Methods in Enzymology

Volume 110

STEROIDS AND ISOPRENOIDS

Part A

METHODS IN ENZYMOLOGY

EDITORS-IN-CHIEF

Sidney P. Colowick   Nathan O. Kaplan

Methods in Enzymology

Volume 110

Steroids and Isoprenoids

Part A

EDITED BY

John H. Law

DEPARTMENT OF BIOCHEMISTRY
UNIVERSITY OF ARIZONA
TUCSON, ARIZONA

Hans C. Rilling

DEPARTMENT OF BIOCHEMISTRY
UNIVERSITY OF UTAH
SALT LAKE CITY, UTAH

1985

ACADEMIC PRESS, INC.

(*Harcourt Brace Jovanovich, Publishers*)

Orlando San Diego New York London
Toronto Montreal Sydney Tokyo

ACADEMIC PRESS, INC.
Orlando, Florida 32887

United Kingdom Edition published by
ACADEMIC PRESS INC. (LONDON) LTD.
24–28 Oval Road, London NW1 7DX

LIBRARY OF CONGRESS CATALOG CARD NUMBER: 54–9110
ISBN 0–12–182010–6

PRINTED IN THE UNITED STATES OF AMERICA

85 86 87 88 9 8 7 6 5 4 3 2 1

Table of Contents

Section I. Early Stages in Isoprenoid Biosynthesis

Section II. Linear Condensations of Isoprenoids

Section III. Cyclization Reactions

Contributors to Volume 110

Article numbers are in parentheses following the names of contributors.
Affiliations listed are current.

WILLIAM S. AGNEW (41), *Department of Physiology, Yale University School of Medicine, New Haven, Connecticut 06510*

D. E. AKIYOSHI (39), *Department of Biochemistry, University of Washington Medical School, Seattle, Washington 98195*

CHARLES M. ALLEN (13, 32), *Department of Biochemistry and Molecular Biology, J. Hillis Miller Health Center, University of Florida, Gainesville, Florida 32610*

TSUNEO BABA (13), *Research Center, Daicel Chemical Industry Ltd., 1239 Shinzaike, Aboshi-ku, Himeji 671-12, Japan*

GRAHAM F. BARNARD (18), *Division of Gastroenterology, Department of Medicine, Stanford University School of Medicine, Stanford, California 94305*

DESIREE L. BARTLETT (20), *Department of Chemistry, University of Utah, Salt Lake City, Utah 84112*

SERGIO BAZAES (9), *Laboratorio de Bioquímica, Pontíficia Universidad Católica de Chile, Casilla 114-D, Santiago, Chile*

JAMES D. BERGSTROM (1), *Department of Biological Chemistry and the Mental Retardation Research Center, UCLA School of Medicine, Los Angeles, California 90024*

PETER BEYER (30, 33), *Institut für Biologie II, Zellbiologie, D-7800 Freiburg i.Br., Federal Republic of Germany*

HENRI BRUNENGRABER (7, 12), *Departments of Nutrition and Medicine, School of Medicine, University of Montreal, Montreal, Quebec H3C 3J7, Canada*

BILAL CAMARA (28, 31), *Laboratoire de Régulations Métaboliques et Differenciation des Plastes, Université Pierre et Marie Curie-Paris 6, 75230 Paris Cédex 05, France*

DAVID E. CANE (44), *Department of Chemistry, Brown University, Providence, Rhode Island 02912*

EMILIO CARDEMIL (10), *Departamento de Química, Facultad de Ciencia, Universidad de Santiago de Chile, Casilla 5659, Santiago, Chile*

ENRIQUE CERDÁ-OLMEDO (27), *Departamento de Genética, Facultad de Biología, Universidad de Sevilla, Sevilla, Spain*

OSVALDO CORI (45), *Departamento de Bioquímica, Facultad de Ciencias Básicas y Farmacéuticas, Universidad de Chile, Casilla 233, Santiago, Chile*

RODNEY CROTEAU (44), *Institute of Biological Chemistry, Washington State University, Pullman, Washington 99164*

V. JO DAVISSON (15), *Department of Chemistry, University of Utah, Salt Lake City, Utah 84112*

JOHN EDMOND (1), *Department of Biological Chemistry and the Mental Retardation Research Center, UCLA School of Medicine, Los Angeles, California 90024*

JAIME EYZAGUIRRE (9), *Laboratorio de Bioquímica, Pontíficia Universidad Católica de Chile, Santiago, Chile*

HIROSHI FUJII (17, 23, 24), *Research Institute for Tuberculosis and Cancer, Tohoku University, Sendai 980, Japan*

ARUN GUPTA (37), *Department of Biological Chemistry, University of Cincinnati, Cincinnati, Ohio 45267*

N. G. HOMMES (39), *Department of Agricultural Chemistry, Oregon State University, Corvallis, Oregon 97331*

YOH IMAI (43), *Department of Biochemistry, Hokkaido University School of Medicine, Sapporo 060, Japan*

KOICHI ISHII (21), *Chemical Research Institute of Non-Aqueous Solutions, Tohoku University, Sendai 980, Japan*

ANA MARIÁ JABALQUINTO (10), *Departamento de Química, Facultad de Ciencia, Universidad de Santiago de Chile, Casilla 5659, Santiago, Chile*

BENJAMIN L. JONES (26), *Campbell Institute for Research and Technology, Camden, New Jersey 08101*

ANDREW A. KANDUTSCH (2), *The Jackson Laboratory, Bar Harbor, Maine 04609*

HIROHIKO KATSUKI (42), *Department of Chemistry, Faculty of Science, Kyoto University, Kyoto 606, Japan*

CHI-HSIN RICHARD KING (20), *The Dow Chemical Company, Pharmaceuticals Process Research, Midland, Michigan 48640*

HANS KLEINIG (30, 33), *Institut für Biologie II, Zellbiologie, D-7800 Freiburg i.Br., Federal Republic of Germany*

RON R. KOPITO (7), *Whitehead Institute, Cambridge, Massachusetts 02142*

TANETOSHI KOYAMA (17, 23, 24), *Chemical Research Institute of Non-Aqueous Solutions, Tohoku University, Sendai 980, Japan*

BERNARD R. LANDAU (12), *Departments of Medicine and Biochemistry, Case Western Reserve University, School of Medicine, Cleveland, Ohio 44106*

E. M. S. MACDONALD (40), *Department of Agricultural Chemistry, Oregon State University, Corvallis, Oregon 97331*

URSULA MITZKA-SCHNABEL (29), *Botanisches Institut der Universität, D-8000 München 19, Federal Republic of Germany*

HENRY M. MIZIORKO (3), *Department of Biochemistry, Medical College of Wisconsin, Milwaukee, Wisconsin 53226*

DANIEL J. MONGER (6), *Clinical Division, Bio-Rad Laboratories, Richmond, California 94804*

R. O. MORRIS (39, 40), *Department of Agricultural Chemistry, Oregon State University, Corvallis, Oregon 97331*

TOKUZO NISHINO (19, 42), *Department of Chemistry, Faculty of Science, Kyoto University, Kyoto 606, Japan*

KYOZO OGURA (17, 19, 21, 22, 23, 24, 25, 36), *Chemical Research Institute of Non-Aqueous Solutions, Tohoku University, Sendai 980, Japan*

NOBUTOSHI OJIMA (36), *Nippon Shinyaku Co., Ltd., Kyoto 601, Japan*

TERUO ONO (43), *Department of Biochemistry, Niigata University School of Medicine, Niigata 951, Japan*

THOMAS S. PARKER (7), *The Rockefeller University, New York, New York 10021*

J. F. PENNOCK (35), *Biochemistry Department, University of Liverpool, Liverpool L69 3BX, England*

JOHN W. PORTER (8, 26), *Lipid Metabolism Laboratory, William S. Middleton Memorial Veterans Hospital, Madison, Wisconsin 53705, and Department of Physiological Chemistry, University of Wisconsin, Madison, Wisconsin 53706*

C. DALE POULTER (15, 20), *Department of Chemistry, University of Utah, Salt Lake City, Utah 84112*

GLENN D. PRESTWICH (46), *Department of Chemistry, State University of New York, Stony Brook, New York 11794*

WERNER RAU (29), *Botanisches Institut der Universität, D-8000 München 19, Federal Republic of Germany*

HANS C. RILLING (14, 16, 38), *Department of Biochemistry, University of Utah, Salt Lake City, Utah 84132*

MARIA CECILIA ROJAS (45), *Departamento de Bioquímica, Facultad de Ciencias Básicas y Farmacéuticas, Universidad de Chile, Casilla 233, Santiago, Chile*

HARRY RUDNEY (37), *Department of Biological Chemistry, University of Cincinnati, Cincinnati, Ohio 45267*

DAVID WARWICK RUSSELL (4), *Department of Biochemistry, Purdue University, West Lafayette, Indiana 47907*

HIROSHI SAGAMI (21, 22, 25), *Chemical Research Institute of Non-Aqueous Solu-*

tions, Tohoku University, Sendai 980, Japan

IKUKO SAGAMI (24, 36), *Research Institute for Tuberculosis and Cancer, Tohoku University, Sendai 980, Japan*

DENNIS M. SATTERWHITE (11), *Institute of Biological Chemistry, Washington State University, Pullman, Washington, 99164*

SHUICHI SETO (19, 36), *Department of Engineering, Nippon University, Koriyama 963, Japan*

TOSHIHIRO SHINKA (19), *Institute of Human Genetics, Kanazawa Medical University, Ishikawa 920-02, Japan*

ABDULLAH SIPAT (5), *Department of Biochemistry and Microbiology, Universiti Pertanian Malaysia, Serdang, Selangor, Malaysia*

IKUKO TAKAHASHI (36), *Chemical Research Institute of Non-Aqueous Solutions, Tohoku University, Sendai 980, Japan*

FREDERICK R. TAYLOR (2), *The Jackson Laboratory, Bar Harbor, Maine 04609*

A. B. WOODSIDE (15), *Department of Chemistry, University of Utah, Salt Lake City, Utah 84112*

HARRY Y. YAMAMOTO (34), *Department of Botany, HITAHR, University of Hawaii, Honolulu, Hawaii 96822*

Preface

Steroids and isoprenoids were last reviewed in *Methods in Enzymology* in 1969 in Volume XV which was edited by R. B. Clayton. Significant articles are also scattered throughout the first six volumes of this series.

Since then remarkable changes have taken place in the field of steroids and isoprenoids. High performance liquid chromatography (HPLC) has revolutionized the preparation and analysis of both substrates and products. Analytical procedures now include high resolution mass spectrometry and gas chromatography–mass spectrometry. Representative enzymes for nearly all of the reactions leading from acetyl-CoA to farnesyl pyrophosphate have been purified to homogeneity, and many other enzymes of polyprenol synthesis and metabolism have been extensively purified from a variety of sources. It is to be noted that several of the enzymes necessary for the conversion of farnesyl pyrophosphate to cholesterol have been solubilized from microsomes, and very significant purifications have been achieved.

An enzyme that has been relatively neglected in Volumes 110 and 111 is 3-hydroxy-3-methylglutaryl-CoA reductase. There are two reasons for this. First, this enzyme was the subject of several articles in Volume 71 of this series. In addition, this area of research is in a period of flux since it is now apparent that the reductase is an intrinsic rather than an extrinsic microsomal protein and much of the earlier work was carried out with a proteolytically degraded molecule.

Volumes 110 and 111 are each divided into three sections. Section I of Volume 110 deals with the enzymes of the early stages of terpenogenesis, and the reactions are common for the biosynthesis of all isoprenoids, i.e., the synthesis of isopentenyl pyrophosphate from acetyl-CoA. Included is a consideration of alternate metabolism of mevalonate. Section II covers the linear head-to-tail (1′-4) condensations as well as the head-to-head (1′-1-4) condensations of terpene biosynthesis. The products of the 1′-4 condensation vary from 3 to 9 isoprene units in length, depending on the system, while squalene and carotenoids are produced in the head-to-head reactions. Reactions between terpenes and nonterpenoid molecules are also described. The final section of this volume covers the cyclization of terpenes in plants as well as the isolation of terpenes from insects, opening the area of biosynthesis of these unusual molecules.

The first section of Volume 111 presents methodology for the isolation and characterization of substrates and products. Several of the articles deal with HPLC and others with mass spectrometry as useful techniques.

Section II on sterol metabolism deals with the transformation of sterols and steroids in a variety of species from vertebrates to insects to plants. Chapters on proteins necessary for the transport and for the acylation of sterols are also included. The final section considers the metabolism of nonsteroidal polyterpenes in vertebrate, plant, and insect systems.

The size and breadth of the area of research represented by the title of these two volumes is far too great to have allowed comprehensive coverage. However, we hope that the material presented is representative. It should provide a reasonable entry into the literature even if it does not present the reader with an "instant" solution to the problem at hand.

We would especially like to thank the authors who contributed to these volumes. The enthusiasm and promptness that they exhibited were most gratifying. Special thanks are due to the staff of Academic Press, both for their encouragement and for knowing what was going on when we did not. Finally, we are indebted to our secretaries, Ellie Moreland and Anne Kidd, for their role in bringing these volumes to fruition.

John H. Law
Hans C. Rilling

METHODS IN ENZYMOLOGY

EDITED BY

Sidney P. Colowick and Nathan O. Kaplan

VANDERBILT UNIVERSITY
SCHOOL OF MEDICINE
NASHVILLE, TENNESSEE

DEPARTMENT OF CHEMISTRY
UNIVERSITY OF CALIFORNIA
AT SAN DIEGO
LA JOLLA, CALIFORNIA

METHODS IN ENZYMOLOGY

EDITORS-IN-CHIEF

Sidney P. Colowick and Nathan O. Kaplan

VOLUME VIII. Complex Carbohydrates
Edited by ELIZABETH F. NEUFELD AND VICTOR GINSBURG

VOLUME IX. Carbohydrate Metabolism
Edited by WILLIS A. WOOD

VOLUME X. Oxidation and Phosphorylation
Edited by RONALD W. ESTABROOK AND MAYNARD E. PULLMAN

VOLUME XI. Enzyme Structure
Edited by C. H. W. HIRS

VOLUME XII. Nucleic Acids (Parts A and B)
Edited by LAWRENCE GROSSMAN AND KIVIE MOLDAVE

VOLUME XIII. Citric Acid Cycle
Edited by J. M. LOWENSTEIN

VOLUME XIV. Lipids
Edited by J. M. LOWENSTEIN

VOLUME XV. Steroids and Terpenoids
Edited by RAYMOND B. CLAYTON

VOLUME XVI. Fast Reactions
Edited by KENNETH KUSTIN

VOLUME XVII. Metabolism of Amino Acids and Amines (Parts A and B)
Edited by HERBERT TABOR AND CELIA WHITE TABOR

VOLUME LX. Nucleic Acids and Protein Synthesis (Part H)
Edited by KIVIE MOLDAVE AND LAWRENCE GROSSMAN

VOLUME 61. Enzyme Structure (Part H)
Edited by C. H. W. HIRS AND SERGE N. TIMASHEFF

VOLUME 62. Vitamins and Coenzymes (Part D)
Edited by DONALD B. MCCORMICK AND LEMUEL D. WRIGHT

VOLUME 63. Enzyme Kinetics and Mechanism (Part A: Initial Rate and Inhibitor Methods)
Edited by DANIEL L. PURICH

VOLUME 64. Enzyme Kinetics and Mechanism (Part B: Isotopic Probes and Complex Enzyme Systems)
Edited by DANIEL L. PURICH

VOLUME 65. Nucleic Acids (Part I)
Edited by LAWRENCE GROSSMAN AND KIVIE MOLDAVE

VOLUME 66. Vitamins and Coenzymes (Part E)
Edited by DONALD B. MCCORMICK AND LEMUEL D. WRIGHT

VOLUME 67. Vitamins and Coenzymes (Part F)
Edited by DONALD B. MCCORMICK AND LEMUEL D. WRIGHT

VOLUME 68. Recombinant DNA
Edited by RAY WU

VOLUME 69. Photosynthesis and Nitrogen Fixation (Part C)
Edited by ANTHONY SAN PIETRO

VOLUME 70. Immunochemical Techniques (Part A)
Edited by HELEN VAN VUNAKIS AND JOHN J. LANGONE

VOLUME 71. Lipids (Part C)
Edited by JOHN M. LOWENSTEIN

VOLUME 72. Lipids (Part D)
Edited by JOHN M. LOWENSTEIN

VOLUME 73. Immunochemical Techniques (Part B)
Edited by JOHN J. LANGONE AND HELEN VAN VUNAKIS

VOLUME 74. Immunochemical Techniques (Part C)
Edited by JOHN J. LANGONE AND HELEN VAN VUNAKIS

VOLUME 75. Cumulative Subject Index Volumes XXXI, XXXII, and XXXIV–LX
Edited by EDWARD A. DENNIS AND MARTHA G. DENNIS

VOLUME 76. Hemoglobins
Edited by ERALDO ANTONINI, LUIGI ROSSI-BERNARDI, AND EMILIA CHIANCONE

VOLUME 77. Detoxication and Drug Metabolism
Edited by WILLIAM B. JAKOBY

VOLUME 78. Interferons (Part A)
Edited by SIDNEY PESTKA

VOLUME 79. Interferons (Part B)
Edited by SIDNEY PESTKA

VOLUME 80. Proteolytic Enzymes (Part C)
Edited by LASZLO LORAND

VOLUME 81. Biomembranes (Part H: Visual Pigments and Purple Membranes, I)
Edited by LESTER PACKER

VOLUME 82. Structural and Contractile Proteins (Part A: Extracellular Matrix)
Edited by LEON W. CUNNINGHAM AND DIXIE W. FREDERIKSEN

VOLUME 83. Complex Carbohydrates (Part D)
Edited by VICTOR GINSBURG

VOLUME 84. Immunochemical Techniques (Part D: Selected Immunoassays)
Edited by JOHN J. LANGONE AND HELEN VAN VUNAKIS

VOLUME 85. Structural and Contractile Proteins (Part B: The Contractile Apparatus and the Cytoskeleton)
Edited by DIXIE W. FREDERIKSEN AND LEON W. CUNNINGHAM

VOLUME 86. Prostaglandins and Arachidonate Metabolites
Edited by WILLIAM E. M. LANDS AND WILLIAM L. SMITH

VOLUME 87. Enzyme Kinetics and Mechanism (Part C: Intermediates, Stereochemistry, and Rate Studies)
Edited by DANIEL L. PURICH

VOLUME 88. Biomembranes (Part I: Visual Pigments and Purple Membranes, II)
Edited by LESTER PACKER

VOLUME 89. Carbohydrate Metabolism (Part D)
Edited by WILLIS A. WOOD

VOLUME 90. Carbohydrate Metabolism (Part E)
Edited by Willis A. Wood

VOLUME 91. Enzyme Structure (Part I)
Edited by C. H. W. HIRS AND SERGE N. TIMASHEFF

VOLUME 92. Immunochemical Techniques (Part E: Monoclonal Antibodies and General Immunoassay Methods)
Edited by JOHN J. LANGONE AND HELEN VAN VUNAKIS

VOLUME 93. Immunochemical Techniques (Part F: Conventional Antibodies, Fc Receptors, and Cytotoxicity)
Edited by JOHN J. LANGONE AND HELEN VAN VUNAKIS

VOLUME 94. Polyamines
Edited by HERBERT TABOR AND CELIA WHITE TABOR

VOLUME 95. Cumulative Subject Index Volumes 61–74 and 76–80
Edited by EDWARD A. DENNIS AND MARTHA G. DENNIS

VOLUME 96. Biomembranes [Part J: Membrane Biogenesis: Assembly and Targeting (General Methods; Eukaryotes)]
Edited by SIDNEY FLEISCHER AND BECCA FLEISCHER

VOLUME 97. Biomembranes [Part K: Membrane Biogenesis: Assembly and Targeting (Prokaryotes, Mitochondria, and Chloroplasts)]
Edited by SIDNEY FLEISCHER AND BECCA FLEISCHER

VOLUME 98. Biomembranes [Part L: Membrane Biogenesis (Processing and Recycling)]
Edited by SIDNEY FLEISCHER AND BECCA FLEISCHER

VOLUME 99. Hormone Action (Part F: Protein Kinases)
Edited by JACKIE D. CORBIN AND JOEL G. HARDMAN

VOLUME 100. Recombinant DNA (Part B)
Edited by RAY WU, LAWRENCE GROSSMAN, AND KIVIE MOLDAVE

VOLUME 101. Recombinant DNA (Part C)
Edited by RAY WU, LAWRENCE GROSSMAN, AND KIVIE MOLDAVE

VOLUME 102. Hormone Action (Part G: Calmodulin and Calcium-Binding Proteins)
Edited by ANTHONY R. MEANS AND BERT W. O'MALLEY

VOLUME 103. Hormone Action (Part H: Neuroendocrine Peptides)
Edited by P. MICHAEL CONN

VOLUME 104. Enzyme Purification and Related Techniques (Part C)
Edited by WILLIAM B. JAKOBY

VOLUME 105. Oxygen Radicals in Biological Systems
Edited by LESTER PACKER

VOLUME 106. Posttranslational Modifications (Part A)
Edited by FINN WOLD AND KIVIE MOLDAVE

VOLUME 107. Posttranslational Modifications (Part B)
Edited by FINN WOLD AND KIVIE MOLDAVE

VOLUME 108. Immunochemical Techniques (Part G: Separation and Characterization of Lymphoid Cells)
Edited by GIOVANNI DI SABATO, JOHN J. LANGONE, AND HELEN VAN VUNAKIS

Section I

Early Stages in Isoprenoid Biosynthesis

[1] Rat Liver Acetoacetyl-CoA Synthetase

By JAMES D. BERGSTROM and JOHN EDMOND

$$\text{Acetoacetate} + \text{ATP} + \text{CoASH} \rightarrow \text{Acetoacetyl-CoA} + \text{AMP} + \text{PP}_i$$

Acetoacetyl-CoA synthetase is a cytosolic enzyme which is found in lipogenic tissues of the rat (and other species) including the infant brain, adult liver, adipose tissue, lactating mammary gland, and adrenal gland. The enzyme-catalyzed reaction produces acetoacetyl-CoA, the first intermediate unique to the pathway of isoprenoid biosynthesis. Its activity level is highly regulated in liver and in many cases, changes in that level are closely coupled to changes in the activity levels of hydroxymethylglutaryl-CoA synthase and hydroxymethylglutaryl-CoA reductase.[1] Hepatic acetoacetyl-CoA synthetase activity is depressed in starvation,[2] by cholesterol feeding,[1] and by mevalonic acid feeding[1] and is enhanced by feeding drugs such as cholestyramine,[2] gemfibrozil,[3] and mevinolin.[1] Its activity has a diurnal variation with the peak in the middle of the dark cycle.[2] Activity levels are higher in females than in males.[3]

Assays

General Comments. The assay of acetoacetyl-CoA synthetase, in cytosolic extracts of lipogenic tissues, is complicated by the presence of high levels of acetoacetyl-CoA thiolase (acetyl-CoA acetyltransferase, EC 2.3.19). During the assay of acetoacetyl-CoA synthetase, the acetoacetyl-CoA produced by the synthetase is rapidly broken down to acetyl-CoA by thiolase. Thus assaying acetoacetyl-CoA synthesis in cytosolic extracts depends upon quantitating the acetyl-CoA produced. This has been accomplished by coupling acetyl-CoA formed to oxaloacetate by the citrate synthase reaction. Oxaloacetate is formed from malate dehydrogenase, and NADH production is followed spectrophotometrically,[2,4] as we will describe later in this chapter. Acetoacetyl-CoA synthesis has been assayed in a fixed time assay by coupling the reaction to endogenous thiolase and adding oxaloacetate and citrate synthase and then measuring

[1] J. D. Bergstrom, G. A. Wong, P. A. Edwards, and J. Edmond, *Fed. Proc., Fed. Am. Soc. Exp. Biol.* **42,** 2082 (1983).

[2] J. D. Bergstrom, K. A. Robbins, and J. Edmond, *Biochem. Biophys. Res. Commun.* **106,** 856 (1982).

[3] J. D. Bergstrom and J. Edmond, unpublished observations.

[4] B. M. Buckley and D. H. Williamson, *FEBS Lett.* **60,** 7 (1975).

ISBN 0-12-182010-6

the citrate produced.[5] Other assays for this enzyme have been used when the enzyme has been purified and they include following the absorption increase at 303 nm due to the absorption of acetoacetyl-CoA in the presence of Mg^{2+},[6] measuring the conversion of NADH to NAD^+ when acetoacetyl-CoA production is coupled to L(+)-3-hydroxyacyl-CoA dehydrogenase[6] and measuring AMP production.[6] However, in our hands these assays only work for the highly purified enzyme and do not work in the presence of thiolase and other enzymes which compete for acetoacetyl-CoA and acetyl-CoA. During the purification of the enzyme a fast and reliable assay is needed that will work both in the presence and absence of thiolase. The second assay we describe is a fixed time assay in which [3-^{14}C]acetoacetate is converted to [3-^{14}C]acetoacetyl-CoA and/or [1-^{14}C]acetyl-CoA and the free [3-^{14}C]acetoacetate is separated from the ^{14}C-labeled CoA esters by instant thin-layer chromatography (ITLC). This assay will work in the presence or absence of thiolase and is used throughout the purification of the enzyme with the exception of the initial cytosolic extract.

Coupled Assay for Use in Cytosolic Extracts

Reagents for Coupled Assay

Tris buffer, 0.2 *M* pH 8.1 containing 0.1 *M* KCl

Substrate mix: 20 m*M* ATP, 40 m*M* malic acid, 100 m*M* $MgCl_2$, 0.4 m*M* CoA, and 15 m*M* NAD^+. Adjusted to pH 6.0 with KOH and stored frozen.

Coupling enzymes: citrate synthase, 35 units/ml and malate dehydrogenase, 140 units/ml are dissolved together in the 0.2 *M* Tris buffer pH 8.1 containing 0.1 *M* KCl and stored frozen

Lithium acetoacetate, 0.1 *M*

Procedure. Two cuvettes are used, one for the reaction and the second a blank that contains all reagents except the acetoacetate. To each cuvette are added 500 μl buffer, 125 μl of substrate mix, 10 μl of coupling enzymes, 10 μl of acetoacetate (reaction cuvette only), and water to bring the volume to 1 ml minus the volume of the enzyme fraction to be added (5 to 100 μl). The cuvettes are brought to 37° and the reaction is initiated by the addition of the enzyme to both cuvettes. The increase in absorbance at 340 nm is followed in a double beam spectrophotometer, subtracting the blank activity from the other. Two moles of NADH is produced for each mole of acetoacetate used. The extinction coefficient for NADH

[5] J. R. Stern, *Biochem. Biophys. Res. Commun.* **44,** 1001 (1971).

[6] T. Fukul, M. Ito, and K. Tomita, *Eur. J. Biochem.* **127,** 423 (1982).

is $6.2 \times 10^3\ M^{-1}\ cm^{-1}$. The final concentrations in the assay mixture are 100 m*M* Tris, 50 m*M* KCl, 12.5 m*M* $MgCl_2$, 2.50 m*M* ATP, 0.050 m*M* CoA, 1.88 m*M* NAD^+, 5 m*M* malate, 1.4 units malate dehydrogenase, 0.35 units citrate synthase, and 1.0 m*M* acetoacetate.

Radiochemical Assay of Acetoacetyl-CoA Synthetase

General Comments. This assay is based on the separation of CoA esters from acetoacetate on silica gel impregnated glass microfiber sheets (ITLC paper). CoA esters remain at the origin while free acids such as acetoacetate, acetate, butyrate, octanoate, and 3-hydroxybutyrate move close to the solvent front. The assay is based on the method that Huang[7] developed for the assay of acetyl-CoA synthetase. The assay is a fixed time assay. It is extremely useful during the purification of the enzyme because it allows large numbers of fractions to be assayed rapidly and works whether or not thiolase is present. In cytosolic extracts this assay gives a lower estimate of the activity of acetoacetyl-CoA synthetase than does the coupled assay presumably because of the presence of acyl-CoA hydrolase activities.

Reagents for Radiochemical Assay

Tris buffer, 0.5 *M*, pH 7.5

Substrate mix: 500 m*M* KCl, 12.5 m*M* ATP, 25 m*M* $MgCl_2$, and 5 m*M* CoA. Adjusted to pH 6.0 with KOH and stored frozen.

[3-^{14}C]Acetoacetate, 2.5 m*M* containing 2.5 μCi/ml of [3-^{14}C]acetoacetate

Preparation of [3-^{14}C]Acetoacetate. Ethyl[3-^{14}C]acetoacetate (40–60 mCi/mmol) in heptane is hydrolyzed by the addition of 2 equivalents of 0.1 *N* NaOH and H_2O is added to bring the solution to a concentration of 5 μCi of [3-^{14}C]acetoacetate/ml. The mixture is heated for 3–4 hr at 50° in a tightly sealed tube. During this period the mixture is vigorously shaken a number of times. At the end of this period, the heptane is removed with a gentle stream of nitrogen and the tube is then heated for one more hour at 50°. One equivalent of HCl is added. The [3-^{14}C]acetoacetate is stable when stored in liquid N_2 at pH 7. The final solution is prepared by making a 1 : 1 dilution of the above solution with a freshly prepared 5 m*M* solution of lithium acetoacetate.

Procedure. A 2 : 1 : 2 mix of buffer : substrate mix : acetoacetate solution is prepared. The assay consists of 50 μl of this mixture, then H_2O and the enzyme fraction are added to bring the final volume of the assay to 100 μl. The assay is initiated by the addition of the enzyme solution and is

[7] K. P. Huang, *Anal. Biochem.* **37,** 98 (1970).

incubated in a water bath at 37°. The reaction is terminated by the addition of 20 μl of glacial acetic acid after 2.5 to 20 min.

Sheets of Gelman's Chromatography medium (ITLC-Type SG, 20 cm × 20 cm) are spotted with 50 μl of the stopped reaction mix (it is not necessary to keep the spots small as the separation is so great). The spots are dried with cool air and the sheets are developed in a mixture of ethyl ether : formic acid (7 : 1). In general 10 samples are applied to each sheet, 2 cm apart. After development the sheets are dried with hot air and a strip 2 cm wide and 4 cm long (in the direction of development) centered at the origins is cut out and counted in 10 ml of a scintillation fluid suitable for aqueous samples. The acetoacetyl-CoA synthetase activity is determined from the fraction of radioactivity that stays at the origin minus a blank (no enzyme) over that of the total radioactivity in 50 μl in the reaction mixture. This fraction is multiplied by the mass of acetoacetate in the assay mixture (50 nmol) and divided by the times of the assay to give micromoles of acetoacetyl-CoA produced/minute.

This assay can be used for any other acyl-CoA synthetase by substituting other ^{14}C substrates such as butyrate, acetate, D(−)-3-hydroxybutyrate, or palmitate. Each of these substances move close to the solvent front in this solvent system while their CoA esters remain at the origin. The specificity of our most purified preparation of acetoacetyl-CoA synthetase was determined using this system.

Definition of Unit. One unit of enzyme activity will produce one μmole of acetoacetyl-CoA per minute at 37°.

Purification

Source. The levels of the enzyme are highly regulated in rat liver and the specific activity of the enzyme in the cytosol can be greatly enhanced by feeding the rats certain drugs. The activity is also higher in livers of females than in livers of males. In female rats the activity level is at its highest at around 50 days of age.[2] The activity also displays a diurnal variation, being higher in the dark cycle than in the light cycle. For a typical purification (results in the table) 15 female rats about 45 days old were fed for 5 days a diet containing 5% cholestyramine and 0.8% gemfibrozil (a lipid lowering drug similar to clofibrate and obtained from Parke-Davis, Morris Plains, NJ). The rats were killed 7 hr into the dark cycle and a total of 106 g of liver was obtained. A 10- to 15-fold increase in the activity level of the enzyme over that in normal males can be achieved with rats treated this way.

Step 1. Cytosolic Extraction. The livers were homogenized in a Potter-Elvehjem homogenizer in 5 volumes of 0.25 *M* sucrose containing 1 m*M*

SUMMARY OF THE PURIFICATION PROCEDURE FOR RAT LIVER ACETOACETYL-CoA SYNTHETASE[a]

Fraction	Volume (ml)	Activity (units)	Recovery (%)	Protein[e] (mg)	Sp. act. (units/mg)
Cytosol	500	109[b]	100	5940	0.018
NH_4SO_4	90.5	84.5[c]	77	3690	0.023
DEAE	28	39.2[c]	36	490	0.080
Blue	4.4	26.4[c]	24	30.8	0.86
S-200	8.4	16.7[c,d]	15	6.35	2.63

[a] Prepared from 106 g of liver taken from 15 female rats 45 days of age fed 5% cholestyramine and 0.8% gemfibrozil for 5 days and killed 7 hours into the dark cycle.
[b] Based on the coupled assay with citrate synthase and malate dehydrogenase.
[c] Based on the radiochemical assay.
[d] Based on the assay coupled to L(+)-3-hydroxyacyl-CoA dehydrogenase.
[e] Protein is determined by the method of M. M. Bradford, *Anal. Biochem.* **72,** 248 (1976).

EGTA, 0.025 m*M* leupeptin, 0.1 unit/ml of aprotinin (trypsin inhibitor units), and 0.005% phenylmethylsulfonyl fluoride at pH 7.5. The homogenates were centrifuged for 15 min at 22,000 *g*. The supernatants were then centrifuged at 100,000 *g* for 45 min. The 100,000 *g* supernatant is the crude cytosolic extract.

Step 2. Ammonium Sulfate Fractionation. The cytosolic extract is brought to 40% saturation with ammonium sulfate at 4° by the addition of 219 g/liter solid ammonium sulfate. The suspension is stirred for 0.5 hr and is centrifuged for 15 min at 22,000 *g* and the supernatant is brought to 80% saturation by the addition of 257 g/liter solid ammonium sulfate. The suspension is centrifuged for 20 min at 22,000 *g* and the pellet is dissolved in 10 m*M* HEPES, 1 m*M* DTT pH 7.5 containing 5% glycerol (Buffer A) that also contained (for this step only) 1m*M* EGTA, 0.025 m*M* leupeptin, and 0.05 units/ml aprotinin. Forty milliliters of this buffer is used to dissolve the pellet per 100 g of liver. The redissolved pellet is dialyzed against 20 volumes of Buffer A containing the EGTA, leupeptin, and aprotinin. The dialyzate is changed once. The total volume (ml) after the dialysis is usually a little less than the original weight (grams) of the liver used. About a 2-fold purification is obtained in this step.

Step 3. DEAE-Sepharose Chromatography. The dialyzed 40–80% ammonium sulfate fraction is applied to a 50 × 2.5 cm column of DEAE-Sepharose (15 meq/100 ml) prewashed with Buffer A. The column is run with a hydrostatic head of 70 to 80 cm which gives a flow rate of about 25 ml/hr. The column is eluted with 200 ml of Buffer A and then with a 800 ml gradient from 0 to 0.4 *M* KCl in Buffer A. Fractions of 20 ml are collected.

The enzyme is eluted in approximately 0.25 *M* KCl. Fractions are assayed for acetoacetyl-CoA synthetase activity and protein measured by absorbance at 280 nm. Fractions containing the peak acetoacetyl-CoA synthetase activity were combined and concentrated to approximately 20 ml in an Amicon Diaflo apparatus with a PM-10 membrane, then the volume is increased to 120 ml with Buffer A and concentrated again to approximately 25 ml. About a 4-fold purification is achieved in this step.

Step 4. Chromatography on Matrix Gel Blue. The concentrated DEAE fraction is applied to a 33 × 1.5 cm column of Matrix Gel Blue A (Amicon Corporation, 2.15 mg of dye/ml) which had been prewashed in Buffer A. The column is eluted with a 500 ml gradient from 0 to 1.4 *M* KCl in Buffer A. A hydrostatic head of about 70 cm gives a flow rate of about 30 ml/hr. Fractions of 15 ml are collected, assayed for acetoacetyl-CoA synthetase activity, and protein is measured by absorbance at 280 nm. The enzyme is eluted in a trailing peak from 0.6 to 0.9 *M* KCl. The fractions of the highest specific activity are combined and concentrated in a Diaflo apparatus using a PM-10 membrane to about 10 ml: after the volume is increased to 100 ml with Buffer A it is then concentrated to approximately 5 ml. About a 10-fold purification is achieved with this step.

Step 5. Gel Filtration on Sephacryl 200. The concentrated fraction from step 4 is applied (in two separate batches of less than 3.0 ml) to a 90 × 2.5 cm column of Sephacryl 200 (Pharmacia Corp.) prewashed with Buffer A. The column is eluted with Buffer A, and 5 ml fractions are collected. The enzyme is eluted in a sharp peak with a V_e/V_0 of 1.37 to 1.40. Most of the contaminating proteins are of a higher molecular weight and the fractions containing the highest specific activity are combined and concentrated in a Diaflo apparatus with a PM-10 membrane to approximately 5 ml. This step gives about a 3-fold purification.

Comments. Thiolase activity is removed by the Matix Blue A column and assays for acetoacetyl-CoA synthetase such as those that utilize the absorbance increase at 303 nm from acetoacetyl-CoA production or couple the reaction to L(+)-3-hydroxyacyl-CoA dehydrogenase may be used. The enzyme appears to be very labile to protease activity particularly during dialysis after the ammonium sulfate fractionation and during DEAE chromatography. Without the protease inhibitors and EGTA present in the buffers used in the early steps of the purification only 5–10% of the original activity is recovered after the DEAE step. The enzyme, after chromatography on matrix gel blue A and Sephacryl 200 is stable for several weeks in Buffer A containing 25–50% glycerol at 4°. All activity is lost with freezing and thawing. The enzyme is not purified to homogeneity by this procedure but this protocol gives about a 150-fold purification from the liver cytosol. The specific activity of the partially purified enzyme after step 5 is about 1500-fold higher than that found in the liver cytosol of male rats.

Properties

Specificity. The enzyme fraction recovered from the Sephacryl column will activate in addition to acetoacetate, L(+)-3-hydroxybutyrate (20% of the rate for acetoacetate), and acetate (45% of the rate for acetoacetate) but not propionate, butyrate, octanoate, or D(−)-3-hydroxybutyrate. We believe the acetate activating activity is due to contamination from a specific acetyl-CoA synthetase because of the report[8] that a purified acetoacetyl-CoA from rat liver activated only acetoacetate and L(+)-3-hydroxybutyrate but not acetate.

GTP and CTP could not replace ATP. The enzyme requires K^+ and Mg^{2+}. Na^+ cannot replace K^+. Other divalent cations may be able to replace Mg^{2+} but have not been tested.

Inhibitors. The enzyme is inhibited by AMP (5 mM gives 70% inhibition) and PP_i (10 mM gives 70% inhibition).

pH Optimum. The pH optimum is 7.5 in Tris buffers (pH 9.0 to 7.5) and in HEPES buffers (pH 8.0 to 7.0) and pH 7.0 in phosphate buffers (pH 7.5 to 6.0). Activity in pH 7.5 Tris is 20% higher than in HEPES pH 7.5 and 40% higher than in phosphate at pH 7.5. Tris apparently stimulates activity.

Molecular Weight. The enzyme has a native molecular weight of 100,000 ± 5000 as determined by gel filtration on Sephacryl 200.

Acknowledgments

This work was supported in part by USPHS Grants HD 06576 and HL 30568 from the National Institutes of Health.

[8] Ito, M., Fukui, T., Kamokari, M., Saito, T., and Tomita, K., *Biochim. Biophys. Acta* **794,** 183 (1984).

[2] Use of Oxygenated Sterols to Probe the Regulation of 3-Hydroxy-3-methylglutaryl-CoA Reductase and Sterologenesis

By FREDERICK R. TAYLOR and ANDREW A. KANDUTSCH

General Introduction

Certain oxygenated derivatives of 5α-cholestan-3β-ol and of 5α-lanostan-3β-ol are potent repressors of hydroxymethylglutaryl-CoA (HMG-CoA) reductase and consequently of the rate of sterologenesis. The bio-

ISBN 0-12-182010-6

chemistry and biological effects of these compounds have been reviewed.[1,2] Structural requirements for high activity are an oxygen function at position 3, and either a hydroxyl at position 20α, 24, 25, or 26 of the side chain or a ketone or α-hydroxyl at position 15 of the D-ring. The positions of double bonds in an oxysterol do not normally have any effect on activity. Some sterols with the second oxygen funtion in other positions, although less potent, are still reasonably good repressors of the reductase. These less potent sterols include a number of natural precursors and metabolites of cholesterol; purified cholesterol itself does not repress HMG-CoA reductase in cultured cells despite substantial uptake.[1,2] Repression of reductase requires intact cells and active protein synthesis and oxysterols do not alter the proportion of reversibly inactivated phosphorylated reductase.[3] There is recent direct evidence that one oxysterol, 25-hydroxycholesterol, represses the synthesis of HMG-CoA reductase[4,5] by altering the level of its mRNA.[6] The effect of oxygenated sterols on the degradation of HMG-CoA reductase is currently in dispute.[5,7] Mutant cell lines selected for resistance to 25-hydroxycholesterol are resistant to all oxysterols tested and to low-density lipoproteins, suggesting that these agents act through a common mechanism. The oxysterols that repress the reductase bind to a cytosolic binding protein with affinities that are proportional to their potencies as repressors of the reductase.[8-10] On this basis it is proposed that it is the oxysterol–protein complex that acts to repress the reductase. The oxysterols are the only pure compounds known to repress the synthesis of the reductase. Two nonsterol fungal metabolites, compactin[11] and mevinolin,[12] are potent competitive inhibitors of reductase activity.

[1] A. A. Kandutsch, H. W. Chen, and H.-J. Heiniger, *Science* **201,** 498 (1978).

[2] G. J. Schroepfer, Jr., *Annu. Rev. Biochem.* **50,** 585 (1981).

[3] S. E. Saucier and A. A. Kandutsch, *Biochim. Biophys. Acta* **572,** 541 (1979).

[4] M. Sinensky, R. Torget, and P. A. Edwards, *J. Biol. Chem.* **256,** 11774 (1981).

[5] J. R. Faust, K. L. Luskey, D. J. Chin, J. L. Goldstein, and M. S. Brown, *Proc. Natl. Acad. Sci. U.S.A.* **79,** 5205 (1982).

[6] K. L. Luskey, J. R. Faust, D. J. Chin, M. S. Brown, and J. L. Goldstein, *J. Biol. Chem.* **258,** 8462 (1983).

[7] M. Sinensky, R. Torget, R. Schnitzer-Polokoff, and P. A. Edwards, *J. Biol. Chem.* **257,** 7284 (1982).

[8] A. A. Kandutsch and E. B. Thompson, *J. Biol. Chem.* **255,** 10813 (1980).

[9] A. A. Kandutsch and E. P. Shown, *J. Biol. Chem.* **256,** 13068 (1981).

[10] F. R. Taylor, S. E. Saucier, E. P. Shown, E. J. Parish, G. J. Schroepfer, Jr., and A. A. Kandutsch, *J. Biol. Chem.,* in press (1984).

[11] A. Endo, M. Kuroda, and Y. Tsujita, *J. Antibiot.* **29,** 1346 (1976).

[12] A. W. Alberts, J. Chen, G. Kuron, V. Hunt, J. Huff, C. Hoffman, J. Rothrock, M. Lopez, H. Joshua, E. Harris, A. Patchett, R. Monaghan, S. Currie, E. Stapley, G. Albers-Schonberg, O. Hensens, J. Hirschfield, K. Hoogsteen, J. Liesch, and J. Springer, *Proc. Natl. Acad. Sci. U.S.A.* **77,** 3957 (1980).

In general oxygenated sterols specifically inhibit sterol synthesis; rates of acetate metabolism to fatty acids and CO_2, and rates of RNA, DNA, and protein synthesis are not immediately affected.[1] HMG-CoA reductase is the only enzyme whose levels have been shown to respond rapidly to the addition or removal of oxysterols in the medium, although slower declines in the levels of some other enzymes have been reported.[2] The increase in cholesterol acyltransferase (acyl-CoA: cholesterol acyltransferase) activity in response to oxysterols in some cell lines has recently been attributed to changes in substrate availability.[13,14] A few oxygenated sterols which repress HMG-CoA reductase also inhibit a late step(s) in the sterol pathway apparently at the level of lanosterol demethylation.[2]

The decline in the sterol content of cells following the block in synthesis has a number of consequences. First appears an inhibition of DNA synthesis followed by the disruption of various membrane functions, the arrest of cell growth, and cell rounding and detachment from the substratum.[1,15] The timing of these events depends on the rate of cell growth. Supplementing the medium with cholesterol or other growth promoting sterol, or mevalonic acid, prevents growth inhibition. However, the long-term effects of oxysterol treatment in the presence of sterol supplements has not been adequately explored. It should also be noted that certain inhibitory oxysterols, such as 20α-hydroxycholesterol, may also serve to some extent as sterol supplements if present in high concentrations.

Use of Oxysterols with Cultured Cells to Investigate the Regulation of HMG-CoA Reductase

Cell Culture

Any established mammalian cell line, or growing primary culture of mammalian cells, may be suitable for studies of the regulation of HMG-CoA reductase, provided that the reductase (and sterol synthesis) is not already prohibitively repressed by substances in the culture medium. Inhibitory substances that may be present in culture media include oxysterols naturally present in serum,[16] or produced there by autoxidation of cholesterol.[17]

[13] S. C. Miller and G. Melnykovych, *Fed. Proc., Fed. Am. Soc. Exp. Biol.* **42,** 1835 (1983).

[14] W. Young, R. Daus, B. Stone, J. Halpern, R. Miller, A. Cooper, and S. Erickson, *Fed. Proc., Fed. Am. Soc. Exp. Biol.* **42,** 2076 (1983).

[15] W. K. Cavenee, H. W. Chen, and A. A. Kandutsch, *Exp. Cell Res.* **131,** 31 (1981).

[16] N. B. Javitt, E. Kok, S. Burstein, B. Cohen, and J. Kutscher, *J. Biol. Chem.* **256,** 12644 (1981).

[17] L. L. Smith, J. I. Teng, Y. Y. Lin, P. K. Seitz, and M. E. McGehee, *J. Steroid Biochem.* **14,** 889 (1981).

Cells growing in serum-free medium or medium containing delipidated[18,19] or lipoprotein-depleted[20] serum are dependent upon *de novo* synthesis of cholesterol and are unable to grow when HMG-CoA reductase activity is repressed. The requirement for cholesterol synthesis and the level of HMG-CoA reductase activity are determined largely by the need for new membranes for cell replication, and by the rate of sterol efflux. The rate of cell replication may be affected by many factors—cell density being an important one.[21] Sterol efflux is highly influenced by the presence of acceptor substances (proteins, phospholipids) in the medium.[22–25]

Studies are most easily carried out with adherent cells, because the medium can be readily changed by pouring it off. The size of the cultures required is determined largely by the methodology used to determine HMG-CoA reductase and sterologenesis. Cultures in 16-mm wells can be adequate.

Addition of Sterols to the Culture Medium

Principle. Oxysterols differ in their solubility in aqueous solutions. However, little definitive information in this regard is available. Methods employed for the addition of sterols to the medium are based largely upon empirical observations and published information regarding the solubility of cholesterol.[26,27] If a trace amount of radioactive cholesterol in a few microliters of ethanol is added to an aqueous solution, time-dependent binding to the plastic or glass walls of the vessel occurs. If the concentration of sterol is high enough (40 n*M* for cholesterol) micelle formation occurs.[26] At concentrations over 2 to 5 μg/ml visible aggregates form. In contrast, if sufficient protein (e.g., delipidated serum) is present in the medium, sterol binds to the protein with little binding to the vessel walls and a stable, relatively homogeneous dispersion of the sterol may be obtained. At high concentrations aggregates and micelles may still be present. If serum proteins are not present in the medium, bovine serum albumin can be used to prepare a stable sterol solution or suspension which can be accurately divided into aliquots for addition to cultures. A procedure for this is as follows.

[18] G. H. Rothblat, L. Y. Arbogast, L. Ouellette, and B. V. Howard, *In Vitro* **12,** 554 (1976).
[19] B. E. Cham and B. R. Knowles, *J. Lipid Res.* **17,** 176 (1976).
[20] C. M. Redding and D. Steinberg, *J. Clin. Invest.* **39,** 1560 (1960).
[21] H. W. Chen, *J. Cell. Physiol.* **108,** 91 (1981).
[22] C. H. Burns and G. H. Rothblat, *Biochim. Biophys. Acta* **176,** 616 (1969).
[23] R. L. Jackson, A. M. Gotto, O. Stein, and Y. Stein, *J. Biol. Chem.* **250,** 7204 (1975).
[24] A. M. Fogelman, J. Seager, P. A. Edwards, and G. Popják, *J. Biol. Chem.* **252,** 644 (1977).
[25] L. C. Bartholow and R. P. Geyer, *Biochim. Biophys. Acta* **665,** 40 (1981).
[26] M. E. Haberland and J. A. Reynolds, *Proc. Natl. Acad. Sci. U.S.A.* **70,** 2313 (1973).
[27] D. B. Gilbert, C. Tanford, and J. A. Reynolds, *Biochemistry* **14,** 444 (1975).

OXYSTEROL REPRESSION OF HMG-COA REDUCTASE AND RELATIVE AFFINITIES FOR THE OXYSTEROL BINDING PROTEIN

Sterol	Repression of HMG-CoA reductase (μM)[a]	Relative binding affinity (μM)[a]
3β-Hydroxy-5α-cholest-8(14)-en-15-one	0.10	0.02
Cholest-5-ene-3β,25-diol	0.15	0.03
(25*S*)Cholest-5-ene-3β,26-diol	0.16	0.20
(25*R*)Cholest-5-ene-3β,26-diol	0.26	0.11
Cholest-5-ene-3β,20α-diol	0.30	0.06
5α-Cholest-7-ene-3β,15α-diol	0.50	0.41
5α-Lanost-8-ene-3β,32-diol	0.70	0.45
5α-Lanost-7-ene-3β,32-diol	1.0	0.2
3β-Hydroxycholest-5-en-7-one	1.7	1.35
Cholest-5-ene-3β,7β-diol	1.9	1.92
Cholest-5-ene-3β,7α-diol	2.5	1.05

[a] The values given are the concentrations necessary for 50% response. Data from F. R. Taylor, S. E. Saucier, E. P. Shown, E. J. Parish, G. J. Schroepfer, Jr., and A. A. Kandutsch, *J. Biol. Chem.*, in press (1984).

Method. To a sterol dissolved in a small amount of distilled ethanol (warming if necessary), add 9 volumes of culture medium containing 5% w/v of bovine serum albumin (Pentex, 3× crystallized, or Sigma, essentially fatty acid free). Mix gently. Low concentrations (<5 μg/ml) will give a clear solution. If high concentrations of sterol are added (10–100 μg/ml) the mixture may contain large sterol aggregates which will settle with time. In this case brief sonication may provide a more stable, more homogeneous dispersion.

Addition of one volume of the sterol suspension to a culture containing nine volumes of culture medium results in a final concentration of 1% ethanol and 0.45% bovine serum albumin. This level of ethanol is not harmful to most cell cultures. However, lower concentrations of ethanol can be used with low concentrations of sterol. This method of solubilizing sterols has been used in investigations of sterol uptake.[28]

Choice of Oxysterol

The most potent oxysterols and those that are known to occur as natural intermediates and metabolites are listed in the table. A wide variety of other sterols have been tested and their synthesis and activities as repressors are referenced in several publications.[1,2,10,29] Of the highly

[28] A. A. Kandutsch and H. W. Chen, *J. Cell. Physiol.* **85,** 415 (1975).
[29] A. A. Kandutsch and H. W. Chen, *Lipids* **13,** 704 (1978).

active sterols only 25-hydroxycholesterol and 20α-hydroxycholesterol are readily obtained in reasonably pure form from commercial sources. 25-hydroxy[26,27-^{3}H]cholesterol is the only radiolabeled, potent repressor with high specific activity (~80 Ci/mmol) available commercially (NEN). The purity of any sterol should be established by recrystallization, thin-layer chromatography, and/or HPLC before use. Radioactive sterols are purified by silica gel thin-layer chromatography shortly before they are used.[19] The purity of sterol preparations that have been allowed to stand for weeks after purification cannot be assumed.

Conditions for Oxysterol Repression of HMG-CoA Reductase

Concentrations of sterols required to repress the reductase to 50% of the control level after an incubation period of 5 hr are shown in the table. These values were obtained with L cell cultures but reasonably similar values for some of the sterols have been obtained with Chinese hamster lung (Dede) and ovary (CHO) cells grown in delipidated serum. The 50% repression values were obtained from linear or logarithmic plots of repression of HMG-CoA reductase activity (as a percentage of the control value) versus the concentration of the sterol in the medium. Four to six concentration points are usually required in order to obtain a reasonably accurate estimate of activity. In our hands, using L cell cultures, the standard error for replicate ($n = 3$) determinations of the 50% repression values for each of three potent sterols (25-hydroxycholesterol, 26-hydroxycholesterol, and 3β-hydroxy-5α-cholest-8(14)-en-15-one) was approximately $\pm$25%. Values obtained may be influenced by the length of the incubation period, by the rate of cell growth, by cell density, and by refeeding with fresh medium. These culture conditions should be standardized. Following refeeding of L cell cultures, HMG-CoA reductase activity increases over a period of about 6 hr. To avoid this change during the time period of an experiment the cultures may be refed about 12 hr before they are used. Generally it is desirable to keep the incubation period as short as feasible in order to avoid autoxidation of the sterols in the medium. Since the half life of HMG-CoA reductase is about 1.5 hr, an incubation period of 5 hr is adequate. If incubation periods are prolonged beyond 5 hr (or as a routine precaution), vitamin E (α-tocopherol) should be added to the culture medium at a concentration of 0.5 μg/ml. If high concentrations of sterol are present in the medium, it is advisable to shake the cultures slowly to avoid settling of aggregates. After incubation, the cells are recovered from the flask, sedimented by centrifugation,and frozen at $-90°$.[30] After thawing, the cells are resuspended in 50 mM potas-

[30] W. K. Cavenee, H. W. Chen, and A. A. Kandutsch, *J. Biol. Chem.* **256,** 2675 (1981).

sium phosphate buffer (pH 7.4), 5 m*M* dithiothreitol, and 1 m*M* EDTA, homogenized by careful sonication, and HMG-CoA reductase activity in an aliquot of the homogenates is determined.[30] To measure fully activated "total" reductase the homogenate should be preincubated for 20 min at 37°[6] or treated with a phosphatase.[31,32] Addition of 50 m*M* sodium fluoride prevents this activation and a combination of these procedures allows measurement of the percentage of active and inactive reductase.[31,32] The effects of oxysterols upon the synthesis of HMG-CoA reductase can be measured directly using antibody precipitation of radiolabeled enzyme.[3,4] The culture medium should be modified to diminish dilution of the radiolabeled amino acid. Recently it has been possible to measure HMG-CoA reductase mRNA levels by *in vitro* translation[33] or through the use of a cDNA probe.[5]

In addition to measuring HMG-CoA reductase, rates of sterol synthesis and the accumulation of pathway intermediates can be determined by the incorporation of radioactive precursors and by other methodologies (see references in 1,2,34).

Oxysterol Binding Protein

A cytosolic protein which binds 25-hydroxycholesterol with high affinity and low capacity has been found in all cell types examined so far.[8,9] This protein can be determined after allowing intact cells to take up the labeled oxysterol, or after incubating a $(NH_4)_2SO_4$-precipitated fraction of the cytosol with the labeled sterol. Since 25-hydroxy[26,27-^{3}H]cholesterol is presently the only radiolabeled sterol of sufficiently high specific activity (80 Ci/mmol) available commercially (NEN), it has been used in all studies so far to detect the binding protein and study its properties. 25-Hydroxycholesterol also binds to other cell fractions and to other cytoplasmic proteins in an apparently nonspecific manner.[8] This is readily deduced by adding a large excess of unlabeled 25-hydroxycholesterol which displaces the radiolabel from the specific binding sites but not from the nonspecific, high capacity material. Relative binding affinities of other unlabeled sterols can be determined by measuring their ability to displace 25-hydroxy[^{3}H]cholesterol. As shown in the table, relative binding affinity correlates well with the ability of the sterols to repress HMG-CoA reduc-

[31] J. L. Nordstrom, V. W. Rodwell, and J. J. Mitschelen, *J. Biol. Chem.* **252,** 8924 (1977).

[32] M. S. Brown, J. L. Goldstein, and J. M. Dietschy, *J. Biol. Chem.* **254,** 5144 (1979).

[33] C. F. Clark, P. A. Edwards, S. F. Lan, R. D. Tanaka, and A. M. Fogelman, *Proc. Natl. Acad. Sci. U.S.A.* **80,** 3305 (1983).

[34] G. F. Gibbons, K. A. Mitropoulos, and N. B. Myant, "Biochemistry of Cholesterol." Elsevier Biomedical Press, New York, 1982.

tase in cultured cells. Some exceptions to this generalization, i.e., sterols that repress HMG-CoA reductase but do not bind to the binding protein, have been shown to be explainable in terms of intracellular metabolism of a sterol with low affinity for the binding protein to produce a sterol with high activity in both tests.[10]

Binding to the Cytosolic Protein in Intact Cells

Binding of exogenous 25-hydroxycholesterol to the cytosolic protein in intact cells reaches a steady-state equilibrium within 30 min.[8] Because the incubation period may be short, it is not necessary to maintain optimum growth conditions for cell cultures. Serum-containing medium can, therefore, be removed, and the cells can be incubated in serum-free medium containing oxysterol dispersed in an ethanol–bovine serum albumin solution as described earlier.

Method. Approximately 1×10^7 cells are required and these may be either attached in a 150-cm flask containing 10 ml of medium or in suspension in 10 ml of medium. Preferentially the cultures are shaken on a gyrating table (40 rpm). 25-Hydroxy[^{3}H]cholesterol (BSA solution) is added at a radioisotope concentration of about 1 μCi/ml of medium. Uptake and binding of the sterol is nearly linear with increasing concentration over a wide range (up to 10 μg/ml). After incubation, the cells are washed three times with 10 volumes of protein-free medium. The cells are then homogenized in 0.5 ml of hypotonic buffer (Buffer A = 20 m*M* Tris containing 1 m*M* MgCl, 2 m*M* $CaCl_2$, 2 m*M* dithiothreitol) in a Potter-Elvehjem homogenizer with 30 strokes followed, after an interval of 5 min, by 20 additional strokes. The homogenate is then made isotonic by the addition of 0.1 ml of a 30% w/w solution of sucrose in buffer A and the cytosolic fraction is isolated by centrifugation, first at 80,000 *g* for 12 min to sediment large particulate material, then at 170,000 *g* for 1 hr to remove microsomes. The supernatant fraction is layered over a 5–20% sucrose gradient buffered with 20 m*M* Tris, pH 7.4, containing 0.15 m*M* EDTA and 0.3 *M* KCl and centrifuged in a SW 60 or SW 50.1 rotor for 16 hr at 257,000 *g* or 220,000 *g*, respectively ($\omega^2 t = 1.5 \times 10^{12}$). The gradients are then fractionated into 20–25 fractions and assayed for ^{3}H. Under these conditions two peaks of radioactivity are found at 7.5 S and 4.5 S, as calculated from the positions of ^{14}C-labeled protein standards. Identification of the 7.5 S fraction as the specific oxysterol binding protein is based on the observations that binding of 25-hydroxycholesterol is saturable and reversible, and that various oxygenated sterols compete for binding in proportion to their ability to repress HMG-CoA reductase.[8-10] On the other hand binding to the 4.5 S fraction is not saturable, nor is 25-hydroxycholesterol displaced by other unlabeled oxysterols; it therefore is considered to be nonspecific binding to substances that sediment with the bulk of the cellular protein.

Assay of the Oxysterol Binding Protein in a $(NH_4)_2SO_4$ Precipitated Fraction of the Cytosol

Ammonium Sulfate Fractionation. The specific binding protein cannot be assayed with any accuracy by incubating the isolated whole cytosol of cells with 25-hydroxy[^{3}H]cholesterol because of a very high background of nonspecific binding.[9] However, a fraction of the cytosol precipitated with $(NH_4)_2SO_4$ between 0.3 and 0.4 saturation contains about 70% of the specific binding protein while nonspecific binding is diminished to a minor component of the total.[9] To obtain the $(NH_4)_2SO_4$ precipitated fraction, cells (approximately 1×10^8) are homogenized in 15 ml of Buffer A and the homogenate is made 5% saturated with sucrose and centrifuged to obtain the cytosolic fraction as described above. The freshly isolated cytosol is stirred on ice and a saturated solution of $(NH_4)_2SO_4$ is added dropwise to give 0.3 saturation. After stirring for 20 min the solution is centrifuged at 80,000 *g* for 15 min, the supernatant is collected and additional saturated $(NH_4)_2SO_4$ solution added, as before, to give 0.4 saturation. After 20 min further stirring on ice the solution is centrifuged and the supernatant is completely decanted. The pellet (designated as the 0.3–0.4 fraction) is then dissolved in the desired buffer. This procedure gives about a 6-fold purification with a 70% recovery. Buffers of pH below 6 do not completely dissolve the pellet and the undissolved material is removed by centrifugation.

Assay by Sucrose Gradient Centrifugation. An aliquot of the 0.3–0.4 fraction (i.e., 0.4 ml containing 0.1 to 1.0 mg protein), is added to a siliconized (Siliclad, Clay Adams Company) test tube containing 25-hydroxycholesterol (saturating concentration approximately 4.6 ng (1 μCi)/ml) dissolved in ethanol and the mixture is incubated at 0°. The final concentration of ethanol should not exceed 10% and is routinely held to 2.5%. The rate of sterol-protein association in Buffer A at pH 7.8 is slow, taking at least a day to reach equilibrium. However, with 25-hydroxycholesterol of high specific activity of binding can be analyzed after as little as an hour of incubation. Binding is then assayed by sedimentation in sucrose gradients as described above.

Assay by a Charcoal-Dextran Competition Method. An alternative method of assaying the partially purified binding protein utilizes dextran-coated charcoal to adsorb unbound sterol. In this procedure duplicate aliquots (0.2 ml) of the 0.3–0.4 fraction are incubated at 0° in 1.5 ml conical polypropylene centrifuge tubes containing 25-hydroxy[^{3}H]cholesterol dissolved in ethanol, one tube with and one without, a 40-fold excess of unlabeled 25-hydroxycholesterol. Unbound 25-hydroxycholesterol is then removed by adding a 50-μl aliquot of a charcoal-dextran solution (described below), incubating at 0° for 15 min with mixing at 5 min intervals. After centrifugation at 12,000 *g* for 5 min replicate 50 μl aliquots of the supernatant fluid are assayed for ^{3}H. Specific binding represents the difference between the ^{3}H measured in the two tubes.

Preparation of the Charcoal-Dextran Mixture. Successful use of this assay requires careful preparation of the charcoal-dextran to remove fine particles. The charcoal-dextran mixture is prepared by adding 1 g Norit A (Pfanstiehl Laboratories) to a test tube and washing with 10 ml of 1 *N* HCl followed by 10 ml H_2O, 10 ml of 5% $NaHCO_3$, 10 ml H_2O, and then five washes with 10 ml 5% Dextran T-500 (Pharmacia) in 10 m*M* Tris–HCl, pH 8. Between washes the charcoal is sedimented by centrifugation at 1000 *g* for 20 min. The dextran is then removed by two washes with 10 ml of 10 m*M* Tris–HCl pH 8 buffer containing 3 m*M* sodium azide and the charcoal is stored at 4° in 20 ml of this buffer. Prior to use 1 g of Dextran T-500 is added and allowed to dissolve overnight. This procedure improves the reproducibility of the assay over that reported previously.[9]

Comments. Under the appropriate conditions, the charcoal-dextran assay gives results similar to those obtained by the sucrose gradient centrifugation assay, although it is linear over a narrower range of protein (0.05–1 mg/ml). It is also somewhat less sensitive, and does not function reliably when levels of nonspecific binding are high. So far the charcoal-dextran assay has not been useful in trials of further purification steps where activity in very dilute fractions must be measured. In these cases sedimentation analysis has continued to give excellent results.

Optimum Binding Conditions

pH. Binding of 25-hydroxycholesterol to the 0.3–0.4 fraction can be measured over the pH range 4.5 to at least 8.0. However, substantial changes in the properties of the binding protein occur within this range.[35] At pH 5.5 the kinetics of binding association and dissociation are relatively fast; measurement of equilibrium binding is possible after only 2 hr of incubation. At pH 7.4 the rates of both association and dissociation are roughly an order of magnitude slower than at pH 5.5 and equilibrium binding requires at least 24 hr. The equilibrium dissociation constants, however, remain nearly the same: 3.9×10^{-9} *M* at pH 5.5 and 7.4×10^{-9} *M* at pH 7.4. Other, perhaps related, pH dependent changes in the molecular form of the protein are described below.

Stability. The binding protein in the cytosolic supernatant or in the partially purified 0.3–0.4 fraction is very unstable in the absence of an

[35] A. A. Kandutsch, F. R. Taylor, and E. P. ·hown, *J. Biol. Chem.*, in press (1984).

oxysterol ligand. In 0.1 *M* KPO_4 buffer, pH 7.4, at 0°, all binding activity is lost within 24 hr. After binding 25-hydroxycholesterol the binding protein remains stable at 0° for up to 10 days. In the cytosolic supernatant, however, a transformation from a 7.5 S form to a 4.2 S form occurs within 24 hr. This is possibly due to proteolysis but attempts to prevent this change with protease inhibitors have so far been unsuccessful. $(NH_4)_2SO_4$ fractionation eliminates this problem when the precipitate is resuspended at pH 7.4 but at pH 5.5 a slow transformation to a 4.2 S form is still sometimes seen. The protein in the 0.3–0.4 fraction can be assayed from 0° to room temperature but at 35° binding activity is lost.

Molecular Forms of the Binding Protein. In studying the binding protein it is important to recognize that its molecular form is dependent upon the pH, the presence of urea and whether it is liganded with an oxysterol.[35] Analyses by sucrose gradients, to determine the sedimentation coefficient, and Sephacryl gel chromatography, to determine the Stokes radius, indicate that the protein consists of three dissociable subunits. When the binding protein is not liganded to an oxysterol the three subunits are associated in a complex with a sedimentation coefficient of 7.7 S and a Stokes radius of 71.6 Å to give a M_r = 236,000. Binding of 25-hydroxycholesterol is associated with a loss of a subunit of M_r 67,000 giving a binding complex with a sedimentation coefficient of 7.5 S, a Stokes radius of 50 Å, and a M_r = 169,000. If this complex is incubated at a pH below 6 and then analyzed in buffers containing 2.5 *M* urea a further, reversible, dissociation of an unliganded subunit of M_r = 72,000 gives a binding complex with a sedimentation coefficient of 4.2 S, a Stokes radius of 53 Å, and a M_r = 97,000. This 4.2 S form may be similar to that which arises, irreversibly, upon incubation of the cytosolic supernatant (see "Stability").

[3] 3-Hydroxy-3-methylglutaryl-CoA Synthase from Chicken Liver

By HENRY M. MIZIORKO

$$\text{Acetyl-CoA} + \text{acetoacetyl-CoA} + H_2O \rightarrow \text{3-Hydroxy-3-methylglutaryl-CoA} + \text{CoA}$$

3-Hydroxy-3-methylglutaryl-CoA is formed in liver as a result of the essentially irreversible reaction of acetyl-CoA with acetoacetyl-CoA,[1] illustrated in the above equation. The enzyme which catalyzes the reaction, 3-hydroxy-3-methylglutaryl-CoA (HMG-CoA) synthase, reacts re-

[1] J. J. Ferguson and H. Rudney, *J. Biol. Chem.* **234,** 1072 (1959).

ISBN 0-12-182010-6

versibly with acetyl-CoA to form an acetyl-enzyme. This species subsequently condenses with the second substrate to form an intermediate which contains HMG-CoA covalently linked to an active site cysteinyl –SH. Hydrolysis of the thioester linkage between enzyme and the condensation product results in release of HMG-CoA and accounts for the irreversible nature of the overall reaction.[2]

The procedure for preparation of the mitochondrial HMG-CoA synthase, which is described here, represents a refinement of the methodology reported by Reed and Lane.[3] Modifications which include stabilization of enzyme in the early stages of purification and elimination of ammonium sulfate precipitation as a method of protein concentration have increased the reproducibility and yield of the preparation.

Assay Method

Principle. HMG-CoA synthase activity can be conveniently measured spectrophotometrically by a modification of the assay developed by Ferguson and Rudney,[1] which monitors the acetyl-CoA-dependent disappearance of acetoacetyl-CoA at 300 nm. An alternative radiochemical assay relies upon the formation of a product ([^{14}C]HMG-CoA) which produces nonvolatile ^{14}C radioactivity under conditions used to volatilize the unreacted [^{14}C]acetyl-CoA substrate (evaporation to dryness from 6 *N* HCl at 95°).

Spectrophotometric Assay

Reagents

Tris–HCl, pH 8.2, 0.2 *M*, containing 0.2 m*M* EDTA
Acetoacetyl-CoA, 1.0 m*M*, pH 4.5
Acetyl-CoA, 10.0 m*M*, pH 4.5

Procedure. The standard assay mixture includes 100 m*M* Tris–HCl, 0.1 m*M* EDTA, 0.2 m*M* acetyl-CoA, 0.050 m*M* acetoacetyl-CoA, and HMG-CoA synthase (3–30 mU) in a final volume of 1.0 ml. The enzyme is incubated for 2 min at 30° in a cuvette containing all components of the assay mix except for acetyl-CoA. Thus, the base line absorbance at 300 nm is established prior to starting the reaction by addition of acetyl-CoA. The acetyl-CoA dependent rate of acetoacetyl-CoA disappearance is a measure of HMG-CoA synthase activity. Under these assay conditions,[3] the apparent extinction coefficient of acetoacetyl-CoA is $3.6 \times 10^3\ M^{-1}$. Increased sensitivity can be achieved by including 20 m*M* $MgCl_2$ in the

[2] H. M. Miziorko and M. D. Lane, *J. Biol. Chem.* **252,** 1414 (1977).
[3] W. D. Reed and M. D. Lane, this series, Vol. 35, p. 155.

assay mix ($\varepsilon_{300\text{ nm}}^{1\text{ cm}} = 13.6 \times 10^3\ M^{-1}$). It should be noted, however, that the avian liver mitochondrial enzyme is partially inhibited by divalent cations,[4] including Mg^{2+}. Due to the high affinity of enzyme for acetoacetyl-CoA,[5] the decrease in $A_{300\text{ nm}}$ is linear with time until this substrate is virtually depleted. It should be noted that $\varepsilon_{300\text{ nm}}$, which reflects the amount of acetoacetyl-CoA in the enolized form, fluctuates in the presence of metal contaminants or with changes in pH.

Radiochemical Assay

Reagents

Tris–HCl, pH 8.2, 0.2 *M*, containing 0.2 m*M* EDTA
[1-^{14}C]Acetyl-CoA, 2.0 m*M*, pH 4.5 (specific activity ~1 mCi/mmol)
Acetoacetyl-CoA, 1.0 m*M*, pH 4.5
Liquid scintillator (compatible with 10% final H_2O content)

Procedure. The standard assay mixture includes 100 m*M* Tris–HCl, 0.1 m*M* EDTA, 0.2 m*M* [1-^{14}C]acetyl-CoA, 0.050 m*M* acetoacetyl-CoA, and HMG-CoA synthase (0.10–1.0 mU) in a total volume of 0.2 ml. The enzyme is incubated for 2 min at 30° with all components of the assay mix except for [1-^{14}C]acetyl-CoA, which is used to initiate the reaction. Aliquots (40 μl) are removed from the reaction mixture at appropriate times and pipetted into glass shell vials (15 × 45 mm; flat bottomed) containing 0.1 ml 6 *N* HCl. The acidified aliquot is taken to dryness at 95° in an aluminum heating block (Lab-Line 2073). Water (0.3 ml) is added to dissolve the dried nonvolatile material. After addition of 3 ml of liquid scintillator, the vials are capped and mounted in 25 mm diameter vial holders (Kimble 74700). The nonvolatile ^{14}C radioactivity (due to [^{14}C]HMG-CoA) is measured by liquid scintillation counting and corrected for any nonvolatile ^{14}C radioactivity in [^{14}C]acetyl-CoA (determined from a control in which enzyme is omitted). Under the conditions outlined here, a linear increase in nonvolatile ^{14}C radioactivity with time is measured until at least 90% of the acetoacetyl-CoA is utilized.

Units. A unit of enzymatic activity is defined as the amount of enzyme necessary to convert 1 μmol of substrates into HMG-CoA per minute, under the conditions described. Prior to elution from phosphocellulose, thiolase contaminates the enzyme preparation. The measured activity is divided by two to correct for thiolase depletion of acetoacetyl-CoA, which is limited by the amount of CoA generated in the synthase catalyzed reaction.[3] Specific activity is expressed in units per milligram of

[4] W. D. Reed, K. D. Clinkenbeard, and M. D. Lane, *J. Biol. Chem.* **250,** 3117 (1975).

[5] L. A. Menahan, W. T. Hron, D. G. Hinckelman, and H. M. Miziorko, *Eur. J. Biochem.* **119,** 287 (1981).

TABLE I
PURIFICATION OF HMG-CoA SYNTHASE FROM CHICKEN LIVER MITOCHONDRIA

Procedure	Total activity[a] (units)	Total protein (mg)	Specific activity (units/mg)	Yield (%)
Mitochondrial matrix	351	9486	0.037	100
DEAE-cellulose chromatography	266	2938	0.091	76
Phosphocellulose chromatography	101	289	0.35	29
Sephadex G-150 chromatography	81	95	0.85	23

[a] Measured using the spectrophotometric assay.

protein. Protein concentration in partially purified samples is estimated by the Lowry method.[6] Protein concentration of homogeneous enzyme samples is calculated using the refractometrically determined extinction coefficient. The relation between A_{280} and refractometrically determined protein concentration is given by $C = (0.88)(A_{280\,\mathrm{nm}}^{1\,\mathrm{cm}})$, where C is protein concentration in mg/ml. To convert protein concentration, estimated by the Lowry method[6] to refractometrically determined protein concentration, the Lowry estimate is multiplied by 0.775.

Purification Procedure

All procedures are performed at 4° (see Table I).

Initial Homogenate. Fresh chicken liver is ground in an Oster meat grinder. One kilogram of ground tissue is homogenized in 2 liters of buffer [0.25 *M* sucrose, 0.1 m*M* EDTA, and 2 m*M* *N*-2-hydroxyethylpiperazine-*N'*-2-ethanesulfonic acid (HEPES), pH 7.2] using a loose fitting Potter-Elvehjem homogenizer with a Teflon pestle. This homogenate is diluted with an equal volume of homogenization buffer and centrifuged at 160 *g* for 10 min and the resulting pellet is discarded. The mitochondria are pelleted by centrifugation at 9000 *g* for 15 min. This pellet is gently resuspended in a volume of 20 m*M* potassium phosphate, pH 7.0, containing 0.1 m*M* EDTA and 0.1 m*M* dithiothreitol (DTT), which is equal to the number of grams of liver processed (e.g., 1000 ml for 1 kg of liver). The original procedure of Reed and Lane[3] involves pelleting the mitochondrial suspension in the low ionic strength phosphate buffer, resuspending in the same buffer and then rupturing the mitochondria by three passes through a precooled Manton-Gaulin laboratory homogenizer (Gaulin Corp.,

[6] O. H. Lowry, N. J. Rosebrough, A. L. Farr, and R. J. Randall, *J. Biol. Chem.* **193,** 265 (1951).

Everett, MA) at 6000 psi. The effluent from the press is centrifuged at 105,000 *g* for 60 min. The supernatant, after filtration through four layers of cheesecloth *must* be immediately subjected to both DEAE-cellulose and phosphocellulose chromatographic procedures (vide infra) without delay or very large losses in enzyme activity will be encountered.

In the modified procedure, the original suspension of mitochondria in low ionic strength phosphate buffer is shell frozen and lyophilized. The freeze dried mitochondria can be stored for several weeks at −60°. The lyophilized powder is extracted with 20 m*M* potassium phosphate, pH 7.0, containing 0.1 m*M* EDTA, 0.1 m*M* DTT, 20 μ*M* acetoacetyl-CoA, and 0.1 m*M* phenylmethylsulfonyl fluoride (PMSF); 1 ml of buffer is used for each gram of liver originally processed. If necessary, the mixture is freed of lumps by homogenizing with a Potter-Elvehjem homogenizer prior to passage through a prechilled Manton-Gaulin laboratory homogenizer at 6000 psi. The effluent from the press is centrifuged at 105,000 *g* for 90 min and the resulting supernatant used for DEAE cellulose chromatography.

DEAE-Cellulose Chromatography. A 7.5 × 80-cm column (Glenco Scientific) is packed with standard DEAE-cellulose (exchange capacity 0.9 ± 0.1 mEq/g; Schleicher & Schuell Co, Keene, NH), washed with 5 column volumes of 0.5 *M* potassium phosphate, pH 7.0, and equilibrated with 10 column volumes of 10 m*M* potassium phosphate buffer, pH 7.0, containing 0.1 m*M* EDTA, 0.1 m*M* DTT, and 0.1 m*M* PMSF. The mitochondrial protein extract, after high-speed centrifugation, is applied to the column, which is then washed with the equilibration buffer. HMG-CoA synthase is not retained by the resin at this ionic strength and pH and elutes slightly after the "breakthrough" fractions. The column effluent is monitored for protein (A_{280}) and HMG-CoA synthase activity (spectrophotometric assay). The active fractions are pooled and *immediately* applied to the phosphocellulose column.

Phosphocellulose Chromatography. A 5.0 × 95-cm column (Glenco Scientific) is packed with phosphocellulose resin (exchange capacity 0.9 ± 0.1 mEq/g; Schleicher & Schuell) in 50 m*M* potassium phosphate, pH 7.0. The column equilibrates slowly with 10 m*M* potassium phosphate buffer, pH 7.0, containing 0.1 m*M* DTT and 0.1 m*M* EDTA. Prior to the actual chromatographic procedure, flow should be initiated through the column and the pH and conductivity of the effluent checked to assure complete equilibration. The active fractions from the DEAE cellulose column are applied to phosphocellulose, followed by one column volume of equilibration buffer. The column is then eluted with a 6 liter linear gradient (5–100 m*M* potassium phosphate, pH 7.0, containing 0.1 m*M* DTT and 0.1 m*M* EDTA). The active fractions, which are free of thiolase contamination at this stage, are pooled and concentrated under an N_2 atmosphere in an Amicon stirred cell using a PM-30 membrane. The enzyme is relatively stable at this point and little activity is lost during the concentration step. In contrast, if the enzyme is concentrated by inward dialysis of ammonium sulfate to 55% saturation,[3] there is a substantial loss of activity. Dissociation of the dimeric enzyme at elevated salt concentrations has been reported[4] and may account for the loss of activity.

Gel Filtration. The concentrated enzyme recovered after phosphocellulose chromatography is loaded on a 2.5 × 90-cm column packed with Sephadex G-150 which is equilibrated with 20 m*M* potassium phosphate, pH 7.0, containing 0.1 m*M* DTT and 0.1 m*M* EDTA. This buffer is used to elute HMG-CoA synthase from the column ($V_e/V_0 \sim 1.9$). The active fractions are pooled and supplemented with glycerol (30%, v/v).

Properties

Stability and Purity. Enzyme purified through the Sephadex G-150 step and stored in the presence of 30% glycerol is stable for several days at 4° and over 1 year at −85°. The procedure yields enzyme which is homogeneous by the criteria of sodium dodecyl sulfate (SDS) gel electrophoresis and analytical ultracentrifugation. The specific activity of the homogeneous enzyme may vary from 0.7 to 1.0 units/mg, depending on the methods employed and the delays encountered in proceeding from the mitochondrial extraction through the phosphocellulose chromatography step.

Molecular Properties. HMG-CoA synthase has a sedimentation coefficient ($s_{20,w}$) of 5.7 S and a molecular weight, determined by equilibrium sedimentation, of 105,000 ± 3000.[4] The enzyme is apparently dimeric, composed of subunits with a molecular weight of 53,000 (determined by SDS gel electrophoresis[4]). The isoelectric point of the chicken liver mitochondrial enzyme is reported[4] to be 7.2. Eight sulfhydryl residues per subunit are detectable upon reaction of SDS denatured samples with 5,5′-dithiobis(2-nitrobenzoic acid). This observation is in good agreement with the number of cysteic acid residues detected upon amino acid analysis of performic acid oxidized samples. The amino acid composition of the purified enzyme is given in Table II.

Catalytic Properties. The enzyme requires an acetylated substrate in order for a productive condensation with acetoacetyl-CoA to occur. Thus, while both acetyl-CoA and propionyl-CoA will react to form a covalent acyl-*S*-enzyme species (H. Miziorko, unpublished), only acetyl-*S*-enzyme reacts further to form product. The enzyme will bind bulkier

TABLE II
AMINO ACID COMPOSITION OF CHICKEN LIVER MITOCHONDRIAL HMG-CoA SYNTHASE

Amino acid	Moles amino acid/mole enzyme subunit
Asx	40
Thr[a]	26
Ser[a]	32
Glx	49
Pro	26
Gly	42
Ala	52
Cys[b]	8
Val	30
Met	8
Ile	14
Leu	47
Tyr	20
Phe	18
Lys	17
His	8
Arg	29
Trp	N.D.[c]

[a] Extrapolated to zero time of hydrolysis.
[b] Measured as cysteic acid after performic acid oxidation and as cysteine by titration of denatured enzyme with 5,5′-dithiobis(2-nitrobenzoic acid).
[c] Not determined.

TABLE III
KINETIC DATA FOR CHICKEN LIVER MITOCHONDRIAL HMG-CoA SYNTHASE

Substrates	K_m (μM)	Inhibitors	K_i (μM)
Acetyl-CoA[a]	100	Acetoacetyl-CoA	10; 6[c]
Acetoacetyl-CoA[b]	0.35	HMG-CoA	12
Oxobutyl-CoA[b]	19	R · CoA[d]	108; 79[e]

[a] Measured at 5 μM acetoacetyl-CoA.[4]
[b] Measured at 200 μM acetyl-CoA.
[c] Dissociation constant measured in ESR experiments involving displacement of the spin-labeled acetyl-CoA analog, R · CoA.
[d] 3-Carboxy-2,2,5,5-tetramethyl-1-pyrrolidinyloxyl-CoA.
[e] Dissociation constant measured in ESR titration experiments.

acyl-CoAs at the acetyl-CoA site[7] (acetoacetyl-CoA; spin labeled R · CoA), but these compounds fail to acylate the active site cysteinyl –SH. Nonetheless, the acyl group of an acyl-CoA bound at the acetyl-CoA site is highly immobilized in the noncovalent complex. The thioester functionality is not an absolute requirement in the substrate which condenses with the acetyl-enzyme intermediate. 3-Oxobutyl-CoA, a thioether, reacts slowly with acetyl-*S*-enzyme to produce the thioether analog of HMG-CoA.[8] The enzyme is not absolutely specific for CoA thioester substrates. Acetyl-3′-dephospho-CoA substitutes for acetyl-CoA and *N*-acetyl-*S*-acetoacetylcysteamine serves as an acetoacetyl-CoA analog.

The reaction proceeds via acetyl-*S*-enzyme and enzyme-*S*-HMG-CoA intermediates; a Ping-Pong kinetic pattern is predicted and has been demonstrated for the ox liver enzyme.[9] The extremely high affinity of the chicken liver enzyme for acetoacetyl-CoA[5] ($K_m = 3 \times 10^{-7}\ M$) has precluded a full kinetic study. However, inhibition patterns are consistent with a Ping-Pong scheme. Competitive inhibition with respect to acetyl-CoA is observed in experiments with the product HMG-CoA ($K_i = 12\ \mu M$) and at elevated levels of the second substrate, acetoacetyl-CoA ($K_i = 10\ \mu M$).[5] A summary of kinetic constants is provided in Table III.

Acknowledgments

Work on enzyme purification and characterization that was performed in the author's laboratory was supported in part by NIH Grants AM-21491 and AM-00645.

[7] H. M. Miziorko, M. D. Lane, and S. W. Weidman, *Biochemistry* **18,** 399 (1979).
[8] H. M. Miziorko, P. R. Kramer, and J. A. Kulkoski, *J. Biol. Chem.* **257,** 2842 (1982).
[9] M. A. Page and P. K. Tubbs, *Biochem. J.* **173,** 925 (1978).

[4] 3-Hydroxy-3-methylglutaryl-CoA Reductases from Pea Seedlings

By DAVID WARWICK RUSSELL

The enzyme HMG-CoA reductase (3-hydroxy-3-methylglutaryl-CoA reductase; EC 1.1.1.34) catalyzes the conversion of HMG-CoA to mevalonic acid, the first intermediate used solely for isoprenoid synthesis. Since this committed step is also rate limiting the properties and regulation of the enzyme are pertinent to the biosynthesis of most of the multitude of primary and secondary isoprenoids synthesized in plants.

ISBN 0-12-182010-6

The reaction catalyzed is

$$\underset{\text{HMG-CoA}}{\mathrm{HO{-}\overset{O}{\overset{\|}{C}}{-}CH_2{-}\underset{OH}{\underset{|}{\overset{CH_3}{\overset{|}{C}}}}{-}CH_2{-}\underset{(1)}{\overset{O}{\overset{\|}{C}}}{-}SCoA}} + 2NADPH + 2H^+ \downarrow$$

$$\underset{\substack{\text{MVA}\\ \text{(mevalonic acid)}}}{\mathrm{HO{-}\underset{(1)}{\overset{O}{\overset{\|}{C}}}{-}CH_2{-}\underset{OH}{\underset{|}{\overset{CH_3}{\overset{|}{C}}}}{-}CH_2{-}\underset{(5)}{CH_2OH}}} + 2NADP^+ + CoASH$$

Thus the enzyme catalyzes the reductive deacylation of HMG-CoA; the thiol-esterified carboxyl group (C-1 of HMG-CoA), is reduced to a primary alcoholic group, on C-5 of mevalonic acid. The stoichiometry shown derives from studies on the enzyme from microorganisms[1] and mammals, and comparison of spectrophotometric and radiochemical assay data for the enzyme from plants indicates similar stoichiometry.[2]

While the procedures described below have been established using pea seedling tissues, the major features probably apply equally well to other tissues. However, minor modifications may be necessary in some cases, and the sections on Compartmentation, Isolation Variables, and enzyme stability, will help to identify isolation problems.

Compartmentation of HMG-CoA Reductases

Many isoprenoid and partly isoprenoid compounds in higher plants occur only in particular organelles. This is well illustrated by the chloroplast (photosynthetic) and primary isoprenoids, and includes for example, the chloroplast location of carotenoids, chlorophyll, and plastoquinone, the mitochondrial ubiquinone and heme A, and the occurrence of sterols mainly in the membranes of the endoplasmic reticulum and those bounding the cytoplasm. This strict subcellular compartmentation gave rise to early speculations that isoprenoid biosynthesis may occur in more than one compartment, and feeding experiments with $^{14}CO_2$ and [^{14}C]mevalonic acid yielded evidence consistent with multiple pathways.[3] More

[1] I. F. Durr and H. Rudney, *J. Biol. Chem.* **235,** 2572 (1960).

[2] J. D. Brooker and D. W. Russell, *Arch. Biochem. Biophys.* **167,** 723 (1975).

[3] K. J. Treharne, E. I. Mercer, and T. W. Goodwin, *Biochem. J.* **99,** 239 (1966).

recent enzyme localization studies[4–6] strongly support the presence of compartmented pathways in the cytoplasm, chloroplasts, and probably mitochondria, which convert HMG-CoA to farnesyl pyrophosphate, from which branch paths presumably lead to the specific isoprenoids characteristic of each compartment.

Clearly, the measurement of HMG-CoA reductase activity can only be validly directed at the measurement of one particular enzyme species, under assay conditions optimal for that particular enzyme, and using defined subcellular fractions with minimal contamination from other sources. Since the reductase activities of the cytoplasm, plastids, and mitochondria are all firmly membrane bound,[5] their individual isolation is quite feasible, but it is essential to rupture the cells under isotonic conditions and without harsh treatment so that the organelles are largely intact. The intact organelles are then isolated, ruptured, and the organelle membranes, containing the reductase activity, are pelleted. Cytoplasmic membranes are sedimented after preliminary centrifugation to remove damaged and fragmented organelles. Thus cross-contamination is minimized and mixed enzyme preparations avoided.

Isolation Variables

There is little point in attempting to assay HMG-CoA reductase activity in crude homogenates for 3 main reasons: (1) three reductase species are present; in the chloroplast, mitochondrial, and cytoplasmic membranes[5] (see above); (2) irreversible inactivation reactions occur, even at ice temperatures; and (3) reversible inactivation or activation may occur. These are briefly discussed below, together with a consideration of isolation buffers.

1. *Multiple Reductase Activities*. The evidence for reductase activities associated with endoplasmic reticulum (cytoplasm), plastids (and chloroplasts), and mitochondria is mentioned above, and studies on the plastid[7,8] and microsomal (cytoplasmic) reductases[2,9–11] show that they have distinctive kinetic and regulatory properties. Assay of a particular reductase

[4] L. J. Rogers, S. P. J. Shah, and T. W. Goodwin, *Biochem. J.* **99,** 381 (1966).

[5] J. D. Brooker and D. W. Russell, *Arch. Biochem. Biophys.* **167,** 730 (1975).

[6] T. R. Green, D. T. Dennis, and C. A. West, *Biochem. Biophys. Res. Commun.* **64,** 976 (1975).

[7] R. J. Wong, D. K. McCormack, and D. W. Russell, *Arch. Biochem. Biophys.* **216,** 631 (1982).

[8] D. W. Russell and J. E. Harvey, *Plant Physiol.* **63,** Suppl., 156 (1979).

[9] J. D. Brooker and D. W. Russell, *Arch. Biochem. Biophys.* **198,** 323 (1979).

[10] D. W. Russell and H. Davidson, *Biochem. Biophys. Res. Commun.* **104,** 1537 (1982).

[11] D. W. Russell and W. D. Dix, *Proc. Int. Bot. Congr., 13th,* p. 316 (1981).

therefore necessitates isolation of the relevant organelle or membrane fraction, requiring isotonic buffers, and homogenization procedures which cause minimal organelle rupture commensurate with adequate cell rupture and enzyme yield. The organelle membranes can be sedimented after organelle rupture. The membrane preparations may also benefit from a wash treatment although this will prolong the isolation time.

2. *Irreversible Inactivation.* Irreversible loss of activity occurs by at least two mechanisms: (i) oxidative inactivation, and (ii) soluble protein-dependent inactivation.

i. *Oxidative inactivation.* This can be prevented by the inclusion of 2-mercaptoethanol in the isolation media; dithiothreitol is as good or better, and can be recommended where financial considerations are unimportant. Although mercaptoethanol is quite satisfactory during isolation, dithiothreitol is essential for enzyme stability in the isolated membrane preparation.[12] Furthermore, dithiothreitol concentrations (10 m*M*) which are optimal for reductase activity in assay mixtures,[2] do not prevent loss of activity in membrane preparations even when stored for short periods (1–2 hr) on ice[12]; enzyme stability in storage requires at least 20 m*M* dithiothreitol (25 or 30 m*M* is better).[12] It is important to note that this high concentration of dithiothreitol is *not recommended* during isolation because it causes activation of a reductase-inactivating principle which is present in the soluble protein fraction[13] (see below, ii).

ii. *Soluble protein-dependent inactivation.* When aliquots of dialyzed soluble protein, containing 25 m*M* dithiothreitol, are added to isolated reductase membrane preparations and incubated in the presence of 25 m*M* dithiothreitol for only 5 min, most of the reductase activity is irreversibly lost.[13] The problem is avoided by using mercaptoethanol during isolation, or lower dithiothreitol concentrations (10–12 m*M*). The evidence suggests the presence of a thiol-activated protease.

3. *Reversible Inactivation and Activation.* Homogenates allowed to stand for 1 hr or more at ice temperature may show increased or decreased reductase activity. Inactivation is observed when mitochondria are present, and activation occurs in the postmitochondrial supernatant.[14] The changes are consistent with the activity of a reductase phosphorylation system. No changes are evident when the media contain EDTA and the extract is processed promptly. The use of NaF may be beneficial, but has no advantages in the pea seedling system since, in the etiolated tissues routinely used, the microsomal reductase is almost completely in the activated state.[10,11]

[12] D. W. Russell and S. Gan, unpublished observations (1981).

[13] J. S. Knight and D. W. Russell, unpublished observations (1982).

[14] D. W. Russell, unpublished observations (1980).

4. *Isolation Buffers.* Some highly acidic tissues require much stronger buffers to maintain the pH: mercaptoethanol or dithiothreitol is necessary (but see Section 2 above), and the use of EDTA and NaF is mentioned in Section 3 above.

Microsomal HMG-CoA Reductase: Assay, Isolation, and Properties

Specific activity of this reductase is highest in young growing tissues of green or etiolated seedlings,[2] reflecting its importance in the biosynthesis of primary isoprenoids necessary for cell division and differentiation. Activity is almost absent in mature tissues.

The apical buds of etiolated seedlings are preferred as a source of enzyme, because the specific activity of the microsomal reductase from these tissues is about 2-fold higher than in analogous preparations from green seedlings.

Assay

Two assay procedures have been used and both are described below. Method 1 employs a final reaction volume of 100 μl and standard apparatus including thin-layer chromatography equipment; this procedure has been extensively used in past work and is recommended for laboratories unfamiliar with microtechniques. Method 2, more recently used,[12,13] employs a reaction volume of 20 μl and is very rapid and economical, but demands greater skills and some specialized apparatus.

Principle. The radiochemical assay used is based on the conversion of *RS*-[3-^{14}C]HMG-CoA to [^{14}C]MVA which is isolated by thin-layer chromatography and counted. The sensitivity of the assay depends on the specific radioactivity of the [^{14}C]HMG-CoA; very high sensitivities can be achieved by using high specific activity [^{14}C]HMG-CoA but under these conditions it is imperative to verify its radiochemical purity.

Assay Method 1 (Microsomal Reductase)

Reagents

Homogenizing medium: 10 m*M* Tris–HCl pH 7; 0.35 *M* sucrose; 30 m*M* EDTA; 10 m*M* 2-mercaptoethanol

Suspension medium: 0.2 *M* K-phosphate buffer pH 6.9; 25 m*M* dithiothreitol

NADPH-generating system: 0.2 *M* glucose 6-phosphate; 40 m*M* NADP$^+$; 20 units/ml glucose-6-phosphate dehydrogenase. The activity of the generating system should be checked regularly

0.2 *M* Dithiothreitol

6 *M* HCl

2 *M* MVA-lactone: mevalonic acid for adding to reaction mixtures is obtained by combining 2 *M* MVA-lac and 0.1 *M* KOH in the ratio 1 : 1 (v/v)

Chloroform and acetone

Thin layer plates, glass: 20 × 5 cm, spread with a 250 μm layer of silica gel G (Merck): when air-dried, these are activated at 110° for 1 hr then stored at 37°, or over desiccant

Scintillation liquid: 9 g butyl-PBD dissolved in 1 liter of toluene, and 500 ml of Triton X-100 added. (Efficiency is much lower in toluene scintillants without Triton)

RS-[3-^{14}C]HMG-CoA 2230 dpm/nmol. Synthesized by the method of Goldfarb and Pitot[15] (reaction vol 0.5 ml/250 μCi), except that the anhydride is added directly to the CoASH solution (0.5 ml) without recrystallizing. The HMG-CoA is purified by paper chromatography.[16] Purity of stock solutions should be checked after 6 months storage at −20°

Procedure. The NADPH-generating system is made up on ice in a test tube (7.5 × 0.8 cm) by combining identical aliquots from separate, concentrated (5-fold), stock solutions of components in K-phosphate buffer, and water is added to give the required concentrations (above); the solution is mixed and kept on ice.

The reaction mixture is prepared on ice, in glass test tubes (10 × 1 cm) and comprises 5 μl of 0.2 *M* dithiothreitol; 5 μl of NADPH-generating system; 25 μl of microsomal enzyme suspension (max 0.8 mg protein) in 0.2 *M* K-phosphate buffer pH 6.9 and 25 m*M* dithiothreitol; water is added to bring the volume to 95 μl. The mixture is warmed 5 min in a 30° shaking waterbath, and the reaction then started by adding 5 μl of *RS*-[3-^{14}C]HMG-CoA containing 70 nmol; the final reaction volume is thus 100 μl.

The tube is capped with Parafilm and the reaction mixture incubated for 20 min at 30° in a shaking waterbath.

The reaction is then stopped by adding 10 μl of 1 *M* MVA and 10 μl 6 *M* HCl with immediate mixing. After standing at room temperature for at least 15 min (for maximum lactonization of mevalonic acid), the mixture is centrifuged (5 min at half speed, bench centrifuge) to pellet the protein, and 60 μl of supernatant (total vol, 120 μl) is streaked onto a thin layer plate of activated silica gel, 250 μm thick (Silica gel G, Merck). The plate is developed with chloroform–acetone (2 : 1, v/v) for 90 min, or until the solvent front is 2 cm from the end of the plate. The plate is dried, scanned

[15] S. Goldfarb and H. Pitot, *J. Lipid Res.* **12,** 512 (1971).

[16] A. I. Louw, I. Bekersky, and E. Mosbach, *J. Lipid Res.* **10,** 683 (1969).

in a radiochromatogram scanner to locate the MVA-lactone (R_f 0.65), which is then scraped into a scintillation vial for counting. A similar area of silica gel in a low background region is also scraped into a vial to provide a background count. (If no scanner is available, the MVA-lactone can be visualized with iodine vapor[17]; in this case additional MVA is added after the reaction is stopped.) Triton-toluene scintillation liquid (10 ml) is added, swirled to disperse the silica gel and dissolve the MVA-lactone, and the samples counted. The specific activity of the reductase is calculated after correcting for background, recovery of MVA-lactone from the thin layer plate, and counting efficiency, and is expressed as units/mg protein (microsomal protein).

Units. One unit of reductase catalyzes the formation of 1 nmol of mevalonic acid per hour, under the above assay conditions.

Protein Assay. The modified Lowry procedure described by Peterson[18] is used: this gives excellent results even with very small samples, and removes molecules (DTT, EDTA etc.) which interfere with the standard Lowry procedure.

Assay Method 2 (Microsomal Reductase)

This is a microassay procedure, similar in principle to Method 1, but requiring 1.5 ml Eppendorf tubes (stoppered polyethylene microcentrifuge tubes), microcentrifuge, micro thin layer plates, and a little practice in dispensing microliter volumes from the micropipets used (Gilson or similar). However, this method is much faster and more economical than Method 1. The reagents are identical to those of Method 1 excepting the higher (5- to 6-fold) specific activity of the [^{14}C]HMG-CoA employed, which gives this assay a similar or better sensitivity than Method 1.

Reagents

Thin layer plates: standard glass microscope slides (7.5 × 2.5 cm) coated with silica gel G (Merck); two slides, held back to back, are dipped into an aqueous slurry of silica gel, the excess is momentarily drained off, the slides are separated, then laid horizontal to produce a uniform layer. Subsequent treatment and activation as in Method 1

Diethyl ether and acetone

[17] T. S. Ingebritsen and D. M. Gibson, this series, Vol. 71, p. 489.

[18] G. L. Peterson, *Anal. Biochem.* **83,** 346 (1977).

RS-[3-^{14}C]HMG-CoA, 12,860 dpm/nmol (for synthesis, see Method 1, microsomal reductase)

See Method 1, microsomal reductase for homogenizing medium; suspension medium; NADPH-generating system; dithiothreitol; HCl; MVA-lactone; and scintillation liquid

Procedure. The NADPH-generating system is made up in a 1.5 ml stoppered polyethylene microcentrifuge tube (Eppendorf) by combining identical aliquots from separate (5× conc.) stock solutions of the different components; after careful mixing this is kept on ice.

The enzyme assay mixture is prepared on ice in 1.5 ml stoppered polyethylene microcentrifuge tubes (Eppendorf) and comprises 1 μl of 0.2 *M* dithiothreitol; 1 μl of NADPH-generating system; 5 μl of microsomal enzyme suspension (max. 0.16 mg protein) in 0.2 *M* K-phosphate buffer pH 6.9, and 25 m*M* dithiothreitol; water is added to give a volume of 17 μl, and after mixing carefully, the tube is warmed for 5 min at 30° in a shaking waterbath. The reaction is then started by adding 14 nmol of *RS*-[3-^{14}C]HMG-CoA in 3 μl, followed immediately by mixing, and the tube is stoppered. Total final reaction volume is thus 20 μl.

The reaction mixture is incubated for 20 min at 30° by floating the tubes, in polystyrene holders, in a 30° shaking waterbath. The tube-holders made from foam-polystyrene sheet (2.25 cm thick) have holes (cork borer) just large enough for a loose fit so that the tubes are held by their rim and their tips project down below the bottom of the polystyrene sheet: these holders are convenient for handling the tubes on the bench, or on ice, as well as in the waterbath.

The reaction is stopped by adding 2 μl of 1 *M* mevalonic acid and 2 μl of 6 *M* HCl. After mixing and standing at room temperature for 15 min (for optimal MVA lactonization), the mixture is centrifuged (~9000 *g*) in an Eppendorf microcentrifuge (or similar) for 1–2 min to pellet the protein. The supernatant is analyzed by applying a 10-μl aliquot of supernatant (total vol, 24 μl) as a streak, 1.3 cm from the base of a micro thin layer plate of activated silica gel. The plate is developed in a standard Shandon tank without preequilibration, using 40 ml of solvent (diethyl ether–acetone, 3 : 1, v/v). When the solvent front reaches the end of the silica gel (5–10 min) the plate is removed, dried, and the top half (R_f 0.5–1.0) is scraped into a scintillation vial for counting. Under these conditions MVA-lactone has an R_f of 0.85[12] and the substrate remains at the origin: background counts should be negligible relative to counts in [^{14}C]MVA. Scintillation liquid (10 ml) is added, swirled to disperse the silica gel and dissolve the MVA-lactone, then counted. Specific activity of the enzyme is calculated after correcting for counting efficiency and recovery of

MVA-lactone from the thin layer plate, and is expressed as units per mg protein (microsomal protein).

Units and Protein Assay. See Method 1, microsomal reductase.

Isolation of Microsomal Membranes

For tissues other than pea seedlings, inspection of the early sections on Compartmentation, and Isolation Variables may be helpful.

Etiolated pea seedlings are grown in darkness in moist vermiculite at 27° and used (green safelight) when 9 days old, at the third internode stage. Green pea seedlings grown in vermiculite under artificial light at 24° are used when 6–15 cm high (10 days to 3 weeks).

All preparative steps are carried out at 0–4°. The tissues (buds of etiolated seedlings; buds and semimature leaves of green seedlings) are harvested into a tared plastic cup on ice and the weight noted (3–20 g). After transfer to a precooled mortar on ice, 3 volumes of homogenizing medium is added, together with purified[19] Polyclar-AT (insoluble polyvinylpyrrolidone, GAF) at 10% of tissue fresh weight, and the tissues then homogenized on ice for 3 min. The mixture is carefully squeezed through muslin (2 layers), and the filtrate is centrifuged at 12,000 *g* for 10 min. The postmitochondrial supernatant is decanted into preparative ultracentrifuge tubes taking care that no loose particulate material at the surface of the pellet is decanted with the supernatant; the pellet and loose surface membrane material is discarded. The postmitrochondrial supernatant is then centrifuged in a Type 65 rotor (Beckman; or similar) at 50,000 *g* for 60 min. The microsomal membranes which sediment under these conditions contain 95% of the reductase activity present in total microsomal preparations sedimented at the standard 105,000 *g* for 60 min; hence the 50,000 *g* pellet contains virtually all the reductase but only about half the protein of standard microsomal preparations, and the specific activity is consequently about double. The supernatant is decanted, and after draining residual liquid from the tube, the walls are wiped dry. The pellet is suspended in suspension medium in the ratio, 0.15 ml of suspension medium per gram of original tissue. Homogeneous suspensions are readily achieved by initially homogenizing (glass rod) the pellet in the residual liquid, then adding an aliquot (~50 μl) of suspension medium, mixing again, then adding the additional suspension medium. Before making to the final volume the homogeneity is checked using a glass micropipet (auto-zero, constriction-type); any visible particulate matter remaining is readily dispersed by 2 or 3 passes through this pipet.

[19] W. D. Loomis and J. Battaile, *Phytochemistry* **5**, 423 (1966).

Properties

Optimal pH. Activity of the microsomal reductase is optimal at pH 6.9[2] and declines sharply at higher or lower values. Higher activities are observed in phosphate than in other buffers: activity is relatively low in Tris–HCl[13] presumably due to perturbation of the membrane by unprotonated molecules of Tris.[20]

K_m Value. The apparent K_m for RS-HMG-CoA is 160 μM.[2,7]

Inhibition by HMG and CoASH. Activity is depressed by low concentrations of HMG or CoASH.[2] The addition of 1.0 m*M* HMG causes 23% inhibition, and 25 μM CoASH inhibits activity by 33%.

Responses to Isoprenoid End Products and Hormones. Addition of sterols, isoprenoid hormones (abscisic acid, gibberellic acid), or auxin (2,4-D), has no effect on activity of the microsomal reductase.[2,10,11]

Inactivation and Activation by Cytosolic Protein

(i) *Irreversible inactivation.* In the presence of cytosolic protein and 25–30 m*M* dithiothreitol, there is a rapid, cofactor-independent, irreversible loss of reductase activity.[13] This inactivation is not observed in 10–12 m*M* dithiothreitol.

(ii) *Reversible inactivation.* Preincubation (5 min) with dialyzed cytosolic protein fractions in the presence of ATP and Mg^{2+} causes a loss (up to 80%) of reductase activity,[13,14] which is restored by incubation with bacterial alkaline phosphatase.[21] Activation also occurs when the reductase preparation is incubated with dialyzed cytosolic protein, and the activation is blocked by NaF.[13] The incorporation of ^{32}P from [γ-^{32}P]ATP accompanies inactivation, and its release parallels restoration of activity.[21] Studies on hormonal control of activity, utilizing quantitative assays for the reductase kinase and phosphatase activities in cytosolic protein fractions, have shown that the hormonal isoprenoid, abscisic acid, at 10 μM, causes up to 86% inhibition of reductase phosphatase activity.[13]

Responses in Vivo. In vivo treatments indicate rapid posttranslational control of activity by phytochrome[9] and abscisic acid[22] (negative control), and by gibberellic acid and zeatin[11] (positive control).

Storage and Stability. The microsomal preparation is best stored at −20° in 25 m*M* dithiothreitol,[12] and the inclusion of glycerol (50%) prevents freezing and enables direct sampling of the stored enzyme without thawing. At this concentration of dithiothreitol, activity is stable for at

[20] N. E. Good and S. Izawa, this series, Vol. 24, p. 63.

[21] D. W. Russell and W. D. Dix, unpublished observations (1981).

[22] D. W. Russell and J. Singh, *Proc. Aust. Biochem. Soc.* **13,** 71 (1980).

least 1 hr at 30°. Both dithiothreitol, and the high concentration of glycerol, are essential for stability; 2-mercaptoethanol is ineffective.

Plastid HMG-CoA Reductase: Assay, Isolation, and Properties

The plastid HMG-CoA reductase, which occurs in the mature chloroplast also, is firmly associated with the envelope membranes. Both the kinetic and regulatory properties of the plastid reductase[7,8] are quite different from those of the microsomal enzyme. The specific activity of the plastid reductase is highest in plastid and chloroplast membrane preparations from young tissues of either green or etiolated seedlings. The higher specific activity of the membrane preparations from young etiolated tissues can be ascribed to the low levels of internal membrane in the plastids (etioplasts) from dark grown seedlings. The assay of activity in chloroplast membrane preparations from mature tissues may require the isolation of envelope-enriched fractions and the use of higher specific activity [^{14}C]HMG-CoA.

Assay

The two assay procedures described below are analogous to those described for the microsomal reductase and similar general comments apply.

Principle. A radiochemical assay is used in which the substrate, *RS*-[3-^{14}C]HMG-CoA, is separated from the product [^{14}C]MVA, by thin layer chromatography, and the isolated product then counted.

Assay Method 1 (Plastid Reductase)

Reagents

Homogenizing medium: 10 m*M* MOPS pH 7.5; 0.35 *M* sucrose; 10 m*M* KCl; 20 m*M* EDTA; 0.1% BSA; 10 m*M* 2-mercaptoethanol

Rupture medium: 10 m*M* MOPS pH 7.5; 10 m*M* KCl; 25 m*M* EDTA; 10 m*M* dithiothreitol

Suspension medium: 0.25 *M* K-phosphate buffer pH 7.9; 25 m*M* dithiothreitol

NADPH-generating system: 0.2 *M* glucose 6-phosphate; 40 m*M* NADP$^+$; 20 units/ml glucose-6-phosphate dehydrogenase; in 0.15 *M* K-phosphate buffer pH 7.9

0.25 *M* K-phosphate buffer pH 7.9

0.2 *M* Dithiothreitol

RS-[3-^{14}C]HMG-CoA, 8300 dpm/nmol (for synthesis, see Method 1, microsomal reductase)

Other reagents as in Method 1 for the microsomal reductase, including HCl, MVA-lactone, chloroform and acetone, thin layer plates, and scintillation liquid

Procedure. The NADPH generating system is made up from separate concentrated (5-fold) stock solutions of the components in K-phosphate buffer; the solution is mixed and kept on ice.

The reaction mixture is made up in test tubes (10 × 1 cm) on ice from the following components: 15 μl of 0.25 *M* K-phosphate buffer, pH 7.9; 5 μl of 0.2 *M* dithiothreitol; 5 μl of NADPH-generating system; 25 μl of the enzyme preparation (max. 0.4 mg protein) in 0.25 *M* K-phosphate buffer pH 7.9 and 25 m*M* dithiothreitol; water is added to bring the volume to 97.5 μl. The mixture is warmed for 5 min with shaking in a 30° waterbath, and the reaction started by adding 2.5 μl of *RS*-[3-^{14}C]HMG-CoA (15 nmol) to give a final reaction volume of 100 μl.

The subsequent incubation conditions, stopping the reaction, analysis by thin-layer chromatography, and calculating specific activity, are identical to the procedures detailed under Method 1 for the microsomal reductase. The specific activity of the plastid enzyme is expressed as units/mg protein (plastid membrane protein).

Units. One unit of plastid HMG-CoA reductase catalyzes the formation of 1 nmol of mevalonic acid per hour under the above assay conditions.

Protein Assay. The modified Lowry procedure described by Peterson[18] is used (for comments, see Method 1, microsomal reductase).

Assay Method 2 (Plastid Reductase)

For general comments about this microassay procedure, see Method 2, microsomal reductase.

Reagents

Refer to Method 1, plastid reductase, for homogenizing medium; rupture medium; suspension medium; NADPH-generating system; K-phosphate; and dithiothreitol

RS-[3-^{14}C]HMG-CoA, 50,000 dpm/nmol (for synthesis, see Method 1, microsomal reductase)

See Method 2, microsomal reductase for thin layer plates, diethylether, acetone

See Method 1, microsomal reductase for HCl, MVA-lac, scintillation liquid

Procedure. The NADPH-generating system is made up in a 1.5 ml stoppered polyethylene centrifuge tube (Eppendorf) on ice, by combining identical aliquots of the separate concentrated (5-fold) stock solutions of the different components made up in K-phosphate buffer pH 7.9; water is added to give the concentrations listed above. This cocktail is carefully mixed and kept on ice.

The reaction mixture is made up on ice in 1.5 ml stoppered polyethylene centrifuge tubes (Eppendorf) by combining 3 μl of 0.25 *M* K-phosphate buffer pH 7.9; 1 μl of 0.2 *M* dithiothreitol; 1 μl of NADPH-generating system; 5 μl of enzyme preparation (max 0.08 mg protein) in 0.25 *M* K-phosphate buffer pH 7.9, and 25 m*M* dithiothreitol; water is added to bring the volume to 18.5 μl and the contents are carefully mixed. The reaction mixture is brought to temperature by warming for 5 min at 30° in a shaking waterbath; the reaction is then started by adding 1.5 μl of [^{14}C]HMG-CoA (3 nmol), the tube is stoppered, and the contents mixed immediately. Thus the final total volume of the reaction mixture is 20 μl.

The reaction mixture is incubated for 20 min at 30° as described under Method 2 for the microsomal reductase. Subsequent procedures, up to and including thin-layer chromatography, scraping into scintillation vials, and counting, are identical to those of Method 2 for the microsomal HMG-CoA reductase. Specific activity of the enzyme is expressed as units/mg protein (plastid membrane protein).

Units and Protein Assay. See Method 1 (plastid reductase).

Isolation of Plastid Membranes

The growth of pea seedlings and the tissues used are described in the section, Isolation of Microsomal Membranes.

All preparative steps are carried out at 0–4°. The tissue is harvested into a tared plastic dish on ice, and after noting the weight, is transferred to a precooled mortar. Homogenizing medium (3 vol) and purified[19] Polyclar-AT (10% of tissue weight) are added and the tissues lightly homogenized for 3 min. The homogenate is squeezed gently through muslin (2 layers), passed through 1 layer of Miracloth (Calbiochem), then centrifuged in a swinging bucket rotor for 2 min at 500 *g* to pellet nuclei. The supernatant is decanted, including loose material at the pellet surface, and centrifuged in a swinging bucket rotor at 3500 *g* for 2 min to pellet the plastids. The plastids are ruptured by resuspending in a minimum of 2 ml of rupture medium per gram of original tissue, and the plastid membranes then sedimented by centrifuging at 50,000 *g* for 16 min. The supernatant is completely decanted and the pellet resuspended in suspension medium, using 30 μl/g of original tissue. Procedures used to obtain a homogeneous suspension are described under Isolation of Microsomal Membranes.

Properties

Optimal pH. Activity of the plastid reductase is optimal at pH 7.9, drops sharply at higher values, and less abruptly at lower pHs. At pH 7 activity is about 30% less than at pH 7.9.[7]

Buffers and Activity. Activity is highest in K-phosphate and the phosphate concentration in the reaction mixture should be about 0.1 *M*.[7] When the presence of phosphate is undesirable, MOPS is the best alternative.[23]

K_m *Value.* The apparent K_m for RS-HMG-CoA is about 0.8 μM,[7] and concentrations of HMG-CoA in excess of saturating levels are not inhibitory.

Response to Isoprenoid Products. In limited testing of plastid pathway end products, addition of gibberellic acid or abscisic acid to reaction mixtures containing only the membrane-bound reductase had little or no effect on activity.[23]

Inactivation and Activation by Stromal Protein. Reversible inactivation occurs when the plastid reductase is incubated for 5 min in the presence of dialyzed stromal protein fractions together with ATP and Mg^{2+}.[23] Activity is restored by adding bacterial alkaline phosphatase; activation also occurs when the inactivated enzyme is incubated with stromal protein fractions only (no cofactors), and the activation is blocked by NaF.[23] Studies on the regulation of the inactivation (kinase) and activation (phosphatase) activities of the stromal protein fraction show that the plastid-synthesized hormonal isoprenoids, abscisic acid and gibberellic acid, stimulate the reductase-inactivating[23] (kinase) activity.

Responses in Vivo. Phytochrome activation causes a rapid post-translational stimulation of plastid reductase activity and the response is blocked by abscisic acid.[7,8,24]

Storage and Stability. When the plastid membrane suspension is quick-frozen and stored at −20° there is no loss of activity for at least several days.[23] Dithiothreitol, at 25 m*M*, is essential for stability; some loss in activity occurs at lower concentrations, and mercaptoethanol is much less effective. Dithiothreitol supplied by Sigma gives satisfactory stability but supplies from cheaper alternative sources have given disastrous results.

Plastid and Microsomal HMG-CoA Reductases: Interrelations

The plastid and microsomal reductases clearly differ in their pH optima, and K_m for RS-HMG-CoA; this, in addition to their characteristic localization, strongly suggests that they are distinctive enzymes rather

[23] T. Wilson and D. W. Russell, unpublished observations (1983).

[24] R. Jackson and D. W. Russell, unpublished observations (1979).

than compartmented forms of the same enzyme, and their differing regulatory properties support this interpretation. The differences between these two reductases suggest that each has an individual and distinctive role in the regulation of isoprenoid biosynthesis in the respective compartments.

Perhaps more striking are their responses to growth regulatory stimuli, and their sophisticated but distinctive phosphorylation control systems. Both enzymes show rapid (1–5 min) posttranslational responses to a variety of growth regulatory stimuli, but the same stimulus may elicit opposite responses in the two enzymes. The nature of the stimuli, together with the speed of the enzyme responses, clearly indicates that the regulation of both enzymes is closely coupled to the regulation of cellular growth and development. It must be concluded that one or more of the primary isoprenoids, other than the isoprenoid hormones, have vital roles in basic molecular reactions of growth and differentiation. Furthermore, the synthesis of hormonal isoprenoids in the plastid, and their regulatory effects on the plastid reductase (feedback), together with their hormonal action on the microsomal (cytoplasmic) reductase, indicates a regulatory coupling between plastid and microsomal reductase which contributes to such coordinated developmental responses as the concomitant light-induced isoprenoid hormone synthesis and development in the chloroplast, and the simultaneous stimulation of leaf growth.

[5] 3-Hydroxy-3-methylglutaryl-CoA Reductase in the Latex of *Hevea brasiliensis*

By ABDULLAH SIPAT

Introduction

Natural rubber is a cis-polyisoprenoid of high molecular weight (10^5 to 4×10^6) obtained from the latex of *Hevea brasiliensis*. The pathway for its biosynthesis up to the formation of isopentenyl pyrophosphate is similar to that for other isoprenoid compounds (Fig. 1). One of the enzymes in this part of the pathway is 3-hydroxy-3-methylglutaryl-CoA (HMG-CoA) reductase (EC 1.1.1.34). This enzyme has been reported to occur in the serum fraction of preserved *Hevea* latex.[1] Hepper and Audley[2] also re-

[1] F. Lynen, *Pure Appl. Chem.* **14,** 137 (1967).

[2] C. M. Hepper and B. G. Audley, *Biochem. J.* **114,** 379 (1969).

METHODS IN ENZYMOLOGY, VOL. 110

ISBN 0-12-182010-6

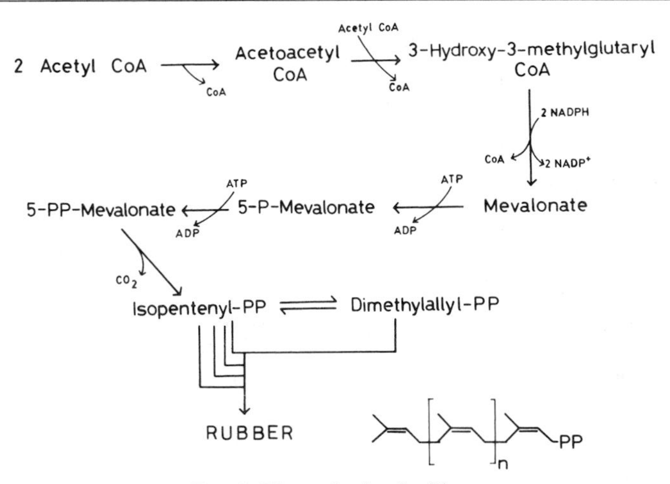

FIG. 1. Biosynthesis of rubber.

ported its occurrence in the bottom fraction of centrifuged (600 *g*) latex obtained from *Hevea* seedlings grown in a glasshouse. More recent work on the enzyme has been carried out in this author's laboratory using the latex from mature *Hevea* trees which are under commercial production.[3-6] This paper describes the methodology that is used to study the reductase and also some of its properties.

Materials and Methods

Plant Material. Previous studies show that the reductase activity varies with the clone from which the latex is obtained. Of the clones that were examined (GT 1, PB 86, Tjir 1, RRIM 501, RRIM 600, and RRIM 701), PB 86 had the highest reductase activity in the bottom fraction suspension as well as the washed bottom fraction suspension (see Fig. 3). RRIM 501 had the lowest reductase activity while the other clones had values between these two. Studies of the reductase reported herein have been carried out using the latex from clone RRIM 600 (approx. 16 years old) grown on the University's Estate. Ten trees were selected and placed under a tapping schedule of once every 2 days in a working week.

Collection of the Latex. The latex is obtained by making a half-spiral incision around the tree trunk (average diameter of 25 cm) using a com-

[3] A. B. Sipat, *Biochim. Biophys. Acta* **705,** 284 (1982).
[4] A. B. Sipat, *Phytochemistry* **21,** 2613 (1982).
[5] R. B. Isa and A.B. Sipat, *Biochem. Biophys. Res. Commun.* **108,** 206 (1982).
[6] A. B. Sipat, *Pertanika* **5,** 246 (1982).

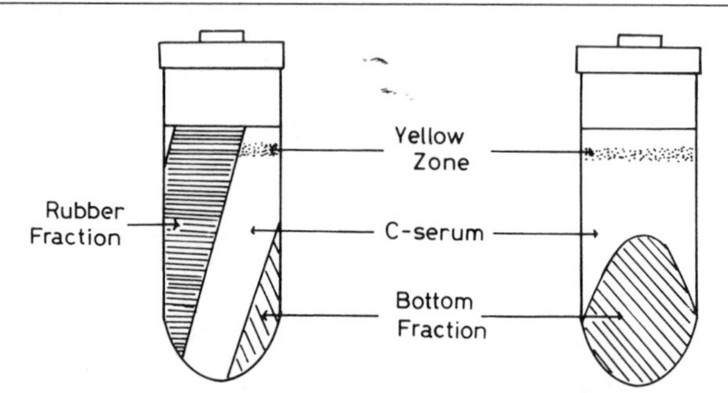

FIG. 2. Separation of latex by centrifugation. The freshly collected latex was centrifuged at 21,000 rpm (59,000 g_{max}) in Rotor 21 using a Beckman L5-65 Ultracentrifuge set to 4°. A polyallomer centrifuge tube is a satisfactory substitute if a cellulose nitrate type is not available.

mercial tapping knife. The flow for the first 3 min is allowed to run to waste since this contains a higher proportion of disrupted subcellular organelles. Thereafter, the latex is collected for the next 30 min into a conical flask surrounded by crushed ice. The collections from the 10 trees are pooled and the yield ranges from 350 to 450 ml.

Fractionation of the Latex. The latex obtained by tapping is a cytosolic suspension of rubber particles (approx. 45%, v/v) and two other nonrubber particles, viz., the lutoids and the Frey-Wyssling complexes. These particles have been described in detail elsewhere.[7-10] The endoplasmic reticulum and the mitochondria—two subcellular organelles with which the reductase of animal tissues are associated—are rarely observed in the fractions of the latex examined by electron microscopy.[9,10]

The latex is fractionated by ultracentrifugation at 21,000 rpm (Rotor 21) or 25,000 rpm (Rotor 65) using a Beckman L5-65 Ultracentrifuge with the temperature set to 4°.[11] This corresponds to a g_{max} value of 59,000 and 54,500, respectively. By treating the latex with neural red before centrifugation, Moir[11] was able to identify at least 11 distinct zones in the centrifuge tube but for the purpose of the present work, only 4 major fractions are considered. These are the rubber fraction, the yellow zone (Moir's zone 4), the C-serum, and the bottom fraction (Fig. 2). This centrifugation scheme is preferred because the ultrastructure of the rubber fraction, the

[7] P. B. Dickenson, *J. Rubber Res. Inst. Malays.* **21,** 543 (1969).

[8] J. B. Gomez and G. F. J. Moir, "The Ultracytology of Latex Vessels in *Hevea brasiliensis,*" Malays. Rubber Res. Dev. Board Monogr. No. 4 Polygraphic Press, Kuala Lumpur, 1979.

[9] S. Bt. Hamzah and J. B. Gomez, *J. Rubber Res. Inst. Malays.* **30,** 161 (1982).

[10] S. Bt. Hamzah and J. B. Gomez, *J. Rubber Res. Inst. Malays.* **31,** 117 (1983).

[11] G. F. J. Moir, *Nature (London)* **184,** 1626 (1959).

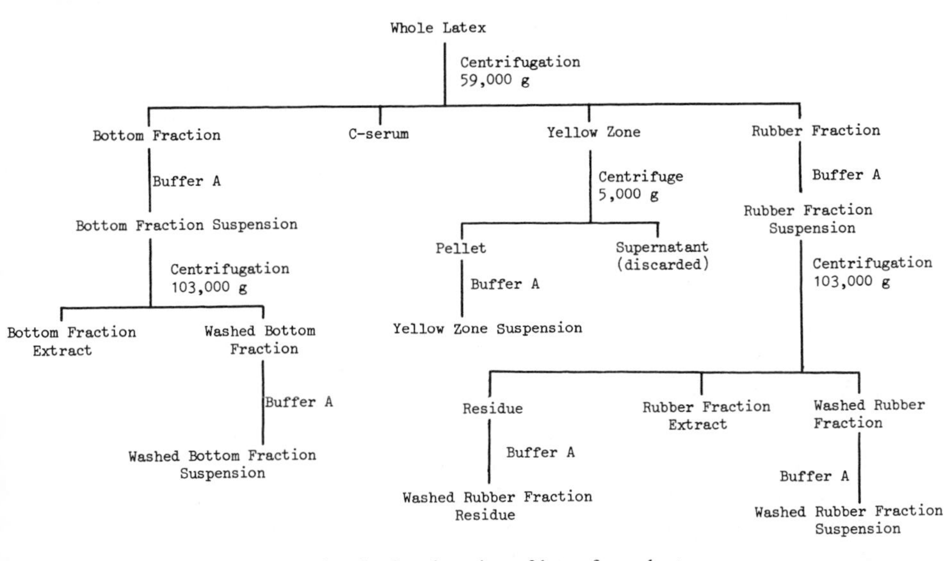

FIG. 3. Scheme for the fractionation of latex for reductase assay.

yellow zone and the bottom fraction has been studied using electron microscopy.[10] Briefly, the rubber fraction consists of rubber particles of sizes ranging from 0.05 to 0.8 μm in diameter with a modal diameter of 0.1 μm. Each rubber particle is enveloped by a film of protein and phospholipids. The yellow zone consists mainly of fragments of the Frey-Wyssling complexes with an occasional intact ones, and also rubber particles. The C-serum is a clear straw-colored liquid which corresponds to the cytosol. Finally, the bottom fraction contains mainly the lutoids and a lesser number of the intact Frey-Wyssling complexes.

The yellow zone is obtained by puncturing the centrifuge tube and carefully drawing it out using a Pasteur pipet. The bottom of the tube is then punctured and the clear C-serum is drained into a chilled beaker. The upper part of the C-serum zone is discarded as it also contains rubber. That part of the tube containing the bottom fraction is cut out completely and rinsed with ice-cold distilled water to remove the residual rubber.

Preparation of the Various Enzyme Fractions for Reductase Assay. The following procedures are to be performed at 0–4°. The outline of the steps employed is shown in Fig. 3.

The Rubber Fraction. This fraction is resuspended in Buffer A (0.1 *M* triethanolamine–HCl containing 2 m*M* dithiothreitol, pH 7.2) in a 1:1 (v/v) ratio using a prechilled mortar and pestle. A gentle action of the pestle is used to minimize the coagulation of the rubber. The milky mixture obtained is referred to as the rubber fraction suspension. This suspension is further centrifuged at 103,000 *g* for 40 min to give 3 subfractions, namely, the washed rubber fraction, the rubber fraction extract, and a small quantity of dark-grey residue at the bottom of the centrifuge tube. The latter is suspended in about 0.5 ml of Buffer A after cutting out that part of the tube containing it. This suspension is referred to as the washed rubber fraction residue. The washed rubber fraction, which now has a much firmer texture compared to the original rubber fraction, can be similarly resuspended in Buffer A although much of it will coagulate during this procedure. The suspension obtained is named the washed rubber fraction suspension.

The Yellow Zone. This is a thick yellowish-white suspension of rubber particles and disrupted Frey-Wyssling complexes in the C-serum. It can be further centrifuged at 5000 *g* to remove much of the rubber and then resuspended in Buffer A to give the yellow zone suspension.

The Bottom Fraction. The bottom fraction is resuspended in Buffer A in a 1:1 (v/v) ratio using a Potter-Elvehjem homogenizer to give the bottom fraction suspension. This suspension is further sonicated for 15 sec twice and then centrifuged at 103,000 *g* for 40 min to give the soluble bottom fraction extract and the washed bottom fraction pellet. The latter is resuspended in Buffer A and is referred to as the washed bottom fraction suspension. The washed bottom fraction pellet can be resuspended in Buffer A containing 1% (w/v) Triton X-100 and after thorough mixing and standing in ice for about 5 min, the suspension is centrifuged at 103,000 *g* for 40 min. The supernatant obtained contains the solubilized reductase.

Reductase Assay. The assay method is based on the formation and the subsequent isolation and measurement of [^{14}C]mevalonate from [^{14}C]HMG-CoA, employing the microassay method described by Shapiro *et al.*[12] Although [3-^{14}C]HMG-CoA is available commercially, it is preferable to synthesize it in the laboratory because of the considerable decomposition which occurs during delivery. The synthesis of the substrate is carried out by reacting hydroxymethyl[3-^{14}C]glutaric acid (HMG) anhydride (prepared as described by Goldfarb and Pitot[13] and with a specific radioactivity of 200 μCi/mmol) with CoA (the lithium salt) under the conditions described by Louw *et al.*[14] except that the pH of the reaction mixture is kept at about 7.6 by the careful addition of small amounts of KOH (5 *N*) as needed. The anhydride is added in several small additions, adjusting the pH as necessary. The final volume of the preparation is about 2 ml. The addition of excess anhydride ensures the absence of free

[12] D. J. Shapiro, J. L. Nordstrom, J. J. Mitschelen, V. W. Rodwell, and R. T. Schimke, *Biochim. Biophys. Acta* **370,** 369 (1974).

[13] S. Goldfarb and H. C. Pitot, *J. Lipid Res.* **12,** 512 (1971).

[14] A. I. Louw, I. Bekersky, and E. H. Mosbach, *J. Lipid Res.* **10,** 683 (1969).

CoA in the substrate preparation. The unreacted [^{14}C]HMG can be removed from the preparation by Sephadex G-15 (Pharmacia, Sweden) column chromatography using 0.5 *M* ammonium formate buffer pH 6.[15] The fractions containing [^{14}C]HMG-CoA are pooled, freeze-dried, and then redissolved in distilled water (pH 6) to the required concentration. An alternative method for the separation of [^{14}C]HMG-CoA from CoA and [^{14}C]HMG using anion exchange column chromatography has also been described.[16]

The reductase reaction mixture contained enzyme protein, 0.2–0.8 mg; NADPH, 0.46 μmol; DL-HMG-CoA, 36 nmol (444 dpm/nmol), and Buffer A in a final volume of 0.13–0.15 ml. The incubation time is 30 min at 30° and the reaction is terminated with 25 μl of HCl (10 *N*). The samples are allowed to lactonize by further incubation at 40° for 1 hr. The mevalonolactone formed, in a 100-μl aliquot of the protein-free supernatant of the reaction mixture, is isolated by TLC on Silica Gel G (5 × 15 cm lane). Unlabeled mevalonolactone is also spotted as the marker and it is located using iodine vapor after the development of the TLC in a solvent of benzene : acetone (1 : 1 v/v). The zone of R_f 0.46–0.73 on the TLC is scraped into a scintillation vial and resuspended in 1 ml distilled water. Radioactivity is measured by liquid scintillation counting in a cocktail of toluene : Triton X-100 (2 : 1,v/v) containing 0.5% (w/v) 2,5-diphenyloxazole. The recovery of mevalonolactone from the TLC is 95–98% and the counting efficiency is 92–94%. Enzyme assays are usually done in triplicates except when a large number of measurements are to be done at a time in which case these are carried out in duplicates. A unit of enzyme activity is arbitrarily defined as the formation of 1 nmol of mevalonate/30 min incubation time.

The rate of mevalonate formation is linear for up to 45 min incubation and up to 1.25 mg enzyme protein under the conditions described above. The enzyme in the washed bottom fraction suspension is saturated at about 300 μM DL-HMG-CoA and the apparent Km, estimated by the Lineweaver–Burk plot, is 56 μM DL-HMG-CoA.[6]

Measurement of Protein. Protein in the various enzyme preparations is determined by the method of Lowry *et al.*[17] after its precipitation with ice-cold trichloroacetic acid (10%, w/v). The protein in the rubber-containing preparations is first solubilized with 0.5% (w/v) Triton X-100 in 2% (w/v) KCl solution (pH 7) overnight and the rubber is separated by centrifugation at 103,000 *g* for 40 min. The solubilized protein is precipitated

[15] A. B. Sipat and J. R. Sabine, *Pertanika* **4,** 35 (1981).

[16] I. P. Williamson and V. W. Rodwell, *J. Lipid Res.* **22,** 184 (1981).

[17] O. H. Lowry, J. J. Rosebrough, A. L. Farr, and R. J. Randall, *J. Biol. Chem.* **193,** 265 (1951).

DISTRIBUTION OF REDUCTASE ACTIVITY IN THE VARIOUS FRACTIONS OF ULTRACENTRIFUGED LATEX

Fraction	Reductase activity	
	Units/mg protein	Units/ml latex[a]
Whole latex	23.01 ± 0.97	165.70
Bottom fraction suspension	16.05 ± 0.44	31.38
C-serum	0.13 ± 0.03	0.53
Yellow zone suspension	8.64 ± 0.16	2.07
Rubber fraction suspension	14.40 ± 1.10	19.50
Washed rubber fraction suspension	11.15 ± 0.58	11.82
Rubber fraction extracts	2.62 ± 0.28	2.97
Washed rubber fraction residue	7.13 ± 1.13	0.30

[a] The values in this column have been corrected for differences in volume and for the quantity of rubber which coagulated.

with trichloroacetic acid and quantitated by Lowry's method. According to Tata[18] the proteins in the rubber fraction and the C-serum are completely precipitated by trichloroacetic acid but the bottom fraction extract (see Fig. 3) contains hevein and pseudo-hevein which are soluble in trichloroacetic acid. These two proteins constitute about 40% of the total proteins in the bottom fraction. Bovine serum albumin is used as the protein standard in this assay.

Some Characteristics of the Reductase

Distribution of Reductase Activity in the Latex. The result of the measurements of reductase activity in the whole latex, and in the 4 fractions obtained after the first centrifugation step, is given in the table. The whole latex shows the highest enzyme specific activity as well as activity/ml latex. Fractionation of this latex results in a loss of enzyme activity. Of the 4 fractions obtained after the first centrifugation step, the bottom fraction has the highest enzyme specific activity as well as activity/ml latex. The yellow zone also contains a high reductase specific activity but its activity/ml latex is low since this fraction constitutes only a small proportion of the latex (approx. 2% by volume). It should be noted that the yellow zone contains particles, or their fragments, which are also present in the rubber fraction and the bottom fraction. The C-serum has neglible reductase activity. Contrary to an earlier report[2] the rubber fraction contains significant enzyme specific activity. Washing this fraction in

[18] S. J. Tata, *J. Rubber Res. Inst. Malays.* **28,** 77 (1980).

Buffer A does not remove the enzyme activity. As shown in Fig. 3 this procedure also gives rise to the rubber fraction extract and the washed rubber fraction residue which is likely to contain membranes from the rubber particles or those which are trapped in the rubber fraction during the first centrifugation. The reductase specific activity in the former subfraction is lower than in the latter subfraction. The reductase activity in the bottom fraction suspension has been shown to be membrane bound.[4] When the bottom fraction is resuspended in Buffer A containing 0.1% (w/v) Triton X-100 and then centrifuged at 103,000 *g*, two subfractions are obtained, the washed bottom fraction and the bottom fraction extract (see Fig. 3). The reductase activity is largely associated with the former subfraction which consists of the membranes of the organelles present in the bottom fraction. It is clear from these results that the bottom fraction should be used as the enzyme source for the study of the reductase in fractionated *Hevea* latex. The enzyme is membrane-bound but it can be solubilized using Triton X-100.[4] The question as to which organelle the enzyme is bound to, ie., the lutoid or the Frey-Wyssling complex or both, still remains to be answered. A major problem here is to obtain a pure preparation of either organelle type.

Cofactor Requirement. The reductase activity in the bottom fraction suspension has a specific requirement for NADPH.[4] NADH is only 6% effective.

Thiol Requirement. The freshly prepared bottom fraction suspension contains appreciable reductase activity even in the absence of added thiol compound. The addition of dithiothreitol, however, increases its activity by about 60% at the optimum concentration of 2 m*M*.[4] The washed bottom fraction enzyme on the other hand requires dithiothreitol for any appreciable activity to be detectable and the optimum concentration is 10 m*M*. The difference in response here is possibly due to the presence of endogenous thiols in the bottom fraction suspension and these are removed after the washing procedure. *p*-Chloromercuribenzoate (1 m*M*) completely inhibits the enzyme activity.

pH Optimum. The reductase in the bottom fraction suspension has an optimum pH of 6.8–7.0 in 0.1 *M* sodium phosphate buffer containing 2 m*M* dithiothreitol.[6]

Effect of Mevalonate, HMG, and CoA. Both mevalonate and CoA are products of the reductase reaction. HMG, and also CoA, can be present in the reaction mixture through the use of impure preparation of the HMG-CoA substrate. The effect of the addition of these compounds on reductase activity has been studied. Mevalonate (2 m*M*) and CoA (2 m*M*) each shows no effect on reductase activity in the bottom fraction suspension.[4] HMG (2 m*M*) inhibits enzyme activity by about 14% in the freshly prepared bottom fraction but when a 4-day-old frozen bottom fraction is used, an activation of approx. 32% is observed.[4]

Enzyme Stability. The bottom fraction which has been stored frozen over 1 day loses about 30% of the reductase activity, and up to 50% when stored over 8 days. Dithiothreitol is present in the buffer used to resuspend the frozen bottom fraction pellet. Latex which has been collected and fractionated at ambient temperature (about 28°) contains no appreciable reductase activity. The stability of the enzyme is likely to be related to the preincubation effect discuss further on.

ATP Inactivation of the Reductase. The reductase in the bottom fraction suspension is inhibited by ATP. The inhibition is concentration-dependent and the addition of 4 m*M* ATP results in about 90% inhibition of the enzyme. Magnesium is not essential for this effect but the bottom fraction is known to contain endogenous magnesium at concentrations ranging from 370 to 520 μM. This effect is also specific for ATP. The other nucleotides tested, ADP, AMP, adenosine, UTP, TTP, GTP, CTP, and 3′,5′-cyclic AMP, have no or a slight (about 10%) activation effect. The inhibition of the reductase activity can be overcome by the addition of excess EDTA to the reaction mixture. Washing the bottom fraction in Buffer A or Buffer A containing 0.1% (w/v) Triton X-100 results in the complete or near complete removal of the inhibition by ATP. The inhibitory effect can be restored by the addition of the C-serum (as low as 5 μl) to the washed bottom fraction suspension. The inhibitory factor appears to be a heat-labile protein present in the C-serum and it is precipitated at 40–60% ammonium sulfate saturation.

The inhibitory effect of ATP on reductase activity in the bottom fraction suspension can be attributed to either the inactivation of the enzyme by phosphorylation as in the case of the rat liver enzyme[19] or to the rapid removal of mevalonate from the reaction mixture by mevalonate kinase, an enzyme known to be present in the C-serum. While the recovery of added [^{14}C]mevalonate from the reductase reaction mixture is virtually complete[4] this evidence does not deny the latter explanation.[20] In fact the data shown in Fig. 4 would support this explanation.

Until recently, one of the major difficulties encountered in studying the ATP inactivation of the reductase is the fact that the enzyme is also inactivated by preincubation at 30° as discussed in the next section. Preincubation of the enzyme with ATP is an essential step in demonstrating its

[19] T. S. Ingebritsen and D. M. Gibson, *in* "Recently Discovered Systems of Enzyme Regulation by Reversible Phosphorylation" (P. Cohen, ed.), p. 63. Elsevier/North-Holland Biomedical Press, Amsterdam, 1980.

[20] G. C. Ness, G. A. Benton, S. A. Deiter, and P. S. Wickham, *Arch. Biochem. Biophys.* **214,** 705 (1982).

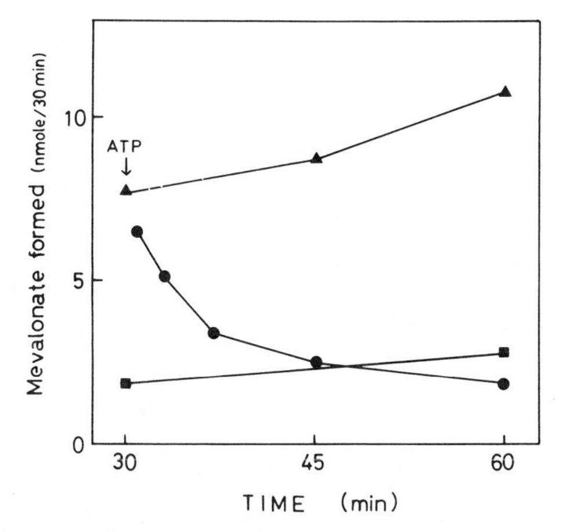

FIG. 4. Effect of ATP on the recovery of mevalonate formed by the reductase reaction. A freshly prepared bottom fraction suspension (1 mg protein) was incubated with the substrate and cofactor in a time-course experiment (▲). ATP (4 m*M*) was added to a parallel series of incubation at time 30 min and the reaction then stopped at the time indicated for the measurement of mevalonate (●). Another set of reaction had ATP (4 m*M*) added together with the substrate and cofactor at the start of the incubation (■).

inactivation by phosphorylation. The sensitivity of the reductase toward preincubation can now be overcome by the addition of a heat-stable protein which has been partially purified from the C-serum.[21] With the addition of this protein to the reductase reaction mixture during the preincubation step, it is now possible to separate the inactivation of the enzyme by preincubation from its inhibition by ATP. Further work on this aspect of the reductase is currently in progress.

Effect of Preincubation. The reductase in the bottom fraction suspension, the washed bottom fraction suspension and the solubilized enzyme preparation loses its activity when preincubated at 30° before starting the reaction proper by the addition of the substrate and NADPH (Fig. 5). Bovine serum albumin (2%, w/v), crude hemoglobin (2%, w/v), NADPH (2 m*M*), and dithiothreitol (up to 50 m*M*) show no stabilizing effect when added individually during the preincubation. The presence of [^{4}C]HMG-CoA during the preincubation reduces the rate of inactivation (57% inactivation as compared to 87% for the control after 1 hr incubation). Lowering the preincubation temperature also reduces the rate of inactivation but even at 0–4°, about 20% loss of activity occurs. The bottom fraction pellet, however, is stable for at least the 1-hr period tested at 0–4°.

[21] A. B. Sipat, *Abstr., Congr. Fed. Asian Oceanian Biochem., 3rd, 1983.*

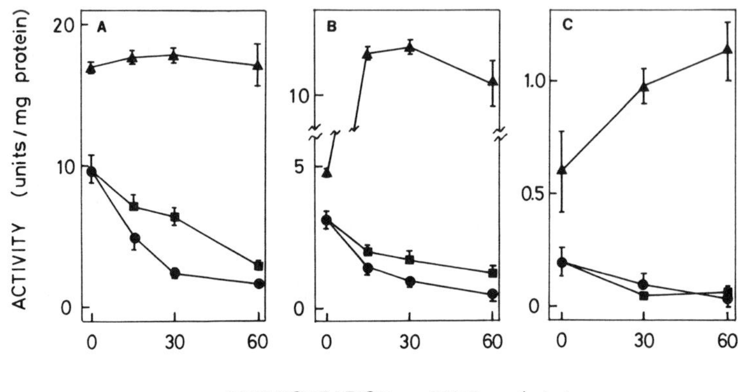

FIG. 5. Effect of preincubation of the reductase in the presence and absence of boiled C-serum on its activity. The above results are obtained using the bottom fraction suspension (A), the washed bottom fraction suspension (B), and the solubilized enzyme preparation (C) as the enzyme source. The enzyme was preincubated for the times indicated either in the absence (●) or the presence of boiled C-serum (40 μl) (▲), and the activity then assayed by the addition of both substrate and cofactor. (■) Boiled C-serum was absent during the preincubation but was added (40 μl) during the assay proper.

So far, the reductase can be stabilized only by the addition of the boiled C-serum, or a protein partially purified from this fraction, during the preincubation step. For the reductase in the washed bottom fraction suspension, and also the solubilized enzyme, the addition of the boiled C-serum during the preincubation results in an increased activity (Fig. 5).

Activation of the Reductase by a Heat-Stable C-Serum Protein. The C-serum has previously been shown to contain a heat-stable factor which increases reductase activity by up to 2-fold.[5] The activation effect is lost upon treatment of the boiled C-serum (the heat-labile proteins have been removed by centrifugation) with trypsin. The activator is also precipitated by ammonium sulfate (25–100% saturation). Partial purification of this activator has been achieved,[21] using the following procedure.

Step 1. The C-serum is boiled in a boiling water-bath for 20 min, cooled, and then centrifuged at 8400 g_{max} to remove the heat-labile proteins.

Step 2. The supernatant from above is concentrated and dialyzed by ultrafiltration using a 10,000 cut-off membrane. The filtrate contains no activator. Alternatively, concentration can also be achieved using 100% ammonium sulfate saturation if an ultrafiltration apparatus is not available.

Step 3. The 10,000 cut-off concentrate is loaded on to a Sephadex G-75 column (1.6 × 60 cm) and eluted with 0.1 *M* potassium phosphate buffer (pH 6.8). Three 280 nm absorbing peaks are obtained and these are arbitrarily named A, B, and C in the order of their elution. The fractions under peak B are pooled and concentrated by ultrafiltration using a 10,000 cutoff membrane. Virtually all the activator is found in this fraction.

Step 4. The concentrated fraction B is then applied on to a DEAE-cellulose (Sigma Chemical Co.) column (1.6 × 2.5 cm). Elution is performed first with the starting buffer (10 m*M* Tris–HCl, pH 7) during which a small 280 nm peak (arbitrarily named B1) is obtained. Further elution is done with the same buffer to which is added NaCl (0.25 *M*) and this results in a second 280 nm peak (B2). Final elution is done with the same buffer containing 0.5 *M* NaCl during which the third peak (B3) is eluted. The activator is found in Fraction B2.

Fraction B (off the Sephadex G-75 column) has the same effect as the boiled C-serum in preventing the inactivation of the enzyme during preincubation (see Fig. 5). Fractions B1, B2, and B3 are not coagulated by boiling. Trichloroacetic acid (10%, w/v) precipitates B2 and B3 but not B1. Thus protein in Fraction B2 is unlikely to be hevein or pseudohevein —proteins which are also heat-stable but are not precipitated by trichloroacetic acid. Partially purified hevein (the pH 5 eluate) has no effect on reductase activity.[5]

While the heat-stable activator in the C-serum increases reductase activity by about 2-fold, its importance lies mainly in its ability to stabilize and to prevent the inactivation of the enzyme during preincubation. Further work on the purification and characterization of this activator protein is still in progress.

[6] Insect 3-Hydroxy-3-methylglutaryl-CoA Reductase

By DANIEL J. MONGER

$$\text{HMG-CoA} + 2\text{NADPH} + 2\text{H}^+ \rightarrow \text{Mevalonate} + 2\text{NADP}^+ + \text{CoASH}$$

Introduction

Unlike mammals all organisms of the class insecta are incapable of sterol biosynthesis,[1] and therefore HMG-CoA reductase is not involved in sterol biosynthetic regulation in insects. Instead, insect HMG-CoA reduc-

[1] J. A. Svoboda, J. N. Kaplanis, W. E. Robbins, and M. J. Thompson, *Annu. Rev. Entomol.* **20,** 205 (1975).

ISBN 0-12-182010-6

tase is involved in the syntheses of juvenile hormones, sesquiterpenoids which regulate larval development and influence adult sexual maturation. This reductase is apparently the rate-limiting step in juvenile hormone production. Its activity is inhibited by F^- and by MgATP,[2] suggesting that the regulatory scheme elucidated for the mammalian enzyme is involved with the insect enzyme as well.

Assay Method

Principle. The assay used is a dual isotope radiochemical assay modified from a procedure used to measure reductase activity in rat liver and L cell fibroblasts.[3]

Reagents

KH_2PO_4 buffer, 50 m*M*, pH 7.4[4]
KCl, 200 m*M*
Dithiothreitol, 5 m*M*
EDTA, 5 m*M*
Kyro EOB, 0.25%[5]
Glucose 6-phosphate, 120 m*M*
Glucose-6-phosphate dehydrogenase (baker's yeast), 0.33 IU/μl
$NADP^+$
DL-[*methyl*-^{3}H]HMG-CoA, 2 m*M*, 5.0 Ci/mol
HCl, 6N
DL-[2-^{14}C]Mevalonolactone, 1 μ*M*, 18 Ci/mol

Procedure. Into a conical-shaped Reacti-Vial (300 μl capacity; Pierce Co.) is placed 50 μl of the enzyme preparation in phosphate buffer containing KCl, dithiothreitol, EDTA, and detergent. Assays are initiated by the 50 μl addition of the NADPH generating system consisting of 38 μl of glucose 6-phosphate, 2 μl of glucose-6-phosphate dehydrogenase, and 0.38 mg of $NADP^+$ together with 10 μl of substrate.

Incubations are performed at 37° for 90 min in the sealed vials housed within a Temp-Blok module heater shaking at 170 rpm. Reactions are terminated with 25 μl of 6 *N* HCl, followed by 25 μl of mevalonolactone. After warming for 30 min at 40°, samples are divided in half and applied to silica gel G thin-layer chromatographic plates (250 μm thickness, 5 × 20 cm) and developed in acetone/benzene (1 : 1). Mevalonolactone is recov-

[2] D. J. Monger and J. H. Law, *J. Biol. Chem.* **257,** 1921 (1982).
[3] D. J. Shapiro, J. L. Nordstrom, J. J. Mitschelen, V. W. Rodwell, and R. T. Schimke, *Biochim. Biophys. Acta* **370,** 369 (1974).
[4] M. S. Brown, S. E. Dana, and J. L. Goldstein, *J. Biol. Chem.* **249,** 789 (1974).
[5] Kyro EOB is a synthetic nonionic detergent supplied by the Procter and Gamble Co.

ered by scraping silica gel from the plates in the region R_f 0.38–0.69 and radioactivity is determined by liquid scintillation in toluene/ethanol/Omnifluor (2 : 1 : 12, v/v/w).

Definition of Units and Specific Activity

One unit is the amount of enzyme that catalyzes the formation of 1 μmol of mevalonate per min at 37°. Specific activity is expressed in units per milligram of protein. Protein is determined by amino acid compositional analysis[6] of hydrolyzed samples using automated ion exchange chromatography. Protein determinations using fluorescamine were found to be impractical due to the relatively small amounts of tissue available. Typical assays measured 5–10 μU of enzyme.

Comments. The assay is linear for 3 hr under the assay conditions employed. No more than 10% of the substrate is converted to product during a 90-min incubation.

Purification Procedure

Rearing of Organism. The insect used is the tobacco hornworm moth, *Manduca sexta* Johannson (Lepidoptera: Sphingidae). Eggs[7] were hatched and larvae reared at 27° in a 15-hr light/9-hr dark photoperiod using the standard diet[8] as modified by Dr. R. A. Bell.

Preparation of Glandular Extracts. Reductase activity was measured in several insect tissues with by far the highest specific activity residing in the corpora allata. Adult *M. sexta* 0–4 days old are sexed, decapitated with a razor blade, and the heads immediately placed on ice. Heads are subsequently rubbed clean of scales, the exoskeleton between the antennae removed with a sharp razor blade, and the heads pinned under a solution consisting of 4 m*M* NaCl, 3 m*M* $CaCl_2$, 40 m*M* KCl, 18 m*M* $MgCl_2$, 0.1 m*M* phenylthiourea, 240 m*M* sucrose, 0.1% (w/v) polyvinylpyrrolidone (Average MW 40,000), and 1.7 m*M* PIPES buffer, pH 6.5. Corpora allata are dissected with the corpora cardiaca and part of the aorta attached.[9] Following two 100 μl washes with assay buffer, the glands (20–30 gland pairs) are homogenized in 100 μl of buffer in micro homogenization vessels. The homogenate is warmed at 37° for 10 min and centrifuged at 10,000 *g* and 23° for 10 min. Supernatants are removed and stored on

[6] D. H. Spackman, W. H. Stein, and S. Moore, *Anal. Chem.* **30,** 1190 (1958).

[7] Eggs were generously provided by J. P. Reinecke of the USDA, Fargo, ND.

[8] R. T. Yamamoto, *J. Econ. Entomol.* **62,** 1427 (1969).

[9] The average wet weight of each corpora allata/corpora cardiaca was determined to be ~20 μg using a Cahn electrobalance equipped with a chart recorder.

ice until assayed.[10] Hence, HMG-CoA reductase activity of *M. sexta* corpora allata resides in the postmitochondrial supernatant. This was also shown to be the case for reductase activity in corpora allata of the viviparous cockroach, *Diploptera punctata,*[11] and may be true for all insects.

Properties

Stability. Enzyme preparations are always stored on ice and assayed for reductase activity within 8 hr from time of glandular dissection. This is to prevent possible losses in enzyme activity resulting from protease contamination. In addition to this, variations in reductase specific activity most likely result from variations in HMG-CoA reductase kinase.[2] Reductase specific activity in corpora allata typically range from 3.5 to 4.5 $\times$ 10^{-4} U/mg.

Stoichiometry and Substrate Specificity. For each mole of HMG-CoA utilized, 1 mol of mevalonate is generated and 2 mol of NADPH is oxidized to $NADP^+$ by HMG-CoA reductase.[12] There is strong evidence to suggest that this same enzyme is also responsible for the conversion of 3-hydroxy-3-ethylglutaryl-CoA (HEG-CoA) to the homolog homomevalonate.[13]

$$\text{HEG-CoA} + 2\text{NADPH} + 2\text{H}^+ \rightarrow \text{Homomevalonate} + 2\text{NADP}^+ + \text{CoASH}$$

This compound, an intermediate in the biosynthesis of the C_{17}, C_{18}, and C_{19} homologs of juvenile hormone, has been demonstrated to be synthesized by corpora allata homogenates prepared from *M. sexta*.[14] In addition, data suggest that mammalian HMG-CoA reductase is capable of reducing HEG-CoA to homomevalonate.[13]

pH and Temperature Optima. The insect reductase exhibits a broad optimum in phosphate buffer between pH 6.8–7.4. Substrate–cofactor preparations are routinely adjusted to pH 7.4 prior to initiating measurement of enzymatic activity. As for temperature, insects maintain their body temperatures to within a fraction of ambient temperature.[15] Hence,

[10] For this purpose, glass Pasteur pipets were heated over a gas flame and drawn out to a fine tip of ~0.1 mm inner diameter.

[11] R. Feyereisen, J. Koener, and S. S. Tobe, *in* "Juvenile Hormone Biochemistry" (G. E. Pratt and G. T. Brooks, eds.), p. 81. Elsevier/North-Holland Biomedical Press, Amsterdam, 1981.

[12] N. Qureshi, R. E. Dugan, and J. W. Porter, *Abstr. Pap., Meet., Am. Chem. Soc.* Biol. No. 57 (1974).

[13] F. C. Baker and D. A. Schooley, *Biochim. Biophys. Acta* **664,** 356 (1981).

[14] E. Lee, D. A. Schooley, M. S. Hall, and K. J. Judy, *J. Chem. Soc., Chem. Commun.* p. 290 (1978).

[15] E. Bursell, "An Introduction to Insect Physiology." Academic Press, New York, 1970.

for entomological systems, experiments are generally conducted at 30°. However, the reductase of *M. sexta* was found to proceed faster at the higher temperature of 37°, the temperature of choice for this assay.

Metal Ion Requirements. There are no apparent requirements for metal ions. On the contrary, insect reductase activity has been shown to be 35% higher in the presence of the divalent metal ion chelator EDTA (5 m*M*) than in its absence.[16] EDTA is needed to inhibit effectively HMG-CoA reductase kinase[2] (see below).

Effectors of Activity. EDTA is needed for measurement of maximal reductase activity. $MgCl_2$ (6 m*M*) and CTP (2 m*M*) have no effect on reductase activity when incubated with enzyme preparations. However, if ATP is substituted for CTP, a 25–30% inhibition in reductase activity is observed. This decreased reductase activity was measured in the presence of EDTA (30 m*M*) and hence was not due to loss of product catalyzed by mevalonate kinase. NaF (50 m*M*) exerts minimal inhibitory activity (<10%) when it is added directly to enzyme preparations. However, if NaF is added to corpora allata prior to dissection, up to 80% inhibition in enzyme activity can be achieved. MgATP and F^- do not appear to inhibit directly HMG-CoA reductase, but rather exert inhibitory effects via other enzymes present in the tissue preparations.[2]

Competitive Inhibition—Compactin. A number of compounds which are analogs of HMG-CoA have been shown to inhibit mammalian HMG-CoA reductase competitively with respect to HMG-CoA.[17,18] Recently, one of these compounds, compactin, has been tested successfully in insects as a potential anti-juvenile hormone agent.[19] An ID_{50} of ~1 n*M* was found for juvenile hormone III biosynthesis *in vivo* by corpora allata of the female cockroach *Periplaneta americana*.[19] Anti-juvenile hormone activity was also manifested in larval *M. sexta* by darkening of the cuticle after injections of compactin *in vivo*.[20] This fungal metabolite has also been demonstrated to inhibit specifically insect HMG-CoA reductase *in vitro*.[20] Reductase assays were performed with corpora allata homogenates from either adult male or female *M. sexta* in the presence of 1×10^{-7} to 1×10^{-10} *M* compactin, using the sodium salt form. Assays were performed with shaking at 37° for 40 min (see Fig. 1). A K_i value for compactin of 0.9 n*M* was found for reductase from corpora allata of both sexes.[20] This is very close to the K_i value of 1.2 n*M* found for compactin

[16] D. J. Monger, unpublished data.

[17] A. Endo, M. Kuroda, and K. Tanzawa, *FEBS Lett.* **72,** 323 (1976).

[18] A. Endo, *TIBS* **6,** 10 (1981).

[19] J. P. Edwards and N. R. Price, *Insect Biochem.* **13,** 185 (1983).

[20] D. J. Monger, W. A. Lim, F. J. Kézdy, and J. H. Law, *Biochem. Biophys. Res. Commun.* **105,** 1374 (1982).

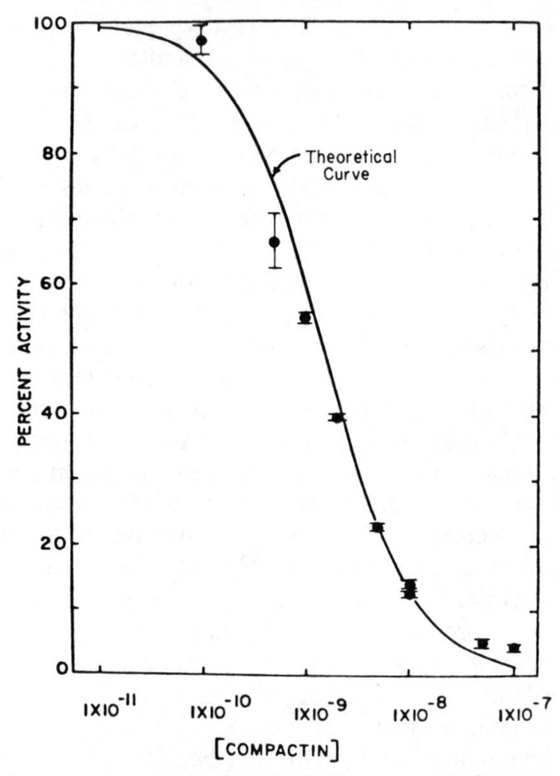

FIG. 1. Reductase assays were performed in the presence of compactin (sodium salt form), using corpora allata homogenates from female *M. sexta*. Each datum point represents the average of four determinations. Percentage activity is the ratio of product formed in the presence of inhibitor versus that measured in its absence (controls). Smooth line is a theoretical curve based on a K_i value of 0.9 nM.[20]

with the HMG-CoA reductase of rat liver microsomes.[21] To put this in perspective, an apparent K_m for the substrate HMG-CoA of 2.7 μM was reported using reductase from corpora allata of adult female *D. punctata*.[11]

Purification of Labeled Substrate

If the radiolabeled HMG-CoA is degraded to any significant extent, purification is suggested. Such decomposition products will not, however, cochromatograph with mevalonolactone when developed in acetone/benzene (1 : 1). Substrate should be applied to a silica gel G thin-layer chro-

[21] A. Endo, *J. Antibiot.* **33,** 334 (1980).

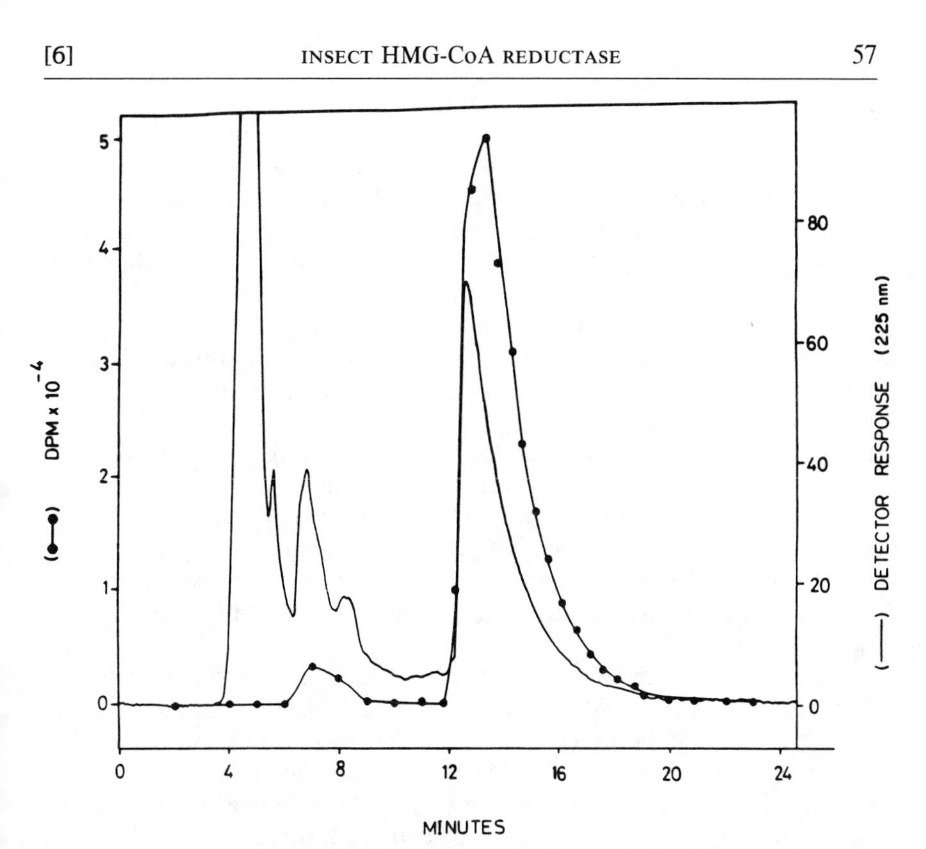

FIG. 2. HPLC chromatogram of DL-[^{3}H]mevalonolactone isolated from a large scale incubation of DL-[*methyl*-^{3}H]HMG-CoA with a 10,000 *g* homogenate prepared from corpora allata from adult female *M. sexta*.

matographic plate (250 μm thickness, 5 $\times$ 20 cm) and developed in *n*-butanol : acetic acid : water (5 : 2 : 3). Scrape the region with R_f 0.68–0.74 and elute through a sintered glass funnel with methanol. The methanolic solution is then concentrated, preferably under reduced pressure, using a rotary evaporator or alternatively under a stream of nitrogen. The labeled substrate is then dissolved in HCl/ethanol (9 : 1) to a concentration of approximately 1 m*M*.

Isolation of Product

Radiolabeled mevalonate which has been recovered from a large scale incubation of corpora allata with high specific activity DL-[*methyl*-^{3}H]HMG-CoA, can be identified as product in either of two ways. Initially, a 10,000 *g* homogenate is prepared from at least 40 corpora allata from adult female *M. sexta* as described under Purification Procedure. Incubate for 90 min with the cofactor–substrate solution containing 70 nmol of DL-[*methyl*-^{3}H]HMG-CoA at approximately 15 times the specific activity as was typically used (~75 Ci/mol). The incubation mixture is worked up as described under Assay Procedure except that the mevalonolactone migrating with an R_f 0.38–0.69 is scraped from the thin-layer chromatographic plates along with the silica gel into acetone. This is filtered through a sintered glass funnel, the filtrate is concentrated under reduced pressure, and dried under nitrogen. This is redissolved in hexane/*n*-butanol (1 : 1) to which carrier DL-mevalonolactone (Fluka AG) is added to a final concentration of 0.5% (g/dl). This material is injected (50 μl) onto a high performance liquid chromatograph using a μPorasil column at a flow rate of 40 ml/hr and pressure of 46 kg/cm^2 using hexane/*n*-butanol (1 : 1) as solvent. UV absorption can be continuously monitored at 225 nm while collecting eluant every 0.5 min for determination of radioactivity by liquid scintillation. Approximately 96% of the total recovered radioactivity comigrated with authentic DL-mevalonolactone with a retention time of 12.7 minutes (Fig. 2).

Alternatively, the labeled incubation product can be diluted with carrier mevalonolactone and subsequently recrystallized to constant specific activity. This is done by converting the mevalonolactone to its benzhydrylamide derivative.[22] Add cold mevalonolactone (1 mmol) to aminodiphenylmethane (1 mmol) and reflux for 1 hr in benzene. Remove the benzene, add dilute phosphoric acid, and wash three times with water. Desiccate over phosphorus pentoxide and recrystallize from benzene or from petroleum ether/benzene (1 : 3); mp 96.5–97.5°. Repeat with recovered labeled mevalonolactone diluted with carrier. Cocrystallizations should be performed repeatedly and should show at least 95% of the recovered labeled product to be authentic mevalonolactone.

[22] J. W. Cornforth, R. H. Cornforth, G. Popják, and I. Y. Gore, *Biochem. J.* **69,** 146 (1958).

[7] Radioenzymatic Assay of Plasma Mevalonate

By THOMAS S. PARKER, RON R. KOPITO, and HENRI BRUNENGRABER

Mevalonate[1] is the product of the rate-limiting step of cholesterol synthesis, catalyzed by hydroxymethylglutaryl-CoA reductase. A small fraction of the mevalonate synthesized by the liver is released into the plasma

[1] The physiological enantiomer *R*-mevalonate is referred to simply as mevalonate.

ISBN 0-12-182010-6

where it circulates at concentrations of 10–100 n*M* in humans and 80–500 n*M* in rats.[2] Plasma mevalonate is taken up by peripheral tissues, mostly by the kidneys.[2–4] About one-fifth of the renal uptake of mevalonate is excreted in urine.[5,6] Interest in an assay for plasma mevalonate arose after the discovery of the shunt pathway of mevalonate metabolism[7] which links cholesterol synthesis with leucine catabolism. Because mevalonate occurs in trace amounts relative to other metabolites (nanomolar vs micro- or millimolar), methods for its quantitation must be extremely sensitive and specific. Gas chromatography combined with mass fragmentography has been used to detect plasma mevalonate[8] or mevalonate formed from HMG-CoA by HMG-CoA reductase in microsome suspensions.[9] However, this approach is technically demanding and requires the dedicated use of expensive instrumentation. By contrast, the radioenzymatic assay, introduced by Popják *et al.*,[2] is a relatively simple and sensitive procedure that is capable of precise quantitation of the picomol amounts of mevalonate found in 0.1 to 1.0 ml of human plasma. The following procedure is taken from the original report of Popják *et al.* with minor modifications.

Principle of the Assay

A set of standards and samples of plasma ultrafiltrates are incubated with [γ-^{32}P]ATP and mevalonate kinase. After stopping the reaction with HCl and addition of a known amount of 5-phospho[^{14}C]mevalonate, the reaction mixture is chromatographed. The ^{32}P/^{14}C ratio in the peak of 5-phosphomevalonate is directly proportional to the amount of mevalonate present in the standards and in the ultrafiltrates. Mevalonate kinase, which is not commercially available, is isolated from pig liver[2,10,11] as described below.

[2] G. Popják, G. Boehm, T. S. Parker, J. Edmond, P. A. Edwards, and A. M. Fogelman, *J. Lipid Res.* **20,** 716 (1979).

[3] K. H. Helstrom, M. D. Siperstein, L. A. Bricker, and L. J. Luby, *J. Clin. Invest.* **52,** 1303 (1973).

[4] R. R. Kopito, S. B. Weinstock, L. E. Freed, D. M. Murray, and H. Brunengraber, *J. Lipid Res.* **23,** 577 (1982).

[5] R. R. Kopito and H. Brunengraber, *Proc. Natl. Acad. Sci. U.S.A.* **77,** 5738 (1980).

[6] H. Brunengraber, S. B. Weinstock, D. L. Story, and R. R. Kopito, *J. Lipid Res.* **22,** 916 (1981).

[7] J. Edmond and G. Popják, *J. Biol. Chem.* **249,** 66 (1974); this volume [12].

[8] L. Hagenfelt and K. Hellstrom, *Life Sci.* **11,** 669 (1972).

[9] C. Cighetti, E. Santiello, and G. Galli, *Anal. Biochem.* **110,** 153 (1981).

[10] G. Popják, this series, Vol. 15 [12].

[11] T. S. Parker, unpublished observations.

Assay of Mevalonate Kinase (EC 2.7.1.36)

The spectrophotometric assay is performed as described extensively by Popják.[10] In the reaction mixture, 1 mM dithiothreitol may conveniently replace cysteine hydrochloride + KOH. Alternatively, the activity can be assayed by incubating the extract with R- or RS-[^{14}C]mevalonate plus MgATP, followed by isolation of 5-phospho-R-[^{14}C]mevalonate by paper chromatography,[10,12] thin-layer chromatography,[10] or high-voltage electrophoresis.[13]

One unit of enzyme activity is defined as that amount of enzyme which, when assayed by the spectrophotometric method, phosphorylates mevalonate at an initial rate of 1 μmol per minute at 25°, unless another temperature is indicated. Specific activities are expressed as units of enzyme activity per milligram protein.

Purification of Mevalonate Kinase

Various procedures have been described to purify mevalonate kinase from yeast[14] or from pig liver.[10,11,13,15] The pig liver enzyme, purified to homogeneity, is reported to have a specific activity of 17.5[15] or 30.4[13] at 30°. For the radioenzymatic assay of mevalonate it is convenient to purify the enzyme to a specific activity of about 5, since purer preparations are not stable. The method for purifying mevalonate kinase, to be described, is a modification of the procedure of Popják.[2] A crude preparation obtained by protamine and ammonium sulfate fractionation[10] is purified by affinity chromatography[11] and hydroxylapatite chromatography.[2,11]

Reagents

a. Sucrose solution, 0.35 M, containing 35 mM $KHCO_3$ and 1 mM EDTA
b. Ammonium sulfate recrystallized from a saturated aqueous solution neutralized with NH_4OH and containing 1 mM EDTA. Alternatively, enzyme-grade neutralized ammonium sulfate can be used if one adds 1 mM EDTA to the extracts prior to adding ammonium sulfate
c. Dilute potassium phosphate buffer: 20 mM, pH 7.5, containing 1 mM EDTA
d. Protamine sulfate from salmon sperm: 10 mg/ml

[12] T. T. Tchen, see this series, Vol. 5 [66], p. 489; Vol. 6 [75], p. 505.

[13] C. S. Lee and W. J. O'Sullivan, *Biochim. Biophys. Acta* **747,** 215 (1983).

[14] T. T. Tchen, *J. Biol. Chem.* **233,** 1100 (1958).

[15] E. Beytia, J. W. Dorsey, J. Marr, W. W. Cleland, and J. W. Porter, *J. Biol. Chem.* **245,** 5450 (1970).

e. Wash buffer for affinity chromatography: 10 m*M* potassium phosphate containing 1 m*M* EDTA and 5 m*M* mercaptoethanol, pH 7.5
f. Buffers for hydroxylapatite chromatography are prepared in CO_2-free deionized distilled water:
   0.5 *M* potassium phosphate, pH 7.5
   1 m*M* potassium phosphate containing 1 m*M* EDTA and 5 m*M* mercaptoethanol, pH 7.5
   200 m*M* potassium phosphate containing 1 m*M* EDTA and 5 m*M* mercaptoethanol, pH 7.5

Method

Initial Extract.[10] Pig liver, obtained as soon as possible after slaughter of the animal and chilled in ice, is finely minced in a rust-free meat grinder. Each 500-g batch of the mince is extracted with 1 liter of the sucrose solution, reagent (a), by gentle stirring for 90 sec. This gentle and brief extraction procedure minimizes the amount of protein (and of ATPase) extracted. The mixture is either filtered through two layers of cheesecloth or centrifuged at 2000 *g* for 30 min, yielding about 1 liter of initial extract.

Protamine Precipitation.[10] Of the protamine solution, reagent (d), 75 ml is added to 1 liter of initial extract to give a final concentration of 0.7 mg/ml. After 2 min of stirring at 0°, the mixture is centrifuged at 2000 *g* for 30 min. Because of variations in the properties of different grades of protamine sulfate, it is recommended to determine in a small scale titration, the maximum amount of protamine sulfate that can be added to the initial extract without precipitating mevalonate kinase. The protamine precipitation step increases the specific activity of the enzyme 3- to 4-fold without loss of activity.

Ammonium Sulfate Precipitation.[10] Finely powdered ammonium sulfate is slowly added to the "protamine supernatant" to a concentration of 30% saturation at 4°. The mixture is stirred gently until the salt dissolves while pH is monitored and maintained at about 7.5 by addition of NH_4OH if necessary. After 30 min, the small precipitate formed is removed by centrifuging at 12,000 *g* for 30 min. More ammonium sulfate is added to the supernatant to a concentration of 45% saturation. The precipitate is collected by centrifuging at 12,000 *g* for 30 min. After decanting the supernatant, the precipitate is suspended in an equal volume of a 80% saturated ammonium sulfate solution containing 20 m*M* potassium phosphate, 1 m*M* EDTA, and 5 m*M* mercaptoethanol, pH 7.5. The suspension can be stored for 1 year with little loss of activity.

Affinity Chromatography.[11] This step is conducted at room temperature. A short glass column (8 × 10 cm i.d.) equipped with a fritted disk

(20–40 μm porosity) at the outlet is packed with 200 ml of Amicon Matrex gel Red A, or an equivalent gel. All solutions are pumped through the column at a rate of 50 ml/min. High affinity sites capable of irreversibly binding mevalonate kinase are inactivated by perfusing the column in closed circuit for 12 hr with 2 liters of 10 m*M* potassium phosphate buffer containing 1 g of bovine serum albumin/liter. Excess albumin is then washed out with 2 liters of 10 m*M* potassium phosphate containing 1 *M* ammonium sulfate and 1 m*M* EDTA, pH 7.5. The same solution is used to regenerate the column after use. Before storage at 4°, the column is washed with 2 liters of 10 m*M* potassium phosphate buffer containing 1 m*M* EDTA and 0.05% sodium azide, pH 7.5. Before use, the column is brought back to room temperature. After gentle stirring of the gel to remove air bubbles, 2 liters of "wash buffer," reagent (e), is pumped through the column.

The slurry of the 30–45% ammonium sulfate precipitate, equivalent to 1 kg of liver, is centrifuged at 12,000 *g* for 30 min. After decanting the supernatant, the pellet is dissolved in 1 liter of "wash buffer." The solution is cycled 5 times through the column. Unbound protein is eluted with 2 liters of "wash buffer." Mevalonate kinase is eluted with 1 liter of 10 m*M* potassium phosphate containing 0.5 *M* ammonium sulfate, 2 m*M* ATP, 1 m*M* $MgCl_2$, and 5 m*M* mercaptoethanol, pH 7.5. The volume of the eluate is reduced to 50 ml by ultrafiltration over a membrane having a cut-off of 10,000 Da.

Hydroxylapatite Chromatography.[2,11] A 2.5 × 60-cm column is packed with defined Bio-Rad hydroxylapatite suspended in 0.5 *M* potassium phosphate. Before each use, the column is equilibrated with 1 liter of 1 m*M* potassium phosphate buffer containing 1 m*M* EDTA and 5 m*M* mercaptoethanol, pH 7.5.

The concentrated preparation from the previous step is centrifuged at about 5000 *g* for 10 min to remove a small amount of insoluble material. The supernatant is diluted with 10 volumes of the buffer used to equilibrate the column. After loading the sample by gravity flow, the column is washed with 100 ml of the same buffer. The column is developed at 30–50 ml/hr with a linear gradient generated with 1 liter each of 1 and 200 m*M* potassium phosphate buffer containing 1 m*M* EDTA and 5 m*M* mercaptoethanol pH 7.5. Mevalonate kinase activity usually appears in the fractions eluted midway through the gradient. The fractions containing mevalonate kinase with specific activity of 0.5 to 2 and greater than 2 are pooled separately. The pools are made 1 mg bovine serum albumin/ml, 2 m*M* EDTA, and 5 m*M* mercaptoethanol. After addition of 4 volumes of saturated ammonium sulfate, the enzyme is concentrated by centrifugation and made up as a 10 units/ml suspension in 100 m*M* potassium phosphate buffer containing 2 m*M* EDTA, 5 m*M* mercaptoethanol, and 60% ammonium sulfate, pH 7.5. The enzyme is stored at 4°.

Assay of Mevalonate

Preparation of Samples. Blood is collected in heparinized centrifuge tubes which are chilled by gentle mixing in an ice + water bath for 30 sec. The tubes are immediately centrifuged at 4° and the plasma is transferred to a centrifuge ultrafiltration device (Amincon CF-50 cone or equivalent). After centrifugation at 1000 *g* and 4° for 30 to 60 min, 1 ml of plasma yields about 0.75 ml of nearly protein-free ultrafiltrate. The concentration of mevalonate in ultrafiltrates kept frozen at −20° is stable for up to 2 years. EDTA should not be used as an anticoagulant because it chelates Mg^{2+} ions which are required for the action of mevalonate kinase. It is essential that the plasma be separated rapidly from the erythrocytes since the latter release glycerate, which interferes with the radioenzymatic assay of mevalonate.

Urine is frozen immediately after collection. Ultrafiltration is not required in the absence of proteinuria. The assay is conducted on a 5-fold dilution of urine.

Reagents. It is imperative that all solutions be made with water freshly distilled from a KOH solution. Stale distilled water, taken from containers contaminated with algae or mold, may contain some mevalonate.

a. Standards of mevalonate. A stock standard solution of 0.1 *M RS*-mevalonate is prepared by dissolving 130 mg of crystalline mevalonolactone in 5.5 ml of 0.2 *N* KOH. The solution is heated at 37° for 30 min to hydrolyze the lactone. After cooling, the pH of the solution is adjusted to 7.3 with 0.1 *N* HCl and its volume is brought to 10 ml with water. The exact concentration of *R*-mevalonate in a 1 : 100 dilution of this stock solution is determined by conducting a spectrophotometric enzymatic assay with mevalonate kinase/pyruvate kinase/lactate dehydrogenase as described by Popják.[10] Just before use in the radioenzymatic assay, aliquots of the stock standard are diluted to 50, 250, and 500 n*M* of *R*-mevalonate.

b. ATP solution, 0.20 *M*, neutralized to pH 7. The ATP must be substantially free of ADP.

c. [γ-^{32}P]ATP purchased with a specific activity of 1000 to 2000 Ci/mol and stored at −80°. Just before use, the stock solutions of unlabeled and labeled ATP are mixed and diluted to 1 m*M* with a specific activity of 30 Ci/mol.

d. $MgCl_2$, 0.3 *M* solution.

e. Potassium phosphate buffer 1 *M*, pH 7.5.

f. Mevalonate kinase, 10 units/ml.

g. 5-Phospho-*R*-[^{14}C]mevalonate, specific activity 10–50 Ci/mol. This derivative is commercially available. The working solution contains 4–5 μCi/ml.

h. Solution containing 0.5 *M* ATP, 0.1 *M* ADP, and 0.5 *M* KH_2PO_4.

i. Absolute ethanol.

j. Triethylammonium carbonate (TEAC) buffer, pH 9.7. A 1 *M* stock solution is prepared by bubbling slowly pure CO_2 into a flask packed in ice containing 1600 ml of water and 278 ml of freshly distilled triethylamine. Bubbling of CO_2 is stopped when all the triethylamine has dissolved and the pH has gone down to 9.7. The volume of the solution is made up to 2 liters. This stock buffer is kept at 4° in a brown bottle.

k. Anion exchange resin, carbonate form. Bio-Rad AG-1-X8 chloride resin, 200–400 mesh, packed in a large column is converted to the carbonate form by 35 bed volumes of 0.1 *M* Na_2CO_3, followed by water until the pH of the effluent is below 9.0, then 10 bed volumes of 20 m*M* TEAC pH 9.7. The resin is removed from the column and kept as a 50% (v/v) slurry in 20 m*M* TEAC buffer.

For each set of determinations, 1 ml of 0.1 *M* TEAC buffer is added to each of 20 Bio-Rad glass Econo-Columns (0.7 × 10 cm) fitted with Luer-type three-way nylon stopcocks at their outflow. Then 4 ml of the slurry of the AG-1-X8 (carbonate) resin is added to each. After the resin has settled, the height of the packing should be about 7 ± 0.25 cm. The columns can be used immediately.

l. Column chromatography apparatus. Simultaneous gradient elution from 20 columns can be accomplished using fraction collectors with either rotating circular or rectangular sample tables. The former requires that an American Optical Co. Model 12063 fraction collector be fitted with a home-made table having five concentric rows for samples. Alternatively, fraction collectors having a rectangular sample table, such as the Buchler Model 400 or equivalent, can be fitted with two banks of column mounts, each holding 10 columns, positioned at opposing corners. Each bank is held together by a 2-cm-thick rectangular Lucite plate drilled with 10 holes, the shape of which matches the bottom of the upper reservoir of each column. Each column holder is supported by a cylindrical metal post passing through one of its corners. The post is itself fixed by a screw on one corner of the fraction collector. This set-up allows each bank of columns to be swung out of the collector and over a waste bucket. The columns are held over the bucket during packing and loading.

Assay Protocol[2]

For the determination of mevalonate in the range of 0–50 pmol, 50 μl of a reaction mixture of the following composition is added to 100 μl of plasma filtrate, and to 100 μl of mevalonate standards containing 0, 5, 25, or 50 pmol of *R*-mevalonate (0, 10, 50 or 100 pmol of *RS*-mevalonate, respectively): 1 *M* potassium phosphate buffer, pH 7.5, 15 μl; 0.3 *M* $MgCl_2$, 5 μl; 1 m*M* [γ-^{32}P]ATP, 7.5 μl, 2 to 3 μCi; mevalonate kinase, 3 μl; and water, 19.5 μl. We usually carry out 20 simultaneous determinations (1 blank, 4 standards, 14 samples, and a reference plasma ultrafiltrate); for such a set we prepare $22 \times 50 = 1100$ μl of the reaction mixture. The assays are carried out in 10×75-mm disposable glass tubes.

Although the final concentration of [^{32}P]ATP in the assay mixture at 50 μM is not optimal for the mevalonate kinase reaction, it is adequate with a relatively large excess of the kinase and a long incubation time. It is unnecessary to determine the specific activity of the ATP, as this value is not needed in the calculations; the 2–3 μCi/7.5 nmol per assay is a convenient value and can be varied within fairly wide limits. If the amounts of mevalonate to be determined are larger than 50 pmol, the samples are diluted. It is essential to include mevalonate standards with each set of determinations; the standards should cover the range of mevalonate values expected in the samples.

The 150 μl (final volume) reaction mixtures are left at room temperature (about 22°) for 2.5 hr. After this, 25 μl of concentrated HCl is added and the tubes are vortexed to stop the reaction.[16] Then, precisely 20 μl of 5-phospho[^{14}C]mevalonate (about 2×10^5 dpm) and 25 μl of reagent (h) (ATP + ADP + KH_2PO_4), in that order, are stirred into each reaction mixture. Absolute ethanol (0.9 ml) is then added and the mixtures are chilled in ice for at least 30 min; the precipitate that is formed is then centrifuged down at 500 *g* for 10 min at 0°. By this procedure, more than 90% of the [^{32}P]ATP and of the [^{32}P]phosphate present in the incubation mixture is precipitated, while doubly labeled 5-phosphomevalonate remains in the 80% ethanol supernatant. Under certain conditions, the 80% ethanol supernatant must be subjected to additional precipitation steps with either ATP + ADP + KH_2PO_4 or with sodium glycerate or both (see below).

The supernatant is aspirated with a Pasteur pipet and is added to 5 ml of 0.1 *M* TEAC buffer, pH 9.7, at room temperature. It is immaterial

[16] It is essential to inactivate mevalonate kinase with HCl before the addition of 5-phospho[^{14}C]mevalonate. Aged preparations of the latter may contain traces of mevalonate, the phosphorylation of which can produce erratic results.

whether or nor precisely the same volume of supernatant is drawn off from each sample tube, as the method is essentially one of isotope dilution.

The 20 diluted samples are simultaneously loaded onto the 20 columns using a 20-channel Technicon model II pump operating at a flow rate of 1.0 ml/min. The proximal end of each of the 20 Tygon tubings is fitted with a 10 cm length of gauge 16 stainless-steel tubing. The 20 steel tubings are put through holes in a rectangular Lucite plate. Thus all tubings can be dipped simultaneously to the bottom of the glass tubes holding the diluted samples. The proximal sections of the Tygon lines pass through two 10-channel Technicon wash valves. These valves are connected via Tygon tubings and a three-way stopcock to (1) a reservoir containing 0.1 *M* TEAC and (2) a linear gradient generator made up of 1000 ml each of 0.1 and 0.7 *M* TEAC buffers. When the tubes holding the samples have emptied, air is allowed to pass through the proximal tubings until it passes through the two 10-channel Technicon wash valves which are then switched. During the following 15 min, 0.1 *M* TEAC buffer is pumped through the column. Then the three-way stopcock is switched, thus starting the gradient development of the 20 columns.

The two banks of columns are swung over the fraction collector which is activated. The first 3 fractions are collected for 15 min each. Subsequently, 2.5-min fractions are collected. Doubly labeled 5-phosphomevalonate is usually eluted between fractions 5 and 9. The development of the column is stopped at this point.[17] Samples (2-ml) are counted briefly to identify the fractions giving the highest ^{14}C counts.[18] These are then counted for ^{32}P and ^{14}C for longer intervals.

The gain and window of two channels of a liquid scintillation spectrometer are adjusted such that the ^{32}P channel excludes ^{14}C counts while the ^{14}C channel records 12–14% of the counts of the ^{32}P channel. Under optimum conditions, the ratio $^{32}P/^{14}C$ is constant in the fractions containing doubly labeled 5-phosphomevalonate (Fig. 1). A standard curve is constructed by plotting the $^{32}P/^{14}C$ ratio of the standards as a function of the amount of *R*-mevalonate present in these standards. The standard curve is linear between 0 and 100 pmol of *R*-mevalonate. The blank of the assay is equivalent to about 1 pmol.

[17] Since residual [^{32}P]ATP is not eluted from the columns with this particular gradient, the column packings are removed after each use, stored for several half-lives of the ^{32}P, regenerated in the usual way, and used again. The glass Econo-Columns and their fittings, after being soaked in detergent and rinsed with water, can be reused immediately.

[18] Alternatively, the fractions can be collected directly into liquid scintillation "minivials" and counted without transfer. The liquid scintillation fluid used must give a clear solution when mixed with the column eluate.

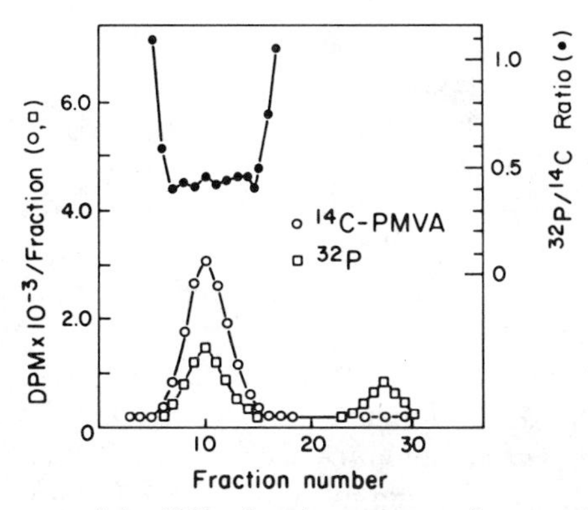

FIG. 1. Chromatogram of the 80% ethanol supernatant from a radioenzymatic assay of plasma mevalonate. Fractions containing [5-^{32}P]phosphomevalonate are identified by the coelution of 5-phospho[2-^{14}C]mevalonate. The radiochemical purity of [5-^{32}P]phosphomevalonate is indicated by the constancy of the ^{32}P/^{14}C ratio throughout the peak.

Occasionally, the ratio ^{32}P/^{14}C is not constant over the fractions containing doubly labeled 5-phosphomevalonate.[11] This may occur when the lot of [γ-^{32}P]ATP used is impure, that is when a substantial fraction of the radioactivity is present in the form of inorganic phosphate or of products of ATP radiolysis. This condition is generally reflected by high ^{32}P counts (>100 cpm) in the fractions eluting before phosphomevalonate, or in more than 100 cpm in the phosphomevalonate peak of the blank. When chromatography of the [γ-^{32}P]ATP confirms such a problem, it is necessary to subject 0.9 ml of the 80% ethanol supernatant to a second precipitation with 0.1 ml of the ATP + ADP + KH_2PO_4 solution and 0.5 ml of absolute ethanol.

Another cause of nonconstancy of the ^{32}P/^{14}C ratio in the fractions containing 5-phosphomevalonate is the presence of a ^{32}P-labeled compound tentatively identified as phosphoglycerate. This compound is eluted slightly after, but is not separated from, 5-phosphomevalonate. The mevalonate kinase preparation is sometimes contaminated with glycerate kinase activity.[19] This interference occurs most often (1) when the plasma was not separated from the erythrocytes immediately after blood sampling, or (2) when hemolysis has occurred, or (3) when mevalonate is assayed in the plasma of uremic patients. This interference can be eliminated by including 1.5 *M* of 3-phospho-*RS*-glycerate in reagent (h).

[19] Glycerate kinase activity is assayed spectrophotometrically using the same assay system as for mevalonate kinase, except that the reaction is started with sodium glycerate instead of mevalonate.

Applications

Urinary Excretion. Mevalonate was identified as a normal constituent of human and rat urine,[5] by (1) radioenzymatic assay conducted on native urine, and (2) spectrophotometric enzymatic assay conducted on an ether extract of 0.5 liter of human urine. The urinary clearance of endogenous mevalonate is 29 and 44% of the glomerular filtration rate in humans and rats, respectively. From studies conducted in perfused rat kidney,[6] it appears that there is no physiological concentration of circulating mevalonate at which the substrate is entirely reabsorbed by the kidney tubule. There is therefore no physiological "renal threshold" for mevalonate.

Kinetics of Plasma Mevalonate. Comparison between the turnover of plasma mevalonate in rats[4] and in humans[20] with the whole-body rate of sterol synthesis shows that the turnover of plasma mevalonate represents about 0.1% of the rate of mevalonate production in the whole body.[21] About two-thirds of the circulating mevalonate is taken up by the kidneys.[2,4] The three main fates of mevalonate in the kidney are incorporation into squalene + sterols, excretion in urine, and degradation via the shunt pathway of mevalonate metabolism.[7] Acute nephrectomy in the rat leads to a 5-fold increase in the level of plasma mevalonate.[5]

Diurnal Rhythm of Plasma Mevalonate. In human subjects put on a feeding schedule of either one[4] or five[20,22] meals per day, the level of plasma mevalonate follows oscillations over a 2- to 5-fold range (Fig. 2). The peak of the rhythm occurs between midnight[22] and 7 AM.[4] The oscillation is abolished by short-term[4] or long-term[22] starvation, as well as by cholesterol feeding.[22] The urinary concentration of mevalonate parallels the variations in plasma concentration.[4]

Plasma Mevalonate as an Index of Cholesterol Synthesis. A linear correlation has been demonstrated between the level of plasma mevalonate (between 7 and 9 AM after an overnight fast) and the rate of whole-body cholesterol synthesis in humans[20,22] (Fig. 3). This relationship applies to normocholesterolemic and hypercholesterolemic subjects, as well as to subjects in whom cholesterol synthesis was either raised or suppressed by cholestyramine or cholesterol feeding respectively. So far, the only technique not requiring the use of radioisotopes that is applicable for measuring whole body cholesterol synthesis in humans is the sterol balance method. As with all metabolic balance methods, synthesis is calcu-

[20] T. S. Parker, D. J. McNamara, C. D. Brown, R. Kolb, E. H. Ahrens, Jr., A. W. Alberts, J. Tobert, J. Chen, and P. J. De Schepper, *J. Clin. Invest.* **74,** 795 (1984).

[21] For an extensive discussion of this topic, see B. R. Landau and H. Brunengraber, this volume [12].

[22] T. S. Parker, D. J. McNamara, C. Brown, O. Garrigan, R. Kolb, H. Batwin, and E. H. Ahrens, Jr., *Proc. Natl. Acad. Sci. U.S.A.* **73,** 3037 (1982).

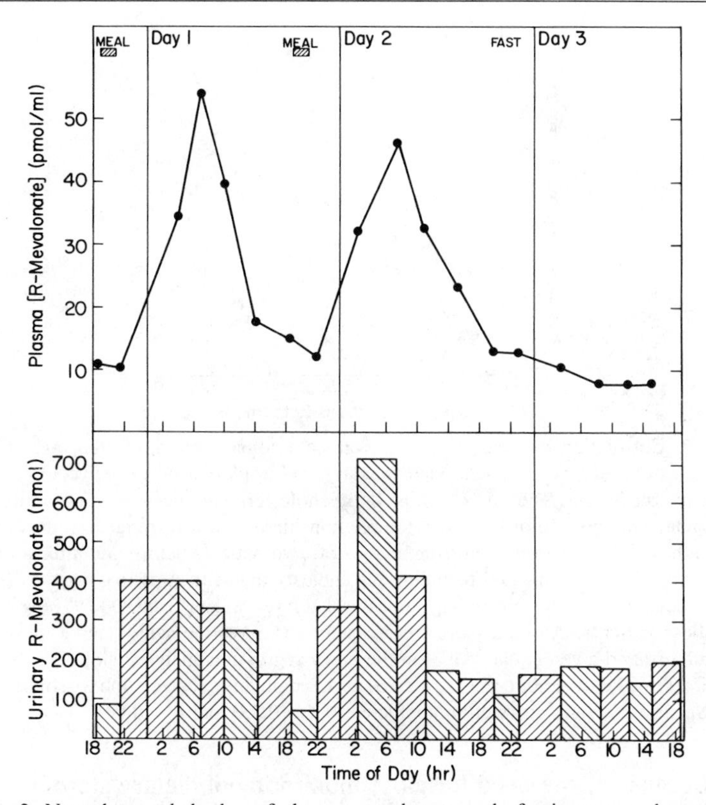

FIG. 2. Nyctohemeral rhythm of plasma mevalonate and of urinary mevalonate excretion in a 40-year-old normocholesterolemic male subject equilibrated on a one-meal-per-day feeding schedule. On day 2, the meal was omitted. The 24-hr urinary excretion of mevalonate amounted to 2.06, 1.55, and 0.85 μmol on days 1, 2, and 3, respectively. Reproduced with permission from Kopito *et al.*[4]

lated as the difference between input and output; in this case requiring quantitative analysis of dietary neural sterols and fecal acidic and neutral sterols. More importantly, this cannot distinguish between sterol synthesis and flux except in the metabolic steady state where net sterol flux is zero by definition. Although it can never be proven that a metabolic steady state exists, it is generally agreed to be a reasonable assumption when the following conditions are met: constant body weight and plasma lipid concentration, constant rate of fecal neutral and acidic sterol excretion, and stable clinical course. The rate of cholesterol synthesis cannot be calculated from sterol balance data in the non-steady state (i.e., growing animals and infants or during the transition between two dietary or therapeutic regimens) without some independent measurement of sterol

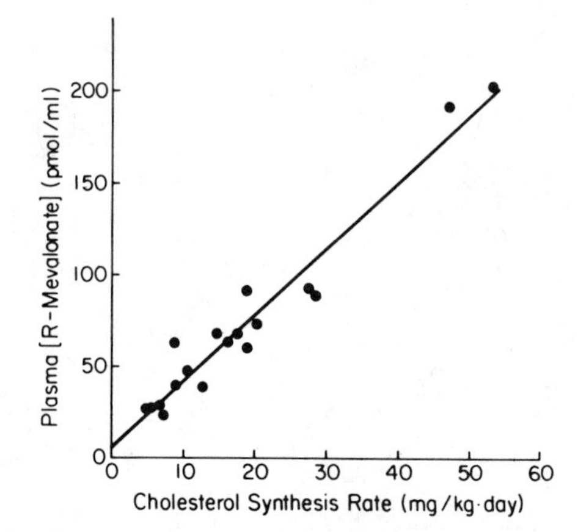

FIG. 3. Correlation between plasma mevalonate concentration and the rate of whole-body cholesterol synthesis. Mevalonate was assayed in plasma taken between 7 and 9 AM after an overnight fast. The rate of whole body cholesterol synthesis was measured by the sterol balance method. In order to test the relationship over a wide range of rates of cholesterol synthesis, data were pooled from studies of: two patients before and after portocaval anastomosis; three patients before and during cholestyramine therapy (8 g twice a day); four patients receiving five diets providing cholesterol at rates ranging from 80 to 981 mg/day; and three obese patients. All data pairs were assumed to have the same degree of statistical independence and given equal weight in the linear regression analysis: slope = 3.63 ± 0.2 (nmol/liter)/(mg/kg/day); intercept = 5.69 ± 4.8 nmol/liter; correlation coefficient (r) = 0.972; Spearman's rank correlation coefficient (r_s) = 0.933 and $p < 0.001$.

flux. Because of the need for close monitoring of dietary sterol intake, a metabolic steady state, and fecal collections, the sterol balance method does not lend itself to large-scale epidemiological studies aimed at identifying individuals at risk for arteriosclerosis. On the other hand, the assay of plasma mevalonate is a fairly simple procedure, once it has been set up. In the future, determination of the level of plasma mevalonate may prove to be a useful alternative to sterol balance as a means of measuring the rate of cholesterol synthesis in certain well defined applications. For example, relatively large changes in cholesterol synthesis, such as those caused by administration of bile acid binding resins or inhibitors of HMG-CoA reductase, can be easily detected and used to adjust dosage. However, it must be pointed out that cholesterol synthesis rates are estimated from mevalonate measurements by extrapolation from a standard curve relating mevalonate data to synthesis as measured by the sterol balance method. This approach assumes that there are no important differences in mevalonate metabolism between the group of patients (or animals) that serves as the reference group and the group under study. This correspondence must be demonstrated for each new case by direct comparison to an accepted reference such as the sterol balance of tritiated water[23] method. Nevertheless, we can foresee that the assay of mevalonate will become a semiroutine clinical test of responsiveness to cholesterol-lowering diets and drugs used to manage patients having hypercholesterolemia or premature atherosclerosis.

[23] J. M. Dietschy, *J. Lipid Res.* (1985), in press.

[8] Mevalonate Kinase

By JOHN W. PORTER[1]

Mevalonate 5-phosphotransferase (mevalonate kinase, EC 2.7.1.36; ATP : mevalonate 5-phosphotransferase) has been found in a wide variety of sources. Some of these are yeast autolysate,[1a] pig[2] and rabbit[3] liver extracts, superovulated rat ovaries,[4] pumpkin seedlings (*Cucurbita pepo*[5]), rubber latex,[6] larva of the flesh fly,[7] green leaves and etiolated cotyledons of French beans,[8–11] *Pinus pinaster* seedlings and extracts of *Agave americana*.[12] Kinase activity has also been found in *Pinus radiata* seedlings,[13] orange juice vesicles,[14] dark-grown cultures of *Kalanchoe crenata*,[15] *Staphylococcus aureus*,[16,17] and melon cotyledons (*Cucumis melo*[18]).

[1] Deceased June 27, 1984.
[1a] T. T. Tchen, *J. Biol. Chem.* **233,** 1100 (1958).
[2] H. R. Levy and G. Popják, *Biochem. J.* **75,** 417 (1960).
[3] K. Markley and E. Smallman, *Biochim. Biophys. Acta* **47,** 327 (1961).
[4] A. P. F. Flint, *Biochem. J.* **120,** 145 (1970).
[5] W. D. Loomis and J. Battaile, *Biochim. Biophys. Acta* **67,** 54 (1963).
[6] I. P. Williamson and R. G. O. Kekwick, *Biochem. J.* **96,** 862 (1965).
[7] R. D. Goodfellow and F. J. Barnes, *Insect Biochem.* **1,** 271 (1971).
[8] L. J. Rogers, S. P. J. Shah, and T. W. Goodwin, *Biochem. J.* **99,** 381 (1966).
[9] L. J. Rogers, S. P. J. Shah, and T. W. Goodwin, *Biochem. J.* **100,** 14C (1966).
[10] J. C. Gray and R. G. O. Kekwick, *Biochem. J.* **133,** 335 (1973).
[11] J. C. Gray and R. G. O. Kekwick, *Biochem. J.* **113,** 37 (1969).
[12] E. Garcia-Peregrin, M. D. Suarez, and F. Mayor, *FEBS Lett.* **38,** 15 (1973).
[13] P. Valenzuela, E. Beytia, O. Cori, and A. Yudelevich, *Arch. Biochem. Biophys.* **113,** 536 (1966).
[14] V. H. Potty and J. H. Bruemmer, *Phytochemistry* **9,** 99 (1970).
[15] D. R. Thomas, *Phytochemistry* **9,** 1443 (1970).
[16] S. Ohnoki, G. Suzue, and S. J. Tanaka, *J. Biochem.* (*Tokyo*) **52,** 423 (1962).
[17] G. Suzue, K. Orihara, H. Morishima, and S. Tanaka, *Radioisotopes* **13,** 294 (1964).
[18] J. C. Gray and R. G. O. Kekwick, *Biochim. Biophys. Acta* **279,** 290 (1972).

METHODS IN ENZYMOLOGY, VOL. 110

ISBN 0-12-182010-6

Mevalonate 5-phosphotransferase has been partially purified from yeast,[1a] hog liver,[2] rabbit liver,[3] *Cucurbita pepo* (pumpkin) seedlings,[5] and *Hevea brasiliensis* (rubber) latex,[6] and it has been purified to homogeneity from hog liver.[19]

This enzyme reacts stereospecifically with 3*R*-mevalonate and MgATP to produce *R*-mevalonate 5-phosphate,[20,21] as indicated in the following reaction. The pure enzyme has an activity of 17 μmol of product formed per minute per milligram of protein.[19]

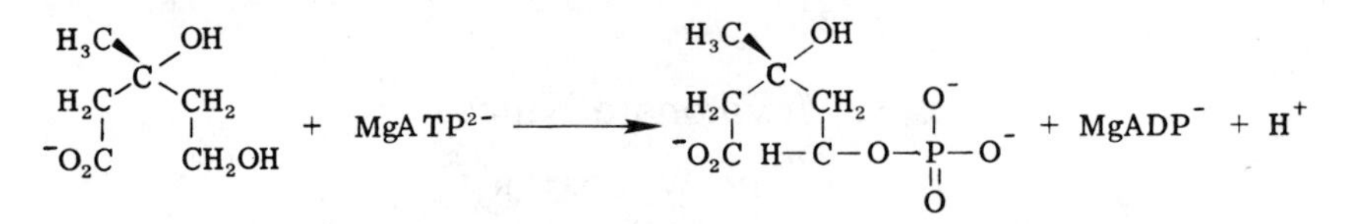

Assay of Enzyme Activity

Spectrophotometric Assay. A coupled spectrophotometric assay procedure that utilizes pyruvate kinase and lactate dehydrogenase is used.[19] In this assay the amount of ADP formed by mevalonate kinase is measured by coupling this reaction with the reactions catalyzed by pyruvate kinase and lactate dehydrogenase. This coupled reaction is monitored spectrophotometrically at 340 nm for NADH oxidation.[1a,2,19,22]

Radiochemical Assay. Mevalonate kinase activity can also be assayed radiochemically.[19] The amount of phosphorylated derivative formed in an incubation mixture from DL-[2-^{14}C]mevalonate can be determined by paper or thin-layer chromatography and subsequent assay for radioactivity.[19,23]

Reagents

Potassium phosphate buffer, 1.0 *M*, pH 7.0, and at the concentrations indicated
2-Mercaptoethanol or dithiothreitol, 200 m*M*
ATP, 200 m*M*
DL-Mevalonate, 300 m*M*
Mevalonate kinase (7.0 units/ml)
$MgCl_2$, 100 m*M*

[19] E. Beytia, J. K. Dorsey, J. Marr, W. W. Cleland, and J. W. Porter, *J. Biol. Chem.* **245**, 5450 (1970).
[20] F. Lynen and M. Grassl, *Hoppe-Seyler's Z. Physiol. Chem.* **313**, 291 (1958).
[21] R. H. Cornforth, J. W. Cornforth, and G. Popják, *Tetrahedron* **18**, 1351 (1962).
[22] J. K. Dorsey and J. W. Porter, *J. Biol. Chem.* **243**, 4667 (1968).
[23] A. deWaard, A. H. Phillips, and K. Bloch, *J. Am. Chem. Soc.* **81**, 2913 (1959).

NADH, 3.0 m*M*
Phosphoenolpyruvate, 50 m*M*
Lactate dehydrogenase (1.0 mg protein and 500 units/ml)
Pyruvate kinase (1.0 mg protein and 400 units/ml)
ADP, 5.0 m*M* solution

Assay Procedures[19,24]

Spectrophotometric Assay. A spectrophotometer such as a Gilford 2400S or a later model that is equipped with a chart recorder is used in the spectrophotometric assay. Normally, several preliminary experiments are performed to determine the optimum concentrations of the reactants listed below and these concentrations are then used in the assays. Initial velocities are calculated from the initial slope of the recorder tracing. The following components are incubated in a 1.0 ml volume in a typical assay[19]: 100 m*M* KH_2PO_4, pH 7.0, 10 m*M* 2-mercaptoethanol or dithiothreitol, 0.16 m*M* NADH, 5 m*M* $MgCl_2$, 4 m*M* MgATP, 3 m*M* DL-mevalonate, mevalonate kinase (about 0.01 unit), 0.5 m*M* phosphoenolpyruvate, lactate dehydrogenase (0.05 mg of protein and 27 units), and pyruvate kinase (0.05 mg of protein and 20 units). The reaction is started by the addition of mevalonate kinase. Appropriate blanks are carried out in the absence of this enzyme.

Radiochemical Assay. A chromatographic assay for mevalonate 5-phosphate is used when it is necessary to identify the product, determine the effect of ADP on the initial velocity of the reaction, or the activity of enzyme in crude extracts is too low to be measured accurately by the spectrophotometric method. Incubation components (buffer, 2-mercaptoethanol, $MgCl_2$, ATP, [^{14}C]mevalonate, and enzyme) are added in a small volume (0.1 to 0.2 ml) and then incubated.[19] At the end of the reaction, enzyme activity is stopped by heating the reaction mixture in boiling water for 2 min. The reaction mixture is centrifuged and the entire supernatant solution is applied to a strip of Whatman No. 1 paper. The small precipitate obtained on centrifugation is washed with 0.075 ml of deionized water, centrifuged again, and the supernatant solution is applied to the paper. Chromatograms are developed by the descending technique in 1-propanol : ammonia : water (60 : 20 : 10) for 12 hr. The papers are scanned for radioactivity and then the sections that contain mevalonate 5-phosphate are cut into small pieces and assayed for radioactivity in a liquid scintillation counter.

Units. A unit of enzyme activity is defined as the amount of enzyme required to catalyze the incorporation of 1 μmol of phosphate from

[24] G. Popják, this series, Vol. 15, p. 393.

MgATP into mevalonate 5-phosphate per minute. Specific activity of the enzyme is expressed as units of enzyme activity per milligram of protein.

Purification of Mevalonate Kinase

Two fresh hog livers (total weight of 2000 g) are homogenized for 15 sec at full speed in a commercial blender with 4 liters of 0.1 *M* K_2HPO_4, pH 7.5, containing 10 m*M* 2-mercaptoethanol and 1 m*M* potassium ethylenediaminetetraacetate acid. This and all subsequent operations are carried out at 0 to 4°. Homogenates are filtered through several layers of alternating cheesecloth and glass wool to remove cell debris and neutral lipids. This extract is then subjected to centrifugation at 15,000 *g* at 0° for 10 min. The supernatant solution is again filtered through a cheesecloth-glass wool mat and then centrifuged at 48,000 *g* for 10 min and at 105,000 *g* for 45 min.

Ammonium Sulfate Precipitation and Calcium Phosphate Gel Adsorption

Protein of the 105,000 *g* supernatant solution is fractionated with solid ammonium sulfate. The fraction precipitating between 15 and 40% saturation is dissolved in 1500 ml of 10 m*M* buffer and dialyzed at 4° for 10 hr against 25 volumes of the same buffer.

The dialyzed protein is treated with calcium phosphate gel[25] as follows. To 10 mg of protein, 1 mg of gel is added; the solution is stirred until homogeneous, and then centrifuged. The precipitate is discarded and additional calcium phosphate gel is added to the supernatant solution (1 mg/1.5 mg of protein in the original dialyzed 15 to 40% ammonium sulfate fraction). The solution is stirred for 5 min, centrifuged, and the precipitate is washed with equal volumes of buffer as follows: once with 10 m*M*, twice with 20 m*M*, and four times with 50 m*M*.

Protein of the combined 50 m*M* eluates from the calcium phosphate gel adsorption is precipitated with solid ammonium sulfate between 0 and 50% of saturation. The precipitate is dissolved in 10 m*M* buffer, and then dialyzed against 10 times its volume of the same buffer for 4 hr, with a change of buffer at 2 hr.

DEAE-Cellulose Chromatography and Sephadex Gel Filtration

The dialyzed protein is divided into four fractions, and each is adsorbed on a column (3.3 × 48 cm) of DEAE-cellulose, previously equili-

[25] K. K. Tsuboi and P. B. Hudson, *J. Biol. Chem.* **224,** 879 (1957).

PURIFICATION OF MEVALONATE KINASE

Purification step	Total protein (mg)	Total units[a] (units)	Specific activity (units/mg)	Volume (ml)	Purification	Yield of units (%)
Extract	136,500	3,969	0.029	4,010		100.0
Ammonium sulfate	60.050	3,120	0.052	1,274	1.78	78.8
Calcium phosphate gel	9,398	2,270	0.242	3,530	8.34	57.4
Ammonium sulfate	7,855	2,035	0.259	335	8.93	51.4
DEAE-cellulose	833	724	0.870	760	30.0	18.3
Ammonium sulfate	562	622	1.105	33	38.10	15.7
Sephadex G-200	140	526	3.766	220	130.80	13.3
DEAE-cellulose	29	258	8.99	153	310.0	6.5
Sephadex G-150	5	87	17.5	30	605.0	2.2

[a] Definitions and conditions of the assay are as reported by E. Beytia, J. K. Dorsey, J. Marr, W. W. Cleland, and J. W. Porter, *J. Biol. Chem.* **245,** 5450 (1970). Reproduced with permission of the *Journal of Biological Chemistry*.

brated with 10 m*M* buffer. The column is washed with 10 m*M* buffer until protein no longer elutes, and then with 20 m*M* buffer. Mevalonate kinase is eluted from the column with 50 m*M* buffer. Enzyme activity from each of the four columns is precipitated between 0 and 60% of saturation with solid ammonium sulfate. The precipitate is dissolved in a minimal volume of 10 m*M* buffer and then divided into two aliquots.

Each of these aliquots is filtered upward through a Sephadex G-200 column (5 × 90 cm), previously equilibrated with 10 m*M* buffer. Protein is eluted with a peristaltic pump at a flow rate of 5 ml/min. Those fractions, 7 ml each, containing the active enzyme are pooled and then adsorbed on a DEAE-cellulose column (1.2 × 20 cm), previously equilibrated with 10 m*M* phosphate buffer. The column is washed with 10 m*M* and then with 20 m*M* buffer. A linear gradient of potassium phosphate buffer (200 ml each of 20 and 80 m*M* buffer) is used to elute the protein. Eluate fractions of 4 ml each are collected.

The enzymatically active fractions are pooled and diluted to a concentration of 10 m*M* buffer and 10 m*M* 2-mercaptoethanol. Protein is concentrated by adsorption on a column of DEAE-cellulose, and then it is eluted with 100 m*M* buffer. The enzymatically active protein from the column is precipitated with solid ammonium sulfate between 0 and 60% of saturation. The precipitate is dissolved in a minimal volume of 10 m*M* buffer, and it is filtered through a Sephadex G-150 column (3 × 55 cm), previously equilibrated, and eluted with the same buffer.

The procedure used in the preparation of mevalonate kinase is summarized in the table.

Properties of the Enzyme

Sephadex Gel Filtration. Mevalonate kinase, prepared as reported in the table, is subjected to Sephadex G-200 gel filtration on a column (1.0 × 90 cm). The column is equilibrated, and the enzyme is eluted with 10 m*M* potassium phosphate buffer, pH 7.5, containing 5 m*M* 2-mercaptoethanol. The enzyme behaves as a single component on this column.

Disc Gel Electrophoresis. Acrylamide disc gel electrophoresis is carried out according to the method of Davis.[26] Acrylamide (5%) is used in the resolving gel, and electrophoresis is carried out at 1.5 mA per gel. After electrophoresis the gels are fixed and stained by the procedure of Chrambach *et al.*[27] A single protein component is observed on the gel.

Sucrose Density Gradient Centrifugation. An aliquot of mevalonate kinase, purified as reported in the table, is concentrated by precipitation with solid ammonium sulfate. The fraction that precipitates between 0 and 80% saturation is collected by centrifugation, dissolved in a minimal amount of 10 m*M* K_2HPO_4, pH 7.5, containing 10 m*M* 2-mercaptoethanol, and then dialyzed against 5000 volumes of the same buffer for 4 hr, with a change of buffer at 1.5 hr. The concentration of the dialyzed protein is adjusted to 15 mg/ml. A 100-μl aliquot of this solution is layered on the top of a linear gradient prepared with equal volumes of 14 and 40% of sucrose, in 10 m*M* K_2HPO_4, pH 7.5, containing 10 m*M* 2-mercaptoethanol. This gradient is centrifuged at 40,000 rpm for 22 hr in a Spinco SW50 rotor, in a model L Spinco preparative ultracentrifuge. The latter enzyme behaves as a single component on sucrose gradient centrifugation.[19]

Molecular Weight Estimation. The gel technique of Whitaker[28] is used to estimate the molecular weight of mevalonate kinase.[19] A column of Sephadex G-100 (3 × 50 cm) is packed and washed with 100 m*M* phosphate buffer, pH 7.0, for several days. The following protein samples are then applied in a volume of 1 ml: fatty acid synthetase, 6.3 mg (molecular weight 450,000[29]); bovine serum albumin (monomer and dimer), 10 mg (molecular weight 70,000[30]); pepsin, 10 mg (molecular weight 35,000[31]); and mevalonate kinase, 340 μg of protein of a specific activity of 9 μmol of

[26] B. J. Davis, *Ann. N.Y. Acad. Sci.* **121,** 404 (1964).

[27] A. Chrambach, R. A. Reisfeld, M. Wyckoff, and J. Zacari, *J. Anal. Biochem.* **20,** 150 (1967).

[28] J. R. Whitaker, *Anal. Biochem.* **35,** 1950 (1963).

[29] P. C. Yang, P. H. W. Butterworth, R. M. Bock, and J. W. Porter, *J. Biol. Chem.* **242,** 3501 (1967).

[30] P. Johnson, *Annu. Rep. Prog. Chem.* **43,** 30 (1946).

[31] F. A. Bovey and S. S. Yanari, *in* "The Enzymes" (P. D. Boyer, H. Lardy, and K. Myrbäck, eds.), 2nd rev. ed., Vol. 4, p. 63. Academic Press, New York, 1960.

product formed/min/mg of protein. Fatty acid synthetase is used to determine the void volume (106.5 ml), and the other proteins are used to construct a standard curve. Solvent flow is directed upward in these separations with the aid of a peristaltic pump at a rate of 16 ml/hr. This enzyme has a molecular weight of 98,000.

Isoelectric Point of Mevalonate Kinase. The isoelectric point of mevalonate kinase is determined with an LKB electrofocusing apparatus and a gradient of sucrose and ampholyte ranging between pH 3.0 and 6.0. The separation is carried out in a No. 1801 column with 25.2 mg of protein purified as reported in the table. Electrofocusing with increasing voltage (up to 650 V) is carried out for 37 hr. The current is kept constant through maintenance of power at 1.5 W. An isoelectric point of 4.7 is obtained for the pure protein.

Mechanism of Enzymatic Reaction

From a combination of studies[19] on the initial velocity of the reaction, inhibition by product, and inhibition by dead end inhibitor, it has been concluded that (1) the phosphorylation of mevalonic acid catalyzed by mevalonate kinase is sequential; that is, all substrates must react with the enzyme before the first product is released, (2) mevalonate reacts with the enzyme first, followed by MgATP, and (3) the order of release of products is mevalonate 5-phosphate and then ADP. This sequence of reactions is shown in diagrammatic form as follows, where A is mevalonate, B is MgATP, P is mevalonate 5-phosphate, and Q is ADP.

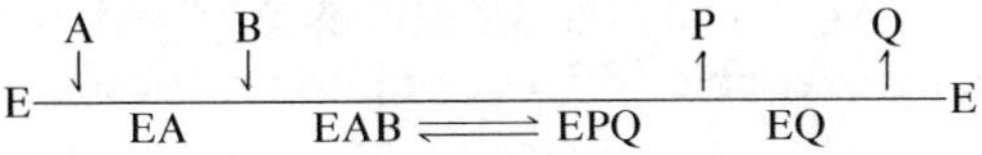

```
     A       B                 P       Q
     ↓       ↓                 ↑       ↑
E──────────────────────────────────────────E
     EA     EAB ⇌ EPQ          EQ
```

Evidence has been presented[19] that an SH group is important in the phosphorylation of mevalonic acid, and that the reactivity of this group is influenced by the presence of potassium mevalonate and MgATP. Presumably, the SH group of the enzyme is present in a cysteine molecule. Evidence has also been presented[32] for the participation of a lysine residue of the enzyme. No evidence exists for the presence of enzyme-bound intermediates in this reaction.

Inhibition of the Enzyme

Evidence has been presented[22] that geranyl and farnesyl pyrophosphates are potent inhibitors of mevalonate kinase. Isopentenyl and di-

[32] M. Soler, A. M. Jabalquinto, and E. Beytia, *Int. J. Biochem.* **10,** 931 (1979).

methyl pyrophosphate and inorganic pyrophosphate do not inhibit this enzyme significantly under the same conditions. The kinetics of this inhibition show that geranyl pyrophosphate inhibition of mevalonate kinase is uncompetitive with respect to mevalonic acid and competitive with respect to MgATP. These results suggest that geranyl and farnesyl pyrophosphates may be physiological regulators of mevalonate kinase activity.

Acknowledgment

This work was supported in part by a Research Grant, HL 16364, from the National Heart and Lung Institute, National Institutes of Health, United States Public Health Service, and by the Medical Research Service of the Veterans Administration.

[9] Phosphomevalonate Kinase from Pig Liver

By JAIME EYZAGUIRRE and SERGIO BAZAES

$$(3R)\text{-Phosphomevalonate} + \text{ATP} \xrightleftharpoons{Mg^{2+}} (3R)\text{-pyrophosphomevalonate} + \text{ADP}$$

Phosphomevalonate kinase (EC 2.7.4.2) catalyzes the phosphorylation of phosphomevalonate (MVAP) by MgATP to pyrophosphomevalonate (MVAPP) and ADP. Only the 3*R* isomer of MVAP is phosphorylated and the reaction is reversible.[1,2] The equilibrium constant is near unity at pH 7.5 and 30°.[3]

Assay Methods

Spectrophotometric Assay[4]

Principle. This method uses a coupled system. ADP generated by the enzyme is transformed into ATP by pyruvate kinase, and the pyruvate formed is reduced to lactate by NADH and lactate dehydrogenase:

[1] K. Bloch, S. Chaykin, A. H. Phillips, and A. deWaard, *J. Biol. Chem.* **234,** 2595 (1959).

[2] H. Hellig and G. Popják, *J. Lipid Res.* **2,** 235 (1961).

[3] D. Valdebenito, E. Cardemil, A. M. Jabalquinto, and J. Eyzaguirre, unpublished work (1983).

[4] S. Bazaes, E. Beytía, A. M. Jabalquinto, F. Solís de Ovando, I. Gómez, and J. Eyzaguirre, *Biochemistry* **19,** 2300 (1980).

ISBN 0-12-182010-6

MVAP + ATP → MVAPP + ADP
PEP + ADP → ATP + pyruvate
Pyruvate + NADH + H^+ → lactate + NAD^+

MVAP + PEP + NADH + H^+ → MVAPP + lactate + NAD^+

The reaction is followed kinetically by measuring the disappearance of NADH at 340 nm.

Reagents

Tris–HCl buffer, 1 *M*, pH 7.5
KCl, 750 m*M*
$MgCl_2$, 50 m*M*
PEP, 25 m*M*
Mercaptoethanol, 100 m*M*
ATP, 10 m*M*
NADH, 24 m*M*
MVAP, 44 m*M*
Pyruvate kinase from rabbit muscle (crystalline suspension in ammonium sulfate
Lactate dehydrogenase from rabbit muscle (crystalline suspension in ammonium sulfate)

Procedure. Enzyme assays are performed at 30° in a spectrophotometer with a thermostatted cell compartment. The assay mixture contains 0.1 ml Tris buffer, 0.1 ml KCl, 0.1 ml $MgCl_2$, 0.1 ml PEP, 0.1 ml mercaptoethanol, 0.1 ml ATP, 0.1 ml NADH, 6 units of pyruvate kinase, 4.6 units of lactate dehydrogenase, enzyme, and water to make a final volume of 1 ml. The change in absorbance at 340 nm is recorded. After registering the base line for 2–3 min the reaction is started by the addition of 10 μl of MVAP and the decrease in absorbance is recorded for 3–5 min. One unit of enzyme is defined as the amount of enzyme which phosphorylates 1 μmol of MVAP per min at 30°.

When assaying crude extracts, 10 m*M* NaF is added to the above reaction mixture in order to inhibit phosphatases.

Radioactive Assay[3]

Principle. This assay is based on the transformation of [^{14}C]MVAP into [^{14}C]MVAPP. Both radioactive compounds are then separated by ion-exchange paper chromatography.

Reagents

Tris–HCl buffer, 1 *M*, pH 7.5
Mercaptoethanol, 1.43 *M*

$MgCl_2$, 100 m*M*
ATP, 100 m*M*
MVAP, 44 m*M*
[^{14}C]MVAP, 10 μCi/0.5 ml, 45.5 μCi/nmol
EDTA, 250 m*M*, pH 7.0

Procedure. The assay is performed at 30°. Reaction mixture for several assays is prepared as follows: 30 μl Tris buffer, 162 μl $MgCl_2$, 24 μl ATP, 1 μl cold MVAP, 24 μl mercaptoethanol, and water up to 300 μl. For one assay, 50 μl of the above mixture is placed in a test tube and 3 μl [^{14}C]MVAP and water is added to a volume of 500 μl. The specific activity in the assay is 6×10^6 cpm/μmol. After preincubating for 2 min, 2 μl phosphomevalonate kinase is added to start the reaction. Aliquots (50 μl) are taken at different times and the reaction is stopped by adding it to tubes containing 40 μl EDTA. A control is run as follows: a separate tube containing the same components except enzyme is prepared; 40 μl EDTA is then added followed by the enzyme.

Quantification of the MVAPP formed is achieved by separating [^{14}C]MVAPP from [^{14}C]MVAP using chromatography in DEAE paper, by a modification of the procedure described by Garcés and Cleland.[5] The reaction mixtures are spotted on 2.5×23 cm DEAE paper strips (Whatman DE-81), which are clipped to a 3×4 cm piece of Whatman 3 MM paper on top, to allow the solvent front to run beyond the DEAE paper. The strips are introduced into a chromatography jar previously saturated with 0.6 *M* ammonium formate pH 3.1 containing 5 m*M* EDTA. The sheets are developed by ascending chromatography at room temperature for 3 hr, removed, dried overnight, and cut in 1-cm pieces. The pieces are counted in a liquid scintillation spectrometer, using 10 ml of the following scintillation mixture: 5 g PPO and 0.25 g POPOP in 1 liter toluene.

Percentage transformation of MVAP to MVAPP is calculated and the values obtained at different times (usually 1, 2, and 3 min) are used for initial velocity calculations.

Other radioassay methods for phosphomevalonate kinase using paper and column chromatography to separate labeled substrates and products are described by Tchen.[6]

Preparation of Phosphomevalonate

The procedures described below is based on the method developed by Levy and Popják[7] and modified by Bazaes *et al.*[4]

[5] E. Garcés and W. W. Cleland, *Biochemistry* **8,** 633 (1969).
[6] T. T. Tchen, this series, Vol. 5 [66].
[7] H. R. Levy and G. Popják, *Biochem. J.* **75,** 417 (1960).

Mevalonic acid lactone is dissolved in 0.1 *N* KOH and incubated for 1 hr at 37° to obtain potassium mevalonate, which is the substrate for MVAP synthesis. After this incubation, the pH is brought to about 7.5, and the concentration of mevalonate is adjusted to nearly 0.1 *M* by diluting with distilled water.

MVAP synthesis is carried out in a reaction mixture containing 20 m*M* potassium phosphate buffer pH 7.4, 5 m*M* potassium mevalonate, 5 m*M* ATP, 10 m*M* mercaptoethanol, 9 m*M* $MgCl_2$, and 80 units of mevalonate kinase in a final volume of 500 ml. Mevalonate kinase of adequate purity for this preparation can be obtained as a by-product during the purification of phosphomevalonate kinase (see below). An alternative reaction mixture, using phosphoenolpyruvate and pyruvate kinase to regenerate ATP, and therefore requiring much less of the nucleotides, is proposed in the article by Cardemil and Jabalquinto.[8]

MVAP can be followed during this preparation either by the spectrophotometric enzyme assay or radiochemically by adding [^{14}C]mevalonate to the initial reaction mixture.

The mixture is incubated at 30° for 3–4 hr and the reaction is stopped by immersing the flask for 4 min in a boiling water bath. After cooling in ice, the precipitated protein is eliminated by centrifugation at 4° in a refrigerated centrifuge for 15 min at 17,000 *g*.

MVAP must now be freed of unreacted mevalonate, nucleotides, and salts. Nucleotides are eliminated by precipitation with 85% ethanol. MVA is converted to its lactone by lyophilization and the lactone is separated from MVAP by precipitating the latter as the Ba salt.[7]

The supernatant obtained above is lyophilized to dryness, resuspended in distilled water, and the pH is adjusted to 4.2 with 6 *N* HCl. Absolute ethanol is added up to a concentration of 85%, and the mixture is left in ice for 2 hr. The white precipitate (containing nucleotides) is centrifuged off for 10 min at 7700 *g* in a refrigerated centrifuge. The precipitate is washed 3 times with 85% ethanol, the washings and supernatant are pooled, and the ethanol is eliminated in a rotary evaporator at a temperature not higher than 50°. The remaining aqueous solution (~30–40 ml) is brought to pH 8.5 with 4 *N* KOH, and the Ba salt of MVAP is precipitated by adding 10 ml 1 *M* $BaCl_2$ and absolute ethanol up to a concentration of 85%. After 2 hr in ice, the precipitate is removed by centrifugation at 27,000 *g* for 10 min. After washing the precipitate twice with 85% ethanol, it is resuspended in 25 ml distilled water and brought to pH 5.2 with 6 *N* HCl. The excess Ba^{2+} is eliminated by adding 10 ml of 1 *M* Na_2SO_4, followed by centrifugation of the insoluble $BaSO_4$. The precipitate is

[8] E. Cardemil and A. M. Jabalquinto, this volume [10].

washed twice with distilled water, and the pooled washings and supernatant are brought to pH 7.5–8.5 with 4 *N* NH_4OH.

Final purification is achieved by the method of Dugan *et al.*[9] using DEAE-cellulose column chromatography. A 3 × 25-cm DEAE-cellulose column (300 ml of resin bed) is equilibrated with 10 m*M* NH_4HCO_3, pH 9.0. The MVAP solution is diluted with distilled water (up to about 1 liter) so as to bring the conductivity below 1.4 mmho, and is then applied to the column at a flow rate of 150 ml/hr. The column is washed (at the same flow rate) with 900 ml of the equilibration buffer and then with 1800 ml of 20 m*M* NH_4HCO_3 buffer, pH 9.0. MVAP is eluted with 1800 ml of 40 m*M* NH_4HCO_3 buffer, pH 9.0, collecting 20 ml fractions. The fractions containing MVAP are pooled and concentrated by lyophilization to a volume of about 150 ml.

The NH_4HCO_3 is eliminated by means of Dowex-50. The MVAP solution is applied to a Dowex 50-X8 column (2.6 × 9 cm) in the H^+ form, at the maximum possible flow rate. The NH_4^+ is retained by the column and the HCO_3^- is eliminated as CO_2. The column is then washed with distilled water until all MVAP is recovered. The fractions containing the substrate are quickly pooled and brought to pH 7.5 with concentrated KOH and lyophilized to a concentration of about 30–50 m*M*. The MVAP solution is kept at −20°. Stability studies show a 1% hydrolysis after 3 months. Nucleotide and mevalonate contamination of this preparation is negligible. A 35–40% yield is obtained. This yield is calculated on the basis of the *R*-mevalonate content of the starting material, which is a racemic mixture.

Purification Procedure

The technique described is based on the method of Bazaes *et al.*[4] All steps are performed at 4°.

Step 1. Preparation of Extract. Pig livers, obtained at the slaughterhouse immediately after killing the animals, are brought to the laboratory in ice. The livers are either processed after no more than 12 hr or kept frozen at −70° for later use. No significant loss in activity of mevalonate kinase or MVAP kinase is observed from frozen liver.

One kilogram of liver free of connective tissue is minced and added is 2 liters of 0.1 *M* potassium phosphate buffer pH 7.5 containing 10 m*M* mercaptoethanol and 1 m*M* EDTA. This mixture is homogenized at top speed in a Waring Blender four times for 30 sec each with 30 sec intervals, and then filtered through 3 layers of cheesecloth. The filtrate is centri-

[9] R. E. Dugan, E. Rasson, and J. W. Porter, *Anal. Biochem.* **22,** 249 (1968).

fuged at 16,000 *g* for 15 min, and the supernatant is centrifuged again at 45,000 *g* for 30 min, discarding the pellets. The supernatant is filtered through glass wool to eliminate lipids.

Step 2. Ammonium Sulfate Fractionation. Ammonium sulfate is slowly added to the supernatant of the previous step up to 30% saturation, with mild stirring, and the solution is then stirred for another 30 min before centrifuging for 15 min at 16,000 *g*. The precipitate is discarded and the supernatant is brought to 60% saturation with ammonium sulfate. After 30 min of slow stirring the precipitate is centrifuged off as above and dissolved in approximately 300 ml 1 m*M* potassium phosphate buffer pH 7.5, 1 m*M* EDTA, 10 m*M* mercaptoethanol (buffer A). The solution is then dialyzed against 10 liters of buffer A for 14 hr.

Step 3. DEAE-Cellulose Chromatography. The pH of the dialyzate is adjusted to 7.5 with 1 *M* potassium phosphate buffer, pH 7.5 and then diluted with buffer A to lower the conductivity below 4 mmho. The enzyme is applied to a DEAE-cellulose column (6 × 23 cm) previously equilibrated with 10 m*M* potassium phosphate buffer, pH 7.5, 10 m*M* mercaptoethanol, 0.1 m*M* EDTA, at a flow rate of 300 ml/hr. The column is then washed with about 10 liters of 30 m*M* potassium phosphate buffer, pH 7.5, 10 m*M* mercaptoethanol, 0.1 m*M* EDTA (flow rate 500 ml/hr) until the absorbance at 280 nm drops below 0.2. If one is interested in obtaining mevalonate kinase for MVAP synthesis (see above), this enzyme can be eluted with 50 m*M* potassium phosphate buffer pH 7.5, 10 m*M* mercaptoethanol, 0.1 m*M* EDTA.

Phosphomevalonate kinase is eluted with a linear gradient of 60–250 m*M* potassium phosphate buffer pH 7.5 containing 10 m*M* mercaptoethanol and 0.1 m*M* EDTA (3 liters total volume) at a flow rate of 150 ml/hr. The active fractions are pooled and the enzyme is precipitated with solid ammonium sulfate added up to 80% saturation. After centrifuging for 15 min at 16,000 *g*, the precipitate is resuspended in about 60 ml of 10 m*M* potassium phosphate buffer, pH 7.5, 10 m*M* mercaptoethanol, 100 m*M* KCl. At this stage the enzyme is free of phosphatase and NADH oxidase activities.

Step. 4. BioGel P-150 Chromatography. The enzyme is applied with inverted flow (18 ml/hr) to a 5 × 75-cm Bio-Gel P-150 column preequilibrated with 10 m*M* potassium phosphate buffer, pH 7.5, 100 m*M* KCl, 10 m*M* mercaptoethanol. The column is further washed with the same buffer until the enzyme is eluted. The active fractions are pooled and concentrated by ultrafiltration to a volume of 15–20 ml. This fraction can be stored indefinitely at −20° in 50% glycerol. At this stage, any remaining mevalonate kinase is separated since due to its higher molecular weight, it elutes earlier from the column.

PURIFICATION OF PHOSPHOMEVALONATE KINASE FROM PIG LIVER[a]

Step	Volume (ml)	Protein (mg)	Units	Specific activity (units/mg)	Purification (x-fold)	Cumulative recovery (%)
Crude extract	1530	85,680	752	0.0088	1	100
30–60% ammonium sulfate	521	49,495	590	0.0119	1.35	78.4
DEAE-cellulose	660	1,716	420	0.24	27.3	55.8
BioGel P-150	340	80.9	290	3.57	406	38.6
Hydroxylapatite	14	13.6	123	9.0	1022	16.3
Blue Dextran-Sepharose	4.6	1.02	70	69.1	7852	9.3

[a] Reprinted with permission from Bazaes *et al.*[4] Copyright 1980 American Chemical Society.

Step 5. Hydroxylapatite Chromatography. The P-150 fraction is diluted 1:3 with 10 mM potassium phosphate buffer, pH 7.5, 1 mM dithiothreitol, and applied to a 1.6 × 10-cm hydroxylapatite column previously equilibrated with the same buffer, at a flow rate of 10 ml/hr. The enzyme does not bind to the column and is washed off with the same buffer.

Step 6. Blue Dextran-Sepharose Chromatography. To the pooled active fractions of the previous step is added 1 M Tris–HCl buffer, pH 7.5 to obtain a 10 mM concentration of the buffer. The solution is then applied to a 1.0 × 5-cm Blue Dextran-Sepharose column previously equilibrated with 10 mM Tris–HCl buffer, pH 7.5, 1 mM dithiothreitol, 10% glycerol, at a flow rate of 17 ml/hr. The column is then washed with 3 volumes of the same buffer, and the enzyme is eluted with the buffer containing 10 mM ATP. The active fractions are pooled and concentrated to a volume of 3–4 ml by ultrafiltration. The solution is brought to 50% glycerol and stored at −20°.

A summary of the results of a typical purification is given in the table.

Properties

Purity and Stability of the Preparation. The enzyme is homogeneous in polyacrylamide gel electrophoresis by the method described by Gabriel.[10] Faint bands of impurities are seen in some preparations when analyzed in sodium dodecyl sulfate gel electrophoresis by the method of Laemmli.[11] The purified enzyme when kept in 50% glycerol at −20° is

[10] O. Gabriel, this series, Vol. 22 [39].

[11] U. K. Laemmli, *Nature* (*London*) **227,** 680 (1970).

highly stable: it maintains 80% of the initial activity after 15 months. The enzyme is very unstable, however, if stored in the absence of thiol compounds.

Substrate and Cofactor Requirements.[4] The enzyme shows absolute specificity for ATP. The best metal cofactor is Mg^{2+}, which can be replaced less effectively by Mn^{2+}, Zn^{2+}, or Co^{2+}.

Kinetic Properties. The enzyme follows hyperbolic kinetics. The kinetic mechanism is ordered, as shown by initial velocity and isotope exchange at equilibrium. MVAP is the first substrate and ADP the last product.[3] This mechanism is similar to that found for mevalonate kinase.[12] True K_m values at pH 7.5 are MVAP 20 μM, ATP 56 μM.[3]

Effect of pH.[4] A plateau of optimum activity ranging from pH 7.5 to 9.5 has been found when studied using TES, MES, and BICINE buffers.

Physical and Chemical Properties.[4] The enzyme is a monomer with a molecular weight of 22,000, as determined by gel filtration, sodium dodecyl sulfate–polyacrylamide gel electrophoresis, and sucrose density gradient centrifugation. Its amino acid composition is Asx_{19}, Thr_9, Ser_{14}, Glx_{33}, Pro_9, Gly_{23}, Ala_{18}, Val_8, Ile_8, Leu_{19}, Tyr_2, Phe_7, Lys_7, His_5, Arg_{11}, Cys_1, Trp_4; no Met has been found.

Other Properties. Chemical modification studies have shown that the cysteine residue is essential for activity.[13] Amino[13] and guanidino[14] groups have also been implicated in the binding of substrates.

Acknowledgments

The authors wish to thank Dr. Emilio Cardemil and Dr. Osvaldo Cori for valuable suggestions.

[12] E. Beytía, J. K. Dorsey, J. Marr, W. W. Cleland, and J. W. Porter, *J. Biol. Chem.* **245,** 5450 (1970).

[13] S. Bazaes, E. Beytía, A. M. Jabalquinto, F. Solís de Ovando, I. Gómez, and J. Eyzaguirre, *Biochemistry* **19,** 2305 (1980).

[14] M. Vergara, M. Alvear, E. Cardemil, A. M. Jabalquinto, and J. Eyzaguirre, *Arch. Biol. Med. Exp.* **15,** 423 (1982).

[10] Mevalonate 5-Pyrophosphate Decarboxylase from Chicken Liver

By EMILIO CARDEMIL and ANA MARÍA JABALQUINTO

$$\text{Mevalonate 5-pyrophosphate} + \text{ATP} \xrightarrow{Mg^{2+}} \text{isopentenyl pyrophosphate} + \text{ADP} + P_i + CO_2$$

Assay Methods

Principle. Three methods are available for the assay of the mevalonate 5-pyrophosphate decarboxylase from chicken liver. In the isotopic assay method, [2-^{14}C]- or [3-^{14}C]mevalonate 5-pyrophosphate is converted to [^{14}C]isopentenyl pyrophosphate, which is hydrolized to [^{14}C]isopentenol with alkaline phosphatase, extracted in hexane, and the radioactivity determined by liquid scintillation spectrometry. In the optical assay method, the formation of ADP is measured by the decrease in absorbancy at 340 nm in the coupled assay of pyruvate kinase plus lactate dehydrogenase. The third method employed in our laboratory is based in the determination of the inorganic phosphate produced in the reaction.

Units. One unit of enzyme activity is defined as the amount of enzyme required to catalyze the decarboxylation of 1 μmol of mevalonate 5-pyrophosphate per min under the assay conditions.

Synthesis of Mevalonate 5-Pyrophosphate. The method employed is based in a modification[1] of the procedure developed by Popják.[2] The method consists in the enzymatic pyrophosphorylation of mevalonic acid using partially purified preparations of pig liver mevalonate kinase and mevalonate 5-phosphate kinase. The product is purified by ion exchange column chromatography.[3]

Mevalonic acid lactone (Sigma Chemical Co.) is transformed into the potassium salt of mevalonic acid as described in the preceding article.[4] If ^{14}C-labeled material is desired, it can be obtained adding a known amount of commercially available [^{14}C]mevalonic acid or [^{14}C]mevalonic acid lactone. Alternatively, unlabeled mevalonate 5-pyrophosphate can be mixed with (*R*)-[2-^{14}C]- or (*R*)-[5-^{14}C]mevalonate 5-pyrophosphate (Amersham International).

[1] A. M. Jabalquinto and E. Cardemil, *Lipids* **15,** 196 (1980).

[2] G. Popják, this series, Vol. 15, p. 393.

[3] R. E. Dugan, E. Rasson, and J. W. Porter, *Anal. Biochem.* **22,** 249 (1968).

[4] J. Eyzaguirre and S. Bazaes, this volume [9].

ISBN 0-12-182010-6

The synthesis of mevalonate 5-pyrophosphate is carried out in a reaction mixture (300 ml final volume) containing 100 m*M* Tris–HCl buffer, pH 8.0, 2.1 m*M* ATP, 100 m*M* KCl, 7 m*M* phosphoenolpyruvate, 4.8 m*M* potassium mevalonate, 10 m*M* 2-mercaptoethanol, 10 m*M* NaF, 5 m*M* $MgCl_2$, 1240 units of rabbit muscle pyruvate kinase, 16 units of pig liver mevalonate kinase, and 47 units of pig liver mevalonate-5-phosphate kinase. The last two enzymes can be obtained by the method described in the preceding article.[4] The enzymes eluted from the DEAE-cellulose column are of adequate purity for this preparation.

The mixture is incubated at 30° for 2 hr and the reaction is stopped by immersing the flask in a boiling water bath for 4 min. After cooling on ice, the precipitated protein is eliminated by centrifugation at 23,500 *g* for 15 min at 4°. The supernatant is adjusted to pH 9.0 with concentrated NH_4OH, diluted to about 7 liters with water to adjust its conductivity to 0.9 mS, and applied (300 ml/hr) to a DEAE-cellulose column (4 × 20 cm) equilibrated with 20 m*M* NH_4HCO_3, pH 9.0. The column is washed (at the same flow rate) with 800 ml of the equilibration buffer and then with 2 liters of the same buffer at 40 m*M*, pH 9.0. Mevalonate 5-pyrophosphate is eluted with 1.5 liters of 60 m*M* buffer, pH 9.0, collecting 20 ml fractions. The fractions containing mevalonate 5-pyrophosphate (located by using the spectrophotometric assay for mevalonate 5-pyrophosphate decarboxylase) are pooled and lyophilized to dryness.

The sample is dissolved in distilled water (~30 ml) and applied to a Dowex 5-X8 column (5 × 10 cm) in the H^+ form. The column is washed with distilled water until all mevalonate 5-pyrophosphate is recovered. The fractions containing the substrate are pooled and brought to pH 7.5 with 10 *M* KOH. Remaining nucleotides are eliminated by adding aliquots of acid washed charcoal until the absorbance at 260 nm is negligible. The charcoal is removed by centrifugation and the supernatant is lyophilized to dryness. The solid residue is resuspended in about 10 ml of water and centrifuged at 40,000 *g* for 5 min to eliminate a small amount of insoluble material. A 45–50% yield is obtained, calculated on the basis of the starting (3*R*)-mevalonic acid.

Isotopic Assay[5]

Procedure. The reaction is carried out in rubber-stoppered 15 ml centrifuge tubes containing (0.6 ml final volume) 0.1 *M* Tris–HCl buffer, pH 7.0, 5 m*M* ATP, 5 m*M* $MgCl_2$, 0.41 m*M* [3-^{14}C]mevalonate 5-pyrophos-

[5] M. Alvear, A. M. Jabalquinto, J. Eyzaguirre, and E. Cardemil, *Biochemistry* **21,** 4646 (1982).

phate (2.45×10^5 cpm/μmol), and mevalonate 5-pyrophosphate decarboxylase. The reaction mixture is incubated at 30° for 10 min and stopped by the addition of 5.5 units of bovine intestine alkaline phosphatase in 0.5 ml of 1 *M* Tris–HCl buffer, pH 8.4, and further incubated for 2 hr at 30°. The incubation mixture is then extracted twice with 1.0 ml of hexane (bp 40–60°). The radioactivity is measured in 1-ml aliquots of the combined hexane phase by conventional liquid scintillation spectrometry. The method is highly reproducible and can be employed without difficulty in crude extracts, but the enzymic activity measured is only about 20% of that detected by the spectrophotometric or the inorganic phosphate assay methods. This is probably due to loss of the volatile [3-^{14}C]isopentenol in the incubation or the extraction procedures. This method has been employed by us for the determination of decarboxylase activity from rat liver and kidney homogenates[1] and for both crude and purified enzyme from chicken liver.[5] An assay method based on the release of $^{14}CO_2$ from [1-^{14}C]mevalonate 5-pyrophosphate has also been used by Popják for the rat liver enzyme.[2]

Spectrophotometric Assay[5]

Procedure. The reaction mixture (1 ml final volume) contains 0.1 *M* Tris–HCl buffer, pH 7.0, 0.1 *M* KCl, 5 m*M* ATP, 6 m*M* $MgCl_2$, 0.5 m*M* phosphoenolpyruvate, 0.23 m*M* NADH, 6.5 units of pyruvate kinase, 11.8 units of lactate dehydrogenase, mevalonate 5-pyrophosphate decarboxylase, and 0.15 m*M* mevalonate 5-pyrophosphate, added to start the reaction. The assay is performed at 30° in a thermostatted spectrophotometer. This method cannot be used at pH values below 5 due to the gradual destruction of NADH, or in crude preparations with high NADH oxidases and phosphatases activities.

Inorganic Phosphate Method[6]

Procedure. The reaction mixture (0.5 ml final volume) contains 0.1 *M* Tris–HCl buffer, pH 7.0, 0.1 *M* KCl, 5 m*M* ATP, 6 m*M* $MgCl_2$, mevalonate 5-pyrophosphate decarboxylase, and 0.44 m*M* mevalonate 5-pyrophosphate added to start the reaction. After 6 min at 30°, the reaction is stopped by the addition of 0.2 ml of 0.1 *M* EDTA, pH 7.0. The inorganic phosphate produced is then measured in a 0.05-ml aliquot according to the micromethod of Lanzetta *et al.*[7] All the glassware used in this method

[6] A. M. Jabalquinto and E. Cardemil, unpublished results.

[7] P. A. Lanzetta, L. J. Alvarez, P. S. Reinach, and O. A. Candia, *Anal. Biochem.* **100,** 95 (1979).

is washed with 1 *N* HCl and rinsed in distilled water before use. This method can be used at all pH values, but when working at low pH, the reaction is stopped by the addition of 0.02 ml of 0.1 *M* EDTA in 1 *M* Tris base. The enzyme activity determined with this method is the same than that detected with the spectrophotometric method. Of the three methods of assay of mevalonate 5-pyrophosphate decarboxylase, this is the only one that does not employ auxiliary enzymes, making it the method of choice when working in conditions that may be inhibitory for the auxiliary enzymes.

Purification of the Enzyme

The method of purifying mevalonate 5-pyrophosphate decarboxylase from chicken liver is essentially that of Alvear *et al.*[5] It provides reproducible and stable preparations of the enzyme with specific activity between 6 and 7 μmol min^{-1}/mg of protein. All steps are performed at 4°, and the buffer used in the purification (unless otherwise indicated) is potassium phosphate pH 7.0 containing 10 m*M* mercaptoethanol plus 0.1 m*M* EDTA.

Crude Extract. Chicken livers, obtained immediately after slaughter, are brought to the laboratory on ice, and frozen at −20°. This material can be kept frozen for several weeks without loss of enzyme. Frozen liver (1 kg) is cut into small pieces and homogenized in a Waring Blender in 3 liters of 0.1 *M* buffer–1 m*M* EDTA. After filtration through cheesecloth and glass wool, a 30 mg/ml solution of protamine sulfate is added to the homogenate to a final concentration of 1.5 mg/ml and is then stirred for 2 min. The slurry is centrifuged for 15 min at 23,500 *g*, and the clear supernatant is carefully decanted and filtered through cheesecloth and glass wool. The pH of the supernatant is adjusted to 5.3 with glacial acetic acid and centrifuged at 23,500 *g* for 5 min. The supernatant is filtered through cheesecloth and glass wool and its pH is readjusted to 7.0 with 10 *M* KOH.

Ammonium Sulfate Fraction. Solid ammonium sulfate is added to the supernatant to 35% saturation, and the pH is maintained at 7.0 with constant addition of 10 *M* KOH. After stirring for 30 min the preparation is centrifuged at 23,500 *g* for 15 min. The supernatant is now brought to 60% saturation with solid ammonium sulfate, stirred for 30 min, and centrifuged as before. The precipitate is resuspended in about 300 ml of 1 m*M* buffer and dialyzed for 18 hr against 10 liters of the same buffer.

DEAE-Cellulose Column Chromatography. The dialyzate is diluted with 1 m*M* buffer to a conductivity of 1.5 mS and applied to a DEAE-cellulose column (Whatman DE-22, 6.5 × 27 cm) equilibrated with 10 m*M* buffer. The column is washed with 4 liters of 50 mM buffer, and the enzyme is eluted with a linear gradient of 2 liters each of 60 and 140 mM buffer; 20-ml fractions are collected. The application of the enzyme and the column wash is carried out with a flow rate of 800–900 ml/hr, while the elution of the enzyme with the gradient is performed at 250 ml/hr. Enzyme activity is measured with the spectrophotometric assay, and those fractions with a ratio of U/ml: A_{280} higher than 0.030 are pooled and EDTA is added to a final concentration of 1 mM. The enzyme is precipitated by addition of ammonium sulfate to 80% saturation and after being centrifuged for 15 min at 23,500 g the precipitate is resuspended in about 60 ml of 10 mM buffer containing 0.2 M KCl. The degree of purification and the yield up to this step is about 51 times and 50%, respectively, as determined by the radioactive assay. The enzyme can be conveniently stored frozen at this stage with the previous addition of glycerol to 20%. Two of these preparations are mixed together and used for the subsequent steps.

Sephadex G-200 Gel Filtration. The pooled fractions from two DEAE-cellulose columns are applied to a Sephadex G-200 column (5 × 95 cm) equilibrated with 10 mM buffer containing 0.2 M KCl and then washed with the same buffer with an upward flow rate of 50 ml/hr. The active fractions (U/ml: $A_{280} \geq 0.040$) are combined and EDTA is added to a final concentration of 1 mM. The enzyme is precipitated by addition of ammonium sulfate to 75% saturation. After centrifugation at 23,500 g for 15 min, the precipitate is resuspended in 5 mM buffer and dialyzed for 18 hr against 2 liters of the same buffer. Glycerol is added to the dialyzate to a final concentration of 20%, the pH of this solution is adjusted to 6.0 with concentrated HCl, and the solution is centrifuged for 10 min at 23,500 g to eliminate a slight precipitate. The supernatant is diluted with 1 mM buffer (pH 6.0)–20% glycerol to a conductivity of 0.7 mS.

Phosphocellulose Column Chromatography. The enzyme is applied to a phosphocellulose column (3.7 × 21 cm) equilibrated with 10 mM buffer (pH 6.0)–20% glycerol. The column is washed with 350 ml of the same buffer, and the enzyme is eluted with a linear gradient of 750 ml each of 20 and 250 mM buffer (pH 6.0)–20% glycerol. Fractions of 18 ml are collected at a flow rate of 60–80 ml/hr. The pH of the pooled active fractions (U/ml: $A_{280} \geq 0.20$) is adjusted to 7.0 with 10 M KOH, and the fractions are concentrated by ultrafiltration. The concentrated enzyme is diluted with 1 mM buffer and again concentrated by ultrafiltration. Glycerol is added to a final concentration of 20%, and the conductivity of the enzyme solution is adjusted with buffer to about 15% above the conductivity of the hydroxylapatite column (see below).

Hydroxylapatite Column Chromatography. The volume of the hydroxylapatite column is calculated according to the following ratio: millili-

PURIFICATION OF MEVALONATE 5-PYROPHOSPHATE DECARBOXYLASE FROM CHICKEN LIVER[a]

Fraction	Protein (mg)	Units (μmol min^{-1})	Specific activity (units mg^{-1})	Purification (x-fold)	Yield (%)
DEAE-cellulose[b]	2213	82.4	0.037		100
Sephadex G-200	516	41.2	0.080	2.1	50
Phosphocellulose	30.5	30.5	0.660	17.6	24.4
Hydroxylapatite	10	13.0	1.3	34.4	15.7
Blue Dextran-Sepharose	1.1	7.4	6.5	175.7	8.0

[a] The enzyme was purified from 1900 g of chicken liver.

[b] The values given are those obtained from two separate preparations from 950 g chicken liver each, purified up this step and then pooled.

ters of hydroxylapatite/milligrams of protein = 0.4. The enzyme is applied to a hydroxylapatite column equilibrated with 13 mM buffer–20% glycerol. The column (10 ml total volume) is washed with the same buffer (flow rate 24 ml/hr), and 1.2 ml fractions are collected. The enzyme is obtained in this washing and fractions with a U/ml: A_{280} ratio higher than 0.80 are pooled.

Blue Dextran-Sepharose Column Chromatography. The enzyme is applied to a column (2.5 × 10 cm) of Blue Dextran-Sepharose prepared according to Ryan and Vestling,[8] equilibrated with 10 mM Tris–HCl buffer (pH 7.0)–20% glycerol–0.1 mM EDTA, without 2-mercaptoethanol. The column is washed with the same buffer (60 ml/hr) until the absorbance at 280 nm is 0. The enzyme is eluted with 75 mM KCl in the same buffer as above, and 5 ml fractions are collected. The active fractions are pooled and concentrated by ultrafiltration. The enzyme is stored frozen at −20°. A summary of the purification procedure is presented in the table.

Properties of Mevalonate 5-Pyrophosphate Decarboxylase

The enzyme can be kept for several months at −20° without loss of activity. The protein obtained is homogeneous by disc gel polyacrylamide electrophoresis but a contaminant protein (less than 20% of the total protein) is observed in the same kind of electrophoresis under denaturing conditions. The preparation contains no mevalonate kinase or mevalonate 5-phosphate kinase activities.

[8] L. D. Ryan and C. S. Vestling, *Arch. Biochem. Biophys.* **160,** 279 (1974).

The decarboxylase is a dimer of molecular weight 85,400 ± 1940, with two identical (or nearly identical) subunits.[5] The purified enzyme does not require the presence of SH-containing reagents for either activity or stability. The enzyme is highly specific for ATP and requires for activity a divalent metal cation, Mg^{2+} and Mn^{2+} being most effective. The optimum pH for the enzyme ranges from 4.5 to 7.0. Inhibitory effects are observed with citrate, phtalate, and other organic carboxylic acids. The isoelectric point, as determined by column chromatofocusing, is 4.8. The kinetics are hyperbolic for both substrates, showing a sequential mechanism; true K_m values of 0.014 and 0.50 mM have been obtained for mevalonate 5-pyrophosphate and ATP, respectively.[5] Evidence has recently been obtained of essential arginyl residues in the enzyme.[9] Preliminary experiments indicate that the enzyme is strongly inhibited by ATPγS but not by the mono- or bidentate complexes of CrATP.[10]

Acknowledgments

We thank Dr. Jaime Eyzaguirre (Universidad Católica de Chile) for his critical reading of the manuscript. Financial assistance from Fondo Nacional de Ciencias and Dirección de Investigación de la USACH is acknowledged.

[9] A. M. Jabalquinto, J. Eyzaguirre, and E. Cardemil, *Arch. Biochem. Biophys.* **225,** 338 (1983).

[10] A. M. Jabalquinto, F. Solís de Ovando, and E. Cardemil, unpublished observations.

[11] Isopentenyldiphosphate Δ-Isomerase

By DENNIS M. SATTERWHITE

Isopentenyldiphosphate Δ-Isomerase

Isopentenyl pyrophosphate isomerase (EC 5.3.3.2, isopentenyldiphosphate Δ-isomerase) catalyzes the reversible reaction shown below providing dimethylallyl pyrophosphate (**2**), the 5-carbon allylic pyrophosphate subunit (isoprene) which initiates the formation of all polyterpenoid molecules by the action of prenyltransferase (EC 2.5.1.1, dimethylallyltransferase). Additionally, **2** may be coupled to nonterpenoid molecules such as tryptophan to form **3**, dimethylallyl tryptophan.

ISBN 0-12-182010-6

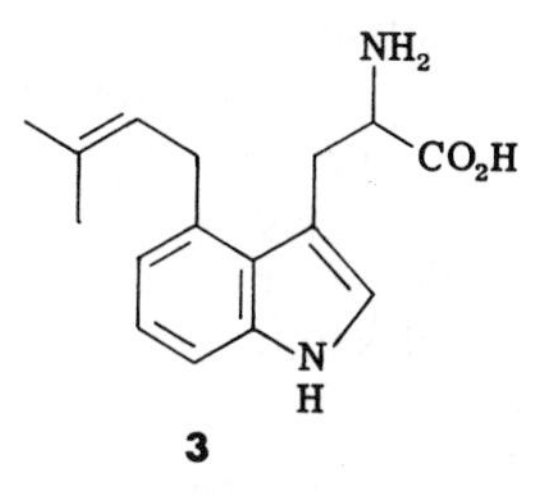

Isomerase was first found in bakers' yeast by Lynen *et al.*[1] Since then isomerase has been studied in pig liver,[1–4] pumpkin,[5] orange peel,[6] pine seedlings,[6] camphor tree,[7] cotton root,[8] *Penicillium cyclopium,*[9] avian liver,[10–12] and *Claviceps purpurea.*[12] One of the purifications from pig liver was extensive, and the enzyme was reported to be nearly homogeneous.[4] The chemistry of this conversion has been recently reviewed in detail by Poulter and Rilling.[13]

Enzyme Assay

The radioassay is dependent on the liberation of alcohols from the labile allylic pyrophosphate under acidic conditions, analogous to the prenyltransferase assay. An aliquot (usually 1–5 μl) of enzyme preparation is added to 0.05 ml of assay buffer (see below) containing 0.04 m*M* (10 Ci/mol) [^{14}C]isopentenyl pyrophosphate (New England Nuclear). The radioactive pyrophosphate obtained from the manufacturer is diluted to this specific activity with isopentenyl pyrophosphate whose preparation is described elsewhere in this volume. The solution is incubated 10 min at

[1] F. Lynen, B. W. Agranoff, H. Eggerer, U. Henning, and F. M. Moslein, *Angew. Chem.* **71,** 657 (1959).

[2] P. W. Holloway and G. Popják, *Biochem. J.* **104,** 57 (1967).

[3] D. H. Shah, W. W. Cleland, and J. W. Porter, *J. Biol. Chem.* **240,** 1946 (1965).

[4] D. V. Banthorpe, S. Doonan, and J. A. Gutowski, *Arch. Biochem. Biophys.* **184,** 381 (1977).

[5] K. Ogura, A. Saito, and S. Seto, *J. Am. Chem. Soc.* **96,** 4037 (1974).

[6] E. Jedlicki, G. Jacob, F. Faini, and O. Cori, *Arch. Biochem. Biophys.* **152,** 590 (1972).

[7] T. Suga, T. Hirata, and K. Tange, *Tennen Yuki Kagobutso Toronkai Koen Yoshishu, 22nd, 1979* p. 251 (1979).

[8] P. Heinstein, R. Widmaier, P. Wegner, and J. Howe, *Recent Adv. Phytochem.* **12,** 313 (1977).

[9] R. M. McGrath, R. N. Nourse, D. C. Neethling, and N. P. Ferreira, *Bioorg. Chem.* **6,** 53 (1977).

[10] D. M. Satterwhite, Ph.D. Dissertation, University of Utah, Salt Lake City (1979).

[11] H. Sagami and K. Ogura, *J. Biochem.* (*Tokyo*) **94,** 975 (1983).

[12] H. C. Rilling, L. T. Chayet, and D. M. Satterwhite, unpublished.

[13] C. D. Poulter and H. C. Rilling, *in* "Biosynthesis of Isoprenoid Compounds" (J. W. Porter and S. L. Spurgeon, eds.), Vol. 1, p. 209. Wiley, New York, 1981.

37°, and the reaction is terminated by the addition of 0.2 ml of 25% concentrated hydrochloric acid in methanol and 0.5 ml deionized water. The sample is again incubated at 37° for 10 min and then is saturated with sodium chloride and extracted twice with 1-ml portions of toluene. The extracts are combined and dried over anhydrous magnesium sulfate. One-half is counted in a standard scintillation solution. At early stages of isolation, phosphatase and transferase activities are contaminants in the preparation that will participate in the assay to give erroneously high values. To check for the former, acid hydrolysis is eliminated. For the latter, 5 μl of 1 m*M* geranyl pyrophosphate is added to the incubation mixture at the onset. Analysis is by acid hydrolysis in the standard manner.

Protein assays are by the biuret procedure or UV absorbance at 280 nm.[14]

Purification of Avian Liver Isomerase

Reproducible procedures for the purification of isomerase from the liver of domestic chicken have been developed by Rilling *et al.*[10,12] These preparations yield up to 200-fold purification from crude tissue extract. At that purification, these preparations invariably show multiple components by SDS–polyacrylamide gel electrophoresis and contain some phosphatase activity and little prenyltransferase activity.

Stock Solutions

Buffer A: 0.1 *M* Tris [Tris(hydroxymethyl)aminomethane, Sigma Chem. Co.], 1 m*M* EDTA (ethylenediaminetetraacetic acid), 1 m*M* dithiothreitol, pH 7.3 at 4°.

Buffer B: 10 m*M* Tris, 1 m*M* dithiothreitol, pH 7.3 at 4°.

Buffer C: 10 m*M* HEPES (*N*-2-hydroxyethylpiperazine-*N*′-2-ethanesulfonic acid, Sigma Chem. Co.), 0.1 *M* potassium chloride, 0.1 m*M* dithiothreitol, pH 7.4 at 4°.

Buffer D: 5 m*M* Tris, 0.1 m*M* EDTA, 1 m*M* dithiothreitol, pH 7.3 at 4°.

Buffer E: 5 m*M* Tris, 1 m*M* dithiothreitol, pH 7.3 at 4°.

Assay buffer: 10 m*M* Tris, 1 m*M* dithiothreitol, 2 m*M* magnesium chloride, pH 7.3 at 4°.

Homogenate of Avian Liver Isopentenyl Pyrophosphate Isomerase

Chicken livers obtained at the slaughterhouse and frozen in 200-g packets after being rinsed with deionized water. One packet is thawed in

[14] S.-L. Lee, H. G. Floss, and P. Heinstein, *Arch. Biochem. Biophys.* **177,** 84 (1976).

400 ml of distilled water containing 1 mM EDTA–1 mM dithiothreitol and homogenized in a Waring blendor for 2 min. The homogenate is centrifuged at 0° for 30 min at 12,000 g in a Sorvall GSA rotor, and the supernatant filtered through 7 layers of cheesecloth to skim the fatty layer away.

Ammonium Sulfate Fractionation

To 300 ml of supernatant, 57.2 g of ammonium sulfate (33%) is added over 15 min at 4°, and followed by stirring for 20 min. The mixture is centrifuged at 0° for 25 min at 12,000 g. The supernatant is filtered through cheesecloth as before. To 240 ml of supernatant, 33.6 g of ammonium sulfate (55%) is added with stirring over 15 min at 4°, followed by stirring for 20 min. The mixture is centrifuged at 0° for 30 min at 23,000 g, and the supernatant is removed by decantation (235 ml). The precipitate is dissolved in Buffer A (152 ml) and dialyzed overnight against 4 liters of Buffer B.

DEAE-Cellulose Fractionation

A 2.5 × 57 cm column of Whatman DE-52 is washed overnight with Buffer B. The enzyme solution from the dialysis step is applied to the column at a rate of 3 ml/min. The column is washed with 750 ml of Buffer B containing 25 mM potassium chloride and then eluted with 3 liters of a linear gradient 25 to 300 mM potassium chloride in Buffer B. Twenty milliliter fractions are collected. Isomerase activity is located in fractions 17–22.

0–80% Ammonium Sulfate Concentration

The column fractions that contained isomerase activity are combined, and 211 g of ammonium sulfate is added (80% saturation) over 35 min at 0° while the pH is maintained at 7.5 with ammonium hydroxide. The suspension is stirred for 30 min and centrifuged at 0° for 30 min at 12,000 g. The supernatant is discarded, and the precipitate dissolved in a minimum volume of buffer.

Gel Filtration

A 2.7 × 81.5 cm column is packed with BioGel A-0.5m, 100–200 mesh (Bio-Rad), washed with 1 liter of 0.1 M potassium chloride, 1 liter of water, and finally with 1 liter of Buffer C. The enzyme from the previous step (~20 ml) is applied to the column and eluted with Buffer C. The flow was controlled at 19 ml/hr and 5-ml fractions collected. The active fractions (85 through 108) are pooled and concentrated with an Amicon concentrator utilizing a PM 10 membrane. The concentrate (17.5 ml) is dialyzed 3 hr

TABLE I
ISOLATION OF ISOMERASE FROM AVIAN LIVER

Purification step	Units[a]	SA[b] ($\times 10^3$)	Yield (%)
Homogenate supernatant	5.37	0.4	100
33–55% ammonium sulfate precipitate	4.03	1.03	75
DEAE-cellulose	1.10	2.91	28
80% ammonium sulfate precipitate	1.01	—	25
Gel filtration	0.76	7.04	14
Electrofocusing, second, after dialysis	0.11	73	2.1

[a] Micromoles per minute.
[b] Specific activity in micromoles per minute per milligram of protein.

against 4 liters of Buffer E and then centrifuged at 0° for 20 min at 12,000 rpm to remove denatured protein.

Isoelectrofocusing

For isoelectric focusing, a 110-ml column is utilized. In addition to enzyme, the column contains 5 m*M* 2-mercaptoethanol and 1% Bio. Lyte (Bio-Rad) pH 5 to 7. Focusing is for 4 days at 400 V. One-ml fractions are collected. Preliminary experiments focusing with wide-range ampholines had shown that no activity is collected outside this narrow range. Fractions containing isomerase activity are pooled and dialyzed overnight against 2 liters of Buffer B at 4°. The protein can be electrofocused a second time to yield a preparation nearly free of prenyltransferase and phosphatase activities. Isomerase elutes from the column as a single activity peak at pH 6.33.

Results of the procedure are summarized in Table I. The final preparation, after dialysis against Buffer E, is stored frozen at −20°. The overall yield of enzyme activity is 2.1% and the preparation had a specific activity of 73 μmol/per/min/mg protein. Gel electrophoresis showed one major band with 5 minor bands.

Purification of Isomerase from Claviceps purpurea

Claviceps sp. strain SD58 are grown in stationary culture as described by Lee *et al.*[14] and modified by Cress *et al.*[15] The hyphae are collected by filtration, lyophilized and stored desiccated at −20° until ready for use.

[15] W. A. Cress, L. T. Chayet, and H. C. Rilling, *J. Biol. Chem.* **256,** 10917 (1981).

Stock Solutions

Buffer A: 20 m*M* Tris–HCl, 20 m*M* mercaptoethanol, 20 m*M* calcium chloride, 10% glycerol, 10 m*M* thioglycolate, 10 m*M* dithiocarbamate, pH 8.2, purged with nitrogen and degassed under vacuum at 4°.

Buffer B: 20 m*M* Imidazole–HCl (recrystallized), 20 m*M* mercapto-ethanol, 10% glycerol, pH 7.3 at 4°.

Homogenate

Lyophilized *Claviceps* are placed in a small ball mill with 20–30 ceramic pellets and milled for 18 hr at 4°. This results in powdered hyphae of variable fineness. The yield of isomerase activity is uniformly 300 units per gram of dry hyphae by the standard assay; this is not corrected for transferase or phosphatase activity although both are low. The powder is suspended in 2 ml Buffer A per gram and blended in a Waring blendor in the cold for 2–3 min. Lower ratios of buffer to hyphae (i.e., 1 ml buffer/g dry weight) have been tried successfully, but 2 ml/g gave optimum unit recovery and pH maintenance. The suspended matter is then centrifuged in the Sorvall GSA rotor at 19,000 *g* for 20 min. The cloudy supernatant is filtered through a minimal amount of glass wool to catch the floating debris.

Ammonium Sulfate Fractionation

While the pH is maintained at 8.2 with 5 *N* sodium hydroxide, sufficient ammonium sulfate is added at 4° to the supernatant to bring the concentration to 45%, stirred 20 min, and the mixture centrifuged as before. This pellet contains DMAT synthetase. The supernatant is filtered through a minimal layer of glass wool to catch floating debris.

With the pH maintained at 8.2, sufficient ammonium sulfate is added to the supernatant to bring the concentration to 70%. After centrifugation as before, the supernatant is discarded and the pellet dissolved in minimal volume of Buffer B. A certain amount of the pellet does not dissolve. The resulting cloudy solution has a reddish brown color and will become somewhat cloudier with dialysis. Dialysis is carried out overnight at 4° into two 2-liter changes of Buffer B.

pH Fractionation

The adjustment of the pH of the enzyme solution to 5.5 with 4 *M* acetic acid or 4 *M* succinic acid results in the precipitation of some proteins which are removed by centrifugation at 29,000 *g* for 30 min. The specific activity is enhanced by 2- to 4-fold by this procedure. Speed is important because exposure to low pH should be as brief as possible. The pH of the supernatant is adjusted with dilute ammonia to 7.3.

Polyethyleneglycol Fractionation (PEG 400)

The supernatant is stirred at 4° and PEG 400 added in the amount of 25% w/v. The mixture is stirred for 20 min and centrifuged at −4° for 45 min at 80,000 *g*. The grey pellet is discarded. Specific activities may increase 2- to 4-fold with 80–100% recovery of units. The PEG supernatant is dialyzed against two 2 liter changes of Buffer B.

DE-52 Chromatography

The dialyzed supernatant is applied to a 2.5 × 35 cm column of DE-52, equilibrated in Buffer B at pH 7.3. The column is then washed with 10 bed volumes of Buffer B and eluted with a linear gradient of 0 to 200 m*M* KCl (500 ml total volume) at 1.5–2 ml/min. The isomerase activity is eluted as a symmetrical peak at 120–140 m*M* KCl at the tail end of the major protein peak. Recoveries range from 40 to 70% with up to a 5-fold increase in specific activity.

Hydroxylapatite Chromatography

The pooled fractions from the DEAE-cellulose column, after dialysis against imidazole buffer, are pumped onto a 2.5 × 15 cm bed of hydroxylapatite previously equilibrated in imidazole buffer. After washing with two volumes of buffer, a linear gradient 0 to 100 m*M* potassium phosphate in 200 ml of the Buffer B is pumped through the column. The isomerase appears at the front of the protein peak at 40–50% (40–50 m*M* PO_4) of elution. The fractions containing isomerase are then concentrated on PM-10 or PM-30 membranes.

BioGel P-150

The concentrate is applied in the sharpest possible zone to a 1.5 cm × 1.2 m column packed with BioGel P-150, 100–200 mesh, previously equilibrated in Buffer B and then eluted with same buffer (void vol = 50 ml) at 10 ml/hr. Isomerase appears in the tail end of the major protein peak (~80 ml). A representative purification is summarized in Table II.

Properties of Isomerase

The instability of this enzyme and presence of phosphatase and prenyltransferase found in these and other preparations have severely

TABLE II
ISOLATION OF ISOMERASE FROM *Claviceps purpurea*[a]

Procedure	Units[b]	SA[c] ($\times 10^3$)	Recovery (%)
Homogenate	20	5	100
45–70% ammonium sulfate ppt.[d]	18	8	90
PEG supernatant	18	20	
DEAE-cellulose	12	50	60
Hydroxylapatite	10	100	50
BioGel P-150	4	200	20

[a] Fifty grams of ball milled *C. purpurea* hyphae in 1 liter Buffer A.
[b] Micromoles per minute.
[c] Specific activity in micromoles per minute per milligram protein.
[d] Corrected for inhibition by ammonium sulfate.

limited the mechanistic characterization of isomerase. Recently, investigators[11] reported multiple forms of avian isomerase prepared by DEAE-cellulose chromatography. However, these may have resulted from proteolysis.[15] The porcine isomerase has been reported[4] to have a molecular weight of 83,000, while gel filtration chromatography of the avian enzyme on BioGel A-0.5m suggests a somewhat smaller molecular weight. Divalent cations, magnesium or manganese, are required for catalysis, with 2 m*M* magnesium providing maximum velocity. The pH optimum is as a broad plateau from pH 7 to 8 in Tris buffer. K_m values are on the order of 10^{-6} *M* for both substrates. A number of organic pyrophosphates inhibit isomerase with K_i values near 10 μM and inorganic pyrophosphate is especially potent with a $K_i = 1.5$ μM. The sulfhydryl directed reagents iodoacetamide, *p*-chloromercuribenzoic acid and *N*-ethylmaleimide are effective inhibitors, while glutathione, mercaptoethanol, and dithiothreitol may help to protect the enzyme during purification and are activators. Potassium fluoride (4 m*M*) and sodium azide have been used in the buffers with no measurable effect on isomerase. Storage at −20° has been achieved for up to 6 months with no loss in catalytic activity.

[12] Shunt Pathway of Mevalonate Metabolism

By BERNARD R. LANDAU and HENRI BRUNENGRABER

The Reactions

In 1970, Popják hypothesized the existence of a pathway linking cholesterol synthesis with leucine catabolism (Fig. 1).[1] His hypothesis was based on the earlier identification of ^{14}C-labeled prenoates[2] (3,3-dimethylacrylate, geranoate, and farnesoate) in tissues from rats injected with [^{14}C]mevalonate. Prenyl pyrophosphates (3,3-dimethylallyl, geranyl, and farnesyl pyrophosphate), synthesized from mevalonate, are hydrolyzed by a microsomal phosphatase[3] to corresponding prenols[2] (3,3-dimethylallyl alcohol, geraniol, and farnesol). The prenols are in turn oxidized to the corresponding prenoates via alcohol dehydrogenase and aldehyde dehydrogenase.[3] Popják hypothesized that 3,3-dimethylacrylate was activated to dimethylacrylyl-CoA, an intermediate in the pathway of leucine catabolism, leading to acetoacetate and acetyl-CoA formation via HMG-CoA.

According to this scheme, carbon 1 of mevalonate is lost as CO_2 at the level of pyrophosphomevalonate decarboxylase; carbons 2, 3, and 6 of mevalonate become carbons 2, 3, and 4 of acetoacetate; carbons 4 and 5 of mevalonate become carbons 2 and 1 of acetyl-CoA. Therefore, following the *in vivo* administration of mevalonate labeled with ^{14}C on any of its carbons except carbon 1, ^{14}C should be incorporated into ketone bodies or acetyl-CoA and into products of their metabolism. Indeed, when Edmond and Popják injected [2-^{14}C]mevalonate into normal suckling rats, significant label was recovered in palmitate isolated from their brains, spinal cords, and skins.[4] When rats were injected with either [2-^{14}C]- or [5-^{14}C]mevalonate, ^{14}C was found in blood 3-hydroxybutyrate.[5] When [5-^{14}C]mevalonate was given to rats and humans, up to 12% of the dose administered appeared as $^{14}CO_2$.[6] The $^{14}CO_2$ formed too rapidly to arise from the oxidation of [^{14}C]cholesterol synthesized from the injected mevalonate.

[1] G. Popják, *Ann. Intern. Med.* **72,** 106 (1970).
[2] G. Popják and R. H. Cornforth, *J. Chromatogr.* **4,** 214 (1960).
[3] J. Christophe and G. Popják, *J. Lipid Res.* **2,** 244 (1961).
[4] J. Edmond and G. Popják, *J. Biol. Chem.* **249,** 66 (1974).
[5] J. Edmond, A. Fogelman, G. Popják, and B. Roecker, *Circ.* **52,** Suppl. II-82 (1975).
[6] A. M. Fogelman, J. Edmond, and G. Popják, *J. Biol. Chem.* **250,** 1771 (1975).

ISBN 0-12-182010-6

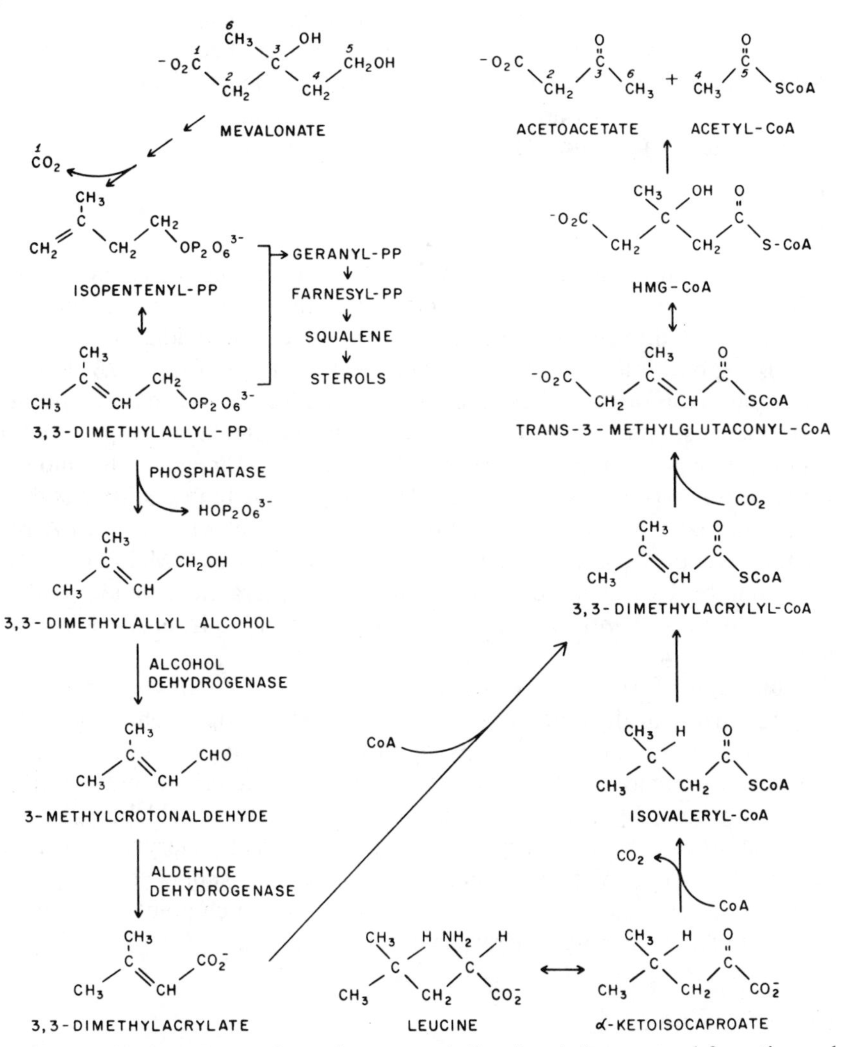

FIG. 1. Shunt pathway of mevalonate metabolism in relation to sterol formation and leucine catabolism.

Further support for the shunt pathway came from determinations by Brady *et al.*[7] of the distributions of ^{14}C in 3-hydroxybutyrates formed by diabetic rats injected with [1,2-^{14}C]- [2-^{14}C]-, [3-^{14}C]-, and [5-^{14}C]mevalonate and more recently from studies with [6-^{14}C]mevalonate.[8] The labeling

[7] P. S. Brady, R. F. Scofield, W. C. Schumann, S. Ohgaku, K. Kumaran, J. M. Margolis, and B. R. Landau, *J. Biol. Chem.* **257,** 10742 (1982).

[8] B. R. Landau, unpublished data (1983).

patterns show that (1) carbon 1 of mevalonate is lost in the formation of acetoacetate, (2) carbons 2, 3, and 6 of mevalonate become respectively carbons 2, 3, and 4 of acetoacetate, and (3) carbon 5 of mevalonate is converted to carbons 1 and 3 of acetoacetate in keeping with the condensation of acetyl-CoA units formed from mevalonate.

Alternate Schemes

An alternate series of reactions for the conversion of mevalonate to HMG-CoA was considered by Popják.[4,9] In this scheme, mevalonate is first dehydrated to Δ^2- or Δ^3-anhydromevalonate. The oxidation of the 5-hydroxy group followed by hydration and formation of the CoA derivative yields HMG-CoA. These reactions are no longer considered a possibility because they are incompatible with (1) the labeling pattern in acetoacetate generated from [4*R*,4-^{3}H]- and [4*S*,4-^{3}H]mevalonate[9] and (2) the lack of incorporation of ^{14}C from [1-^{14}C]mevalonate into 3-hydroxybutyrate and fatty acids.[10] Reversal of the hydroxymethylglutaryl-CoA reductase reaction could also result in the formation of HMG-CoA from mevalonate.[4] However, reversal is unlikely on energetic grounds[4] and is also not in accord with the lack of incorporation of ^{14}C from [1-^{14}C]mevalonate.

Schroepfer[11] proposed a branching of the shunt pathway from the sterol pathway at the level of geranyl or farnesyl pyrophosphate or both, rather than at 3,3-dimethylallyl pyrophosphate. He hypothesized a series of reactions, analogous to those described in the bacterial degradation of allylic alcohols, by which geranoate and farnesoate would be progressively shortened by two carbon units to yield dimethylacrylyl-CoA and hence HMG-CoA. While there are no experiments to support this hypothesis, data to date cannot differentiate between a branch point at the level of dimethylallyl, geranyl, or farnesyl pyrophosphate.

Compartmentation

The shunt appears to generate HMG-CoA, the immediate precursor of mevalonate. For this reason, the pathway was originally viewed as "mevalonate recycling."[4] However, there are two pools of intracellular HMG-CoA: (1) a mitochondrial pool, the precursor of ketone bodies,

[9] G. Popják, personal communication; also quoted in J. W. Bardenheier, Ph.D. Dissertation, University of California, Los Angeles California (1979).

[10] S. B. Weinstock, R. R. Kopito, G. Endemann, J. F. Tomera, E. Marinier, D. M. Murray, and H. Brunengraber, *J. Biol. Chem.* **259,** 8944 (1984).

[11] G. J. Schroepfer, Jr., *Annu. Rev. Biochem.* **50,** 585 (1981).

fueled by fatty acid oxidation and by leucine catabolism, and (2) an extramitochondrial pool, the precursor of mevalonate. Evidence that HMG-CoA generated in the shunt is mitochondrial has been obtained through the use of (−)-hydroxycitrate. This inhibitor of ATP citrate lyase[12] blocks the transfer of acetyl groups from the mitochondria to the cytosol via citrate. In the perfused liver, (−)-hydroxycitrate decreases by half the incorporation of ^{14}C from [5-^{14}C]mevalonate into fatty acids.[10] Since alcohol dehydrogenase is a cytosolic enzyme, the steps of the shunt from isopentenyl pyrophosphate to methylcrotonaldehyde are extramitochondrial. Since leucine catabolism occurs in mitochondria, the steps beyond dimethylacrylyl-CoA are mitochondrial. Since aldehyde dehydrogenase is present in both compartments, the intermediate of the shunt that crosses the mitochondrial membrane is methylcrotonaldehyde or dimethylacrylate.

Methods of Measurement

Principles. The rate of shunting can be measured using mevalonate labeled with ^{14}C or ^{3}H. Compounds generated directly or indirectly from the shunt are isolated and assayed for radioactivity. The rate of operation of the sterol pathway is usually assessed, at the same time as that of the shunt, from the incorporation of label into nonsaponifiable lipids (squalene plus sterols). The percentage ratio of the rates [shunt pathway/(shunt + sterol pathways)] × 100 may be regarded as the extent to which the flux of mevalonate entering the synthetic pathway for cholesterol is diverted, i.e., shunted from cholesterol synthesis.

The simplest method for measuring the shunt is based upon the yield of $^{14}CO_2$ from either [5-^{14}C]- or [4,5-^{14}C]mevalonate. The $^{14}CO_2$ comes from the oxidation in the tricarboxylic acid cycle of [1-^{14}C]- or [1,2-^{14}C]acetyl-CoA formed via the shunt. Essentially, all that is required is the quantitative collection of $^{14}CO_2$ and an accurate determination of its radioactivity.

[3-^{14}C]Mevalonate, which is converted in the shunt to [3-^{14}C]acetoacetate, yields radioactivity not only in acetoacetate, 3-hydroxybutyrate and acetone, but also in CO_2, since many organs including liver[13] oxidize ketone bodies. From [2-^{14}C]mevalonate, there are two sources of $^{14}CO_2$: (1) the shunt pathway through the formation of [2-^{14}C]acetoacetate; (2) the sterol pathway which generates $^{14}CO_2$ at the demethylation of lanosterol.

[12] J. A. Watson and J. M. Lowenstein, *J. Biol. Chem.* **245**, 5993 (1970).

[13] G. Endemann, P. G. Goetz, J. Edmond, and H. Brunengraber, *J. Biol. Chem.* **257**, 3434 (1982).

For every 6 molecules of [2-^{14}C]mevalonate converted to a 27 carbon sterol, one atom of ^{14}C evolves as $^{14}CO_2$. Then using [2-^{14}C]mevalonate[14]:

$$\text{Rate of shunting} = \frac{\text{total dpm in } ^{14}CO_2 - (\text{dpm in C-27 sterols}/5)}{\text{mevalonate specific activity}}$$

When the bulk of $^{14}CO_2$ production from [2-^{14}C]mevalonate is generated in the sterol pathway, the calculated production of $^{14}CO_2$ in the shunt is inherently imprecise, i.e., it is the difference between two similar quantities.

From [5-^{3}H]mevalonate, ^3H_2O is generated by two mechanisms: (1) in the shunt pathway, alcohol dehydrogenase and aldehyde dehydrogenase catalyze the formation of NAD^{3}H from which ^{3}H is incorporated into H_2O through the respiratory chain; (2) in the sterol pathway, one atom of ^{3}H is liberated in the formation of the central carbon–carbon bond of squalene and one in the conversion of Δ^7-cholestanol to $\Delta^{5,7}$-cholestadienol.[15,16] Therefore, using [5-^{3}H]mevalonate[17]:

$$\text{rate of shunting} = \frac{\text{total dpm in } ^3H_2O - (\text{dpm in cholesterol})/5 - (\text{dpm in squalene} + \text{lanosterol})/11}{\text{mevalonate specific activity (S.A.)}}$$

If [5-^{14}C]mevalonate is used in addition to [5-^{3}H]mevalonate, the nonsaponifiable lipids (NSL) do not require fractionation. Then

$$\text{rate of shunting} = \frac{^3\text{H dpm in } ^3H_2O}{[5\text{-}^3\text{H}]\text{mevalonate S.A.}} - \left(\frac{^{14}\text{C dpm in NSL}}{[5\text{-}^{14}\text{C}]\text{mevalonate S.A.}} - \frac{^3\text{H dpm in NSL}}{[5\text{-}^3\text{H}]\text{mevalonate S.A.}}\right)$$

Procedures. Measurement of $^{14}CO_2$ produced by a live small animal after a single injection of [^{14}C]mevalonate is as follows. Immediately after injection, the animal is introduced into a closed container (such as an Erlenmeyer flask), the atmosphere of which is suctioned through a gas dispersion device (fritted glass disk) into 1 *N* NaOH. There should be a 5-fold excess of NaOH over the CO_2 to be trapped. If a constant infusion of tracer is required, the animal, fitted with an intravenous catheter (in a tail vein, or in a jugular vein), is kept in a closed metabolic cage, the atmosphere of which is suctioned into NaOH.

[14] M. Righetti, M. H. Wiley, P. A. Murrill, and M. D. Siperstein, *J. Biol. Chem.* **251,** 2716 (1976).

[15] G. Popják, D. S. Goodman, J. W. Cornforth, R. H. Cornforth, and R. Ryhage, *J. Biol. Chem.* **236,** 1934 (1961).

[16] C. R. Childs, Jr. and K. Bloch, *J. Biol. Chem.* **237,** 62 (1962).

[17] R. R. Kopito, D. M. Murray, D. L. Story, and H. Brunengraber, *J. Biol. Chem.* **259,** 372 (1984).

Since partially carbonated NaOH solutions are difficult to count in scintillation fluid, transfer of the $^{14}CO_2$ from the NaOH solution to an amine such as ethanolamine is recommended. A suitable volume of the NaOH solution is pipetted into an Erlenmeyer flask containing a few drops of methyl orange solution. The volume of the flask is at least 3 times larger than the sample. The flask is closed by a rubber stopper (or serum cap) through which a small well containing ethanolamine is suspended (Kontes #882320 plastic well can contain up to 0.3 ml of ethanolamine; for larger volumes, a counting vial containing up to 2 ml of ethanolamine is suspended in a 0.5- to 1-liter Erlenmeyer). After closing, the pressure inside the flask is lowered to avoid losses of $^{14}CO_2$ after acidification. This is achieved by a brief suction via a needle inserted through the stopper. Then 1 *N* H_2SO_4 is injected through the stopper until the pH indicator changes color. When large Erlenmeyer flasks are used, it is convenient to insert permanently a long needle through the rubber stopper. The needle extends lower than the vial containing ethanolamine. The needle is fitted with a 3-way plastic valve which facilitates application of partial vacuum and injection of acid. Standards of $NaH^{14}CO_3$ dissolved in a model trap solution (0.08 *M* NaOH + 0.01 *M* Na_2CO_3) are run in parallel with the samples to assess the recovery of the transfer of label from NaOH solution to ethanolamine. For a 25- to 100-ml Erlenmeyer flask, almost complete recovery of $^{14}CO_2$ (>95%) in ethanolamine is achieved in 1–3 hr. When 1-liter flasks are used, this may require 20–24 hr.

If $^{14}CO_2$ production is to be measured in a cell or tissue suspension incubated in the presence of 95% O_2 + 5% CO_2, an empty hanging well is suspended from the stopper of the flask. At the end of the incubation, 0.3 ml of ethanolamine is carefully injected through the stopper into the well. Then 1 *N* H_2SO_4 is injected through the stopper into the suspension. After 3 hr of incubation in a shaker, the flasks are opened and the ethanolamine wells are transferred into liquid scintillation vials. The counting fluid used must be compatible with the amine. For ethanolamine, Oxyfluor-CO_2 from New England Nuclear is adequate. When isolated organs are perfused with [^{14}C]mevalonate, most of the $^{14}CO_2$ generated is found in the effluent gas mixture of the oxygenator (see scheme 1 in ref. 18). This gas mixture is suctioned into a NaOH trap. At the end of the perfusion, the $^{14}CO_2$ present as bicarbonate in the final perfusate is first stabilized by making a 50 ml sample of perfusate 10 m*M* in NaOH. Then this solution and that of the NaOH trap are treated as described above.

To measure the production of ^3H_2O from [5-^{3}H]mevalonate, a sample of plasma, incubation medium or perfusate is ultrafiltered in an Amincon

[18] H. Brunengraber, M. Boutry, Y. Daikuhara, L. Kopelovich, and J. M. Lowenstein, Vol. 35 [56].

CF-50 ultrafiltration centrifuge cone or equivalent. The ultrafiltrate is made 0.1 *N* in NaOH and is distilled under atmospheric pressure. The first fraction of distillate is discarded because it may contain some labeled impurity (when a mixture of [5-^{14}C]- and [5-^{3}H]mevalonate has been used in the experiment, the first fraction of distillate usually contains a small amount of ^{14}C). In the subsequent fractions of distillate, the ^{3}H dpm per ml are constant.

If the volume of sample available is too small for distillation and if the amount of $^{3}H_2O$ dpm is too little for the sample to be diluted, the following procedure may be used. It applies to samples containing [^{3}H]- and [^{14}C]mevalonate. A sample (0.2–0.5 ml) of ultrafiltrate is pipetted into two counting vials containing 0.01 ml of 10% acetic acid. After 40 min of incubation at room temperature to eliminate $^{14}CO_2$, counting fluid is added to one of the vials. Two milliliters of methanol is added to the second vial which is evaporated under a gentle stream of air. After dissolution of the residue on 0.5 ml of water, counting fluid is added, and both vials are counted for ^{3}H and ^{14}C. The counting procedure chosen should yield absolute dpm ^{3}H and ^{14}C in both samples (preferably using internal standards of [^{3}H]- and [^{14}C]toluene). If the ^{14}C dpm is identical in both samples, the loss of ^{3}H dpm in the evaporated sample is ascribed to $^{3}H_2O$ present in the original ultrafiltrate. If evaporation has not been gentle, some dpm [^{14}C]- and [^{3}H]mevalonate can be lost. The loss of [^{3}H]mevalonate may be calculated from the loss of dpm ^{14}C and the $^{3}H/^{14}C$ ratio in the mixture of tracers used.

Regardless of the procedure used to assess the production of $^{3}H_2O$, the amount of [^{3}H]volatile radioactivity present in the stock solution of doubly labeled mevalonate should be determined. A suitable dilution of this solution is processed in the same manner as the samples.

Method Limitations

Limitations may be considered general as they relate to the use of any labeled compound and specific as they relate directly to the metabolism of mevalonate.

General Limitations. Underestimates are caused through the loss of label in exchange and synthetic processes. Neither $^{14}CO_2$ nor $^{3}H_2O$ is generated directly from labeled mevalonate by the shunt, but rather is derived from the shunt products, [^{14}C]acetyl-CoA and NAD^{3}H. Because of the exchange and synthetic reactions these shunt products undergo, collections of $^{14}CO_2$ and $^{3}H_2O$ are not quantitative nor fixed portions of the quantities of [^{14}C]acetyl-CoA and NAD^{3}H formed. Synthetic processes (ketogenesis and lipogenesis) divert the ^{14}C of [^{14}C]acetyl-CoA from the tricarboxylic acid cycle. Some of the label is lost in exchange processes with unlabeled amino acids and glycolytic intermediates. When the cycle operates as a nonsynthetic pathway, these exchange processes remove equal fractions of label from both carbons of acetyl-CoA.[19] When the cycle operates as a synthetic pathway, a greater portion of label from carbon 2 than carbon 1 of acetyl-CoA is lost.[19,20] In addition, a sizeable fraction of the $^{14}CO_2$ formed can be reincorporated by exchange processes into compounds that are nonvolatile in acid[21] and therefore not apparent in the yield of $^{14}CO_2$. The amounts of ^{14}C in all the products formed via the shunt could be summed, e.g., in CO_2, acetoacetate, 3-hydroxybutyrate fatty acids, urea, etc., but obviously this can be tedious, subject to errors in recoveries, and a significant product may be overlooked.

Brady *et al.*[22] provided evidence that production of $^{14}CO_2$ from [5-^{14}C]- or [2-^{14}C]mevalonate significantly underestimates the rate of the shunt pathway. The yield of $^{14}CO_2$ from [1-^{14}C]mevalonate reflects the utilization of exogenous mevalonate by all pathways (shunt, sterol, ubiquinone, dolichol), since $^{14}CO_2$ is formed before any branching of mevalonate metabolism occurs. Therefore, when [1-^{14}C]- and either [2-^{14}C]- or [5-^{14}C] mevalonate are used in parallel experiments, $^{14}CO_2$ from [1-^{14}C]mevalonate can be compared to the production of labeled products from the [2-^{14}C]- or [5-^{14}C]mevalonate. In liver and kidney yields of $^{14}CO_2$ from [1-^{14}C]mevalonate markedly exceed the sum of the yield of ^{14}C in CO_2 and nonsaponifiable lipids from either [5-^{14}C]- or [2-^{14}C]mevalonate. Thus, either one or more pathways of mevalonate metabolism, other than the shunt and sterol pathways, is operative to a marked degree or estimates of the rate of the shunt pathway based on the yields of $^{14}CO_2$ from [2-^{14}C]- or [5-^{14}C]mevalonate are significantly underestimated.[22]

There are methods for correcting for underestimations caused by these processes. The yield of label from [1-^{14}C]acetyl-CoA to $^{14}CO_2$ can be estimated by presenting the preparation under study with α-ketoisocaproate (KIC) labeled alternatively with ^{14}C in carbons 1 and 2. KIC is used at a low concentration, so as not to perturb metabolic patterns through its presence.[23] Carbon 1 of KIC is liberated as CO_2 by mitochondrial branch-

[19] E. O. Weinman, E. H. Strisower, and I. L. Chaikoff, *Physiol. Rev.* **37,** 252 (1957).

[20] J. Katz and I. L. Chaikoff, *Biochim. Biophys. Acta* **18,** 87 (1955).

[21] J. F. Tomera, P. G. Goetz, W. M. Rand, and H. Brunengraber, *Biochem. J.* **208,** 231 (1982).

[22] P. S. Brady, W. C. Schumann, S. Ohgaku, R. F. Scofield, and B. R. Landau, *J. Lipid Res.* **23,** 1317 (1982).

[23] J. F. Tomera, R. R. Kopito, and H. Brunengraber, *Biochem. J.* **210,** 265 (1983).

chain ketoacid dehydrogenase and C-2 of KIC becomes C-1 of acetyl-CoA (see Fig. 1). Thus, from the yield of $^{14}CO_2$ from [1-^{14}C]KIC the quantity of acetyl-CoA formed is obtained. From the yield of $^{14}CO_2$ from [2-^{14}C]KIC the portion of C-1 of the acetyl-CoA forming CO_2 is then calculated. When dealing with a single cell type this approach is likely to be valid. When the preparation under study has more than one cell type, the assumptions must be made that mevalonate is metabolized via the shunt in the same cell type(s) and to the same relative extent as KIC. A correction for the quantity of $^{14}CO_2$ reincorporated can be made from the recovery of $^{14}CO_2$ when $H^{14}CO_3^-$ is presented to the system under study.[21] It must then be assumed that endogenously formed $^{14}CO_2$ is treated in the same manner as the exogenously administered $H^{14}CO_3^-$.

Underestimations due to these processes can be large. Thus, in perfusions with [5-^{14}C]mevalonate of livers from fed normal and diabetic rats the quantity of [1-^{14}C]acetyl-CoA formed was calculated to be 6 times and 28 times the quantity of $^{14}CO_2$ evolved.[23]

Specific Limitations. *RS*-Mevalonate has been employed in many studies, but only the R-isomer is utilized. Considerations have been given to the purity of the [^{14}C]mevalonates used.[24,25] Certain lots of *RS*-[^{14}C]mevalonate were found to contain impurities which, in the presence of kidney tissue, even if dead, yield acid-volatile radioactivity which, like $^{14}CO_2$, is trapped in alkali.[24] These impurities can be removed by chromatography on Celite.[24] Alternatively, one can purify and resolve the RS mixture by treating it with MgATP and mevalonate kinase.[26,27] 5-Phospho-*R*-[^{14}C]mevalonate is isolated, hydrolyzed with alkaline phosphatase, and the resulting *R*-[^{14}C]mevalonate is chromatographed.[24,26] Mevalonolactone is taken up faster than mevalonate salt by *in vitro* preparations.[28] At physiological pH, 95% of plasma mevalonate is in the salt form.[29] The equilibration of the lactone and salt forms of mevalonate is catalyzed in plasma by a lactonase.[29,30]

The yield of ^3H_2O from [5-^{3}H]mevalonate has a major advantage over the yield of $^{14}CO_2$ from [5-^{14}C]mevalonate in that it does not involve

[24] J. W. Bardenheier and G. Popják, *Biochem. Biophys. Res. Commun.* **74,** 1023 (1977).

[25] M. H. Wiley, P. A. Murrill, M. M. Howton, S. L. Huling, D. C. Cohen, and M. D. Siperstein, *Biochem. Biophys. Res. Commun.* **79,** 1023 (1977).

[26] G. Popják, G. Boehm, T. S. Parker, J. Edmond, P. A. Edwards, and A. M. Fogelman, *J. Lipid Res.* **20,** 716 (1979).

[27] T. S. Parker, R. R. Kopito, and H. Brunengraber, this volume [7].

[28] P. A. Edwards, J. Edmond, A. M. Fogelman, and G. Popják, *Biochim. Biophys. Acta* **488,** 493 (1977).

[29] T. S. Parker, *Fed. Proc., Fed. Am. Soc. Exp. Biol.* **38,** 632 (1979).

[30] M. H. Wiley, S. Huling, and M. D. Siperstein, *Biochem. Biophys. Res. Commun.* **88,** 605 (1979).

transfer of label through the tricarboxylic acid cycle.[7,17] Studies with [^{3}H]ethanol are in keeping with water being the major fate for the ^{3}H of NAD^{3}H.[31] Incorporation of ^{3}H into lactate can be used as a convenient measure of the quantity of ^{3}H from NAD^{3}H incorporated into compounds formed using NADH as cofactor. However, incorporation of ^{3}H can occur into all other compounds formed by reduction with NADH as cofactor. In experiments of such a duration that steady state is not achieved, yields of label in products will be less than at steady state.

True tracer experiments cannot be performed *in vivo* with [^{14}C]mevalonate because the maximum specific activity of [^{14}C]mevalonate achievable is not sufficient to yield adequate label in a product unless the quantity of the mevalonate administered raises the plasma concentrations of mevalonate far above those existing in animals (10–100 n*M* in humans; 80–500 n*M* in rats).[26,32–34] Since [5-^{3}H]mevalonate can be prepared with a much higher specific activity than [^{14}C]mevalonate, studies can be performed *in vivo* using [5-^{3}H]mevalonate with only minimal increases in plasma mevalonate concentrations.[33] When isolated cells and organs are incubated or perfused with synthetic media, [^{14}C]mevalonate can be added to the media at physiological concentrations and yields of labeled products prove to be sufficient. However, hepatocytes and isolated liver preparations release endogenous mevalonate into the media so that the integrated specific activity of the mevalonate must be determined to estimate shunt activity.[10]

Mevalonate, *in vivo,* is released by the liver into blood and is cleared by peripheral tissues, mostly by the kidneys. The turnover rate of plasma mevalonate, calculated from the kinetics of a tracer quantity of [5-^{3}H] mevalonate injected into the blood, is 2.3 nmol/kg $\times$ min in a 200 g rat.[10,33] The rate of mevalonate synthesis in the whole body, calculated from the incorporation of ^{3}H from ^3H_2O into squalene + sterols,[35] is 2000 times greater.[10] In humans, based upon sterol balance studies, the rate of mevalonate synthesis is approximately 1000-fold more than the rate of turnover of plasma mevalonate.[36] Weinstock *et al.*[10] perfused livers with a

[31] P. Havre, M. A. Abrams, R. J. M. Corrall, L. C. Yu, P. A. Szczepanik, H. B. Feldman, P. Klein, M. S. Kong, J. M. Margolis, and B. R. Landau, *Arch. Biochem. Biophys.* **182,** 14 (1977).

[32] T. S. Parker, D. J. McNamara, C. Brown, O. Garrigan, R. Kolb, H. Batwin, and E. H. Ahrens, Jr., *Proc. Natl. Acad. Sci. U.S.A.* **79,** 3037 (1982).

[33] R. R. Kopito, S. B. Weinstock, L. E. Freed, D. M. Murray, and H. Brunengraber, *J. Lipid Res.* **23,** 577 (1981).

[34] R. R. Kopito and H. Brunengraber, *Proc. Natl. Acad. Sci. U.S.A.* **77,** 5738 (1980).

[35] J. M. Dietschy, *J. Lipid Res.,* in press (1985).

[36] T. S. Parker, D. J. McNamara, D. C. Brown, R. Kolb, E. H. Ahrens, Jr., A. W. Alberts, J. Chen, and P. J. De Schepper, *J. Clin. Invest.* **74,** 795 (1984).

medium containing ^3H_2O and physiological levels of [^{14}C]mevalonate. Extracellular mevalonate contributed only about 0.13% of the mevalonate metabolized by the liver. The reason that the turnover of plasma mevalonate is not the measure of turnover of body mevalonate is that there is a lack of equilibration of the labeled mevalonate in plasma with total body mevalonate. The reason that the metabolism of [^{14}C]mevalonate perfused through liver is not the measure of total mevalonate utilization in liver is that there is a lack of equilibration between the labeled mevalonate in the perfusate and mevalonate in the liver.

Evidence for this nonequilibration is to be found in the studies of Lakshmanan and Veech.[37] They injected ^3H_2O into rats to measure sterol synthesis, and hence mevalonate production, and at the same time injected increasing amounts of [^{14}C]mevalonate to determine the dose at which endogenous production of mevalonate ceased. From the ^{3}H/^{14}C ratios in cholesterol as a function of the dose of mevalonate injected, they concluded that 1 to 5 mmol *R*-mevalonate/kg is required to suppress endogenous mevalonate production during a short-term (1 hr) experiment. The pool of extracellular mevalonate had to be increased by more than 3×10^4 before the specific activities of extracellular and intracellular mevalonate were the same.

In studies conducted *in vivo* or in perfused kidney, attention must be given to the amount of mevalonate excreted in urine. This can be considerable when the tubular reabsorption of mevalonate is saturated.[17,34,38] When relatively large amounts of [^{14}C]mevalonate are administered, labeled prenoates, mainly farnesoate, form in significant quantities, and are found in fatty acid fractions after petroleum ether extraction.[4] Therefore, when incorporation of [^{14}C]mevalonate into fatty acids is used as a measure of shunt activity, it is important to separate n-fatty acids, i.e., palmitate and stearate, from the prenoates.[4]

Experimental Results

In all estimates made of shunt activity *in vivo*, [5-^{14}C]mevalonate was injected intravenously. The yield of expired $^{14}CO_2$ was the measure of the shunt and the incorporation of ^{14}C into NSL the measure of the sterol pathway. The kidney has been concluded to be responsible for most of

[37] M. R. Lakshmanan and R. L. Veech, *J. Biol. Chem.* **252,** 4667 (1977).

[38] H. Brunengraber, S. B. Weinstock, D. L. Story, and R. R. Kopito, *J. Lipid Res.* **22,** 916 (1981).

circulating mevalonate metabolized by the shunt.[39,40] Shunt activity has been reported to be reduced in uremia,[41] inhibited by phosphate,[42] greater in females than males,[43,44] inhibited by testosterone,[45] stimulated by estrogen,[43,45] altered during pregnancy,[46] essentially unaffected by thyroid states,[47] and decreased in diabetes.[48]

In these studies the doses of [^{14}C]mevalonate administered markedly elevated the blood concentrations of mevalonate, to levels 1 to 4 orders of magnitude higher than physiological. When mevalonate is injected intravenously it essentially bypasses the liver because of lack of equilibration, but it is metabolized in the proximal and distal convoluted tubules of the kidney[39] from which it is also excreted.[34,38] These *in vivo* protocols are then in essence intrarenal infusions of label giving data on alteration in the pathways of metabolism for the relatively small amount of mevalonate utilized by kidneys (less than 0.1% of mevalonate metabolized in the body).[17] Furthermore, when estimates for shunt activity under two conditions are compared, as male versus female[43,44] or normal versus diabetic,[48] the assumption must be made that the fraction of ^{14}C metabolized by the shunt that is recovered as $^{14}CO_2$ is the same for both conditions.

A summary of estimates of shunt activity made *in vitro* is presented in the table. Only the studies in perfused rat kidney[17] and liver[10] have been done using physiological concentrations of mevalonate. Except for those two studies, no account was taken of endogenous dilution of the labeled mevalonate nor a correction made for the portion of ^{14}C entering the shunt pathway that was not recovered in $^{14}CO_2$. Assuming that about 5% of mevalonate utilized by rat liver *in vivo* is via the shunt, as is the case *in vitro,* since the liver is the major site of mevalonate utilization, it is the

[39] J. Edmond, A. M. Fogelman, and G. Popják, *Science* **193,** 154 (1976).

[40] M. H. Wiley, M. M. Howton, and M. D. Siperstein, *J. Biol. Chem.* **252,** 548 (1977).

[41] K. R. Feingold, M. H. Wiley, G. MacRae, G. Kaysen, P. Y. Schoenfeld, and M. D. Siperstein, *Metab., Clin. Exp.* **32,** 215 (1983).

[42] M. H. Wiley, K. R. Feingold, G. L. Searle, P. Y. Schoenfeld, G. A. Kaysen, and M. D. Siperstein, *Clin. Res.* **28,** 524A (1980).

[43] M. H. Wiley, M. M. Howton, and M. D. Siperstein, *J. Biol. Chem.* **254,** 837 (1979).

[44] K. R. Feingold, M. H. Wiley, G. L. Searle, B. K. Machida, and M. D. Siperstein, *J. Clin. Invest.* **66,** 361 (1980).

[45] P. S. Brady, R. F. Scofield, S. Mann, and B. R. Landau, *J. Lipid Res.* **24,** 1168 (1983).

[46] K. R. Feingold, M. H. Wiley, G. MacRae, and M. D. Siperstein, *Metab., Clin. Exp.* **29,** 885 (1980).

[47] K. R. Feingold, M. H. Wiley, G. MacRae, and M. D. Siperstein, *J. Clin. Invest.* **66,** 646 (1980).

[48] M. H. Wiley, K. R. Feingold, M. M. Howton, and M. D. Siperstein, *Diabetologia* **22,** 118 (1982).

QUANTITATIONS OF MEVALONATE METABOLISM VIA THE SHUNT PATHWAY IN TISSUE *in Vitro*

Preparation	Labeled mevalonate; *R*-isomer concentration	Product(s) assayed	Conclusions from references
Slices rat tissues	2-^{14}C,5-^{14}C; 0.05 m*M*	CO_2; NSL; long chain fatty acids	Shunt significant in kidney, ileum, spleen, lung, testes; minor or undetectable in liver, brain, skin, adipose tissue; kidney 21 times more active than any other tissue per unit weight[a]
Slices calf tissues, calf villi and crypt cells	2-^{14}C; 4.2 m*M*	CO_2; NSL	Shunt in kidney cortex 15–80 times more active than ileum, muscle, kidney medula, liver; not detected in adipose tissue[b]
Slices liver chick	2-^{14}C; 0.02–4.0 m*M*	CO_2; NSL	13–20% of metabolism in kidney via shunt at 0.05–4.0 m*M*; 5% at 0.04 m*M*; quantitatively insignificant in liver[c]
Rat villous cells	1-^{14}C,2-^{14}C; 0.25 m*M*	CO_2	Suggest shunt absent in intestinal epithelial cells[d]
Slices liver and kidney from normal and diabetic rats treated and untreated with insulin	5-^{14}C; 0.05 m*M*	CO_2; NSL	Shunt decreased in liver and kidney in diabetes; insulin restores activity[e]
Slices liver and kidney of normal rats and rats in diabetic ketosis	1-^{14}C,2-^{14}C, 5–^{14}C, 5-^{3}H; 0.05 m*M*	CO_2; NSL	Shunt greater in kidney than liver; no difference between normal and diabetic[f]
Slices of kidney from male and female rats	5-^{14}C; 0.05 m*M*	CO_2	Kidney slices from female form $^{14}CO_2$ at more than twice the male rate[g]

QUANTITATIONS OF MEVALONATE METABOLISM VIA THE SHUNT PATHWAY IN TISSUE *in Vitro* (*continued*)

Preparation	Labeled mevalonate; *R*-isomer concentration	Product(s) assayed	Conclusions from references
Slices liver and kidney from male and female hyper- and hypothyroid rats	5-^{14}C; 0.07 m*M*	CO_2; NSL	Thyroid hormone status changes produce only minor change in shunt activity in liver and kidney; kidney slices from female and male form $^{14}CO_2$ at same rate[h]
Cultured human lymphocytes	2-^{14}C; 0.07 m*M*	NSL; fatty acids	Shunt (incorporation into fatty acids + neutral alphatic components) 20–31% of total metabolism (total radioactivity incorporated)[i]
Perfused rat kidneys	2-^{14}C,4,5-^{14}C,5-^{14}C, 5-^{3}H; 250 n*M*	Ketone bodies; fatty acids; CO_2; NSL	Shunt activity higher in kidney from female than male rats; metabolism by shunt as much as 1/5 that by sterol-forming pathway[j]
Perfused rat livers	2-^{14}C,3-^{14}C,4,5-^{14}C, 5-^{14}C; 215–380 n*M*	CO_2; ketone bodies fatty acids; NSL	Shunt accounts for about 5% of mevalonate metabolism in liver[k]

[a] M. Righetti, M. H. Wiley, P. A. Murrill, and M. D. Siperstein, *J. Biol. Chem.* **251,** 2716 (1976).

[b] J. R. Linder and D. C. Beitz, *J. Lipid Res.* **19,** 836 (1978).

[c] J. A. Aguilera, A. Linares, V. Arce, and E. García-Peregrín, *Comp. Biochem. Biophys., B* **71B,** 617 (1982).

[d] F. Malki, K. Badjakian, and I. F. Durr, *Int. J. Biochem.* **13,** 187 (1981).

[e] M. H. Wiley, K. R. Feingold, M. M. Howton, and M. D. Siperstein, *Diabetologia* **22,** 118 (1982).

[f] P. S. Brady, W. C. Schumann, S. Ohgaku, R. F. Scofield, and B. R. Landau, *J. Lipid Res.* **23,** 1317 (1982).

[g] M. H. Wiley, M. M. Howton, and M. D. Siperstein, *J. Biol. Chem.* **254,** 837 (1979).

[h] K. R. Feingold, M. H. Wiley, G. MacRae, and M. D. Siperstein, *J. Clin. Invest.* **66,** 646 (1980).

[i] C. Tabacik, S. Aliau, B. Serrou, and A. C. de Paulet, *Biochem. Biophys. Res. Commun.* **101,** 1987 (1981).

[j] R. R. Kopito, D. M. Murray, D. L. Story, and H. Brunengraber, *J. Biol. Chem.* **259,** 372 (1984).

[k] S. B. Weinstock, R. R. Kopito, G. Endemann, J. F. Tomera, E. Marinier, and D. M. Murray, *J. Biol. Chem.* **259,** 8944 (1984).

major site of shunt activity in the whole animal. About one-fifth of the small amount of mevalonate utilization by kidney is via the shunt.

Conclusions

Isopentenyl pyrophosphate can be converted to sterols or shunted to the formation of mitochondrial HMG-CoA. The importance, if any, of this shunt pathway is unknown. It might divert, under certain conditions and over an extended period, isoprene units from cholesterol formation so as to significantly alter the quantity of cholesterol present *in vivo*.[1,44] Intermediates in the pathway could participate in the regulation of cholesterol formation.

The quantity of mevalonate metabolized by the shunt has been estimated from the fate of the carbons or hydrogens of mevalonate in their metabolism by the pathway. The yield from a tissue of label from labeled mevalonate into a product of the shunt depends on the quantity of the labeled mevalonate that enters that tissue, the dilution of that labeled mevalonate by endogenous unlabeled mevalonate and the portion of the label entering the shunt that is recovered in the product, as well as on the activity of the shunt.

When labeled mevalonate is injected into the circulation of an animal, the yield from the shunt of labeled product can be expected to be greater for those tissues into which the mevalonate can most readily enter and in which endogenous mevalonate formation is least. There is as yet no satisfactory method for estimating the extent of metabolism of endogenous mevalonate by the shunt pathway *in vivo*.

Methods are available for estimating the rate of shunting in *in vitro* preparations. The closer the *in vitro* preparation is to one containing a single cell type, the less the assumptions required in making the estimation. The simplest methods are based upon the measurement of the yield of $^{14}CO_2$ from [5-^{14}C]mevalonate and the yield of $^{3}H_2O$ from [5-^{3}H]mevalonate. The fraction of the label entering the shunt that is recovered in the product is determined as well as the extent of dilution by endogenous mevalonate. The latter is determined from the incorporation of ^{3}H from $^{3}H_2O$ into sterols.

Section II

Linear Condensations of Isoprenoids

[13] Photolabile Analogs of the Allylic Pyrophosphate Substrate of Prenyltransferases

By CHARLES M. ALLEN and TSUNEO BABA

The synthesis of photolabile analogs of isoprenoid compounds provides a tool for probing the structure of the substrate binding sites of a variety of prenyltransferases. Brems and Rilling[1] first described the synthesis and utilization of *o*-azidophenethyl pyrophosphate, a photolabile analog of the homoallylic substrate isopentenyl pyrophosphate, as an effective inhibitor of the avian liver prenyltransferase, *trans,trans*-farnesylpyrophosphate synthetase. Recently,[2] we have synthesized two photolabile analogs of the allylic pyrophosphate substrates, geranyl and *trans,trans*-farnesyl pyrophosphate, which bear the 2-diazo-3-trifluoropropionyloxy (DATFP) moiety. The use of a substrate with this DATFP moiety as a photoaffinity probe was first described by Chowdhry *et al.*[3] Other DATFP containing lipids have been used subsequently as effective probes of lipid–lipid and lipid–protein interactions.[4] The isoprenoid analogs described here have been tested as substrates of the bacterial enzyme undecaprenylpyrophosphate synthetase. One of these photolabile analogs, DATFP-geranyl pyrophosphate, the *trans,trans*-farnesylpyrophosphate analog, is both a substrate and an irradiation-dependent inactivator of the bacterial enzyme. The synthesis of these analogs and their utilization as substrates and inactivators are described below.

Preparation of 2-Diazo-3-trifluoropropionyloxyprenyl Analogs and Their Respective Synthetic Intermediates (Scheme I)[2]

(E)-3,7-Dimethyl-1-chloroacetoxy-2,6-octadiene (Geranyl Chloroacetate, ***2b****)*. Geraniol (0.77 g, 5 mmol) and chloroacetic anhydride (2.57 g, 15 mmol) are reacted together in pyridine (9.78 g, 124 mmol) at 0° for 2 hr by the method of Cook and Maichuk.[5] Water is added to stop the

[1] D. N. Brems and H. C. Rilling, *Biochemistry* **18,** 860 (1979).
[2] T. Baba and C. M. Allen, *Biochemistry* **23,** 1312 (1984).
[3] V. Chowdhry, R. Vaughan, and F. Westheimer, *Proc. Natl. Acad. Sci. U.S.A.* **73,** 1406 (1976).
[4] R. Radhakrishnan, C. M. Gupta, B. Erni, R. J. Robson, W. Curatolo, A. Majumdar, A. H. Ross, Y. Takagaki, and H. G. Khorana, *Ann. N.Y. Acad. Sci.* **346,** 165 (1980).
[5] A. Cook and D. T. Maichuk, *J. Org. Chem.* **35,** 1940 (1970).

ISBN 0-12-182010-6

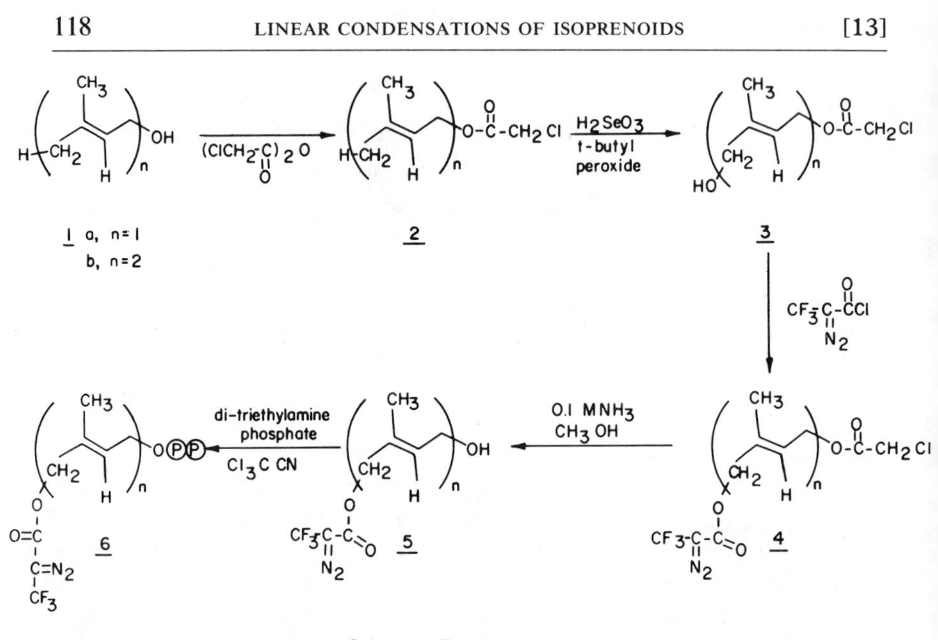

Scheme I

reaction and then the water and pyridine are removed by evaporation under vacuum. The oily residue is dissolved in ethyl ether and this solution is washed extensively with 1 *M* $NaHCO_3$. The ethyl ether extract is then concentrated and applied to a silica gel column in petroleum ether–benzene (5:1, v/v). The ester is isolated by elution with petroleum ether–benzene (2:1, v/v). Yield (81%, 0.93 g).

3-Methyl-1-chloroacetoxy-2-butene (3,3-Dimethylallyl chloroacetate, ***2a****).* 3-Methyl-2-butenol (1.46 g, 17 mmol), which is prepared by the reduction of 3,3-dimethylacrylic acid with $LiAlH_4$,[6] is reacted with chloroacetic anhydride (5.80 g, 34 mmol) in pyridine (13.4 g, 170 mmol). The resulting ester is isolated as described above for the geranyl derivative. Yield (78%, 2.15 g).

(E,E)-3,7-Dimethyl-1-chloroacetoxy-2,6-octadien-8-ol *(**3b**).* Geranyl chloroacetate (2.30 g, 10 mmol) is oxidized with *t*-butyl hydroperoxide (4.0 ml, 36 mmol) in the presence of catalytic quantities of H_2SeO_3 (26 mg, 0.2 mmol) and salicylic acid (140 mg, 1 mmol) in CH_2Cl_2 for 20 hr at room temperature by an adaption of the method of Umbreit and Sharpless.[7] Dichloromethane is removed by rotary evaporation. The amount of *t*-butyl hydroperoxide is reduced by repeated (2–3 times) addition of ben-

[6] C. Yuan and K. Bloch, *J. Biol. Chem.* **234,** 2605 (1959).

[7] M. A. Umbreit and K. B. Sharpless, *J. Am. Chem. Soc.* **99,** 5526 (1977).

zene and subsequent evaporation. The resulting residue is dissolved in ethyl ether and this solution is washed thoroughly with 1 *M* $NaHCO_3$ to remove H_2SeO_3. The ether phase is then concentrated, dissolved in benzene, and applied to a silica gel column. The trans ω-allylic alcohol is eluted with benzene–ethyl acetate (10 : 1, v/v). Yield (47%, 0.99 g).

(E)-3-Methyl-1-chloroacetoxy-2-buten-4-ol (***3a***). Dimethylallyl chloroacetate (1.65 g, 10.1 mmol) is treated with *t*-butyl hydroperoxide (4.1 ml, 36.5 mmol), H_2SeO_3 (132 mg, 1.01 mmol), and salicylic acid (142 mg, 1.01 mmol) for 93 hr at room temperature as described above. The trans alcohol was obtained in 37% yield (0.71 g).

The 2-diazo-3-trifluoropropionyloxy (DATFP) derivatives of the diol monoesters are prepared with 2-diazo-3-trifluoropropionyl chloride. The latter is prepared by reacting trifluorodiazoethane with phosgene.[3] Trifluorodiazoethane is prepared by mixing 50 ml aqueous solutions of 2-trifluoroethylamine hydrochloride (25 g) and $NaNO_2$ (15 g) over 100 ml of CH_2Cl_2, then distilling the CH_2Cl_2–trifluorodiazoethane mixture at 50° while collecting the distillate in a flask cooled by a dry ice–methanol bath. The distillate is filtered to remove ice and the organic phase is dried and stored over $CaCl_2$ in the freezer. 2-Diazo-3-trifluoropropionyl chloride is prepared by mixing dichloromethane solutions of trifluorodiazoethane and phosgene with freshly ground K_2HPO_4 according to the method of Chowdry *et al.*[3]

(E,E)-8-DATFP-3,7-Dimethyl-1-chloroacetoxy-2,6-octadiene (***4b***). (*E,E*)-3,7-Dimethyl-1-chloroacetoxy-2,6-octadien-8-ol (2.30 g, 9.32 mmol) is esterified with 2-diazo-3-trifluoropropionyl chloride (1.92 g, 11.2 mmol) in pyridine (15 g, 186 mmol) at room temperature for 2 hr. The pyridine is removed at the end of the reaction by rotary evaporation. Ethyl ether is added to the resulting residue and the mixture is filtered. The ether extract is then extensively washed sequentially with 0.1 *N* HCl, 1 *M* $NaHCO_3$, and a saturated solution of NaCl. The extract is concentrated, dissolved in benzene–petroleum ether (1 : 5, v/v) and applied to a silica gel column. The diester is isolated by elution in benzene–petroleum ether (1 : 2, v/v). Yield (74%, 2.63 g).

(E)-4-DATFP-3-Methyl-1-chloroacetoxy-2-butene (***4a***). (*E*)-3-Methyl-1-chloroacetoxy-2-buten-3-ol (1.94 g, 11 mmol) is esterified with 2-diazo-3-trifluoropropionyl chloride (2.27 g, 13.2 mmol) in pyridine (1.74 g, 22 mmol) for 2 hr at room temperature. The diester is isolated as described above for the geranyl analog. Yield (71%, 2.13 g).

(E,E)-8-DATFP-3,7-Dimethyl-2,6-octadien-1-ol (DATFP-Geraniol) (***5b***). (*E,E*)-8-DATFP-3,7-Dimethyl-1-chloroacetoxy-2,6-octadiene (1.61 g, 4.21 mmol) is hydrolyzed with 0.1 *M* NH_3 (84.2 ml, 8.42 mmol) in aqueous methanol (90% MeOH, v/v) at room temperature for 90 min by an adap-

tion of the method of Reese and Stewart[8] to give the DATFP ester. Methanol is removed by rotary evaporation, then water and ethyl ether is added. The ether extract is washed with a solution of saturated NaCl and the organic phase dried over Na_2SO_4. The ether extract is concentrated, dissolved in benzene and applied to a silica gel column. The ester is isolated by elution in benzene–ethyl acetate (5 : 1, v/v). Yield (87%, 1.12 g).

*(E)-4-DATFP-3-Methyl-2-buten-1-ol (DATFP-Dimethylallyl Alcohol, **5a**).* (*E*)-4-DATFP-3-Methyl-1-chloroacetoxyl-2-butene (2.13 g, 6.78 mmol) is hydrolyzed to the monoester with 0.1 *M* NH_3 (136 ml, 13.6 mmol) in aqueous methanol at room temperature for 60 min and purified as described above for the geranyl derivative. Yield (79%, 1.27 g).

Each of the DATFP-isoprenols is phosphorylated by the method of Popják *et al.*[9] Purification of the pyrophosphate esters (**6a,b**) is accomplished by an adaption of the procedures of Holloway and Popják[10] where Amberlite XAD-2 and DEAE-cellulose chromatographies are used. The phosphorylated products are applied to a 20 × 210-mm Amberlite XAD-2 (20–30 mesh) column previously equilibrated with 1 m*M* NH_4OH. Pure DATFP-geranyl pyrophosphate (DATFP-GPP) or DATFP-dimethylallyl PP, free of any inorganic phosphate, is eluted with 1 m*M* NH_4OH–50% methanol or with 1 m*M* NH_4OH, respectively. The phosphate content is determined by the method of Chen *et al.*[11]

Enzyme Preparations and Assays

Undecaprenyl pyrophosphate ($C_{55}PP$) synthetase is prepared from *L. plantarum* through the hydroxylapatite purification step as described elsewhere in this volume [32]. The enzyme assays contain in a final volume of 0.5 ml: 9–22 μg of protein, 100 m*M* Tris–HCl buffer (pH 7.5), 0.5% Triton X-100, 200 μM $MgCl_2$, 8.7 μM Δ^3-[^{14}C]isopentenyl pyrophosphate (IPP) (3.1 μCi/μmol, 30,000 dpm), 10 μM *trans,trans*-farnesyl pyrophosphate (FPP) and/or photolabile analog. The reaction mixtures are incubated at 35° for 30 min and the products are analyzed after acid hydrolysis as described elsewhere in this volume [32].

Photolysis Conditions[2]

Photolysis is conducted at 4° for 5 min in a 1-cm path-length quartz cuvette situated at 1 cm distance from the surface of a 8 W GE ger-

[8] C. B. Reese and J. C. M. Stewart, *Tetrahedron Lett.* pp. 4273 (1968).

[9] G. Popják, J. W. Cornforth, R. H. Cornforth, R. H. Ryhage, and D. S. Goodman, *J. Biol. Chem.* **237,** 56 (1962).

[10] P. W. Holloway and G. Popják, *Biochem. J.* **104,** 57 (1967).

[11] P. S. Chen, Jr., T. Y. Toribara, and H. Warner, *Anal. Chem.* **28,** 2756 (1956).

micidal lamp (G8T5). Ninety-five percent of the output of the lamp is at 254 nm.

Photolysis mixtures, containing 125 m*M* Tris–HCl buffer, 250 μ*M* $MgCl_2$, 11 μ*M* [^{14}C]IPP (3.1 μCi/μmol) and different concentrations of photoanalog and/or *trans,trans*-FPP in a volume of 1.2 ml, are cooled to 4° in the cuvettes. A solution of the C_{55}PP synthetase at 0° is added to the contents of each cuvette and mixed well. This mixture is then irradiated for the desired time. Triton X-100 is omitted from the irradiation mixtures to prevent both the absorption of irradiating light and product formation. Following irradiation, Triton X-100 and *trans,trans*-FPP are added to bring the concentration of all components to the level that is used in the standard assay conditions. Two 0.5-ml aliquots from each of these mixtures are then transferred to an assay tube and assayed by the standard assay procedure.

The enzyme exhibits a gradual loss in enzymic activity in dilute solutions in the absence of Triton X-100, therefore, to ensure reproducible and meaningful results two alternative procedures must be followed: (1) minimize the time between enzyme addition and photolysis to less than 2 min and add Triton X-100 immediately after irradiation is completed or (2) maintain a constant time interval (10–15 min) between the time of enzyme addition and time of Triton X-100 addition for each test mixture.

Absorption Spectrum and Photolability of DATFP-GPP

DATFP-GPP (**6b**) has an absorption maximum at 236 nm (ε = 14,000) and a shoulder at 340 nm (ε = 27). Irradiation of this compound in 1 m*M* NH_4OH in a quartz cell at room temperature results in its decomposition with a $t_{1/2}$ of 1.5 min as measured by the loss in absorbance at 236 nm.

Photolabile Analogs as Prenyltransferase Substrates and Inhibitors[2]

The DATFP derivatives were compared with *trans,trans*-FPP as substrates for the *L. plantarum* C_{55}PP synthetase. DATFP-GPP, the *trans,trans*-FPP analog, was about 47% as reactive as a substrate as FPP, when the enzyme was saturated with the substrate analog. The K_m value for DATFP-GPP was 0.17 μ*M*. This value compares favorably with that for *trans,trans*-FPP (K_m of 0.13 μ*M*). DATFP-dimethylallyl-PP (**6a**), a GPP analog, was not a substrate, which is consistent with the poor reactivity of GPP as a substrate.[12]

DATFP-Dimethylallyl PP was also not reactive as a substrate with the chicken liver prenyltransferase, *trans,trans*-FPP synthetase, and gave no

[12] T. Baba and C. M. Allen, *Biochemistry* **17,** 5598 (1978).

inhibition of enzymic activity when the analog was in 100-fold excess of the natural substrate GPP.

(The photolabile analogs of IPP, *o*- and *p*-azidophenethyl pyrophosphate were not inhibitors of $C_{55}PP$ synthetase, even when they were present at 10–100 times the concentration of IPP in the reaction mixture.)

Analysis of Enzymatic Products

TLC. The photolabile product obtained from the action of the $C_{55}PP$ synthetase on DATFP-GPP and IPP may be extracted from the reaction mixture with *n*-butanol instead of subjecting it to acid hydrolysis. Furthermore, it may be hydrolyzed to the monophosphate and free polyprenol as described elsewhere in this volume [32] for further analysis. These radiolabeled photolabile products may then be characterized by TLC. The DATFP-polyprenyl pyrophosphate migrated with an $R_f = 0.35$ on Silica Gel 60 plastic-backed TLC sheets (E. Merck) in diisobutyl ketone–acetic acid–H_2O (60 : 37.5 : 7.5, v/v/v), Solvent A. Treatment of the enzymatic product with wheat germ phosphatase at pH 6.2 for 4 hr yielded products, whose principal component, the monophosphate, chromatographed on silica gel with an $R_f = 0.48$ in Solvent A. These mobilities are comparable to the R_f values of 0.35 and 0.51 for $C_{55}PP$ and $C_{55}P$, respectively, in the same TLC system. Complete hydrolysis of the enzymic product to the free polyprenyl was obtained by treatment with potato acid phosphatase in 60% methanol at pH 5.0 for 3 hr. The R_f values of the resulting DATFP-polyprenol and undecaprenol on silica gel were 0.85 and 0.89, respectively, in Solvent A and 0.51 and 0.53, respectively in benzene–ethyl acetate (10 : 1, v/v). Reverse phase chromatography of the DATFP-polyprenol on paraffin coated Kieselguhr G plates in acetone–H_2O (80 : 20, v/v) gave a major component with an R_f value to 0.35 and minor components with R_f values of 0.46, 0.50, and 0.65. Undecaprenol chromatographed in this system with an $R_f = 0.04$.

Product Chain Length. The ^{14}C-labeled DATFP-polyprenols were also acetylated with [^3H]acetic anhydride and the esters separated by reverse phase TLC in acetone–H_2O (92 : 8, v/v). This provides a method of determining the product chain length. An evaluation of the $^{14}C/^3H$ ratios observed for the major esterified polyprenol analog showed that the analog was formed by the addition of seven isoprene units.

Photoinactivation of $C_{55}PP$ Synthetase[2]

$C_{55}PP$ synthetase was incubated with and without 0.63 μM DATFP-GPP in the presence of 11 μM IPP and 250 μM $MgCl_2$ and irradiated with

UV light at 4°. The time dependent loss of enzymic activity was determined. A $t_{1/2}$ of 2.6 min was calculated for the process of photoprobe-dependent inactivation. However, the enzyme was inactivated by only about 50% even when it was saturated with inactivator. This partial inactivation may be accounted for by the nonproductive loss of the irradiation-generated carbanion of the inactivator by the reaction of the carbanion with the solvent or the Tris buffer instead of its reaction with enzyme. The GPP analog, DATFP-dimethylallyl PP, which had no activity as a substrate, did not significantly inactivate the enzyme even at photoanalog concentrations 50-fold higher than those which were shown to inactivate when DATFP-GPP was used.

Comments

It was the purpose of this work to design, prepare and test photolabile analogs of the allylic pyrophosphate substrate of prenyltransferases as substrates, inhibitors or covalent modifying agents for the long-chain prenyltransferase, C_{55}PP synthetase.

There are several lines of evidence that show that the irradiation-dependent inactivation of the enzyme with DATFP-GPP was directed toward a specific functional domain of the prenyltransferase.

First, the diazo derivative, DATFP-GPP, was a substrate for the C_{55}PP synthetase with a K_m value similar to that of the natural substrate *trans,trans*-FPP. On the other hand, DATFP-dimethylallyl PP, a substrate analog of GPP, failed to serve as a substrate or inactivate the enzyme on irradiation. The observation that the farnesyl analog, DATFP-GPP, was a substrate illustrates the considerable tolerance the enzyme permits toward the character of the long chain allylic substrate, particularly toward the nature of the ω-terminal residue.

Second, the irradiation-induced enzyme inactivation showed saturation kinetics with increasing DATFP-GPP concentration. The K_i determined from kinetic analysis was 0.22 μM. This was similar to the K_m value for DATFP-GPP.

Third, the same K_m values of the enzyme for FPP were observed with enzyme UV-irradiated either in the presence or the absence of the inactivator. This indicates no major change in the catalytic activity of the surviving enzyme but complete inactivation of some of the enzyme.

Fourth, the substrate *trans,trans*-FPP protects the enzyme from inactivation in the presence of its photolabile analog, DATFP-GPP, in a concentration dependent manner.

Fifth, both cosubstrate, IPP, and divalent cation, Mg^{2+}, must be present to enable DATFP-GPP to inactivate the enzyme on irradiation.

Therefore, it can be inferred that both IPP and Mg^{2+} are required for the binding of the natural substrate, FPP, to the active site. This situation is different than the observations made with the avian liver FPP synthetase, where it was shown directly, that either the allylic or homoallylic substrate bound to the FPP synthetase in the absence of the other substrate.[13] Furthermore, this binding occurred even in the absence of the divalent cation.[14]

Apparent differences in the binding of substrates and their analogs to the C_{55}PP synthetase and FPP synthetase may be partly explained in terms of the differences in the chain length and the stereochemistry of the substrate and the final product. The large chain length of the product of the C_{55}PP synthetase may require a larger active site domain and therefore more tolerance to alterations in the substrate size and shape. It has been suggested that the lack of strict geometric specificity in the binding of the allylic PP substrate to the C_{55}PP synthetase may be caused by a requirement of the allylic PP binding site to accommodate both cis and trans residues during polymerization.[12] The present observations, taken in conjunction with the previous results, indicate that the binding of the allylic PP substrate to the active site of the C_{55}PP synthetase may require a more extensive cooperation of the homoallylic substrate and divalent cation than that observed with FPP synthetase.

The availability of these photolabile prenyltransferase substrates now offers a number of other opportunities for their application in studying prenyltransferases. The ability to prepare long chain photolabile polyprenyl phosphates such as photoactivatable undecaprenyl- and dolichylmonophosphates would be of significant value in providing a tool to probe enzymes or enzyme complexes, which use the parent prenyl phosphates as glycosyl carriers in bacterial cell wall biosynthesis or mammalian glycoprotein biosynthesis.

Acknowledgments

This work was supported by NIH Grant GM-23193.

[13] B. C. Reed and H. C. Rilling, *Biochemistry* **15,** 3739 (1976).

[14] H. L. King, Jr. and R. C. Rilling, *Biochemistry* **16,** 3815 (1977).

[14] Photoaffinity Substrate Analogs for Eukaryotic Prenyltransferase

By HANS C. RILLING

The instant chemical reactivity of photochemicals on irradiation has made them extremely useful tools for probing the catalytic sites of proteins.[1,2] This technology has been exploited for identifying the catalytic sites of several prenyltransferases (dimethylallyltransferases). Two independent and different approaches have been taken. In Allen's laboratory an ingenious procedure was utilized to prepare substrate analogs.[3,4] A photoreactive moiety was attached to the end of the isoprenoid chain. The procedure is general and can be applied to almost any polyisoprenoid molecule that has a terminal isopropylidene group. Brems,[5,6] on the other hand, took advantage of the broad substrate specificity of the eukaryotic prenyltransferases, a property that had been demonstrated in the laboratories of both Popják and Ogura.[7-9] One of the analogs that had been found to be inhibitory was phenylethyl pyrophosphate. This suggested that an aryl azide bearing the ethyl pyrophosphate moiety might provide a useful photolabel for this enzyme, and it was to this end that the following analogs were synthesized.

Preparation of o-Azidophenylethanol. Because of the unavailability of starting material for synthesizing *o*-azidophenylethanol as prepared earlier,[5] a different and incidentally simpler synthesis was devised for this compound.[6] *o*-Nitrophenylethyl alcohol (1 g) (Aldrich) was stirred with anhydrous pyridine (150 ml) at 15°. Dipyridine-chromium(VI) oxide complex[10] (6 *M* excess) dissolved in 290 ml of dichloromethane was added in small portions and allowed to react for 90 min. The extent of reaction was monitored by thin-layer chromatography on silica gel plates (Eastman, with fluorescent indicator) with 5% ethyl acetate in toluene as solvent. *o*-Nitrophenylacetaldehyde has an R_f of 0.63, while the alcohol has one of

[1] H. Bayley and J. R. Knowles, this series, Vol. 46, p. 69.
[2] J. R. Knowles, *Acc. Chem. Res.* **5,** 155 (1972).
[3] T. Baba and C. M. Allen, *Biochemistry* **23,** 1312 (1984).
[4] C. M. Allen and T. Baba, this volume [13].
[5] D. N. Brems and H. C. Rilling, *Biochemistry* **18,** 860 (1979).
[6] D. N. Brems and H. C. Rilling, *Biochemistry* **20,** 3711 (1981).
[7] G. Popják, J. L. Rabinowitz, and J. M. Baron, *Biochem. J.* **113,** 861 (1969).
[8] T. Whinka, K. Ogura, and S. Seto, *J. Biochem.* (*Tokyo*) **78,** 1177 (1975).
[9] A. Saito, Ph.D. Thesis, Tohoku University (1976).
[10] J. C. Collins, *Tetrahedron Lett.* **30,** 3363 (1968).

ISBN 0-12-182010-6

0.08. The reaction was terminated by removal of solvent by rotary evaporation. The products were extracted into diethyl ether and were separated on a column of silica gel (Woelm-ICN) with 5% ethyl acetate in toluene as solvent. A yield of 30% was obtained. An excess of *o*-nitrophenylacetaldehyde (0.60 mmol) was reduced by 0.11 mmol of NaB^3H_4 (25 mCi). *o*-Nitrophenyl[1-^{3}H]ethyl alcohol thus obtained was reduced to *o*-aminophenyl[1-^{3}H]ethyl alcohol by stirring with concentrated hydrochloric acid (1.33 mL) and stannous chloride (0.62 g) at room temperature for 15 hr. NaOH (4 *N*) was added until the white tin complex just dissolved. The solution was extracted with ether, and the extract was washed with water, dried (Na_2SO_4), filtered, and evaporated to give a yellow oil (80 μmol). This was immediately converted into the azide (75 μmol) by treatment with $NaNO_2$.[11,12] Spectral data (IR and NMR) were consistent with the structure.

o-[1-^{3}H]Azidophenylethanol. One gram of *o*-aminophenylethanol was reacted with an equivalent of acetic anhydride. After 0.5 hr, the solution solidified. The product was recrystallized from absolute ethanol with a yield of 95%. The crystalline solid has a melting point of 103°. Spectral data were consistent with the anticipated product. The aldehyde was then formed by oxidation with 3 equivalents of chromium trioxide in water which was added to *N*-acetyl-*o*-aminophenylethanol in pyridine. The reaction was monitored and products were detected by fluorescence quenching on silica gel thin-layer chromatography. A mixture of at least three products was obtained.

One hundred milligrams of this mixture was chromatographed on preparative plates of silica gel G with diethyl ether as solvent. The 75 mg of *N*-acetyl-*o*-aminophenylacetaldehyde (R_f 0.5) thus obtained was reduced to *N*-[1-^{3}H]acetyl-*o*-aminophenylethanol by NaB^3H_4. This material was then hydrolyzed with 6 *N* HCl at 100° for 1 hr. Excess HCl was then removed under vacuum, and the product was neutralized with NH_4OH. After excess NH_4OH was removed under vacuum, *o*-[1-^{3}H]aminophenylethanol was dissolved in ethyl ether and the borates removed by filtration. The pyrophosphate ester of *o*-[1-^{3}H]azidophenylethanol was prepared as described below. The radioactive pyrophosphate (sp. act. 31 Ci/mol) had an identical R_f value in 2-propanol–ammonia–water (6 : 3 : 1) and extinction coefficient as the nonradioactive counterpart.

p-Azidophenylethanol. p-Aminophenylacetic acid was purchased from Aldrich Chemical Co. and was reduced to the corresponding alcohol with $LiAlH_4$. Spectral data for the product (IR and NMR) were consistent with

[11] G. Smolinsky and B. L. Feuer, *J. Am. Chem. Soc.* **86,** 3085 (1964).

[12] R. O. C. Norman and G. K. Rada, *J. Chem. Soc.* p. 3030 (1961).

the desired product. The remaining synthesis was identical with that for the ortho isomer.

3-Azido-1-butanol. 1,3-Butanediol was purchased from Aldrich Chemical Co. One gram of the diol was reacted with 1 equivalent of acetic anhydride in 8 equivalents of pyridine for 24 hr. After the excess pyridine was neutralized with HCl, the products were extracted into ethyl ether. The primary acetylated alcohol was separated from other products by chromatography on a Woelm (ICN) silica gel column, with 30% ethyl acetate in hexane as solvent. A yield of 50% was obtained. Spectral data (IR and NMR) were consistent with the anticipated product. 1-Acetoxy-3-hydroxybutanol was reacted at reflux with 1.5 equivalents of carbon tetrabromide and 1 equivalent of triphenylphosphine in dichloromethane. 1-Acetoxy-3-bromobutanol (70%) was isolated by distillation. Again the spectral data were consistent with the product. 1-Acetoxy-3-bromobutanol was hydrolyzed with 1 equivalent of Na_2CO_3 in water–methanol. 3-Bromo-1-butanol (30%) was resolved from other products on a silica gel column (solvent system 10% ethyl acetate in hexane). 3-Bromo-1-butanol was refluxed for 10 hr with a 10-fold excess of potassium azide in acetonitrile and a catalytic amount of 18-crown-6 ether. 3-Azido-1-butanol (98%) was used without further purification. Spectral data including a mass spectrum were consistent with the structure.

The pyrophosphate esters of the above alcohols were prepared by the method of Cramer as described by Cornforth and Popják[13] and purified by ion-exchange chromatography on Dowex AG-1-X8 formate from Bio-Rad by use of a linear ammonium formate gradient in methanol.[14] Improved procedures for preparation of pyrophosphate esters are reported elsewhere in this volume.[15] Purity of the pyrophosphates was verified by thin-layer chromatography on ammonium sulfate impregnated silica gel H plates with $CHCl_3$–CH_3OH–H_2O (5 : 5 : 1) as solvent or on Whatman No. 1 paper with 2-propanol–ammonia–water (6 : 3 : 1) as solvent.

Concentrations of the esters were determined by analysis of total phosphate as described earlier.[16] Presence of the azide after phosphorylation was confirmed by applying the compounds to a thin-layer plate of silica gel H containing fluorescent indicator. After irradiation of several of the spots by UV light, the plate was developed with 2-propanol–ammonia–water (6 : 3 : 1). The plate was then visualized under UV light. If the compounds had not been irradiated prior to chromatography, single spots

[13] R. H. Cornforth and G. Popják, this series, Vol. 15, p. 385.

[14] S. S. Sofer and H. C. Rilling, *J. Lipid Res.* **10,** 183 (1969).

[15] D. L. Bartlett, R. H. S. King, and C. D. Poulter, this volume [20].

[16] B. C. Reed and H. C. Rilling, *Biochemistry* **15,** 3739 (1976).

that had chromatographed up the plate were obtained. If they had been irradiated, the majority of the material remained at the origin.

Inhibition Constants of I, II, and III. Kinetically obtained inhibition constants for the ortho ester are 18 and 9 μM, for the para compound are 60 and 85 μM, and for the butanol derivative are 125 and 60 μM against geranyl and isopentenyl pyrophosphate, respectively. K_m values for natural substrates isopentenyl pyrophosphate and geranyl pyrophosphate for the synthesis of farnesyl pyrophosphate are less than 0.5 μM.[17,18] Since the kinetic data suggest that *o*-azidophenylethyl pyrophosphate binds 5–10 times more tightly than the others, only it was used in the following experiments.

Half-Life of Photodecomposition. *o*-Azidophenylethyl pyrophosphate has a UV absorption maximum at 250 nm with an extinction coefficient of 2230 liters mol^{-1} cm^{-1} in water. The half-life for photolytic formation of the nitrene from the corresponding azide was determined by loss of absorbance at 250 nm. A maximum loss of absorbance at 250 nm was obtained after 60 sec, while the absorbance after irradiating for 30 sec showed that one-half the azide had disappeared.

Photoaffinity Labeling. Crystalline prenyltransferase is stored in saturated ammonium sulfate. To remove this salt, it was chromatographed on a Sephadex G-25 column equilibrated with 50 m*M* *N*-tris(hydroxymethyl)methyl-2-aminoethanesulfonic acid buffer, pH 7.0, containing 1 m*M* $MgCl_2$, 10 m*M* 2-mercaptoethanol, and 100 m*M* KCl or was dialyzed against the same buffer at 4°. Dithiothreitol, which is usually included to stabilize the enzyme, was omitted since aryl azides are reduced to the corresponding amines by this reagent.[19] Solutions of the enzyme and analog were photolyzed at room temperature under a nitrogen atmosphere in a 1-cm quartz cuvette closely situated between two Mineralights (UVS-11, maximum emission 254 nm). Typically, the enzyme was irradiated with 30-fold molar excess of the aryl azide for four 1-min intervals. After each irradiation, the amount of affinity label was restored to the original concentration and the process repeated. The extent of photoaffinity labeling was monitored by the loss of enzyme activity; under these conditions, labeling resulted in 80 ± 5% inactivation of enzymatic activity. One cycle of irradiation led to 40% inactivation, while irradiation of enzyme under nitrogen in the absence of affinity reagent resulted in essentially no loss of enzymatic activity even after four sequential 1-min irradiations.

[17] F. M. Laskovics, J. M. Krafick, and C. D. Poulter, *J. Biol. Chem.* **254,** 9458 (1979).

[18] B. C. Reed and H. C. Rilling, *Biochemistry* **14,** 50 (1975).

[19] J. V. Staros, H. Bayley, D. N. Standing, and J. R. Knowles, *Biochem. Biophys. Res. Commun.* **80,** 568 (1978).

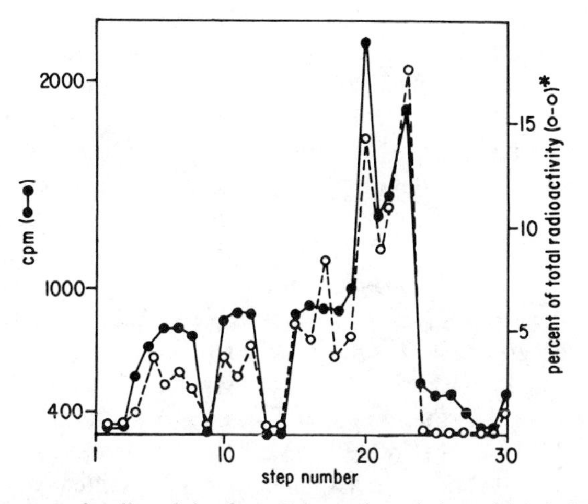

FIG. 1. Recovery of radioactivity from Edman degradation of labeled CNBr peptide. Seventy nanomoles (1.6×10^6 cpm) of photolabeled CNBr peptide was subjected to automated sequence analysis. Pth amino acids liberated at each step of Edman degradation were analyzed for radioactivity (solid line) and for percentage of the total radioactivity expected after correction for the initial and repetitive yields and carryover from the previous step. Counts per minute (solid line) represent 20% of the total radioactivity liberated at each step. Reproduced with permission from D. N. Brems and H. C. Rilling, *Biochemistry* **20,** 3711 (1981).

To localize the part of the protein that was labeled, the protein was carboxymethylated and then subjected to tryptic digestion. The tryptic maps of enzyme affinity labeled to the extent of 0.6 mol of analog per catalytic site had at least three radioactive spots, one at the origin. The other two migrated in the first dimension (chromatography) but did not move during the electrophoresis. The two radioactive spots which migrated did not coincide with any fluorescamine-positive areas and were diffuse. These results were interpreted to indicate that the affinity label had reacted with more than one amino acid. The absence of fluorescamine-positive staining associated with the radioactive spots may result from the diffuse nature of these regions. When geranyl pyrophosphate was included during the labeling process, the radioactive spot which migrated furthest away from the origin was greatly diminished. Thus, apparently this peptide is principally associated with the allylic site. All other attempts to isolate radioactive tryptic peptides failed. Ion-exchange chromatography resulted in broad radioactive elution profiles, confirming the heterogeneity suggested by the tryptic maps.

Consequently, procedures that would yield larger peptides were utilized. After cleavage with CNBr, analysis of the number of peptides with attached affinity reagent revealed that only one of the eight peptides was significantly labeled. Over 80% of the recovered radioactivity was located in this peptide. Isolation of this radioactive peptide was not easily achieved, chiefly due to the insolubility and aggregating nature of the CNBr peptides. Key to the isolation of that peptide was maleylation, which rendered the peptides soluble in aqueous buffers and consequently more amenable to purification by conventional chromatographic techniques.

Sequence Determination. Sequence analysis by automated Edman degradation of native CNBr peptides yielded all 30 residues shown by amino acid analysis. The results showed the sequence to be Leu-Asp-Leu-Ile-Gly-Ala-Pro-Val-Ser-Lys-Val-Asp-Leu-Ser-Thr-Phe-Gln-Glu-Glu-Arg-Tyr-Lys-Ala-Phe-Val-Pro-Tyr-Lys-Ala-Met.

Sequence analysis of peptide from affinity labeled protein also gave 30 residues. The results show the sequence to be the same. An aliquot of the Pth-amino acid liberated at each step of the Edman degradation was analyzed for radioactivity. The results are illustrated in Fig. 1. A total of 16 individual steps released significant amounts of radioactivity. After 30 steps, 60% of the total radioactivity was recovered. Thus, these results indicate that it is possible to label the catalytic site of prenyltransferase with moderate selectivity.

Acknowledgments

This research was supported in part by a Grant (AM13140) from the National Institutes of Health. The skillful participation of Dr. D. N. Brems in this is gratefully acknowledged.

[15] Synthesis of Allylic and Homoallylic Isoprenoid Pyrophosphates

By V. Jo Davisson, A. B. Woodside, and C. Dale Poulter

Allylic pyrophosphates are important metabolites in the isoprenoid pathway. Along with isopentenyl pyrophosphate, these compounds are substrates and products for the prenyltransferases that catalyze the basic chain elongation reactions in the pathway. With the exception of [^{14}C]isopentenyl pyrophosphate,[1] none of the isoprenoid pyrophosphates can be

[1] Available from Amersham, Arlington Heights, Ill. 60005.

ISBN 0-12-182010-6

purchased nor are there dependable high yield syntheses of these compounds from the corresponding alcohols, many of which are available commercially.

Difficulties encountered in the synthesis and subsequent purifications can be traced to two reinforcing factors. The 3,3-dialkylallylic moiety found in the substrates is highly reactive, and phosphate and pyrophosphate residues are superb leaving groups, especially when they are protonated.[2] The procedure commonly used to synthesize allylic pyrophosphates was first reported in 1959[3] and has not been altered significantly since then. The reaction involves treatment of a mixture of the alcohol and inorganic pyrophosphate with trichloroacetonitrile to generate a complex mixture of organic and inorganic mono-, di-, and triphosphates. Yields of the desired products rarely exceed 30%, and further losses are usually encountered during purification. In addition, the procedure becomes difficult to manage if more than 5–10 mg of material is involved.

This chapter is devoted to the synthesis and purification of four commonly used compounds—isopentenyl pyrophosphate (**1**), dimethylallyl pyrophosphate (**2**), geranyl pyrophosphate (**3**), and farnesyl pyrophosphate (**4**). A synthesis of **1** is published in an earlier volume of this series.[4]

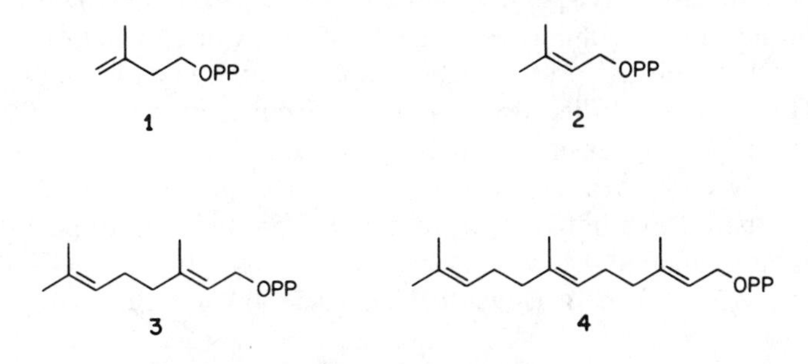

We find the new procedure described herein to be easier to execute, especially since reagents and intermediates can be prepared in bulk and stored until needed for the pyrophosphorylation. Many of the problems associated with highly reactive allylic pyrophosphates **2–4** are circumvented by introduction of the labile carbon–oxygen bond in the final step by a direct displacement with inorganic pyrophosphate and by a chromatographic purification on cellulose. The procedure has been used to prepare a variety of pyrophosphates in addition to those listed above and can be scaled-up to produce useful quantities of material.

[2] B. K. Tidd, *J. Chem. Soc. B* p. 1168 (1971).

[3] F. Cramer and W. Bohm, *Angew. Chem.* **71,** 775 (1959).

[4] R. H. Cornforth and G. Popják, this series, Vol. 15, p. 359.

Materials

Reagents

N-Chlorosuccinimide, copper(I) bromide, dimethyl sulfide, *p*-toluenesulfonyl chloride, tetra-*n*-butylammonium hydroxide, and sulfosalicylic acid are available from Aldrich Chemical Co. *N*-Chlorosuccinimide is recrystallized from refluxing benzene and stored over phosphorus pentoxide at room temperature. Dimethyl sulfide is distilled from calcium hydride under a nitrogen atmosphere prior to use. Disodium dihydrogen pyrophosphate is obtained from Stauffer Chemical Co. Pyridine, ammonium bicarbonate, and ferric chloride (ACS certified grade) are available from Fisher Scientific. Dichloromethane and acetonitrile are distilled from phosphorus pentoxide under a nitrogen atmosphere just prior to use. All other solvents are routinely purified by passage through neutral alumina followed by distillation.

Isoprenoid Starting Materials

Isoprene, 3-methyl-3-buten-1-ol (isopentenol), and (*E*)-2,7-dimethyl-2,6-octadien-1-ol (geraniol) are available from Aldrich Chemical Co. Isoprene is distilled just prior to use. Isopentenol is used without purification. Geraniol is distilled (bp 52–53°, 0.05 mm Hg), and the distillate is further purified by flash chromatography on silica gel[5] by elution with 50 : 48 : 2 (v/v/v) hexanes, diethyl ether, and isopropanol. *E,E*-Farnesol is prepared synthetically from geranyl bromide.[6] The purity of this material is evaluated by capillary gas chromatography on an OV-1 WCOT column (12.5 m) using a Hewlett Packard 5880A gas chromatograph.

Ion Exchange Resin

Dowex AG 50W-8X cation exchange resin (hydrogen form) is available from Bio-Rad Laboratories and is converted to the ammonium form by treatment with concentrated ammonium hydroxide. The resin is recycled by elution with 0.5 liter of concentrated hydrochloric acid following exchange of the tetra-*n*-butylammonium counterion.

[5] This reference gives a complete description of the procedures and apparatus required to conduct flash chromatography successfully [W. C. Still, M. Kahn, and A. Mitra, *J. Org. Chem.* **43,** 2923 (1978)].

[6] F. W. Sum and L. Weiler, *J. Am. Chem. Soc.* **101,** 4401 (1979).

Chromatography

Whatman CF11 fibrous cellulose powder, available from Whatman Inc., is used for flash chromatography. Cellulose thin layer plates are manufactured by E. Merck and are available from American Scientific Products. Silica gel, Grade 60 (230/400 mesh), is available from Aldrich Chemical Co.

NMR of Organic Pyrophosphates

Samples of organic pyrophosphates for ^{1}H NMR, ^{13}C NMR, and ^{31}P NMR are prepared just prior to analysis and may be stored at 0° for short periods of time. It is necessary to adjust the pH of deuterium oxide to pH 8 with ammonium deuteroxide-d_5 (available from Aldrich Chemical Co.) to prevent solvolysis of the allylic pyrophosphates. All of the samples are prepared in concentrations of 35–50 mg/ml with sodium 2,2-dimethyl-2-silapentane 5-sulfonate (DSS) added as an internal reference. One or two drops of a 1% w/v EDTA in D_2O solution is added to each sample just prior to obtaining its ^{31}P NMR spectrum. ^{31}P NMR spectra are referenced to external phosphoric acid.

Synthesis of Pyrophosphates

The general strategy employed to synthesize pyrophosphates involves preparation of activated precursors susceptible to nucleophilic displacement with inorganic pyrophosphate.[7] Several activated precursors can be used. Chloride or bromide derivatives are suitable for the allylic systems, with the slightly less reactive chloride being preferred for the geranyl and farnesyl moieties. The homoallylic isopentenyl system requires a better leaving group, and one that is not susceptible to elimination under the reaction conditions. These criteria are met by the *p*-toluene sulfonate (tosylate) derivative. The displacements are performed in dry acetonitrile with tris(tetra-*n*-butyl)ammonium hydrogen pyrophosphate, which is both a soluble and reactive form of nucleophilic pyrophosphate.

*Tris(tetra-n-butyl)ammonium Hydrogen Pyrophosphate (**5**)*

A 2.0 by 30 cm column is slurry packed with Dowex AG 50W-8X cation exchange resin (hydrogen form) in deionized water. After washing the column with two column volumes of deionized water, 3.13 g (14

[7] V. M. Dixit, F. M. Laskovics, W. I. Noall, and C. D. Poulter, *J. Org. Chem.* **46,** 1967 (1981).

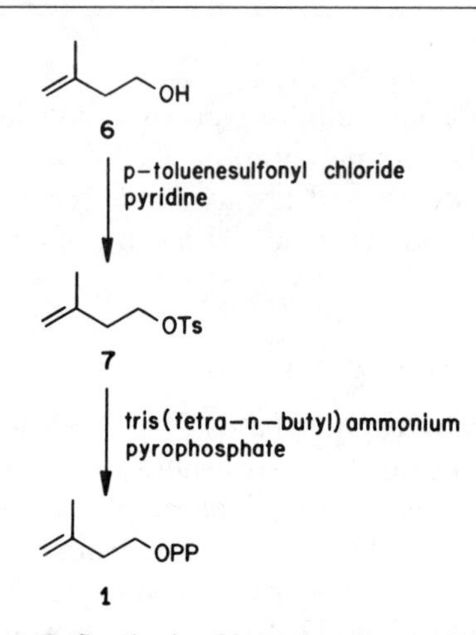

SCHEME I. Synthesis of isopentenyl pyrophosphate.

mmol) of disodium dihydrogen pyrophosphate is applied to the resin in 30 ml of deionized water. The pH of the eluent is monitored, and the eluent is collected when the pH becomes acidic. Collection (approximately 100 ml) is halted when the pH returns to that of deionized water. The solution of pyrophosphoric acid is then *immediately* titrated to pH 7.3 with tetra-*n*-butylammonium hydroxide. The salt is dried by lyophilization to yield 12.53 g (98%) of a hydroscopic, white solid. This preparation is suitable for most purposes. However, it contains a small amount of material that does not dissolve in acetonitrile. The insoluble residue can be removed by centrifugation prior to a displacement reaction.

The water content of **5** can be reduced by dissolving the solid in dry acetonitrile and removal of the solvent on a rotary evaporator at aspirator vacuum. This process is repeated until no further reduction is seen in the signal for water in acetonitrile ($\delta \approx 4.5$ ppm) in the ^{1}H NMR spectrum. The reagent is dried in the appropriate reaction vessel just prior to the pyrophosphorylation reaction.

*Isopentenyl Pyrophosphate (**1**)*

The reactions used to synthesize **1** are shown in Scheme I. The major variation from the syntheses of allylic pyrophosphates discussed later in this chapter is the use of the tosylate leaving group. Attempts with halides were unsuccessful because of competing elimination reactions. The procedure described in this section is useful as a general method for synthesizing nonallylic primary pyrophosphates.

3-Methyl-3-butenyl-p-Toluene Sulfonate (Isopentenyl Tosylate, **7***).* The synthesis and physical date for this compound are described in the literature.[8] The reaction is run under a blanket of nitrogen in an oven-dried 25 ml two-neck round bottom flask equipped for magnetic stirring. One gram (11.6 mmol) of 3-methyl-3-buten-1-ol (**6**) and 1.83 g (23.2 mmol) of pyridine are added by syringe. The flask and its contents are cooled in an ice bath before the direct addition of *p*-toluenesulfonyl chloride (2.21 g, 11.6 mmol). After 15 min, the ice bath is removed and the reaction is allowed to stir at room temperature for 5 hr before the contents are transferred to a 125-ml separatory funnel. The reaction mixture is diluted with 50 ml of water and extracted three times with 25 ml portions of diethyl ether. The combined ether extracts are washed with 25 ml of 0.5 *N* sulfuric acid followed by 25 ml of water. Finally the organic layer is washed with 15 ml of saturated sodium chloride solution and dried over magnesium sulfate. Solvent is removed by rotary evaporation. The resulting yellow oil is purified by flash chromatography[5] on silica gel using an 87 : 13 (v/v) mixture of hexanes and ethyl acetate as the eluent to yield 2.12 g (76%) of a colorless oil, R_f 0.4 (TLC on silica gel, visualized with iodine); ^{1}H NMR 90 MHz ($CDCl_3$) δ 1.07 [3H, s, methyl at C(3)], 2.35 [2H, t, $J = 6.6$ Hz, methylene at C(2)], 2.45 [3H, s, methyl at C(4′) of aromatic], 4.14 [2H, t, $J = 6.6$ Hz, methylene at C(1)], 4.75 [2H, d, $J = 9.6$ Hz, vinyl at C(4)], 7.35 [2H, d, $J = 6.6$ Hz, C(3′) and C(5′) of aromatic], 7.82 [2H, d, $J = 6.6$ Hz, C(2′) and C(6′) of aromatic]. Tosylate **7** is stable and can be stored without decomposition at 0°.

3-Methyl-3-butenyl Pyrophosphate (Isopentenyl Pyrophosphate, **1***).* The reaction is run under a blanket of nitrogen in a flame-dried 50-ml round bottom flask equipped for magnetic stirring. Tris(tetra-*n*-butyl)ammonium hydrogen pyrophosphate (3.25 g, 3.60 mmol) is dissolved in 4.5 ml of acetonitrile. Tosylate **7** (288 mg, 1.20 mmol) dissolved in 0.5 ml of acetonitrile is added, and the resulting mixture is allowed to stir for 2 hr at room temperature. Solvent is then removed by rotary evaporation, and the resulting opaque residue is dissolved in 3.0 ml of 1 : 49 (v/v) isopropanol : 25 m*M* ammonium bicarbonate (ion exchange buffer). This clear, colorless solution is passed through a column containing 63.5 ml (108 meq) of Dowex AG 50W-X8 cation exchange resin (ammonium form) previously equilibrated with two column volumes of ion exchange buffer. The column is eluted with 127 ml of the same buffer. The clear, colorless eluent is lyophilized to dryness to yield 1.13 g (99%) of a fluffy white solid.

[8] B. M. Trost and R. A. Kunz, *J. Am. Chem. Soc.* **97,** 7152 (1975).

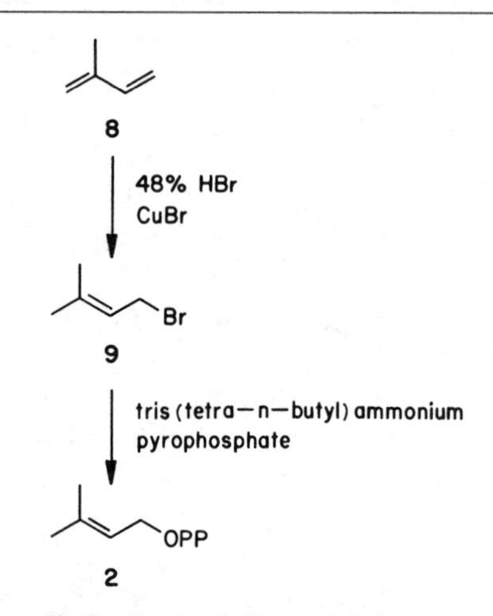

SCHEME II. Synthesis of dimethylallyl pyrophosphate.

Dimethylallyl Pyrophosphate (**2**)

The synthesis for dimethylallyl pyrophosphate shown in Scheme II utilizes allylic bromide **9** in the displacement step. This material is available from Aldrich Chemical Co. or can be synthesized from isoprene as described in this chapter. Alternatively, dimethylallyl alcohol can be converted to the bromide using phosphorus tribromide[7] or the chloride by the procedures described later for pyrophosphates **3** and **4**.

1-Bromo-3-methyl-2-butene (Dimethylallyl Bromide, **9**). The synthesis of **9** is briefly described in a previous communication.[9] Copper(I) bromide is purified by Soxhlet extraction with acetone for 24 hr followed by a thorough rinse with benzene. The powder is dried at 75° for 12 hr. A 2.0 g portion of copper(I) bromide is dissolved in 100 ml of 48% hybrobromic acid and transferred to a 250 ml separatory funnel. Freshly distilled isoprene (10 ml) is added to this deep purple solution, and the reaction mixture is vigorously shaken for 5 min. The organic layer is removed and dried over magnesium sulfate. After filtration, the mixture is distilled at aspirator pressure from a small amount of potassium carbonate. Bromide **9** distills at 35–37° to yield 1.36 g (20%) of a colorless liquid; ^{1}H NMR 90 MHz ($CDCl_3$) δ 1.77 (3H, s, methyl), 1.80 (3H, s, methyl), 4.00 [2H, d, J = 8.4 Hz, methylene at C(1)], 5.53 [1H, t, J = 8.4 Hz, vinyl at C(2)]. The compound decomposes upon thin-layer chromatography on silica gel; however, it can be stored for up to a year at −20°.

[9] P. J. R. Neederlof, M. S. Moolenaar, E. R. deWaard, and H. D. Huisman, *Tetrahedron* **33,** 579 (1977).

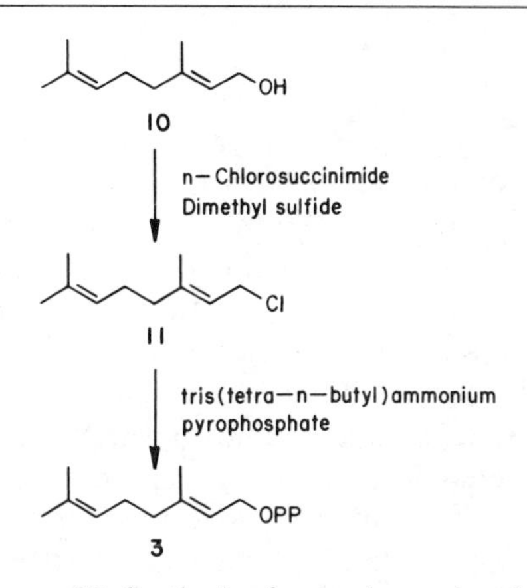

SCHEME III. Synthesis of geranyl pyrophosphate.

3-Methyl-2-butenyl Pyrophosphate (Dimethylallyl Pyrophosphate, ***2****).* The reaction is run under a blanket of nitrogen in a flame-dried 50-ml round bottom flask equipped for magnetic stirring. Tris(tetra-*n*-butyl)ammonium hydrogen pyrophosphate (2.27 g, 2.52 mmol) is dissolved in 4.5 ml of acetonitrile. Dimethylallyl bromide (**9**) (179 mg, 1.20 mmol) dissolved in 0.5 ml of acetonitrile is added to this solution. The reaction mixture is stirred for 2 hr at room temperature before solvent is removed by rotary evaporation. The resulting opaque residue is dissolved in 3.0 ml of 1 : 49 (v/v) isopropanol : 25 m*M* ammonium bicarbonate. The resulting clear colorless solution is passed through a column containing 44.5 ml (75.5 meq) of Dowex AG 50W-X8 cation exchange resin (ammonium form) which has been equilibrated with two column volumes of ion exchange buffer. The column is eluted with 89 ml of the same buffer. The clear, colorless eluent is lyophilized to dryness to yield 780 mg (99%) of a fluffy white solid.

Care must be taken during lyophilization. The sample decomposes if left under vacuum at room temperature for extended periods. Presumably ammonia slowly evolves from the salt giving more reactive partially protonated allylic pyrophosphates.

*Geranyl Pyrophosphate (**3**) and Farnesyl Pyrophosphate (**4**)*

Geranyl (**3**) and farnesyl (**4**) pyrophosphate are synthesized from the corresponding alcohols using identical reagents. Only the procedure for pyrophosphate **3** outlined in Scheme III is described in detail. Allylic chlorides **11** and **13** are used in the phosphorylation reaction because these derivatives are easier to handle than the corresponding bromides but are sufficiently reactive for the displacement by inorganic pyrophosphate.

(E)-1-Chloro-3,7-dimethyl-2,6-octadiene (Geranyl Chloride, ***11****).* The method employed for the synthesis of these compounds is based upon a general procedure for the synthesis of allylic chlorides and bromides from the corresponding alcohols.[10] A flame-dried, three-neck, 100-ml round bottom flask equipped with a rubber septum, a low temperature thermometer, and a magnetic stirrer is employed for these reactions. All of the glassware utilized for the various manipulations is oven dried at 110°, and the reaction is run under a blanket of nitrogen. *N*-Chlorosuccinimide (1.47 g, 11 mmol) is dissolved in 45 ml of dry dichloromethane. The contents of the flask are cooled to −30° in an acetonitrile/dry ice bath. Dimethyl sulfide (0.88 ml, 0.74 g, 12 mmol) is added dropwise by syringe to the cold, well-stirred heterogeneous reaction mixture. The contents of the flask are briefly allowed to warm to 0° before the temperature is lowered to −40°. Geraniol (1.73 ml, 1.54 g, 10 mmol) in 5 ml of dry dichloromethane is added by syringe to the milky white suspension over a 3-min period. The reaction is allowed to warm to 0° over 1 hr and maintained at that temperature for an additional hour. During this period, the mixture becomes a clear, colorless solution. The ice bath is then removed, and the reaction is stirred at room temperature for 15 min before it is poured into a 250-ml separatory funnel which contains 25 ml of cold saturated sodium chloride. The aqueous layer is extracted with two 20 ml portions of pentane. The organic layers are combined with an additional 20 ml of pentane and washed with two 10 ml portions of cold saturated sodium chloride. The organic layer is then dried over magnesium sulfate for 15 min and filtered by gravity. Volatiles are removed by rotary evaporation at aspirator vacuum. Traces of volatile impurities are removed under vacuum (1.0 mm Hg) at room temperature for 4 hr to yield **11** as a colorless oil, 1.63 g (94%); ^{1}H NMR 90 MHz ($CDCl_3$) δ 1.61 (3H, s, methyl), 1.70 (3H, s, methyl), 1.75 (3H, s, methyl), 2.10 [4H, br m, methylene at (C4) and (C5)], 4.09 [2H, d, $J = 8.9$ Hz, methylene at (C1)], 5.1 [1H, br m, vinyl at (C6)], 5.47 [1H, t, $J = 8.9$ Hz vinyl at (C2)]. Chloride **11** can be stored at −20° for several months in an inert atmosphere.

(E,E)-1-Chloro-3,7,11-trimethyl-2,6,10-dodecatriene (E,E-Farnesyl Chloride, ***13****).* Following a procedure identical to that described for synthesis of **11**, 0.33 g (1.5 mmol) of farnesol (**12**) is treated with 0.22 g (1.65 mmol) of *N*-chlorosuccinimide and 0.12 ml (0.10 g, 1.65 mmol) of dimethyl sulfide. After work-up, **13** is obtained as a colorless oil, 0.35 g (97%); ^{1}H NMR 90 MHz ($CDCl_3$) δ 1.63 (3H, s, methyl), 1.68 (3H, s,

[10] E. J. Corey, C. U. Kim, and M. Taheeda, *Tetrahedron Lett.* p. 4339 (1972).

methyl), 1.72 (6H, s, methyls), 2.02 [8H, br m, methylenes at C(4), C(5), C(8) and C(9)], 4.11 [2H, d, J = 8.9 Hz, methylene at C(1)], 5.14 [2H, br m, vinyl at C(6) and C(10)], 5.48 [1H, t, J = 8.9 Hz, vinyl at C(2)]. Chloride **13** can also be stored at −20° in an inert atmosphere.

*(E)-3,7-Dimethyl-2,6-octadienyl Pyrophosphate (Geranyl Pyrophosphate, **3**).* The reaction is run under nitrogen in a flame-dried 50-ml round bottom flask equipped for magnetic stirring. Tris(tetra-*n*-butyl)ammonium hydrogen pyrophosphate (2.27 g, 2.52 mmol) is dissolved in 4.5 ml of dry acetonitrile before addition of 206 mg (1.20 mmol) of geranyl chloride in 0.5 ml of acetonitrile. The resulting mixture is stirred for 2 hr at room temperature. Solvent is then removed by rotary evaporation, and the resulting opaque residue is dissolved in 3.0 ml of 1 : 49 (v/v) isopropanol : 25 m*M* ammonium bicarbonate. The resulting clear, colorless solution is passed through a column containing 44.5 ml (75.7 meq) of Dowex AG 50W-X8 cation exchange resin (ammonium form) which has been equilibrated with two column volumes of ion exchange buffer. The column is eluted with 89 ml of the same buffer. The clear, colorless eluent is lyophilized to dryness to yield 780 mg, 99% of a fluffy white solid. The same precautions described for allylic pyrophosphate **2** must be observed.

*(E,E)-3,5,7-Trimethyl-2,6,10-dodecatrienyl Pyrophosphate (E,E-Farnesyl Pyrophosphate, **4**).* Following a procedure identical to that described for **3**, 144 mg (0.6 mmol) of chloride **13** is treated with 1.10 g (1.20 mmol) of **5** in 2.5 ml of dry acetonitrile. Ion exchange is accomplished using 28 ml (48 meq) of Dowex AG 50W-X8 resin (ammonium form). The resulting milky suspension is lyophilized to yield 422 mg (98%) of a fluffy white solid.

Purification of Isoprenoid Pyrophosphates

Procedures are described for purification of isoprenoid pyrophosphates obtained from the displacement reactions by flash chromatography on cellulose. The apparatus and procedures are patterned after a description by Still and co-workers[5] for flash chromatography on silica gel. The columns used in our work are similar to those described by Still except the standard taper joints used to join the nitrogen valve to the column are replaced by more rugged O-ring seals.

Fresh cellulose powder (Whatman CF11) is pretreated by gentle washing for 30 min with two volumes of 0.2 *N* hydrochloric acid followed by three rinses with deionized water. The cellulose is then washed for 30 min with a double volume of 0.2 *N* sodium hydroxide solution followed by three rinses with deionized water. This material is stored in 1 : 1 (v/v) isopropanol : water.

Columns are packed using a slurry technique. The bottom is plugged with glass wool and covered by a bed of clean sand 1 cm thick. The sand is covered with acetonitrile before addition of a slurry consisting of one volume of cellulose in two volumes of 1 : 1 (v/v) acetonitrile : 0.1 *M* ammonium bicarbonate. The cellulose is allowed to settle by gravity flow and packed under nitrogen pressure until no further settling occurs. Three column volumes of acetonitrile are passed through the column to remove any pockets of air. The column is equilibrated at the desired flow rate with four column volumes of the solvent to be used for chromatography. Following chromatography, the column is regenerated by washing with 0.1 *M* ammonium bicarbonate (two column volumes) followed by deionized water. If the solvent contains chloroform, the column is washed with increasing amounts of 0.1 *M* ammonium bicarbonate in acetonitrile to prevent phase separation. The regenerated column is equilibrated as described above just before use. Before equilibration an intermediate wash with 1 : 1 (v/v) isopropanol : acetonitrile is necessary for elution with solvents containing chloroform. All of the separations are performed at room temperature.

A partial purification of the organic ammonium pyrophosphates is achieved by extraction. This step increases the capacity and resolution of the chromatography. Complete suspension of the reaction mixtures before extraction is critical for efficient recovery. This process is accelerated by rapid vortexing and ultrasonication. The overall efficiency of the extraction step is determined by analysis of the residue and the extract by thin-layer chromatography. The recovery of organic pyrophosphate is determined by weighing the residue after drying by lyophilization.

Thin layer chromatography is performed on 0.1-mm cellulose plates. The solvent front routinely is allowed to travel a distance of 8 cm. There are two separate techniques employed for visualization. The first uses a stain specifically for phosphate-containing materials. The system utilizes two separate solutions: (1) 0.2% ferric chloride in 4 : 1 (v/v) ethanol : water and (2) 1.0% sulfosalicylic acid in 3 : 2 (v/v) ethanol : water. The plates are sprayed with sulfosalicylic acid, air dried, and sprayed with ferric chloride. Pyrophosphate-containing material appears as white spots on a pink background.[11] The second procedure consists of staining plates with iodine vapor, a method sensitive for residues with unsaturation. All the solvent systems are selected to ensure that ammonium pyrophosphate and ammonium phosphate have lower R_f values than the desired organic pyrophosphate.

[11] Z. Gunter and J. Sherma, eds. "CRC Handbook of Chromatography," Vol. 2, p. 143. CRC Press, Cleveland, Ohio, 1972.

The elution buffers are removed from pyrophosphates **1–4** by rotary evaporation followed by lyophilization. There is no evidence for contamination from bicarbonate by ^{13}C NMR. It is appropriate to emphasize again that prolonged lyophilization of the allylic salts is deleterious.

*3-Methyl-3-butenyl Pyrophosphate (Isopentenyl Pyrophosphate, **1**)*

The white solid obtained from cation exchange of the products from the displacement reaction is dissolved in 4 ml of 0.1 *M* ammonium bicarbonate and transferred to a 25 ml test tube. This solution is then treated with 10 ml of a 1 : 1 (v/v) mixture of isopropanol : acetonitrile. A white precipitate forms after 1–2 min of vigorous vortexing. The precipitate is allowed to settle, and the solution is decanted into a centrifuge tube. The precipitate is dissolved in 3 ml of 0.1 *M* ammonium bicarbonate and extracted with 7 ml of 1 : 1 isopropanol : acetonitrile as described above. A final extraction is performed by dissolving the second precipitate in 2 ml of 0.1 *M* ammonium bicarbonate and extraction with 5 ml of 1 : 1 isopropanol : acetonitrile. The extracts are clarified by centrifugation in a bench top centrifuge and combined. The volume is reduced by rotary evaporation at aspirator vacuum to yield 0.50 g of a thick oil. This material is either stored overnight at −20° or chromatographed directly.

The concentrate obtained from the extraction is dissolved in 2 ml of isopropanol, acetonitrile, and 0.1 *M* ammonium bicarbonate in a ratio of 4.5 : 2.5 : 3 (v/v/v). This solution is applied to a 3.5 by 15 cm column of cellulose equilibrated with the same solution. After allowing the material to be absorbed by gravity flow, 10 ml of elution buffer is applied under pressure to ensure complete absorption. The reservoir is then filled with buffer, and products are eluted at a flow rate of 30 ml/min. After a 140 ml forerun, 30 30-ml fractions are collected. A 10-μl aliquot of every other fraction is assayed by thin-layer chromatography on cellulose. The plates are developed with the solvent used for flash chromatography. Those fractions showing a single spot for **1** by TLC (R_f = 0.35) are combined. The total volume in which **1** elutes is 280 ml (two column volumes). The volume is reduced to one-third by rotary evaporation at aspirator vacuum. The remainder of the organic solvent is then removed by rotary evaporation at 4 mm Hg to yield approximately 40 ml of a clear aqueous solution. During the course of the concentration step, the pH is maintained between 7.2 and 7.5 by the addition of 0.1 *N* ammonium hydroxide or by bubbling carbon dioxide through the solution. The aqueous residue is dried by lyophilization to yield 0.24 g (80%) of a finely divided white solid which is stored over calcium sulfate at −78°; ^1H NMR 300 MHz (D_2O/ND_4OD) δ 1.77 [3H, s, methyl at C(3)], 3.11 [2H, t, $J_{^1H,^1H}$ = 6.6 Hz, methylene at C(2)], 4.78 [2H, dt, $J_{^1H,^1H}$ = 6.6 Hz, $J_{^1H,^{31}P}$ = 3.3 Hz, methylene at C(4)], 4.86 [2H, s, vinyl at C(4)]; ^{13}C NMR 75 MHz (D_2O/ND_4OD) δ 24.51, 40.72 (d, $J_{^{13}C,^{31}P}$ = 7.2 Hz), 67.02 (d, $J_{^{13}C,^{31}P}$ = 4 Hz), 114.62, 147.39; ^{31}P NMR 32 MHz (D_2O/ND_4OD) δ −11.03 [1P, J_{P_1,P_2} = 20 Hz, P(1)], −7.23 [1P, P(2)].

The ^{31}P chemical shifts and, to a lesser extent, ^{31}P–^{31}P coupling constants are dependent on pH, counterion, and concentration. Although AB quartets are typically observed, we occasionally see only a single peak in the ^{31}P spectrum because of fortuitous chemical shift equivalence.

*3-Methyl-3-butenyl Pyrophosphate (Dimethylallyl Pyrophosphate, **2**)*

The procedure used to purify **2** is identical in all aspects to that described for **1**. Chromatography of 0.47 g of concentrated extract yields 0.24 g (80%) of a finely divided white solid, R_f 0.35; ^1H NMR 300 MHz (D_2O/ND_4OD) δ 1.72 (3H, s, methyl), 1.76 (3H, s, methyl), 4.45 [2H, dd, $J_{^1H,^1H}$ = 7 Hz, $J_{^1H,^{31}P}$ = 7 Hz, methylene at (C1)], 5.46 [1H, t, J = 7 Hz, vinyl at (C2)]; ^{13}C NMR 75 MHz δ 19.98, 27.81, 65.43 (d, $J_{^{13}C,^{31}P}$ = 4 Hz), 123.00 (d, $J_{^{13}C,^{31}P}$ = 7.2 Hz), 143.34; ^{31}P NMR 32 MHz (D_2O/ND_4OD) δ −11.03 [1P, J_{P_1,P_2} = 20 Hz, P(1)], −9.02 [1P, P(2)].

*(E)-3,7-Dimethyl-2,6-octadienyl Pyrophosphate (Geranyl Pyrophosphate, **3**)*

The solid obtained from cation exchange of the products from the displacement reaction is dissolved in 3 ml of 0.1 *M* ammonium bicarbonate and transferred to a 25 ml test tube. This solution is then treated with 10 ml of a 1 : 1 (v/v) mixture of isopropanol : acetonitrile. A white precipitate forms after 1–2 min of vigorous vortexing. The precipitate is allowed to settle, and the solution is decanted into a centrifuge tube. The solid is dissolved in 3 ml of 0.1 *M* ammonium bicarbonate, and the extraction is repeated. The remaining solids are dissolved in 2 ml of 0.1 *M* ammonium bicarbonate and extracted with 6 ml of the isopropanol and acetonitrile mixture. The extracts are clarified by centrifugation in a bench top centrifuge and combined. The volume is reduced by rotary evaporation at aspirator vacuum to yield 0.41 g of a thick, pale yellow oil. This material is either stored overnight at −20° or chromatographed directly.

The concentrate obtained from the extraction is dissolved in 2 ml of isopropanol, acetonitrile, and 0.1 *M* ammonium bicarbonate in a ratio of 5 : 2.5 : 2.5 (v/v/v). The pale yellow solution is applied to a 3.5 by 10.5 cm column of cellulose that has been equilibrated in the same solution following the procedure described for **1**. Products are eluted at a flow rate of 30 ml/min. After a 100 ml forerun, 25 20-ml fractions are collected. A 10-μl aliquot of every other fraction is assayed by thin-layer chromatography on cellulose with the same solvent used to elute the column. Those fractions showing a single spot for **3** (R_f = 0.30) are combined. The total volume in which 3 elutes is 200 ml (two column volumes). The volume is reduced to one-third by rotary evaporation at aspirator pressure. The remainder of the organic solvent is removed by rotary evaporation at 4 mm Hg to yield 30 ml of a clear aqueous solution. During the course of the concentration, the pH of the solution is maintained between 7.2 and 7.5 as described for **1**. The aqueous residue is dried by lyophilization to yield 0.30 g (78%) of a white, flocculent solid which is stored over calcium sulfate at −78°; ^{1}H NMR 300 MHz (D_2O/ND_4OH) δ 1.62 (3H, s, methyl), 1.68 (3H, s, methyl), 1.72 (3H, s, methyl), 2.11 [4H, m, methylenes at (C4) and (C5)], 4.47 [2H, dd, $J_{^1H,^1H}$ = 6.5 Hz, $J_{^1H,^{31}P}$ = 6.5 Hz, methylene at (C1)], 5.22 [1H, t, $J_{^1H,^1H}$ = 6.5 Hz, vinyl at (C6)], 5.47 (1H, t, $J_{^1H,^1H}$ = 6.5 Hz, vinyl at (C2)]; ^{13}C NMR 75 MHz (D_2O/ND_4OD) δ 19.69, 27.55, 28.33, 41.55, 65.42 (d, $J_{^{13}C,^{31}P}$ = 4.0 Hz), 122.85 (d, $J_{^{13}C,^{31}P}$ = 7.5 Hz), 127.10, 136.68, 145.76; ^{31}P NMR 32 MHz (D_2O/ND_4OD) δ −11.23 [1P, $J_{P_1P_2}$ = 20 Hz, P(1)], −9.10 [1P, P(2)].

*(E,E)-3,7,11-Trimethyl-2,6,10-dodecatrienyl Pyrophosphate (E,E-Farnesyl Pyrophosphate, **4**)*

The white solid obtained from the cation exchange from the products of the displacement reaction is suspended in 5 ml of 0.1 *M* ammonium bicarbonate and transferred to a 50 ml test tube. This suspension is then treated with 16 ml of a 1 : 1 (v/v) mixture of isopropanol and acetonitrile. A white precipitate forms after rapid vortexing for 1–2 min. The precipitate is allowed to settle, and the clear extract is decanted into a centrifuge tube. The solid is dissolved in 3.0 ml of 0.1 *M* ammonium bicarbonate, and the resulting suspension is extracted with 11 ml of 1 : 1 isopropanol : acetonitrile. The remaining solids are dissolved in 2.0 ml of 0.1 *M* ammonium bicarbonate and extracted with 9 ml of 1 : 1 isopropanol : acetonitrile. The extracts are clarified by centrifugation and combined. The volume is reduced by rotary evaporation at aspirator vacuum to yield 0.27 g of a thick golden oil. This material is either stored at −20° overnight or used directly in the chromatography. Analysis by thin-layer chromatography on cellulose with isopropanol, chloroform, acetonitrile, and 0.1 *M* ammonium bicarbonate in a ratio of 5.5 : 2 : 1 : 1.5 (v/v/v/v) shows a major spot for **4** (R_f = 0.53) and one minor contaminant (R_f = 0.4).

The concentrate obtained from the extraction is dissolved in 2 ml of isopropanol, chloroform, acetonitrile, and 0.1 *M* ammonium bicarbonate in a ratio of 5 : 3 : 1 : 1 (v/v/v/v). This solution is applied directly to a 4.5 by

11 cm column of cellulose that has been equilibrated with the same solution following the procedure described for **11**. Products are eluted at a flow rate of 35 ml/min. After a 300 ml forerun, 25 20-ml fractions are collected. A 10-μl aliquot of every second fraction is analyzed by thin-layer chromatography on cellulose with the same solvent used to elute the column. Those fractions showing a single spot for **4** ($R_f = 0.31$) are combined. The total volume in which **4** elutes is 180 ml (1.1 column volumes). The sample is concentrated as described for **3**. During the concentration step, the pH of the solution is maintained at 7.2 by the addition of a 25 m*M* ammonium bicarbonate solution. When the total volume is reduced to 20 ml, the aqueous residue is lyophilized to yield 0.19 g (72%) of a flocculent white solid. This material is stored over calcium sulfate at $-78°$; ^{1}H NMR 300 MHz (D_2O/ND_4OD) δ 1.58 (3H, s, methyl), 1.60 (3H, s, methyl), 1.65 (3H, s, methyl), 1.71 (3H, s, methyl), 2.06 [8H, m, methylenes at (C4, C5, C8, and C9)], 4.46 [2H, dd, $J_{^1H,^1H} = 6.0$ Hz, $J_{^1H,^{31}P} = 6.0$ Hz, methylene at (C1)], 5.15 [2H, m, vinyls at (C6 and C10)], 5.46 [1H, t, $J_{^1H,^1H} = 6.4$ Hz, vinyl at (C2)]; ^{13}C NMR 75 MHz (D_2O/ND_4OD) δ 18.20, 18.52, 19.90, 27.91, 28.91, 29.13, 42.04, 42.14, 65.17 (d, $J_{^{13}C,^{31}P} = 4.0$ Hz), 122.89 (d, $J_{^{13}C,^{31}P} = 7.5$ Hz), 127.09, 127.43, 134.43, 138.55, 145.30; ^{31}P NMR 32 MHz (D_2O/ND_4OD) δ -11.45 [1P, $J_{P_1,P_2} = 20$ Hz, P(1)], -10.53 [1P, P(2)].

Conclusion

The procedures described in this chapter for synthesis of pyrophosphate esters from primary alcohols offer several advantages. The scale of the reactions and purifications can be varied to produce submilligram to multigram quantities of material without difficulty. The reagents and intermediates can be prepared in bulk and stored at $-20°$ in a desiccator until needed. The purification procedures we report are suitable for the highly reactive allylic systems and yield material that is free of contaminants as judged by TLC and ^{1}H, ^{13}C, and ^{31}P NMR spectroscopy.

Acknowledgments

The work described in this chapter was supported by grants from the National Institutes of Health, GM 25521 and GM 21328.

[16] Eukaryotic Prenyltransferases

By HANS C. RILLING

The 1′-4 prenyl transfer reaction is the polymerizing condensation of polyterpenoid biosynthesis (see ref. 1 for a review). It is a condensation of an allylic pyrophosphate with isopentenyl pyrophosphate, and the products are the next higher homolog of the allylic substrate and inorganic pyrophosphate. There are many different prenyltransferases which produce every size of product from a dimer (geranyl pyrophosphate) to the high molecular weight polymers, gutta percha and rubber. Cis as well as trans isomers may be produced. Bacterial prenyltransferases which synthesize C_{55} polymers are covered elsewhere in this volume by Allen [32]. This chapter will consider the prenyltransferase (dimethylallyltransferase, EC 2.5.1.1) of sterol biosynthesis in eukaryotes. This enzyme condenses either a C_5 or a C_{10} allylic pyrophosphate with the homoallylic pyrophosphate to give, as the ultimate product, farnesyl pyrophosphate which then serves as a substrate for squalene and sterol synthesis.

$$\text{Dimethylallyl pyrophosphate} + \text{isopentenyl pyrophosphate} \xrightarrow{Mg^{2+}} \text{geranyl pyrophosphate}$$
$$\text{Geranyl pyrophosphate} + \text{isopentenyl pyrophosphate} \xrightarrow{Mg^{2+}} \text{farnesyl pyrophosphate}$$

It is also possible that this enzyme produces farnesyl pyrophosphate for dolichyl pyrophosphate and nonaprenyl pyrophosphate (ubiquinone) synthesis. However, it is not clear if it is this or other prenyltransferases that participate in those pathways. A chapter by Barnard on the human liver farnesyl pyrophosphate synthetase is also included in this volume [18].

Assay Procedure

Prenyltransferase has been assayed from the onset by measuring the conversion of the acid-stable substrate (isopentenyl pyrophosphate bearing a radiolabel) to the acid-labile allylic product. There are many minor variations in procedure, but in general, after incubation a mineral acid is used for hydrolysis and the nonpolar products thus formed are extracted into an organic solvent for the determination of radioactivity.

[1] C. D. Poulter and H. C. Rilling, *in* "Biosynthesis of Isoprenoid Compounds" (J. W. Porter and S. L. Spurgeon, eds.), Vol. 1, p. 163. Wiley, New York, 1981.

ISBN 0-12-182010-6

Specifically, in this laboratory the assay mixture contains 10 m*M* HEPES buffer pH 7.0, 1 m*M* $MgCl_2$, 1 m*M* dithiothreitol, 40 μM [1-^{14}C]isopentenyl pyrophosphate and 200 μM geranyl pyrophosphate. The final volume is 50 μl and the incubation is for 10 min at 37°. Substantially lower concentrations of substrates can be used since the K_m values are in the 0.1 μM region.[2] Following the incubation, 0.2 ml of a 4 : 1 methanol : concentrated HCl mixture is added, and the incubation is continued for another 10 min at the same temperature. One milliliter of hexanes is added along with 0.5 ml of water. After thorough mixing with a vortex mixer, 0.5 ml of the organic solvent is transferred to a scintillation vial for counting. When dilution of the enzyme is required, buffer containing 10 m*M* $KHPO_4$ (pH 7.), 1 m*M* EDTA, and 10 m*M* 2-mercaptoethanol is used. This assay is linear to nearly complete substrate conversion. However, when crude extracts or partially purified fractions are analyzed, nonlinearity can be encountered. Radioisotopic isopentenyl pyrophosphate can be obtained from Amersham/Searle or New England Nuclear. It is diluted to a specific activity of 10 μCi/μmol with isopentenyl pyrophosphate prepared as described elsewhere.[3,4]

Units

A unit is defined as 1 nmol product formed min^{-1} at 37°.

Protein Determination

Protein was determined by the biuret method or by absorbance at 280 nm. The homogeneous enzyme from avian liver had an absorbance of 1.0 mg^{-1} ml^{-1} at 280 nm.[5]

Enzyme Purification

This protein has been purified to homogeneity from several eukaryotic sources. In general, the procedures are straightforward and entail extraction, ammonium sulfate fractionation, and chromatography on DEAE and then hydroxylapatite. At this stage, the protein can usually be crystallized.

[2] F. M. Laskovics, J. M. Krafcik, and C. D. Poulter, *J. Biol. Chem.* **254,** 9458 (1979).
[3] D. N. Brems and H. C. Rilling, *Biochemistry* **18,** 860 (1979).
[4] C. M. Allen, this volume [32].
[5] E. Layne, this series, Vol. 3, p. 447.

Enzyme Preparation from Yeast. The enzyme from yeast was partially purified in Lynen's laboratory in 1959.[6] The first purification to homogeneity was by Eberhardt in this laboratory with yeast as the source of the enzyme.[7] All procedures were performed at 4°. Two pounds of yeast were suspended in 500 ml of 10 m*M* phosphate, pH 6.0, and passed through a French press at 18,000 to 20,000 psi. This was a laborious procedure, and rupturing the yeast with a Manton-Gaulin mill or a Bead-Beater (Biospec Products) would undoubtedly suffice. The supernatant obtained after centrifugation (14,000 *g* for 10 min) was diluted to 2.0 liters with 10 m*M* phosphate buffer, pH 6.9, adjusted to pH 7.0 with 7 *N* ammonium hydroxide, and subjected to ammonium sulfate fractionation. The protein precipitating between 50 and 75% saturation of ammonium sulfate was collected by centrifugation (14,000 *g* for 30 min). The pellet thus obtained was dissolved and exhaustively dialyzed against 1 m*M* potassium phosphate buffer pH 6.0. The dialyzate was applied to a column (5 × 31 cm) of DE-52 cellulose (Whatman) previously equilibrated with this buffer. The column was washed with the starting buffer (about 2 liters) until the absorbance of the eluate dropped to less than 0.05. The column was developed with a linear gradient of 1 m*M* potassium phosphate to 20 m*M* phosphate, pH 6.0 (volume = 5 liters). Fractions with a specific activity of 200 or greater were combined and the protein precipitated with ammonium sulfate (75% saturation, pH 7). After collection by centrifugation, the protein was dissolved and dialyzed overnight against several changes of 10 m*M*, potassium phosphate, buffer pH 6.0. The protein was then applied to a column (0.9 × 22 cm) of hydroxylapatite previously equilibrated with this buffer. After a preliminary wash with 100 ml of starting buffer, enzyme was eluted with a linear gradient of 10 m*M* phosphate, pH 6.0, to 125 m*M* phosphate, pH 6.9 (volume = 200 ml). Fractions with specific activities greater than 3200 were combined and collected by ammonium sulfate precipitation. After exhaustive dialysis against 1 m*M* phosphate, pH 6.0, enzyme was incorporated into a sucrose gradient (0 to 40%) with 1% carrier Ampholine (pH 5 to 7). The protein was subjected to electrofocusing in a 115-ml column (LKB 8101) at a constant voltage of 400 V for 46 hr. Active enzyme fractions were pooled and stored as an ammonium sulfate suspension (75% saturation, pH 7) in the presence of 1 m*M* $MgCl_2$ and 20 m*M* mercaptoethanol.

Enzyme Storage and Lability. Enzyme stored in the above fashion was not stable, and the half-life of loss of activity was approximately 15

[6] F. Lynen, B. W. Agranoff, H. Eggerer, U. Henning, and E. M. Moeslein, *Angew. Chem., Int. Ed. Engl.* **71,** 657 (1959).

[7] N. L. Eberhardt and H. C. Rilling, *J. Biol. Chem.* **250,** 863 (1975).

days. Other methods such as storing samples in varying concentrations of glycerol at 0 to −20° were unsuccessful. Freezing enzyme solutions was found to completely abolish the activity. Concentrated enzyme solutions could be stored for a few days at 0°; however, the rate of loss of activity exceeded that of the ammonium sulfate suspensions.

Crystallization. Several preparations yielded crystalline enzyme after chromatography on hydroxylapatite when the protein was dialyzed against 65% saturated ammonium sulfate. The crystals, appearing as long, thin needles under the microscope, were enzymatically active after repeated washings. The crystals were unstable and reverted to amorphous material within several days with a concomitant loss of activity.

Table I gives a synopsis of the purification procedure.

A similar procedure was developed for purifying this enzyme to homogeneity from the fungus *Phycomyces blakesleeanus*.[8] The specific activity of the pure enzyme was equivalent to that from yeast. Since neither of these preparations was especially stable and therefore unsuitable for mechanistic studies, other sources of the enzyme were sought. Chicken liver proved to be suitable and the enzyme has also been purified to homogeneity from chicken livers.

Enzyme Purification from Avian Liver. All procedures were at 4° unless otherwise stated. The standard buffer was 5 m*M* potassium phosphate (pH 7.0) containing 1 m*M* EDTA and 10 m*M* 2-mercaptoethanol. Initially the procedure called for homogenizing the liver in phosphate buffer.[9] However, if fatty chicken livers are used, the pellets obtained on ammonium sulfate fractionation are very loose. To avoid this, the following procedure is now used.[10] Livers (1.5–2 kg) are thawed in 50 m*M* imidazole buffer pH 7.0 containing 10 m*M* 2-mercaptoethanol. Livers are then homogenized in a Waring Blendor in the same buffer using 2.35 liter of buffer per kg liver. After homogenization, 1 *M* $CaCl_2$ is added to 40 m*M* final concentration. The preparation is then centrifuged at 13,000 *g* for 30 min. The supernatant is filtered through cheesecloth and then made 40 m*M* in potassium phosphate, pH 7.0. Ammonium sulfate was added to obtain 36% saturation. After centrifugation at 13,200 *g* for 45 min, ammonium sulfate was added to 57% saturation, and after 15 min, the mixture was centrifuged at 13,200 *g* for 30 min. The pellet was suspended in standard buffer to a final volume of approximately 750 ml. This solution was dialyzed against several 10-liter changes of buffer until the conductivity of the protein solution approached that of the buffer (approximately 1.2 mmho). After centrifugation at 27,000 *g* for 30 min, the protein solu-

[8] H. C. Rilling, unpublished.

[9] B. C. Reed and H. C. Rilling, *Biochemistry* **14,** 50 (1975).

[10] D. N. Brems, E. Bruenger, and H. C. Rilling, *Biochemistry* **20,** 371 (1981).

TABLE I
YEAST PRENYLTRANSFERASE PURIFICATION SCHEME[a]

Step	Total protein (mg)	Specific activity	Total units ($\times 10^{-4}$)	Recovery	Purification (fold)
French press supernatant	5.25×10^4	15.3	80.3	100	
Ammonium sulfate	1.09×10^4	30.8	33.6	41.9	2.01
DEAE-cellulose	250	326	8.40	10.5	21.3
Hydroxylapatite	9.43	3640	3.44	4.3	238
Isoelectric focusing	6.28	5220	3.29	4.07	341

[a] Reproduced from Eberhardt and Rilling,[7] with permission from the *Journal of Biological Chemistry*.

tion was applied to a 7.5 × 30 cm column of DE-52 cellulose previously equilibrated with standard buffer. After washing, the column was developed with a linear gradient of 10 to 60 m*M* potassium phosphate buffer (pH 7.0) containing 1 m*M* EDTA and 10 m*M* 2-mercaptoethanol (4 liter total volume). Fractions of specific activity above 100 were combined and precipitated with ammonium sulfate at 50% saturation. After centrifugation, the pellet was dissolved in 10 ml of 10 m*M* Tris, pH 7.0, containing 10 m*M* 2-mercaptoethanol and dialyzed against the same buffer. The protein solution was applied to a 2.5 × 27 cm hydroxylapatite column which was washed with two column volumes of 6 m*M* potassium phosphate buffer. The enzyme was eluted with a 1-liter gradient of 6–100 m*M* potassium phosphate containing 10 m*M* 2-mercaptoethanol, pH 7.0. Solid ammonium sulfate was added to 50% saturation and the solution centrifuged. The pellet thus obtained was extracted successively with 5 ml each of 40, 35, and 30% saturated ammonium sulfate in 0.1 *M* potassium phosphate, pH 7.0, containing 1 m*M* EDTA and 2 m*M* dithiothreitol. The protein in the extracts crystallized on standing at room temperature; the main crop of crystals appeared in the 35% extract. The procedure is summarized in Table II. After crystallization, the enzyme crystals were suspended in saturated neutral ammonium sulfate and stored at 4°. As a crystalline suspension, the enzyme is stable for as long as a year. The enzyme has also been stored free of ammonium sulfate at −20° in the absence of thiols. If thiols are included during freezing, the protein upon thawing is an inactive white precipitate.

Interconvertible Forms of Prenyltransferase

Several tissues have yielded separable prenyltransferases that are apparently very similar in chemical properties as well as substrate and prod-

TABLE II
PRENYLTRANSFERASE FROM AVIAN LIVER[a]

Step	Total protein	Specific activity	Units	Recovery (%)	Purification (fold)
Extract	306 g	2	6×10^5	—	—
Ammonium sulfate	63 g	6.6	4×10^5	66	3
DE-52	1.6 g	210	3.4×10^5	56	100
Hydroxylapatite	180 mg	1400	2.6×10^5	43	700
First crystals	50 mg	1500	7.5×10^4	13	750

[a] Typical preparation from 2 kg liver.

uct specificity. Ogura and collaborators found this with the enzyme from pig liver, and they postulated that these forms resulted from oxidation and reduction of disulfide bonds.[11] Popják and co-workers found a similar phenomenon.[12] In any event, it has been possible to purify this enzyme to homogeneity from another mammalian source.[12,13]

Preparation of Prenyltransferase from Pig Liver.[14] Pig livers were obtained fresh at the slaughterhouse, iced immediately, and then frozen within an hour. Frozen liver, 1500 g, was cut into small pieces and then homogenized in 1 vol of cold, distilled water in a large Waring Blendor. The pH of the homogenate was adjusted to 5.5 with 3.5 *N* acetic acid, and 2-mercaptoethanol was added to a final concentration of 10 m*M*. The thick slurry was centrifuged for 30 min at 10,000 rpm and the clear supernatant decanted. The pH of the supernatant was adjusted to 6.0 with alkaline DE-52 cellulose which had been prepared by adjusting the pH of a slurry of DE-52 to approximately 10, after which the gel was collected on a sintered glass funnel and thoroughly washed with deionized water. Usually 100 g of damp gel was required for adjusting the pH. DE-52 cellulose, 600 g wet, previously equilibrated to pH 6.0 with 10 m*M* potassium acetate, was added to the preparation, which was then stirred for 30 min. After the gel had settled, the supernate was assayed for prenyltransferase. If significant levels of prenyltransferase were detected in the supernate, additional DE-52 cellulose was added and the mixture was stirred for 30 min, after which the supernate was again assayed to assure that the enzyme had been adsorbed to the gel. The gel was then collected

[11] T. Koyama, Y. Saito, K. Ogura, and S. Seto, *J. Biochem.* (*Tokyo*) **82,** 1585 (1977).

[12] G. F. Barnard, B. Langton, and G. Popják, *Biochem. Biophys. Res. Commun.* **85,** 1097 (1978).

[13] G. F. Barnard, this volume [18].

[14] L.-S. Yeh and H. C. Rilling, *Arch. Biochem. Biophys.* **183,** 718 (1977).

on a sintered glass funnel and washed with 10 m*M* potassium acetate, pH 6.0, containing 10 m*M* 2-mercaptoethanol, until the washes were colorless. The gel was then transferred as slurry in the same buffer to a 5.0-cm-diameter column which already contained approximately 8 cm of DE-52 equilibrated to pH 6.0 with the acetate buffer. When the column of DE-52 cellulose was packed, the height of the gel was approximately 50 cm. The column was eluted with a linear gradient of 0 to 100 m*M* ammonium sulfate in 10 m*M* potassium acetate, pH 6.0, containing 10 m*M* mercaptoethanol, at a rate of 240 ml/hr. The total volume of the gradient was 4 liters.

Active fractions, specific activity greater than 16, were combined and concentrated on a Diaflo PM-30 (Amicon Corp.) membrane. Frequently we observed a 2-fold increase in apparent specific activity on combining and concentrating fractions obtained by chromatography. The protein solution was then dialyzed against 10 m*M* Hepes buffer, pH 7.8, containing 10 m*M* mercaptoethanol, and applied to a 2.5 × 40-cm column of DE-52 cellulose equilibrated to pH 7.8 with the same buffer. The column was washed with 300 ml of this buffer containing 20 m*M* ammonium sulfate, and then a linear gradient of 20 to 150 m*M* ammonium sulfate in the same buffer was applied. The total volume of the gradient was 3 liters and 20-ml fractions were collected. The enzyme activity eluted as a single symmetrical peak from this column. Fractions of specific activity greater than 50 were combined and concentrated as before.

The enzyme was dialyzed against 10 m*M* TES buffer, pH 7.0, containing 10 m*M* mercaptoethanol, and was then applied to a 2.5 × 20-cm column of hydroxylapatite previously equilibrated with the same buffer. The protein was eluted with a linear gradient (450-ml total volume) of 0 to 100 m*M* potassium phosphate in 10 m*M* TES, pH 7.0, containing 10 m*M* mercaptoethanol. Prenyltransferase activity eluted with the first band of protein to emerge from the column. Individual fractions were analyzed by electrophoresis in alkaline gels and for specific activity. Fractions of high specific activity containing a single protein were combined and concentrated on a Diaflo PM 30 membrane. The protein could be crystallized from ammonium sulfate solution at this stage, but the yields were never high, nor was there an increase in specific activity. The enzyme was stored as a suspension in 60% ammonium sulfate. The purification scheme is summarized in Table III.

Properties of Prenyltransferase

All of the eukaryotic farnesyl pyrophosphate synthetases are dimeric proteins with a molecular weight of about 80,000. Electrophoresis has

TABLE III
PIG LIVER PRENYLTRANSFERASE PURIFICATION SCHEME[a]

Fraction	Protein (mg) (× 10^{-4})	Units (× 10^{-3})	Specific activity	Yield (%)	Purification[b] (*n*-fold)
Acid, supernatant[c]	6.4×10^4	19	3	100	
DE-52, pH 6.0	4.1×10^3	15	37	79	12
DE-52, pH 7.8	6.3×10^2	7.4	117	39	39
Hydroxylapatite	55	5	914	26	305

[a] Reproduced with permission of Academic Press.[14]
[b] Calculated from the acid supernatant.
[c] Preliminary experiments showed that this step removed half of the protein without loss of enzyme units. The crude homogenate was not routinely assayed for enzyme activity or protein.

shown the subunits to be of the same molecular weight in all instances. In a series of binding studies, Reed[15] demonstrated that the avian liver enzyme was comprised of two subunits of identical binding properties. Brems,[10] working with the same protein, derivatized the catalytic site with a photoreactive substrate analog. Both subunits were modified and the CNBr fragment that bore the labeled analog was successfully sequenced.[10] Thus, all evidence indicates that the subunits are identical.

The specific activities of the homogeneous prenyltransferases vary from 900 for the pig liver enzyme to 5200 for the enzyme from yeast. All of these enzymes accept geranyl pyrophosphate as well as dimethylallylpyrophosphate as substrates, with Michaelis constants between 0.1 and 1 μM for all substrates. Farnesylpyrophosphate was also shown to be a substrate for the avian liver enzyme as well as the yeast enzyme, but geranyl pyrophosphate was formed at only a few percent of the rate of farnesylpyrophosphate synthesis. The farnesylpyrophosphate synthetase from *Phycomyces blakesleeanus,* an organism that synthesizes carotenoids, does not synthesize geranylgeranylpyrophosphate from farneslypyrophosphate. This indicates that there must be a different prenyltransferase for carotene synthesis.

Acknowledgments

This research was supported in part by a grant (AM 13140) from the National Institutes of Health. It is a pleasure to recognize the contribution of Drs. Eberhardt, Reed, Yeh, and Brems, as well as that of E. Bruenger.

[15] B. C. Reed and H. C. Rilling, *Biochemistry* **15,** 3739 (1976).

[17] Enzymatic Hydrolysis of Polyprenyl Pyrophosphates

By TANETOSHI KOYAMA, HIROSHI FUJII, and KYOZO OGURA

In the structural study of prenyl pyrophosphates formed by prenyltransferase reaction, it is important to hydrolyze the pyrophosphate esters to the corresponding prenols. Although short-chain prenyl pyrophosphates such as farnesyl and geranylgeranyl pyrophosphate are easily cleaved with alkaline phosphatase, polyprenyl pyrophosphates with a carbon chain longer than C_{20} are resistant to the usual phosphatase treatment.[1] Efficient enzymatic cleavage of such polyprenyl pyrophosphates to the corresponding primary alcohols is achieved when they are treated with potato acid phosphatase in the presence of a large amount of methanol.[2]

Reagents

Acetate buffer, 1.0 *M*, pH 5.6
Triton X-100, 1% (w/v)
Acid phosphatase from potato (approx. 2 units/mg of protein, grade II, Boehringer), 22 mg/ml
Methanol

Procedure. The polyprenyl pyrophosphates (0.001 ~ 4 μmol) formed by the action of solanesylpyrophosphate synthetase are extracted into 1-butanol (more than 3 times) from the biosynthetic incubation mixture. The 1-butanol extracts are combined and washed with water. Then the solvent is removed on a rotary evaporator at less than 40° until the total volume is ~1 ml. About 2 ml of methanol is added to the concentrated 1-butanol extracts to make the final volume of the solution 3 ml. To the resultant solution are added 0.5 ml of 1% (w/v) Triton X-100, 0.5 ml of 1.0 *M* acetate buffer, pH 5.6, and 1 ml of the potato acid phosphatase solution (44 units). The mixture is incubated at 37° for more than 12 hr. The polyprenols, octaprenol and nonaprenol, liberated are extracted with pentane, and the extract is then washed with water.

Applicability. Solanesylpyrophosphate synthetase catalyzes the synthesis of all-*trans*-polyprenyl pyrophosphates with carbon chains of C_{25}, C_{30}, C_{35}, C_{40}, and C_{45} in a ratio variable depending on the conditions of

[1] C. M. Allen, W. Alworth, A. Macre, and K. Bloch, *J. Biol. Chem.* **242,** 1895 (1967).
[2] H. Fujii, T. Koyama, and K. Ogura, *Biochim. Biophys. Acta* **712,** 716 (1982).

METHODS IN ENZYMOLOGY, VOL. 110

ISBN 0-12-182010-6

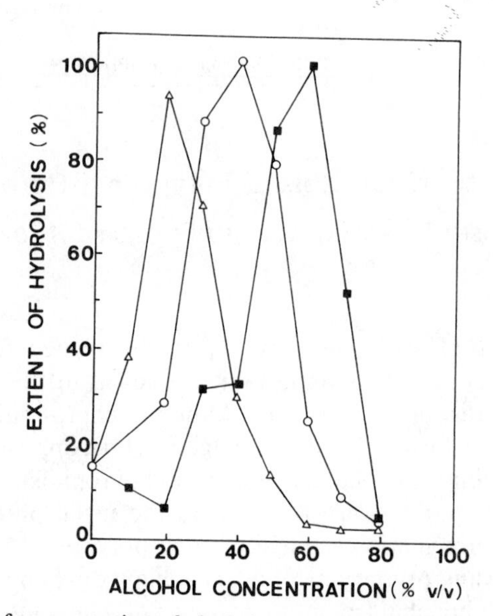

FIG. 1. Effect of concentration of alcohol on the hydrolysis of polyprenyl pyrophosphates. The indicated amount of alcohol is added to the reaction mixture containing 0.1% Triton X-100, 22 units of acid phosphatase and ^{14}C-labeled polyprenyl pyrophosphates. Alcohols added: methanol, ■; ethanol, ○; 1-propanol, △.

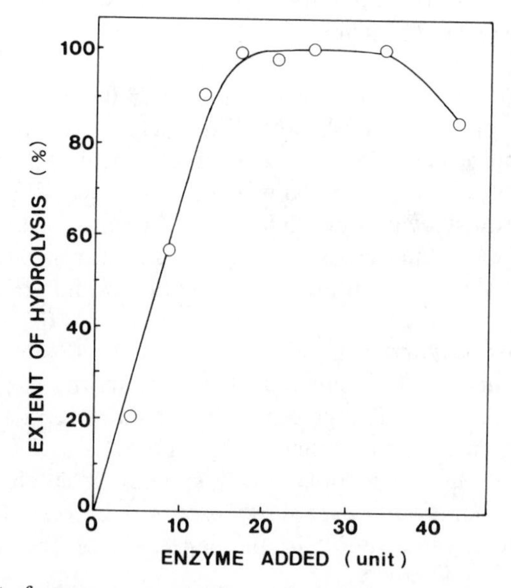

FIG. 2. Effect of enzyme concentration on hydrolysis. The indicated amount of acid phosphatase is added to the reaction mixture containing 0.1% Triton X-100, 60% methanol, and ^{14}C-labeled polyprenyl pyrophosphates.

incubation.[3,4] This method cleaves all of these products effectively. The method is also applicable for the hydrolysis of the products of hexaprenylpyrophosphate synthetase,[4] heptaprenylpyrophosphate synthetase,[5] decaprenylpyrophosphate synthetase,[6] and undecaprenylpyrophosphate synthetase[7] reactions. The corresponding monophosphate esters are also hydrolyzed by this method.

Remarks

1. As the acid phosphatase is strongly inhibited by fluorides, contamination with fluorides should be prevented. When polyprenyl pyrophosphate are extracted from an incubation mixture containing fluoride ions, the 1-butanol solution should be washed thoroughly with water to remove the fluorides.
2. Ethanol (40%, v/v) and 1-propanol (20%, v/v) are also effective instead of 60% (v/v) methanol (Fig. 1).
3. Since a large amount of the phosphatase is used to complete the hydrolysis (Fig. 2), a considerable amount of protein will precipitate during a long incubation.
4. In some cases Triton X-100 can be omitted.

[3] H. Fujii, H. Sagami, T. Koyama, K. Ogura, S. Seto, T. Baba, and C. M. Allen, *Biochem. Biophys. Res. Commun.* **96,** 1648 (1980).
[4] H. Fujii, T. Koyama, and K. Ogura, *J. Biol. Chem.* **257,** 14610 (1982).
[5] I. Takahashi, K. Ogura, and S. Seto, *J. Biol. Chem.* **255,** 4539 (1980).
[6] K. Ishii, H. Sagami, and K. Ogura, *Biochem. Biophys. Res. Commun.* **116,** 500 (1983).
[7] I. Takahashi and K. Ogura, *J. Biochem. (Tokyo)* **92,** 1527 (1982).

[18] Prenyltransferase from Human Liver

By GRAHAM F. BARNARD

Introduction[1]

Prenyltransferase (dimethylallyltransferase, EC 2.5.1.1; dimethylallyl diphosphate : isopentenyl diphosphate dimethylallyltransferase) catalyzes two sequential, irreversible 1′–4 condensations of isopentenyl pyrophos-

[1] Abbreviations: IPP, isopentenylpyrophosphate (= 3 methylbut-3-en-1-yl pyrophosphate); DMAPP, 3,3-dimethylallylpyrophosphate; GPP, *trans*-geranylpyrophosphate; FPP, *trans,trans*-farnesylpyrophosphate; PP_i, inorganic pyrophosphate ($HOP_2O_6^{3-}$); -PP, pyrophosphoryl moiety; TES, *N*-tris(hydroxymethyl)methyl-2-aminoethanesulfonic acid; SDS, sodium dodecyl sulfate.

ISBN 0-12-182010-6

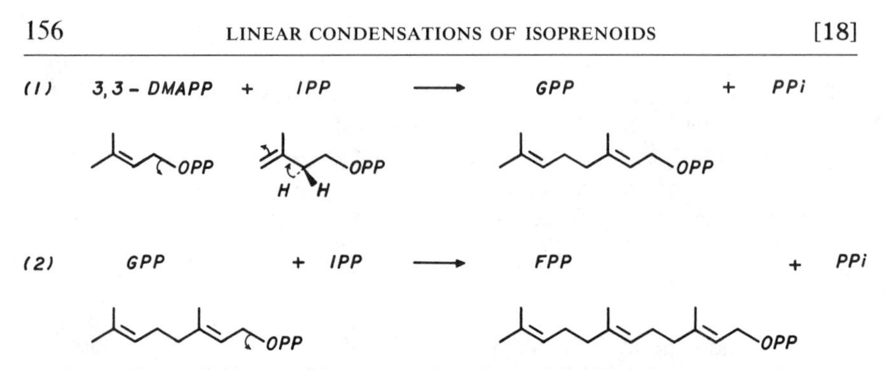

FIG. 1. Sequential irreversible 1′–4 condensations of dimethylallyl pyrophosphate.

phate first with dimethylallyl pyrophosphate and then with the resultant geranylpyrophosphate to produce the C_{15} *trans,trans*-farnesyl pyrophosphate [Eqs. (1) and (2) in Fig. 1]. These bimolecular nucleophilic substitution (S_N2) reactions in the pig liver enzyme proceed with an inversion of configuration at the pyrophosphate-bearing carbon of the allylic substrate and removal of the pro-*R* hydrogen from the C-2 of IPP in these *trans*-polyprenol synthetic reactions.[2]

Prenyltransferase has been purified to homogeneity from yeast,[3] pig liver,[4,5] and in crystalline form from chicken liver.[6] The homogeneous human liver enzyme, the first sterol synthesizing enzyme to be purified from a human tissue, has properties distinct from the enzyme from these sources.[7]

Enzymatic Assay

Principle. The radioassay of prenyltransferase activity with ^{14}C-labeled IPP of known specific activity as radiolabeled substrate is based on the hydrolysis of the allylic products in acid at 37°, the extraction of the resultant alcohols into light petroleum, and their quantitative estimation by liquid scintillation counting.[8] For example, the allylic pyrophosphates dimethylallyl pyrophosphate, geranyl pyrophosphate, and farnesyl pyrophosphate will be hydrolyzed to alcohols derived from their tertiary and primary carbonium ion intermediates in a ratio of about 4 : 1, DMAPP →

[2] J. W. Cornforth, R. H. Cornforth, G. Popják, and L. Yengoyan, *J. Biol. Chem.* **241,** 3970 (1966).

[3] N. L. Eberhardt and H. C. Rilling, *J. Biol. Chem.* **250,** 863 (1975).

[4] L-S. Yeh and H. C. Rilling, *Arch. Biochem. Biophys.* **183,** 718 (1977).

[5] G. F. Barnard and G. Popják, *Biochim. Biophys. Acta* **617,** 169 (1980).

[6] B. C. Reed and H. C. Rilling, *Biochemistry* **14,** 50 (1975).

[7] G. F. Barnard and G. Popják, *Biochim. Biophys. Acta* **661,** 87 (1981).

[8] G. Popják, this series, Vol. 15 [12].

methylvinyl carbinol + dimethylallyl alcohol; FPP → nerolidol + farnesol.

Geranyl-PP is customarily used as the allylic substrate. [1-^{14}C]Isopentenyl-PP is used as the homoallylic substrate and it is stable in acid at 37°. Incidentally, any produced ^{14}C-labeled isopentenol would not be extracted from an aqueous-ethanolic solution by light petroleum.

When assaying crude enzyme preparations it is important to correct for contaminating isopentenylpyrophosphate isomerase activity by subtracting the value obtained with control incubations which contain only ^{14}C-labeled IPP and no allylic substrate. The isomerase catalyzes the interconversion of IPP with DMAPP, therefore any ^{14}C-labeled IPP can be converted to the acid-hydrolyzable DMAPP without the mediation of prenyltransferase.

Assay Reagents. The stock solutions (stored frozen) are

Tris–HCl buffer, pH 7.8, 100 m*M*
$MgCl_2 \cdot 6H_2O$, 100 m*M*
Dithiothreitol, 100 m*M*
Geranyl-PP, 0.010 m*M*
(Dimethylallyl-PP), 0.010 m*M*
[1-^{14}C]Isopentenyl-PP (56 μCi/μmol), 0.010 m*M*

Substrates.[9] GPP is prepared by the phosphorylation of geraniol (pure trans isomer, "Meranol" from Bush, Boake & Allen, London, England) as described by Holloway and Popják,[10] and purified as described by Parker *et al.*[11] DMAPP is prepared by phosphorylation of 3,3-dimethylallyl alcohol by the method of Cramer and Bohm[12] as modified by Popják *et al.*[13] IPP is synthesized in a 3-step sequence: isopentenol → isopentenyl monophosphate → isopentenyl phosphomorpholidate → IPP as detailed by Donninger and Popják.[14] The product is isolated and converted to the triammonium salt as detailed by Parker *et al.*[11] [1-^{14}C]IPP (3-methyl [1-^{14}C]but-3-en-1-yl pyrophosphate) of specific activity 56 μCi/μmol is obtainable from the Radiochemical Centre, Amersham, Bucks, UK. For routine assays using 16 μM substrates a sample of [1-^{14}C]IPP of specific activity 0.5 μCi/μmol is prepared by admixture of the radioactive and nonradioactive samples. Purity of the substrates can be verified by thin

[9] R. H. Cornforth and G. Popják, this series, Vol. 15 [11].
[10] P. W. Holloway and G. Popják, *Biochem. J.* **104,** 57 (1967).
[11] T. S. Parker, G. Popják, K. Sutherland, and S.-M. Wong, *Biochim. Biophys. Acta* **530,** 24 (1978).
[12] F. Cramer and W. Bohm, *Angew. Chem.* **71,** 775 (1959).
[13] G. Popják, J. W. Cornforth, R. H. Cornforth, R. Rhyage, and D. S. Goodman, *J. Biol. Chem.* **236,** 56 (1962).
[14] C. Donninger and G. Popják, *Biochem. J.* **105,** 545 (1967).

layer chromatography on silica gel H plates 250 μm thick developed with 1-propanol–conc. ammonium hydroxide–water (6 : 3 : 1 v/v). The substances were visualized by exposure to I_2 vapor, followed by spraying with a molybdate solution in 10% H_2SO_4 and then ascorbic acid[15] to reveal phosphorylated compounds. The R_f IPP = 0.15–0.20.

Assay Procedure.[5,7,10] Routine assays to locate prenyltransferase activity during purification contained 16 μ*M* substrates and ^{14}C-labeled IPP of specific activity 0.5 μCi/μmol. Studies with the purified enzyme used 0.5 μ*M* substrates and ^{14}C-labeled IPP of specific activity 56 μCi/μmol. The latter method will be described as illustration.

Incubations of 0.5 ml final volume are performed at 37° for 10 min. They contain 10 m*M* Tris–HCl buffer, pH 7.8, 2 m*M* $MgCl_2$, 2 m*M* dithiothreitol, 0.2 n*M* purified prenyltransferase (15.5 ng/ml, ~8 mU), 0.5 μ*M* ^{14}C-labeled IPP (56 μCi/μmol), 0.5 μ*M* GPP. A mixture containing Tris–HCl, Mg, dithiothreitol, and IPP sufficient for 20 assays is prepared fresh and then aliquots are added to prewarmed tubes containing prenyltransferase and GPP to start the reaction. To ensure linear product formation with time the assays are terminated when not more than 25% of the substrates have been consumed, by the addition of 1 ml of 2 *N* HCl in 80% ethanol and 0.1 ml of a carrier prenol solution (1 mg each of geraniol and farnesol in ethanol). After incubation at 37° for 10 min the mixtures are neutralized with 0.7 ml of 10% NaOH and the prenols are extracted with 1 ml of light petroleum (bp 40–60°). The single extraction facilitates multiple assays, it consistently removes 85% of the ether extractable counts (as determined by repeat extractions), and there is no necessity for centrifugation to separate the ether and aqueous phases. For routine assays the value obtained in the first extraction is used in the calculation of enzymic activity without correction for incomplete extraction. ^{14}C radiolabel content is measured by directly adding the light petroleum extract to 10 ml of RPI-3a70B scintillation fluid (Research Products International, Elk Grove, IL). Samples were counted in a Packard Tri-Carb spectrometer model 3320 at an efficiency of about 70% as determined by recounting in the presence of an internal ^{14}C-labeled standard. Control incubations contain either no enzyme or no allylic substrate. Activity is expressed as nmol allylic products produced/min/mg protein after correcting for contaminating isomerase activity by substraction as discussed above.

Units. One unit of prenyltransferase activity catalyzes the conversion of one nmol of ^{14}C-labeled IPP to farnesyl pyrophosphate/min in the pres-

[15] C. S. Hanes and F. A. Isherwood, *Nature* (*London*) **164,** 1107 (1949).

ence of GPP (or DMAPP). The specific activity of the enzyme is units/mg protein.

Purification[7]

The method is modified from those used previously for the porcine enzyme[4,5] and utilizes a hydrophobic chromatography step.[7]

First appropriate legal permission must be obtained for the use of human liver obtained at autopsy. Donors who had not known active or past liver disease are most suitable and samples of liver should be obtained preferably not later than 5 hr postmortem; the liver can be frozen at −70° with solid CO_2 and stored frozen until needed.

Buffers

Buffer A: 10 m*M* sodium acetate buffer, pH 6.0, + 10 m*M* 2-mercaptoethanol

Buffer B: 10 m*M* sodium acetate buffer, pH 5.0, + 5 m*M* 2-mercaptoethanol

Buffer C: 10 m*M* potassium phosphate buffer, pH 7.2, + 10 m*M* 2-mercaptoethanol

Buffer D: 60 m*M* potassium phosphate buffer, pH 7.2, + 10 m*M* 2-mercaptoethanol

Buffer E: 10 m*M* TES-KOH buffer, pH 7.0 + 10 m*M* 2-mercaptoethanol

Buffer F: 90 m*M* potassium phosphate buffer, pH 7.0, + 10 m*M* TES-KOH, pH 7.0, and 10 m*M* 2-mercaptoethanol

Step 1. Crude Extract. The frozen liver sample is broken into small pieces and 100 *g* batches are homogenized in 200 ml of 10 m*M* 2-mercaptoethanol (initially at 22°) for 30 sec in a Waring blender. All subsequent operations are at 4°. The pH of the stirred homogenate is adjusted from 7.1 to 5.2 with 3.5 *N* acetic acid. The homogenate is stirred for another hour and then centrifuged at 3000 rpm for 45 min. The supernatent is decanted through glass wool to remove floating lipid and is kept. The pellet is resuspended in one third the volume of the original homogenate in 10 m*M* 2-mercaptoethanol, pH 5.2 and the suspension is recentrifuged. The second supernatant is also passed through glass wool and combined with the first supernatant.

Step 2. Ammonium Sulfate Fractionation. Powdered ammonium sulfate is added with stirring over 30 min to the pooled supernatant from step 1 to 47.5% saturation (295 g/liter). The suspension is stirred for 1 hr and the precipitate then sedimented at 16,000 *g* for 45 min. The supernatant is discarded and the precipitate resuspended in buffer A and then dialyzed overnight against the same buffer. Small amounts of precipitated protein which do not go back into solution upon dialysis have low specific activity and can be removed by centrifugation of the solution at 16,000 *g* for 30 min.

Step 3. Batch Treatment with Carboxymethyl Cellulose (CM-52). Prenyltransferase remains unbound to CM-52 cellulose in this step. The pH of the dialysed preparation from step 2 (32 mg protein/ml and conductivity 0.87 mmho) is adjusted to 5.0 by the addition with stirring, of 3.5 *N* acetic acid. Whatman CM-52 cellulose (375 mg dry resin/30 mg protein) is equilibrated with buffer B and excess buffer removed by decantation after brief centrifugation at 1000 rpm. The equilibrated resin is added to the enzyme preparation with stirring and after 15 min the suspension is centrifuged at 6000 rpm for 10 min and the supernatant decanted and kept. The resin is washed twice with buffer B by resuspension and centrifugation; the supernatants from each centrifugation are pooled and the pH adjusted to 6.0 with 1 *N* NaOH.

Step 4. DEAE-Cellulose Column Chromatography. DEAE-cellulose (Whatman DE-52, 167 mg wet weight/10 mg protein, equilibrated with buffer A) is added to the enzyme sample from step 3 (0.75 mmho conductivity and 3.3 mg protein/ml). The suspension is stirred for 15 min. After the resin has settled the supernatant can be discarded if an assay verifies that approximately 95% of the enzyme units are bound to the resin. The resin is washed three times with buffer A by resuspension and decantation and then it is layered onto a 1.6 × 75-cm column of fresh DEAE-cellulose equilibrated with buffer A (final height 87 cm). The column is washed with 20 ml of equilibrating buffer A and then a linear gradient of 1800 ml total volume from 0 to 140 m*M* ammonium sulfate in the same buffer is applied at a flow rate of 1.5 ml/min. Fractions of 14 ml are collected and assayed for enzymatic activity, conductivity, and protein content (by absorbance at 280/260 nm). Prenyltransferase elutes in a single Gaussian peak centered at 5.3 mmho conductivity. The enzyme fractions are pooled and concentrated to about 5 ml by pressure filtration through a PM-10 membrane (Amicon Corp., Lexington MA) at 50 lb/square inch. This solution is then dialyzed against buffer C.

Step 5. Butyl-agarose Column Chromatography. The transferase is retarded but not bound by butyl-agarose at pH 7.2. Longer chain alkyl-agaroses (C_5–C_{10}) bind prenyltransferase increasingly strongly but with progressively poorer recovery of enzymic activity. A 1 × 60-cm column of butyl-agarose (Sigma Chemical Co., St. Louis, MO) is equilibrated with buffer C and the protein sample from step 4 applied to the top. The column is washed with 80 ml of buffer C and then an 800 ml total volume

PURIFICATION OF HUMAN LIVER PRENYLTRANSFERASE[a]

Purification step	Volume (ml)	Total enzyme units[b]	Total protein (mg)	Specific activity[c]	Degree of purification	Yield (%)
1. Supernatant of homogenate[d]	3740	26910	52950	0.5	2.8	100
2. $(NH_4)_2SO_4$ precipitate	238	19515	7567	2.6	14.3	73
3. CM-52 supernatant	429	16961	1650	10.3	57.0	63
4. DE-52 column	40	12748	232	54.9	305.0	47
5. Butyl-agarose column	11.2	9690	44	220.0	1215.0	36
6. Calcium phosphate column	3.6	7373	8	922.0	5122.0	27

[a] Data from G. F. Barnard and G. Popják, *Biochim. Biophys. Acta* **661,** 87 (1981).

[b] Enzyme activity is corrected for contamination by isopentenylpyrophosphate isomerase (less than 2% after the first column chromatography).

[c] nmol FPP synthesized from GPP + IPP, nmol/min/mg protein; that of whole liver homogenate before centrifugation was 0.18.

[d] Prepared from 1.32 kg of liver.

linear gradient of buffer C–buffer D is applied. Fractions of 8 ml are collected at a flow rate of 1 ml/min and assayed for enzymatic activity, conductivity, and protein content. Those containing prenyltransferase are pooled and concentrated to about 6 ml by pressure filtration as above.

Step 6. Calcium Phosphate Column Chromatography. A 1.5 × 85-cm column of freshly prepared calcium phosphate gel[16] is equilibrated with buffer F. The enzyme sample from step 5 is applied to the column which is then washed with 200 ml of buffer F. A 1600 ml total volume linear gradient of buffer E–buffer F is then applied and fractions of 13 ml are collected at a flow rate of 1.5 ml/min and are assayed for enzymatic activity, conductivity, and protein content. The pooled fractions of purified enzyme are stored at 4° as a precipitate in ammonium sulfate solution at 60% saturation (390 g/liter). The enzyme in this form is quite stable, 30–50% of the original activity remained after 12 months. Typical results of purification are shown in the table.

Properties

Purity. SDS–polyacrylamide gel electrophoresis is performed as described by Weber *et al.*[17] using 10% gels at pH 7.2 and with a constant

[16] C. K. Mathews, F. Brown, and S. S. Cohen, *J. Biol. Chem.* **239,** 2957 (1964).

[17] K. Weber, J. R. Pringle, and M. Osborn, this series, Vol. 26 [1].

current of 8 mA/tube for 4 hr. Gels are stained for 3 hr with Coomassie brilliant blue R-250 (1.25 g + 227 ml methanol + 40 ml glacial acetic acid + 229 ml water, filtered). They are destained with a solution containing 50 ml methanol, and 75 ml glacial acetic acid per liter water and AG 50-1X8 resin. The purified enzyme gives a single sharply stained protein band whether the gel is lightly or heavily loaded, i.e., from 4 to 20 μg. For the native enzyme, 10% polyacrylamide gels polymerized with ammonium persulfate are made as described by Maurer[18] (gel system 3) and preelectrophoresed overnight at 1 mA/tube. After application of the experimental samples electrophoresis is at 4 mA/tube for 3 hr. Gels are stained with Coomassie G-250 prepared fresh by adding 1–2 ml of 1% G-250 in water to 50 ml of 12.5% trichloroacetic acid, filtering the suspension, and staining the gels with the filtered solution for 2 hr. Gels are then washed in water and destained in 7% acetic acid until the background is clear. The single stained protein band of the native purified preparation is somewhat diffuse, but it coincides precisely with the transferase activity in duplicate unstained gels which are sliced and assayed in the usual assay medium.

Molecular Weight.[7] The mean subunit molecular weight of human liver prenyltransferase is 38,400 ± 1220 as determined by polyacrylamide gel electrophoresis using two sets of molecular weight markers: (1) polymers of a cross-linked monomer M_r 14,300–57,700; (2) a set of natural protein standards (Sigma, SDS-6). An M_r = 74,000 ± 1400 is obtained for the native enzyme by gel exclusion chromatography on a Sephacryl S-200 column using glucose-6-phosphate dehydrogenase, bovine serum albumin, and hen ovalbumin as standards, and is consistent with a dimeric protein structure.

Amino Acid Composition.[7] The amino acid composition of reduced and carboxymethylated human prenyltransferase following hydrolysis with 3.5 *N* mercaptoethanesulfonic acid and correcting for decomposition of each amino acid, in residues/subunit is Cys(CM) 5.8, Asx 32.8, Thr 10.0, Ser 10.3, Glx 45.2, Pro 12.0, Gly 23.7, Ala 27.2, Val 20.9, Met 6.4, Ile 16.5, Leu 37.4, Tyr 17.4, Phe 14.2, Trp 2.0, Lys 21.8, His 4.4, Arg 17.5. This composition with a high content of hydrophobic residues and few half-cysteines is similar to that of the pig liver prenyltransferase.[5]

pH Optimum and Effects of Ionic Strength on Enzymatic Activity.[7] Of various buffers tested, activity is maximal with Tris–HCl. A broad pH optimum from 7.3 to 8.8 is obtained. Both phosphate buffers and high ionic strength buffers are reversibly inhibitory. For example, activity in 100 m*M* Tris–HCl buffer, pH 7.8 or in 100 m*M* KCl/10 m*M* Tris–HCl

[18] H. R. Maurer, *in* "Disc Electrophoresis and Related Techniques of Polyacrylamide Gel Electrophoresis," p. 50. W. de Gruyter Berlin, 1971.

buffer, pH 7.8 is 76% of that obtained in 10 mM buffer alone; in 250 mM buffer or KCl it is 44% and in 500 mM buffer or KCl in dilute buffer activity is only about 10% of the maximum in 10 mM Tris–HCl buffer alone. Routine assays are performed in 10 mM Tris–HCl buffer, pH 7.8.

Divalent Metal Requirement.[7] Human liver prenyltransferase has an absolute requirement for divalent magnesium or manganese ions for activity. Not more than background activity (equivalent to 0.7% of the activity seen with 2 mM Mg^{2+}) is observed in the absence of divalent metals. No activity above background is present with up to 20 mM $CaCl_2$. Although magnesium and manganese ions give equal maximum activity, Mn^{2+} has a much higher affinity than Mg^{2+} for the enzyme, with a half-maximal activation at 3.7 μM for Mn^{2+} vs 89.0 μM for Mg^{2+}.

Thiol Requirement. The effect of thiol-reducing agents on human prenyltransferase is intriguing albeit not completely understood. In terms of activity thiol-reducing agents are essential. If the purified enzyme is dialyzed against Tris–HCl buffer, pH 7.8 assayable activity becomes completely dependent on the presence of thiol-reducing agents, and is proportional to the logarithm of the thiol concentration up to a maximum activation. Dithiothreitol produced full activation of the enzyme in less than 3 min and was the most effective thiol-reducing agent tested with a half-maximal activation at 0.48 mM; dithioerythritol (0.63 mM), L-cysteine (6.7 mM), glutathione (12.0 mM), and 2-mercaptoethanol (13.9 mM) give a similar activation at higher (cited) concentrations.[7] Figure 2A, a double reciprocal plot of 1/enzymatic activity vs 1/thiol concentration emphasizes this higher affinity for dithiols and the fact that a similar maximum activity is achieved by the various reducing agents.

In terms of enzyme stability thiol-reducing agents are detrimental as shown in Fig. 2B. If the purified enzyme as an ammonium sulfate precipitate is dialyzed against 10 mM Tris–HCl buffer, pH 7.3 and kept at 4° the activity loss is about 5% over 24 hr (curve 1). If the suspension is simply diluted in buffer and assayed in the absence of thiols then the thiol, presumably carried over from the purification procedure, supports a reduced but stable enzymatic activity (curve 2). If dithiothreitol is present then 95% of maximal activity is lost in 4–5 hr (curve 3). There is partial protection against this activity loss by Mn^{2+} or Mg^{2+} (curve 4) and by GPP (curve 5.)[19] Considering Fig. 2A and B together with the kinetic data ± thiols (see below) it is reasonable to suggest that thiols produce a rapid

[19] Because of this activity loss it is probably best to store the purified enzyme in buffers which do not contain thiols. I am informed that some batches of the enzyme do not exhibit thiol-induced instability. The apparent instability could therefore result from an inhibitor produced by the interaction between thiols and components or contaminants of the dialysis tubing.

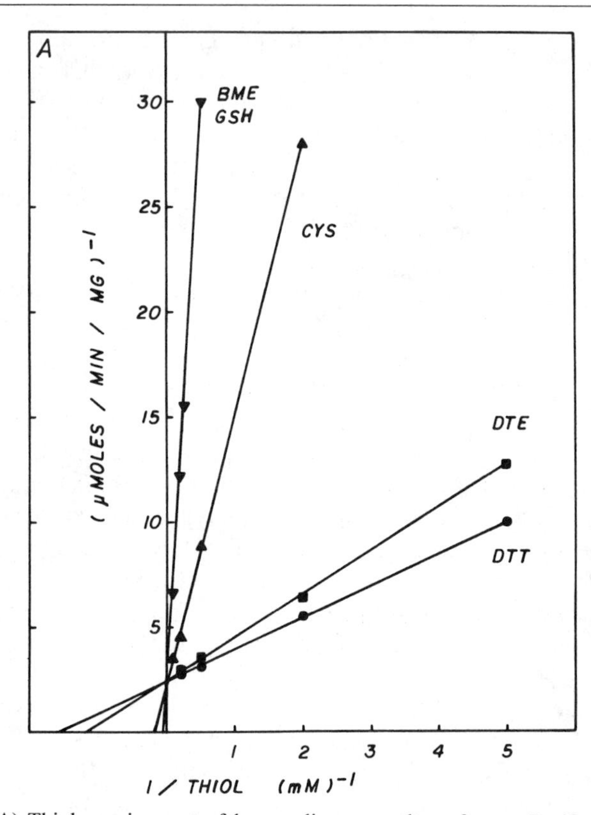

FIG. 2. (A) Thiol requirement of human liver prenyltransferase. Purified human prenyltransferase (15.5 μg/ml) is dialyzed against Tris–HCl buffer, pH 7.8 for 2.5 hr at 4°. Samples of 5 μl are assayed for 4 min in assay mixtures containing 0.5 μ*M* substrates and different concentrations of various thiol-reducing agents. Control incubations contain no thiol. Reciprocal enzyme activity as (μmol/min/mg)$^{-1}$ is plotted as a function of the reciprocal thiol concentration, mM^{-1}; ●, dithiothreitol, DTT; ■, dithioerythritol, DTE; ▲, L-cysteine, CYS; ▼, 2-mercaptoethanol, BME, or reduced glutathione, GSH. (B) Thiol-induced instability of human prenyltransferase activity. At time zero, samples of human prenyltransferase ammonium sulfate precipitated are diluted to 7.75 μg/ml and dialyzed against 10 m*M* Tris–HCl buffer, pH 7.3 with 1 ■, no additions; 2 ▲, no additions; 3 ●, 0.5 m*M* dithiothreitol; 4 ◆, 0.5 m*M* dithiothreitol *plus* 0.2 m*M* Mg^{2+}, or 0.2 m*M* Mn^{2+}; 5 ▼, 0.5 m*M* dithiothreitol *plus* 0.1 m*M* geranyl-PP. At various times 2-μl aliquots are assayed, in the presence (curves 1, 3–5) or absence (curve 2) of 2 m*M* dithiothreitol. Activity is expressed as percentage of the initial activity at time zero in the presence of dithiothreitol. The addition of sample aliquots (2 μl) containing GPP does not significantly interfere with the assays for enzymatic activity.

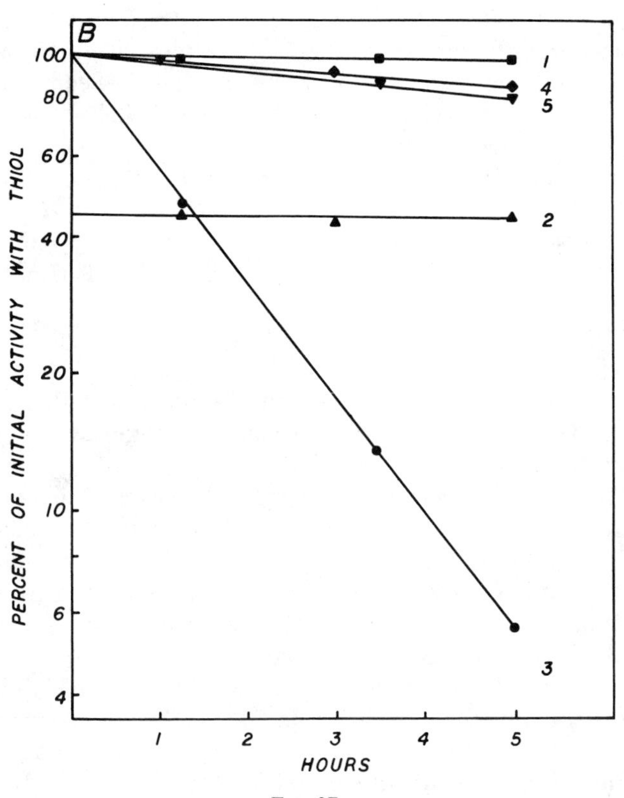

FIG. 2B.

enzyme thiol-reduction affecting the allylic substrate binding site and also a time-dependent instability perhaps caused by reduction of additional disulfide bond(s). Thiol reduction of the porcine enzyme affects enzyme structure and the number of titratable enzyme thiol groups.[5,20]

During the purification of human prenyltransferase only one form is noted. The pig liver enzyme exists in two forms, A and B, which are interconvertible by altering the state of thiol-oxidation/reduction of the protein.[20] Using crude, partially purified or fully purified preparations of the human enzyme under conditions which separately favor the thiol-reduced A-form or the thiol-oxidized B-form of the porcine enzyme, we detect only one form of human prenyltransferase.[7] This form elutes from DEAE-cellulose columns in a position identical to the B-form of porcine prenyltransferase.

[20] G. F. Barnard, B. Langton, and G. Popják, *Biochem. Biophys. Res. Commun.* **85,** 1585 (1978).

Kinetic Properties.[7] Double reciprocal plots of initial velocity data with isopentenyl-PP and geranyl-PP as substrates give a pattern consistent with an ordered sequential reaction mechanism. Secondary plots of the slopes and intercepts of the primary graphs permit calculation of the kinetic constants (mean ± SD):

K_a (Michaelis constant for S_1, GPP) = 0.44 ± 0.07 μM
K_b (Michaelis constant for S_2, IPP) = 0.94 ± 0.27 μM
K_s (Dissociation constant for S_1, K_{-1}/K_{+1}) = 0.14 ± 0.03 μM
$V_{1'}$ (Maximum reaction velocity, using 1/GPP) = 1.07 ± 0.19 μmol/min/mg
$V_{1''}$ (Maximum reaction velocity, using 1/IPP) = 1.10 ± 0.23 μmol/min/mg

The value of $V_{1'}$ is obtained from secondary plots using GPP as the varied substrate and agrees closely with the value of $V_{1''}$ obtained from secondary plots using IPP as the varied substrate.

Isopentenyl-PP in concentrations above 2 μM result in a pronounced inhibition, competative with geranyl-PP, and with a K_i (dissociation constant for the ES_2 complex) of 0.70 μM. Conversely, no significant substrate inhibition is seen with geranyl-PP as the varied substrate.

As mentioned above, purified human liver prenyltransferase is dependent on thiol-reducing agents for enzymic activity. Kinetic analysis of partially thiol-reactivated samples of the enzyme ± 2 mM dithiothreitol reveal a quasi-competitive pattern with GPP as the varied substrate but a noncompetitive pattern with IPP as the varied substrate. The apparent K_m for GPP in the absence of dithiothreitol is 2.15 ± 0.07 μM but in its presence it is only 0.26 ± 0.02 μM. The apparent K_m for IPP is the same ± dithiothreitol at 0.35 ± 0.04 μM. The change caused by dithiothreitol thus appears to affect only the binding of the allylic substrate GPP and not that of the homoallylic substrate IPP.

Inactivation by Phenylglyoxal.[7] Human liver prenyltransferase is sensitive to inactivation by the arginine modifying reagent phenylglyoxal. Partial protection against inactivation is afforded by the substrates DMAPP and GPP implying that arginine residues may be involved in the binding of the pyrophosphorylated substrates to the enzyme.

Immunological Properties.[7] Rabbit antiserum raised against porcine prenyltransferase shows cross-reactivity but nonidentity with the human enzyme in double-immunodiffusion experiments. Therefore the human enzyme shares at least some of the antigenic determinants of the porcine enzyme. For a quantitative comparison, 10 μl of antiserum precipitated 413 ng of the porcine enzyme but only 196 ng of the human enzyme in immunotitration experiments.

Acknowledgments

This work was supported by USPHS Research Grant HL-12745 to Dr. George Popják at the department of Biological chemistry, University of California at Los Angeles during the authors tenure as assistant research biochemist. I thank Drs. S. K. Erickson, L. H. Hohberger, and G. Popják for critical review during the preparation of this manuscript.

[19] Prenyltransferases of Pumpkin Fruit

By Kyozo Ogura, Tokuzo Nishino, Toshihiro Shinka, and Shuichi Seto

Farnesylpyrophosphate synthetase (dimethylallyltransferase, EC 2.5.1.1) and geranylgeranylpyrophosphate synthetase can be obtained from pumpkin fruit and separated from each other.[1,2] These enzymes catalyze the consecutive condensation of isopentenyl pyrophosphate with allylic pyrophosphates as follows.

Farnesyl-PP synthetase

$$\text{Dimethylallyl-PP} + \text{isopentenyl-PP} \xrightarrow{Mg^{2+}} \text{geranyl-PP} + \text{PP}_i$$

$$\text{Geranyl-PP} + \text{isopentenyl-PP} \xrightarrow{Mg^{2+}} \text{farnesyl-PP} + \text{PP}_i$$

Geranylgeranyl-PP synthetase

$$\text{Dimethylallyl-PP} + \text{isopentenyl-PP} \xrightarrow{Mn^{2+}} \text{geranyl-PP} + \text{PP}_i$$

$$\text{Geranyl-PP} + \text{isopentenyl-PP} \xrightarrow{Mn^{2+}} \text{farnesyl-PP} + \text{PP}_i$$

$$\text{Farnesyl-PP} + \text{isopentenyl-PP} \xrightarrow{Mn^{2+}} \text{geranylgeranyl-PP} + \text{PP}_i$$

Under usual conditions, however, the product of the farnesylpyrophosphate synthetase reaction starting with dimethylallyl pyrophosphate and isopentenyl pyrophosphate is a mixture of geranyl pyrophosphate and all-*trans*-farnesylpyrophosphate, whereas all-*trans*-geranylgeranyl pyrophosphate is the only product in the reaction catalyzed by geranylgeranylpyrophosphate synthetase regardless of the chain length of the allylic substrate with which the reaction is initiated.

Assay Methods

Principle. Isopentenyl pyrophosphate is resistant to treatment with acid while allylic prenyl pyrophosphates yield a mixture of products solu-

[1] K. Ogura, T. Nishino, and S. Seto, *J. Biochem.* (*Tokyo*) **64,** 197 (1968).
[2] K. Ogura, T. Shinka, and S. Seto, *J. Biochem.* (*Tokyo*) **72,** 1101 (1972).

ISBN 0-12-182010-6

ble in nonpolar solvents. The assay method determines the amount of ^{14}C-labeled isopentenyl pyrophosphate which reacts with allylic substrates to form ^{14}C-labeled compounds which, after acid treatment, are extractable with hexane. In the assay of crude preparations, it is recommended to use geranyl pyrophosphate plus [^{14}C]isopentenyl pyrophosphate in the presence of Mg^{2+} and farnesyl pyrophosphate plus [^{14}C]isopentenyl pyrophosphate in the presence of Mn^{2+} for the assay of farnesylpyrophosphate synthetase and geranylgeranylpyrophosphate synthetase, respectively. Potassium fluoride also is added to inhibit phosphatase activity present in crude preparations. Incorporation of ^{14}C-labeled isopentenyl pyrophosphate into the hexane-extractable fraction in the absence of any added allylic substrate indicates contamination of the enzyme preparation by isopentenylpyrophosphate isomerase. Iodoacetamide can be used to inhibit the isomerase, if necessary.

For product analysis, incubation mixtures are treated with alkaline phosphatase to hydrolyze prenyl pyrophosphates to the corresponding primary alcohols.

Reagents. [1-^{14}C]Isopentenyl pyrophosphate. It is convenient to use ^{14}C-labeled material with specific activity of about 1 Ci/mol prepared by diluting commercially available [1-^{14}C]isopentenyl pyrophosphate of higher specific activity, 50 ~ 60 Ci/mol (Amersham) with unlabeled isopentenyl pyrophosphate.

Prenyl Pyrophosphates. Dimethylallyl alcohol, geraniol, and all-*trans*-farnesol are pyrophosphorylated by the method of Poulter *et al.*[3]

Tris–HCl buffer, 0.5 *M*, pH 7.0
$MgCl_2$, 0.1 *M*
$MnCl_2$, 0.1 *M*
[1-^{14}C]Isopentenyl pyrophosphate (sp. act., 1 Ci/mol), 0.5 m*M*
Geranyl pyrophosphate, 1 m*M*
Farnesyl pyrophosphate, 1 m*M*
Iodoacetamide, 0.1 *M*
KF, 0.5 *M*
HCl, 6 *M*
NaOH, 6 *M*
Hexane

Procedure. In a 10-ml screw-capped test tube are placed 0.1 ml of Tris buffer and 0.05 ml each of $MgCl_2$ (for farnesylpyrophosphate synthetase) or $MnCl_2$ (for geranylgeranylpyrophosphate synthetase), ^{14}C-isopentenyl pyrophosphate, geranyl pyrophosphate (for farnesylpyrophosphate synthetase), or farnesyl pyrophosphate (for geranylgeranylpyrophosphate

[3] D. L. Bartlett, R. H. S. King, and C. D. Poulter, this volume [20].

synthetase). In the assay with crude preparations 0.1 ml each of KF and iodoacetamide are also included. Enzyme and water are added to give a final volume of 1.0 ml. The mixture is incubated at 37° for 1 hr. The reaction is stopped by the addition of 0.1 ml of HCl, and the mixture is kept at 37° for 30 min to complete the hydrolysis of prenyl pyrophosphate. The mixture is then made alkaline by the addition of 0.15 ml of NaOH and extracted with 4 ml of hexane. The hexane solution is washed with water and a 2-ml aliquot of the solution is counted for radioactivity in toluene scintillator. Enzyme activity is given by the radioactivity in the hexane extract.

Product Analysis. Acid hydrolysis is convenient for routine assay of prenyltransferase activity but unsuitable for identifying the products of the enzymatic reaction because of the complex nature of the decomposition products. For product analysis, treatment with alkaline phosphatase is recommended to hydrolyze prenyl pyrophosphates to the corresponding primary alcohols as exemplified in the following procedure: A mixture containing the products of enzymatic reaction carried out under the standard conditions is adjusted to pH 9.0 with Tris–HCl buffer, and 0.01 ml of a solution of intestinal alkaline phosphatase (Boehringer Mannheim GmbH, 10 mg/ml) is added to the mixture. The mixture is incubated at 37° for 3 hr and then allowed to stand at room temperature overnight. The mixture is extracted with light petroleum, and the petroleum extract is subjected to radiochromatography (TLC, HPLC, or GC) with authentic prenols. For example, the extract may be chromatographed on a plate coated with cellulose powder (Avicel plate, Funakoshi Chemicals, Japan) which has been impregnated with paraffin oil by dipping the plate once into a 5% solution of liquid paraffin in light petroleum. The solvent system is acetone–water (13 : 7, v/v) saturated with paraffin oil. The chromatography of the extract on a silica gel plate with a system of benzene–ethyl acetate (4 : 1, v/v) is also useful. Authentic prenols are located by exposing the developed plate to iodine vapor.

Purification of Enzymes

All steps are carried out at 0–4° unless otherwise mentioned.

Extraction. Pumpkin (about 1 kg) obtained locally is sliced, seeds having been removed and pulverized in a homogenizer. The homogenate is suspended in 250 ml of 0.05 *M* phosphate buffer, pH 6.8. The resulting mixture is squeezed through 8 layers of gauze to remove debris. The filtrate is centrifuged at 77,000 *g* for 1 hr.

First Ammonium Sulfate Fractionation. The supernatant is fractionated by slow addition of solid ammonium sulfate with stirring, and the fraction precipitating between 20 and 80% saturation is collected by centrifugation at 26,000 *g* for 30 min. The precipitate is dissolved in a minimum volume of 0.05 *M* phosphate buffer, pH 6.8, and the solution is desalted by passing through Sephadex G-25 equilibrated with the same buffer.

Second Ammonium Sulfate Fractionation. The desalted solution is subjected to a second ammonium sulfate fractionation, and the fraction precipitating between 35 and 55% saturation is collected and dissolved in 4 ml of 0.05 *M* of Tris–HCl buffer pH 7.0 containing 0.1 *M* NaCl.

DEAE Sephadex Chromatography. The resulting solution is filtered through a column (2.5 × 30 cm) of Sephadex G-50 with the same buffer containing 0.1 *M* NaCl. The protein fraction eluted from the column is concentrated to about 5 ml by ultrafiltration using a Diaflo apparatus with a UM-10 membrane (Amicon) of an exclusion limit of M_r = 10,000, and is applied to a DEAE Sephadex A-50 column (1.5 × 28 cm). The elution is performed with a linear concentration gradient established between 250 ml of 0.05 *M* Tris buffer, pH 7.0 containing 0.1 *M* NaCl and 250 ml of 0.05 *M* Tris buffer, pH 7.0 containing 0.4 *M* NaCl. The effluent is collected in 5 ml portions. Geranylgeranylpyrophosphate synthetase is recovered in fraction numbers 40–50, and farnesylpyrophosphate synthetase is recovered in fraction numbers 48–56. The geranylgeranylpyrophosphate synthetase fractions thus obtained contain 110 mg of protein capable of catalyzing the synthesis of 88 nmol of geranylgeranyl pyrophosphate per minute. The farnesylpyrophosphate synthetase fractions contain 157 mg of protein catalyzing the synthesis of 854 nmol of farnesyl pyrophosphate per minute. At this stage these enzymes are stable for at least 1 week when kept frozen at −20°.

Properties

Activators. Farnesylpyrophosphate synthetase requires Mg^{2+} or Mn^{2+} ions (the former preferred) whereas geranylgeranylpyrophosphate synthetase is activated by Mn^{2+} ions much more effectively than Mg^{2+} ions. A similar geranylgeranylpyrophosphate synthetase is contained in crude preparations from carrot root,[4] and pumpkin seedlings.[5]

Effect of Substrate Concentration. Apparent K_m values of farnesylpyrophosphate synthetase for geranyl pyrophosphate and of geranylgeranylpyrophosphate synthetase for farnesyl pyrophosphate are 1.3×10^{-6} and 1.7×10^{-6} *M*, respectively.

[4] D. L. Nandi and J. W. Porter, *Arch. Biochem. Biophys.* **105,** 7 (1964).

[5] T. Shinka, K. Ogura, and S. Seto, *Phytochemistry* **13,** 2103 (1974).

pH Optimum. Farnesylpyrophosphate synthetase and geranylgeranylpyrophosphate synthetase show maximum activities at pH 7.5 and 7.0 in Tris–HCl buffer, respectively.

Substrate Specificity. A number of 3-methyl-2-alkenyl pyrophosphates ranging in carbon number from 6 to 13 act as substrate to react with isopentenyl pyrophosphate in the reactions catalyzed by farnesylpyrophosphate synthetase[6-8] and geranylgeranylpyrophosphate synthetase.[9]

[6] K. Ogura, T. Nishino, T. Koyama, and S. Seto, *J. Am. Chem. Soc.* **92,** 6036 (1970).
[7] T. Nishino, K. Ogura, and S. Seto, *J. Am. Chem. Soc.* **94,** 6849 (1972).
[8] T. Nishino, K. Ogura, and S. Seto, *Biochim. Biophys. Acta* **302,** 33 (1973).
[9] T. Shinka, K. Ogura, and S. Seto, *J. Biochem.* (*Tokyo*) **78,** 1177 (1975).

[20] Purification of Farnesylpyrophosphate Synthetase by Affinity Chromatography

By DESIREE L. BARTLETT, CHI-HSIN RICHARD KING, and C. DALE POULTER

The fundamental building reaction in the isoprenoid pathway is a 1′-4 prenyl transfer which attaches C-1 of an allylic pyrophosphate to C-4 of isopentenyl pyrophosphate to generate a larger five-carbon homolog of the allylic substrate.[1] Beginning with dimethylallyl pyrophosphate, a variety of products which differ in the length of the isoprenoid chain and the stereochemistry of the double bonds can be formed. A family of enzymes catalyze 1′-4 prenyl transfers. Individual members show different substrate specificities based on chain length and double bond stereochemistry of the allylic substrate and produce five carbon homologs with exclusively *E* or *Z* trisubstituted double bonds. In principle it should be possible to use the different substrate specificities to purify selectively individual members of the family from a crude homogenate by affinity chromatography.

Farnesylpyrophosphate synthetase (dimethylallyltransferase, EC 2.5.1.1) is a 1′-4 prenyltransferase that produces a key intermediate in the isoprenoid pathway which is the precursor for a variety of essential metabolites, including sterols, ubiquinones, dolichols, and some hemes. The enzyme synthesizes (*E,E*)-farnesyl pyrophosphate from dimethylallyl pyrophosphate and two molecules of isopentenyl pyrophosphate in

[1] C. D. Poulter and H. C. Rilling, *Acc. Chem. Res.* **11,** 307 (1978).

ISBN 0-12-182010-6

two steps. The product of the first step, geranyl pyrophosphate, binds to the enzyme more tightly than the other substrates or the final product[2] and is, therefore, a logical candidate for the ligand in an affinity column to purify the enzyme. This chapter describes the synthesis of an affinity column for farnesylpyrophosphate synthetase based on the geranyl moiety and a rapid purification of the enzyme from avian liver and yeast.

General Methods

Infrared, Mass, and Nuclear Magnetic Resonance Spectra

Infrared (IR) spectra were recorded on a Perkin-Elmer 299 Infrared Spectrophotometer and were calibrated to the 1601 cm^{-1} absorption of polystyrene. Solid samples were analyzed either as potassium bromide pellets or as 10% solutions in spectrograde chloroform. Liquids or oils were analyzed neat as a thin film between two salt plates. All absorptions are reported in wave numbers (cm^{-1}). Nuclear magnetic resonance (NMR) spectra were recorded on Varian EM-390, FT-80, and SC-300 spectrometers. Proton spectra are reported in parts per million downfield from internal tetramethylsilane. Phosphorus-31 spectra are reported in parts per million as negative ppm if downfield from external 85% phosphoric acid or as positive ppm if upfield from the external reference. Mass spectra (chemical ionization and electron impact) were obtained on a Varian MAT 1125 mass spectrometer.

Liquid Scintillation Spectrometry and Electrophoresis

Radioactivity was measured using a Packard TRI-CARB 4530 liquid scintillation counter, and the samples were analyzed in 10 ml of INSTA-FLUOR (Packard) liquid scintillation cocktail. Polyacrylamide gel electrophoresis (in sodium dodecyl sulfate) was conducted in a Bio-Rad Protean Dual Vertical Slab Gel Electrophoresis Cell using a Buchler 3-1500 Constant Power Supply.

Solvents and Reagents

All solvents were reagent grade and distilled. Anhydrous solvents were prepared by heating at reflux under nitrogen over a drying agent followed by distillation under a nitrogen atmosphere. Tetrahydrofuran was heated at reflux over sodium metal with benzophenone as an indicator until the blue color persisted. *N*,*N*-Diisopropylamine, dichlorome-

[2] B. C. Reed and H. C. Rilling, *Biochemistry* **15**, 3739 (1976).

thane, and methanol were heated at reflux over calcium hydride for several hours and then distilled. Methanol was redistilled over magnesium turnings. Dimethylformamide was warmed (80°) over calcium hydride for several hours and then distilled under vacuum (10 mm Hg). Ethylene glycol was distilled under vacuum (10 mm Hg).

Dimethyl methylphosphonate and 1,1′-carbonyldiimidazole were purchased from Aldrich Chemical Co. Dimethyl methylphosphonate was distilled under vacuum (35° at 0.3 mm Hg). Sodium boro[^{3}H]hydride was purchased from New England Nuclear. [^{14}C]Isopentenyl pyrophosphate was purchased from Amersham. Geranyl pyrophosphate was prepared according to the procedures presented in chapter 15 of this volume. Deuterium oxide, sodium 2,2-dimethyl-2-silapentane 5-sulfonate (DDS), chloroform-*d*, isopropyl alcohol-d_8, and tetramethylsilane (TMS) were purchased from MSD Isotopes.

Chromatography

Dowex AG 50W-8X cation exchange resin (hydrogen form) and Affi-Gel 10 agarose beads were purchased from Bio-Rad Laboratories. Silica gel (Merck grade 60, 230–400 mesh and grade 62, 60–200 mesh) was purchased from Aldrich Chemical Co. and preparative TLC plates (silica gel F-254; 0.5 mm, 20 × 20 cm) from EM reagents.

Reactions were routinely monitored by TLC (thin-layer chromatography) with 7.5 × 2.5 cm Baker-flex silica gel IB-F sheets (J. T. Baker), and the spots were visualized with iodine. Silica gel columns for chromatography (flash or gravity) were packed with dry silica gel (230–400 mesh, unless otherwise noted) and then equilibrated with solvent.

Affinity chromatography with derivatized agarose beads was run in a Glenco precision bore 25 × 0.6 cm medium pressure liquid chromatography column. Elution of protein was monitored at 280 nm using a LKB 2138 Uvicord S UV monitor (LKB-Produkter AB). Proteins were concentrated in a Micro-ProDiCon (model no. MPDC-115; Bio-Molecular Dynamics) negative pressure micro protein dialysis concentrator using a ProDiMem membrane (model no. PA-15; Bio-Molecular Dynamics; molecular weight cut off at 15,000).

Construction of the Affinity Gel

The strategy employed to synthesize the affinity gel is shown in Scheme I. The phosphonate moiety used to link the geranyl chain to phosphorus differs from the normal substrate by substitution of the oxygen attached to C-1 by carbon. This alteration gives a stable linkage which

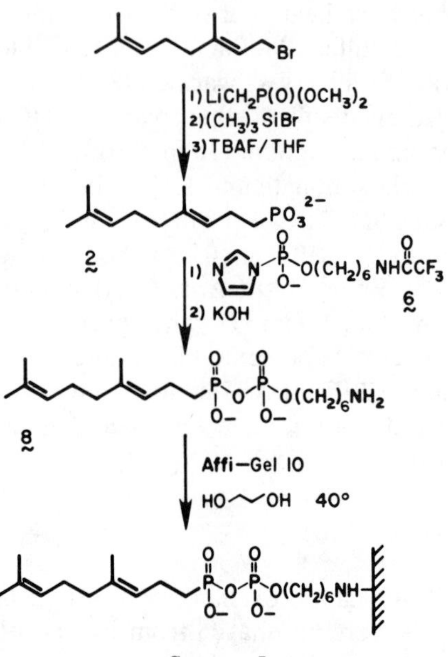

SCHEME I.

is not susceptible to decomposition by solvolysis, a reaction responsible for the notorious instability of allylic pyrophosphates.[3] Furthermore, similar compounds are excellent inhibitors of farnesylpyrophosphate synthetase.[4] The phosphate linkages were placed between the geranyl moiety and the hexamethylene spacer to afford resistance against nonspecific phosphatases, all of which require terminal phosphates as substrates.[5] In addition, tritium was incorporated into the hexamethylene spacer so the progress of the coupling of the ligand to the gel could be monitored. The yield of the reaction was calculated by measuring residual radioactivity in the supernatant.

*Dimethyl Geranylmethylphosphonate (**1**)*

To a solution of 8.57 g (69 mmol) of dimethyl methylphosphonate in 120 ml of anhydrous tetrahydrofuran at −78° was added 32 ml (76 mmol) of 2.4 *M* *n*-butyl lithium in hexane over a period of 10 min. The resulting mixture was stirred at −78° for 30 min before dropwise addition of a

[3] V. J. Davisson, A. B. Woodside, and C. D. Poulter, this volume [15].

[4] E. J. Corey and R. P. Volante, *J. Am. Chem. Soc.* **98,** 1291 (1976).

[5] C. C. Richardson, *Annu. Rev. Biochem.* **38,** 708 (1969).

solution of 15 g (69 mmol) of geranyl bromide[6] in 20 ml anhydrous tetrahydrofuran over 15 min. The reaction mixture was allowed to stir at $-78°$ for 2 hr before 10 ml of brine was added. The organic layer was washed with 50 ml of 5% aqueous ammonium chloride and 50 ml of brine and was then dried over anhydrous magnesium sulfate. Solvent was removed at reduced pressure to give 16.07 g of crude dimethyl geranylmethylphosphonate **1**.

Purification of the crude product by flash chromatography[7] on silica gel by elution with ethyl acetate (R_f 0.23) gave 14.04 g (54 mmol, 78%) of **1** as a clear oil; IR (neat) 3020, 2950, 2915, 2850, 1650, 1445, 1375, 1250, 1180, 1058, 1030, and 810 cm^{-1}; ^1H NMR ($CDCl_3$) δ 1.60 (6H, s, two methyls at C-4 and C-8), 1.67 (3H, s, methyl at C-8), 1.60–2.50 (8H, m, H at C-1, C-2, C-5, and C-6), 3.73 (6H, d, $J_{^1H,^{31}P} = 11$ Hz, two ester methyls), and 5.11 (2H, br t, vinyl H at C-3 and C-7); ^{31}P NMR ($CDCl_3$, ext. ref. 85% H_3PO_4) δ 34.8; mass spectrum (CI, CH_4), m/z 261 ($M^+ + 1$), 218, 205, 191, 179, 137, 124.

Tetra-n-butylammonium Monohydrogen Geranylmethylphosphonate (**2**)

To 0.895 g (3.44 mmol) of dimethyl geranylmethylphosphonate **2** at 0° was added dropwise, 1.08 g (0.93 ml, 7.06 mmol) of trimethylbromosilane. The reaction mixture was warmed to room temperature and stirring was continued for 2 hr. Excess trimethylbromosilane and by-product methyl bromide were removed under vacuum (1 hr at 0.03 mm Hg) to give 1.29 g (3.44 mmol, 100%) of bis(trimethylsilyl)geranylmethylphosphonate. A proton NMR in $CDCl_3$ showed 18 H's at 0.30 ppm for the six silyl methyl groups. The bis(trimethylsilyl) ester was dissolved in 15 ml of tetrahydrofuran, cooled to $-78°$ in a dry ice/acetone bath, and 3.44 ml of 1 *M* tetra-*n*-butylammonium fluoride (3.44 mmol) in tetrahydrofuran was added. The reaction mixture was allowed to stir at $-78°$ for 15 min, at room temperature for 45 min, and then concentrated *in vacuo*. The residue was dried by addition of anhydrous acetonitrile and removal of solvent under reduced pressure (3 times) and subjected to high vacuum to give 1.63 g (3.44 mmol, 100%) of **2** as a pale yellow solid; IR ($CHCl_3$) 3450–3020, 3000–2890, 2880, 1645, 1655, 1480, 1380, 1240 cm^{-1}; ^1H NMR ($CDCl_3$) δ 1.0 (12 H, br m, *n*-butyl methyls), 1.50 (6H, s, two methyls at C-4 and C-6), 1.65 (3H, s, methyl at C-8), 1.30–2.50 (8H, m, H at C-1, C-2, C-5, and C-6), 3.30 (H, br m, methylene H α to N), 5.10 (2H, br m, vinyl H at C-3 and C-7) and 10.50

[6] R. M. Coates, D. A. Ley, and P. L. Cavander, *J. Org. Chem.* **43,** 4915 (1978).
[7] W. C. Still, M. Kahn, and A. Mitra, *J. Org. Chem.* **43,** 2923 (1978).

(1H, br m, hydroxyl proton); ^{31}P NMR (DMF, ext. ref. 85% H_3PO_4) δ −22.67.

*O-(6-N-Trifluoroacetylamino-1-hexyl)-P-geranylmethyl Phosphonophosphate (**7**)*

A sample of 2.295 g (7.83 mmol) of *N*-trifluoroacetyl-6-amino-1-hexylphosphate[8] **5** and 1.34 g (8.22 mmol) of 1,1′-carbonyldiimidazole was dried under vacuum (1 hr at 0.03 mm). The solids were dissolved in 7 ml of anhydrous dimethylformamide, and the reaction mixture was stirred at room temperature for approximately 5 hr.[9] After the reaction was complete (formation of **6**), 3.33 g (7.05 mmol) of geranylmethylphosphonate **2** in 8 ml of anhydrous dimethylformamide was added. The resulting solution was allowed to stir at room temperature for 20–24 hr. Solvent was removed under high vacuum (short path distillation) to give a thick yellow oil. The oil was dissolved in water (100 ml) and slowly eluted through Dowex AG 50W-X8 (ammonium cycle), and the column was washed with two additional 50-ml portions of water. The eluants were combined and lyopholized to give 3.70 g of crude **7** as a pale yellow solid. The solid was purified by flash chromatography.[7] Silica gel (600 ml) was preequilibrated with acetonitrile. A 1.5 g sample of **7** in 7 ml of 10% aqueous acetonitrile was introduced onto the column which was then eluted successively with 600, 1000, and 800 ml of 10, 15, and 20% aqueous acetonitrile (containing 0.5% of 58% ammonium hydroxide), respectively.

Fractions containing product (R_f 0.37 in 15% aqueous CH_3CN containing 0.5% of 58% NH_4OH) were combined, and acetonitrile was removed under aspirator vacuum (compound **7** is a detergent and the concentration process must be watched closely). The concentrate was lyopholized to give 3.046 g (5.64 mmol, 80%) of **7** as a white solid[10]; mp, decomposes at 220°; IR (KBr) 3300, 2925, 1700, 1555, 1440, 1250–1050, 950 cm^{-1}; NMR (isopropyl alcohol-d_8/D_2O) δ 1.30–2.38 (16H, m, H at C-1, C-2, C-5, C-6, C-2′, C-3′, C-4′, and C-5′), 1.59 (3H, s, methyl at C-8), 1.63 (3H, s, methyl at C-8), 1.66 (3H, s, methyl at C-4), 3.30 (2H, t, J = 6Hz, methy-

[8] R. Barker, I. P. Trayer, and R. L. Hill, this series, Vol. 34 [56].

[9] The progress of the reaction was followed by ^{31}P NMR. A 0.5 ml sample of the DMF mixture was transferred to an NMR tube, and a smaller NMR tube containing D_2O was inserted and used as an external lock for the spectrophotometer. Starting material **5** and the imidazole adduct **6** have chemical shifts in DMF at 0.0 and −10.7 ppm, respectively, using 85% phosphoric acid as an external reference.

[10] Compound **7**, as a pure white solid, is insoluble in H_2O, CH_3CN, DMF, DMSO, $CHCl_3$, EtOH, *i*-PrOH, *t*-BuOH, and *n*-BuOH. It is slightly soluble in *i*-PrOH, *t*-BuOH, and *n*-BuOH water mixtures (approximately 8 : 2 alcohol–water) and completely soluble in ethylene glycol.

lene α to N), 3.90 (2H, m, methylene α to O), 5.09 (1H, m, vinyl H at C-7), and 5.19 (1H, m, vinyl H at C-3).

*O-(6-Amino-1-hexyl)-P-geranylmethyl Phosphonophosphate (**8**)*

To a suspension of 225 mg (0.416 mmol) of compound **7** in *t*-butyl alcohol was added 18.7 ml of 0.1 *M* potassium hydroxide (1.87 mmol) at room temperature. The suspension was then cooled to 0° and stirred for 5 to 6 hr. TLC analysis (1 : 4 H_2O/CH_3CN containing 0.5% of 58% NH_4OH) of the reaction mixture showed only one compound, *O*-(6-amino-1-hexyl)-*P*-geranylmethyl phosphonophosphate **8** (R_f 0.23), when the hydrolysis was complete (R_f 0.64 for starting material **7**). Acidification at 0° with 0.2 *N* hydrochloric acid to pH 7.5 and lyophilization gave **8** as a pale yellow solid. For spectral data **8** was chromatographed on silica gel (60–200 mesh) using conditions outlined under the synthesis of [1-^{3}H]-**8**; IR (KBr) 3700–2300 (br), 1630, 1535, 1445, 1380, 1215 (br), 1110, 1060, 925 cm^{-1}; NMR (isopropyl alcohol-d_8/D_2O) δ 1.40–2.40 (16H, m, H at C-1, C-2, C-5, C-6, C-2′, C-3′, C-4′, and C-5′), 1.59 (3H, s, methyl at C-8), 1.64 (3H, s, methyl at C-8), 1.66 (3H, s, methyl at C-4), 3.02 (2H, br t, J = 7 Hz, methylene α to N), 3.98 (2H, m, methylene α to 0), 5.10 (1H, t, J = 7 Hz, vinyl H at C-7), and 5.19 (1H, t, J = 7 Hz, vinyl H at C-3).

*[1-^{3}H]-O-(6-Amino-1-hexyl)-P-geranylmethyl Phosphonophosphate (**8**)*

To 27 mg (127 μmol) of *N*-trifluoroacetyl-6-amino-1-hexanal[11] **3** in 1 ml of anhydrous methanol was added, at room temperature, 18 μmol (6.25 mCi) of sodium boro[^{3}H]hydride (347.8 mCi/mmol). After 12 hr at room temperature, 5 mg sodium borohydride was added to ensure that the reduction was complete. After 30 minutes, 1 ml of brine was added, and the resulting mixture was extracted twice (with vortexing) with 1 ml portions of ether. The ether extracts were combined and filtered through a plug of anhydrous magnesium sulfate. The solvent from a portion of the ether extract (approximately 600 μl in a 5-ml vial) was evaporated under a stream of nitrogen to give 7.7 mg (36 μmol) of alcohol **4**. The alcohol was phosphorylated using the method of Ramirez.[12] Thus, 15 mg (43 μmol) of solid 1,2-dibromo-1-phenylethyl phosphonic acid was added, and the reaction vial was thoroughly flushed with nitrogen and stoppered. Following the addition of 0.77 ml of anhydrous dichloromethane and 14 μl (10 mg, 81 μmol) of anhydrous diisopropylethylamine, the reaction mixture

[11] Aldehyde **3** was prepared by Swern oxidation of alcohol **5**; A. J. Mancuso, S.-L. Huang, and D. Swern, *J. Org. Chem.* **43,** 2480 (1978).

[12] F. Ramirez, J. F. Marecek, and S. S. Yemul, *J. Org. Chem.* **48,** 1417 (1983).

was allowed to stir for 12 hr at room temperature. One milliliter of water was added, and the aqueous phase was extracted twice (with vigorous stirring) with 2 ml portions of ether to remove organic soluble byproducts. The aqueous layer which contained the phosphorylated product was passed through a small column of Dowex AG 50W-8X (hydrogen cycle), and the column was washed twice with 0.5 ml portions of water. The combined eluants were evaporated to dryness under a stream of nitrogen with warming (approximately 40°). Residual water was removed by addition of acetonitrile and evaporation under a stream of nitrogen with warming (approximately 40°) followed by drying over anhydrous magnesium sulfate to give 10 mg (34 μmol, 96%) of *N*-trifluoroacetyl-6-amino-1-[1^{3}H] hexylphosphate **5**. No starting alcohol **4** was present as determined by TLC (R_f 0.42 for alcohol in 15% aqueous CH_3CN containing 0.5% of 58% NH_4OH).

To approximately 2 mg of [^{3}H]-**5** and 15 mg unlabeled **5** (58 μmol total) was added, under nitrogen, 11 mg (62 μmol) of 1,1′-carbonyldiimidazole in 160 μl of anhydrous dimethylformamide. The reaction vial was sealed and agitated to ensure mixing. After 7 hr at room temperature, 32 mg (68 μmol) of solid geranylmethylphosphonate **2** was added. The reaction vial was again flushed with nitrogen, agitated, and allowed to set for 24 hr at room temperature. Solvent was removed overnight under a stream of nitrogen, and the residue purified by thin-layer chromatography [two 500 μm 20 × 20 cm silica gel plates; 15% aqueous CH_3CN containing 0.5% of 58% NH_4OH; R_f = 0.44 for product; bands were visualized by developing the edges of the plate with iodine; product was extracted with 3 : 7 H_2O/*t*-BuOH]. After chromatography, 20 mg (37 μmol, 60%, SA 2.6 μCi/μmol) of *O*-(6-*N*-trifluoroacetylamino-[1-^{3}H]hexyl)-*P*-geranylmethyl phosphonophosphate **7** was obtained as a white solid.

To 108 mg (200 μmol) of **7** and 2 mg (3.7 μmol, 9.6 μCi) of [^{3}H]-**7** was added a mixture containing 4.0 ml of 0.1 *N* potassium hydroxide (400 μmol) and 6 ml of *t*-butanol at room temperature. The mixture was cooled to 0° in an ice bath and 0.5 ml of 1 *N* potassium hydroxide (500 μmol) was added. After 5 hr the reaction mixture was acidified to pH 7.5 with 0.2 *N* hydrochloric acid, and the resulting solution was concentrated under a stream of nitrogen to give a pale yellow oil. The oil was chromatographed on silica gel (35 ml, 60–200 mesh) by successive elution with 10, 80, and 90 ml portions of 15, 20, and 25% aqueous CH_3CN (containing 0.5% of 58% NH_4OH), respectively. The fractions (20–31, 5 ml/fraction) contraining product (R_f = 0.23) were combined and lyophilized to give 55 mg (124 μmol, 62%; SA 0.031 μCi/μmol) of *O*-(6-amino-[1-^{3}H]hexyl)-*P*-geranylmethyl phosphonophosphate **8** as a white solid.

*Coupling of **8** to Affi-Gel 10*

Ten milligrams of [^{3}H]-**8** and 84 mg of **8** (212 μmol total) were dissolved in 14 ml of dry ethylene glycol, and the resulting solution was analyzed for total radioactivity. Approximately 14 ml (210 μmol) of Affi-Gel 10[13] (15 μmol/ml maximum capacity) was filtered, washed twice with 25 ml of dry ethylene glycol, and added to **8** as a slightly moist gel. The reaction flask was flushed with nitrogen, sealed, and shaken overnight (18 hr) at 40° (do not stir!). The solution was analyzed for residual radioactivity, and the extent of coupling of **8** to the gel was determined to be 40%. Ethanolamine (13 μl, 212 μmol) was added to derivatize the remaining reactive sites on the gel, and the reaction flask was flushed with nitrogen and shaken vigorously for 5 hr at 40°. The resulting suspension was filtered, and the gel was washed in succession with two 20 ml portions of ethylene glycol, 20 ml isopropanol, 20 ml deionized water (0°), and two 15 ml portions of 10 m*M* PIPES[14] buffer (1 m*M* $MgCl_2$, 10 m*M* 2-mercaptoethanol, 0.25 m*M* sodium azide, pH 7.0, 0°). The moist gel was suspended in 14 ml of PIPES buffer and stored at 4° until needed.

Affinity Chromatography

Buffer Solutions

Standard buffer: 10 m*M* PIPES, 1 m*M* $MgCl_2$, 10 m*M* 2-mercaptoethanol, pH 7.0
Salt buffer: standard buffer containing 30 m*M* KCl (or 70 m*M* KCl)
Elution buffer: standard buffer containing 1 m*M* disodium pyrophosphate
Wash buffer: standard buffer containing 0.5 *M* KCl and 5 m*M* disodium pyrophosphate
Dialysis buffer: 40 m*M* phosphate, 1 m*M* EDTA, 10 m*M* 2-mercaptoethanol, pH 7.0

Assays and Protein Determinations

All assays were at 37° in a buffer consisting of 20 m*M* BHDA (bicyclo-[2.2.1]hept-5-ene-2,3-dicarboxylic acid),[15] 1 m*M* magnesium chloride, 10

[13] Affi-Gel 10 (Bio-Rad) is an *N*-hydroxysuccinimide ester of a derivatized crosslinked agarose gel. The gel has a neutral 10-atom spacer arm which contains the active ester functionality at the end. Ligands with a primary amino group can be coupled to the gel in aqueous and nonaqueous solvents.

[14] PIPES [piperazine-*N*,*N'*-bis(2-ethanesulfonic acid)].

[15] M. F. Malette, *J. Bacteriol.* **94,** 283 (1967).

mM 2-mercaptoethanol, and 0.01% bovine serum albumin. The acid lability assay[2] was used to determine the extent of the reaction. One unit of activity represents the formation of 1 μmol of product per min. Specific activities are given as μmol product produced/min/mg of protein. The concentrations of proteins were determined by the method of Lowry *et al.*[16] The concentration of pure FPP synthetase was determined by measuring the absorbance at 280 nm using an extinction coefficient of 1.03 ml mg^{-1} cm^{-1}.[2]

Purification of Farnesylpyrophosphate Synthetase

General Procedures. In the experimental procedure described below protein was chromatographed on a 21 × 0.6 cm medium pressure column (agarose bed volume of 6 ml, 36 μmol maximum capacity). A scrubber column located immediately before the affinity column contained 2 ml of agarose derivatized with 2-aminoethanol. The scrubber did not bind FPP synthetase but served to prolong the life of the affinity column. Buffer was passed through the system with a peristaltic pump, and elution of protein was monitored continuously at 280 nm. Six milliliter fractions were collected in 13 × 100-cm glass test tubes, except for those fractions known to contain farnesylpyrophosphate synthetase where plastic test tubes were used. All solutions were maintained at 4° and chromatographies were run in a cold room at 4°. In a typical purification, the crude sample from an ammonium sulfate precipitation was diluted to 10–20 mg/ml with standard buffer and loaded onto the column slowly (usually overnight) at a rate of 8 ml/hr. The column was then eluted with standard buffer until the absorbance at 280 nm returned to baseline. Subsequent elution protocols were carried out at a flow rate of 30 ml/hr. When the chromatography was complete and the column had been washed thoroughly, the gel was equilibrated with standard buffer containing 0.25 mM sodium azide and stored at 4°. The gel was left in the Glenco column between runs and care was taken to insure it remained covered with buffer.

Avian Liver Farnesylpyrophosphate Synthetase. Livers were collected fresh at a local slaughter house, washed, and stored at −70° until needed. Samples for affinity chromatography were prepared by following the first steps of the procedure of Reed and Rilling[17] for purification of the enzyme by standard chromatographic techniques. Protein that precipitated between 35 and 50% saturation with ammonium sulfate was dialyzed against 4 changes (1 l apiece) of 10 mM BHDA buffer (1 mM $MgCl_2$, 10

[16] O. H. Lowry, N. J. Rosebrough, A. L. Farr, and R. J. Randall, *J. Biol. Chem.* **193,** 265 (1951).

[17] B. C. Reed and H. C. Rilling, *Biochemistry* **14,** 50 (1975).

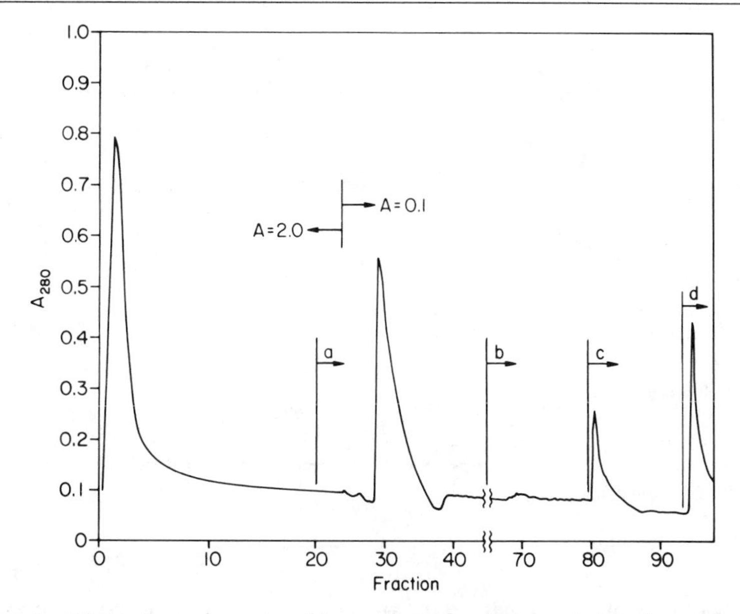

FIG. 1. Affinity chromatography of farnesylpyrophosphate synthetase from avian liver on geranylmethylphosphonate-agarose. Eighty milligrams of crude protein in 4 m of standard buffer (10 m*M* PIPES, 1 m*M* $MgCl_2$, 10 m*M* 2-mercaptoethanol, pH 7.0) was applied to the column (21 × 0.6 cm) and eluted overnight (fractions 1–20) in the same buffer at a rate of 8 ml/hr. The flow rate was increased (fraction 21) to 30 ml/hr and the column was eluted with (a) a linear 0–30 m*M* gradient of KCl followed by 30 m*M* KCl in standard buffer, (b) standard buffer, (c) 1 m*M* pyrophosphate in standard buffer, and (d) 0.5 *M* KCl, 5 m*M* pyrophosphate in standard buffer. Fraction volumes were 6 ml and protein was monitored by the absorbance at 280 nm. Farnesylpyrophosphate synthetase activity eluted from the column as a tight band (fractions 80 and 81) directly following the front for 1 m*M* pyrophosphate in standard buffer.

m*M* 2-mercaptoethanol, pH 7.0). An 80 mg portion was diluted and loaded onto the column as described in the preceding section. After elution (overnight) with standard buffer, the affinity column was eluted in succession with a 0–30 m*M* gradient of KCl (50 ml total volume) and 100 ml of 30 m*M* salt buffer to remove nonspecifically bound proteins. The column was then eluted with 100 ml of standard buffer, followed by 80 ml of elution buffer. Wash buffer (100 ml) was passed through the column to remove residual protein, and the gel was equilibrated with standard buffer containing 0.25 m*M* sodium azide before storage at 4°.

A typical chromatogram is shown in Fig. 1. Individual fractions were assayed for farnesylpyrophosphate (FPP) synthetase, and the enzyme was found in a tight band (10–15 ml) beginning just after the voild volume for the elution buffer. Minor amounts of activity were found in the first

TABLE I
PURIFICATION OF AVIAN LIVER FPP SYNTHETASE

Fraction	Units	Protein (mg)	Specific activity	Yield (%)	Purification (fold)
Crude supernatant	1.58	395	4.0×10^{-3}	100	0
35% ammonium sulfate supernatant	1.16	221	5.23×10^{-3}	75	1.3
35–50% ammonium sulfate precipitate after dialysis	1.11	80	13.8×10^{-3}	70	3.45
Affinity chromatography	0.83	0.55	1.50[a] (2.20[b])	52[a] (35[b])	375[a] (550[b])
Dialysis-concentration	0.76	0.41	1.85[a]	48[a]	463[a]

[a] 30 m*M* salt wash.
[b] 70 m*M* salt wash.

few column volumes where most of the other protein eluted. If too much crude protein was applied to the column or if the column was loaded too rapidly, a greater percentage of FPP synthetase activity was found in these fractions. Small amounts of activity were also found in the protein fraction that eluted with 30 m*M* salt buffer. If a 70 m*M* KCl salt wash was used instead, slightly more activity was found in this protein fraction. In either case, the majority of activity was found in the tight band of protein that eluted with the 1 m*M* pyrophosphate elution buffer. These fractions contained FPP synthetase of high specific activity at concentrations of 0.01–0.04 mg/ml. Protein that eluted with the high salt buffer wash had no farnesylpyrophosphate synthetase activity.

The fractions containing farnesylpyrophosphate synthetase were combined and concentrated-dialyzed (ProDiCon Concentrator) against dialysis buffer at 0°. The resulting dialyzates contained protein at concentrations of 0.2 to 0.4 mg/ml. The avian liver enzyme from the affinity chromatography was concentrated with greater than 90% recovery of activity. Fractions containing protein which eluted with 30 and 70 m*M* salt buffer and with the wash buffer were also concentrated in dialysis buffer to final concentrations of 0.5 to 1.5 mg/ml.

These samples and FPP synthetase were analyzed by sodium dodecyl sulfate (SDS)–polyacrylamide gel electrophoresis. The 30 and 70 m*M* KCl washes both contain a protein, MW 53,000 ± 10%, which was a major band in the crude preparation applied to the column. The protein had a

TABLE II
PURIFICATION OF FPP SYNTHETASE FROM *S. cerevisiae*

Fraction	Units	Protein (mg)	Specific activity	Yield (%)	Purification (fold)
Crude supernatant	1.30	309	4.2×10^{-3}	100	0
50–75% ammonium sulfate precipitate after dialysis	1.25	39	31.8×10^{-3}	96	7
Affinity chromatography[a]	0.82	0.314	2.62	63	624
Dialysis-concentration[a]	0.58	0.25	2.33	45	555

[a] 30 m*M* salt wash.

considerable affinity for Affi-Gel and was even found in the protein fraction that eluted with wash buffer. When a larger scrubber column (6 ml) was used during the initial protein elution and then by-passed during the remainder of the chromatography, much of the MW 53,000 protein remained absorbed on the underivatized gel. Farnesylpyrophosphate synthetase, purified using a 70 m*M* salt wash (180 ml) before the elution step, gave a single band upon SDS gel electrophoresis. Under these conditions approximately 50% of the activity originally loaded on the column was recovered as homogeneous protein of high specific activity (2.2 μmol/min/mg). However, those samples of farnesylpyrophosphate synthetase purified using a 30 m*M* salt wash gave two bands, a major band for the prenyltransferase, MW 43,000 ± 10%, and a smaller band for the MW 53,000 protein. The latter band corresponded to about 30% of the total intensity. Under these conditions 74% of the activity originally loaded on the column was recovered. The purification of the enzyme is summarized in Table I.

Yeast Farnesylpyrophosphate Synthetase. Fresh yeast (*Saccharomyces cerevisiae*) was obtained from a local bakery. Samples for affinity chromatography were prepared by following the initial steps in the purification of the enzyme by Eberhardt and Rilling.[18] Protein that precipitated between 50 and 75% saturation with ammonium sulfate was prepared for the affinity column as described in the previous section. The yeast enzyme gave a chromatographic profile similar to that shown in Fig. 1 for the avian liver prenyltransferase. SDS gel electrophoresis of the active

[18] N. L. Eberhardt and H. C. Rilling, *J. Biol. Chem.* **250,** 863 (1975).

fractions and the washes indicated that the MW 53,000 protein contaminant found in liver was also present in yeast. The purification of yeast prenyltransferase is summarized in Table II.

Conclusions

Farnesylpyrophosphate synthetase can be purified to homogeneity rapidly by affinity chromatography using an affinity ligand based on geranyl pyrophosphate. The single affinity step separates the enzyme from hundreds of other proteins present after ammonium sulfate precipitation of the crude homogenate with a 460- to 600-fold purification. The recovery of farnesylpyrophosphate synthetase activity from the affinity column was 74% for the chicken liver enzyme, although recoveries as low as 50% and as high as 90% were also obtained in other runs depending on elution conditions. The yeast enzyme, however, is less stable,[18] and recovery of activity from the column was only 66%.

The derivatized gel is stable to prolonged storage at 4°, and a single packed column was used by us repeatedly for 5 months without detectible degradation. It is apparent that the geranyl pyrophosphate linkage is stable to both chemical and enzymatic degradation. We anticipate similar ligands will prove useful in purification of other prenyltransferases.

Acknowledgments

This work was supported by NIH Grants GM 25521 and GM 21328 and by NIH postdoctoral fellowship GM 09198 to Desiree L. Bartlett.

[21] Geranylgeranylpyrophosphate Synthetase of Pig Liver

By HIROSHI SAGAMI, KOICHI ISHII, and KYOZO OGURA

Geranylgeranylpyrophosphate synthetase obtained from pig liver in a form free of farnesylpyrophosphate synthetase (dimethylallyltransferase, EC 2.5.1.1) catalyzes the following two reactions, and does not catalyze the condensation between dimethylallyl pyrophosphate and isopentenyl pyrophosphate.[1]

$$\text{Geranyl-PP} + \text{isopentenyl-PP} \xrightarrow{Mn^{2+}} \text{farnesyl-PP} + \text{PP}_i$$
$$\text{Farnesyl-PP} + \text{isopentenyl-PP} \xrightarrow{Mn^{2+}} \text{geranylgeranyl-PP} + \text{PP}_i$$

[1] H. Sagami, K. Ishii, and K. Ogura, *Biochem. Int.* **3**, 669 (1981).

ISBN 0-12-182010-6

Assay Method

Principle. The method is based on the same principle as described earlier, determining the extent to which ^{14}C-labeled isopentenyl pyrophosphate reacts with nonlabeled geranyl or farnesyl pyrophosphate in the presence of enzyme and manganese ions[2] to form ^{14}C-labeled geranylgeranyl pyrophosphate which is acid-labile.

Reagents

Potassium phosphate buffer, 0.5 *M*, pH 7.0
$MnCl_2$, 0.01 *M*
[1-^{14}C]Isopentenyl pyrophosphate (specific activity, 1 Ci/mol), 0.5 m*M*
Farnesyl pyrophosphate, 1 m*M*
Iodoacetamide, 0.2 *M*
HCl, 6 *M*
NaOH, 6 *M*

Procedure. In a 10-ml capped test tube are placed 0.1 ml of 0.5 *M* phosphate buffer, pH 7.0, 0.1 ml of $MnCl_2$, 0.1 ml of iodoacetamide, 0.05 ml of [^{14}C]isopentenyl pyrophosphate, and 0.05 ml of farnesyl pyrophosphate. Water and enzyme are added to give a final volume of 1.0 ml. After the mixture is incubated at 37° for 12 hr, 0.1 ml of HCl is added to the mixture to terminate the enzymatic reaction, and the mixture is kept at 37° for 30 min. The mixture is then made alkaline by the addition of 0.35 ml of NaOH, and is shaken with 4 ml of hexane. The hexane layer is washed with water and an aliquot of the hexane extract is counted for radioactivity in toluene scintillator. The enzyme activity is given by the radioactivity in the hexane extract. Identification of the product is carried out with the free alcohol obtained by alkaline phosphatase treatment by means of thin-layer chromatography with an authentic sample of all-*trans*-geranylgeraniol. For chromatography, a plate coated with cellulose powder impregnated with paraffin in petroleum ether is used in a system of acetone–water (13 : 7, v/v) saturated with paraffin oil.

Purification of Enzyme

All steps are carried out at 0–4° unless otherwise stated.

Pig liver, obtained from a slaughterhouse, is cut into small pieces and homogenized for 1 min with a Potter-Elvehjem homogenizer in a 5-fold volume of 0.1 *M* phosphate buffer, pH 7.4 containing 125 m*M* sucrose, 4 m*M* $MgCl_2$, 1 m*M* EDTA, 30 m*M* nicotinamide, and 2.5 m*M* glutathione. The homogenate is filtered through double layers of gauze, and the filtrate

[2] D. L. Nandi and J. W. Porter, *Arch. Biochem. Biophys.* **105,** 7 (1964).

PURIFICATION OF GERANYLGERANYLPYROPHOSPHATE SYNTHETASE FROM PIG LIVER

Step	Protein (mg)	Activity		Specific activity (units/mg)	Purity (fold)
		Units[a]	Yield (%)		
100,000 *g* supernatant	38,415	32,197	100	0.8	1.0
0–60% $(NH_4)_2SO_4$ fraction	14,007	19,292	59.9	1.4	1.8
DEAE Sephadex A-50	1,493	9,470	29.4	6.3	7.9
Hydroxylapatite	74	2,067	6.4	27.9	34.9

[a] One unit of enzyme activity is defined as the activity required to convert 1 pmol of isopentenyl pyrophosphate into geranylgeranyl pyrophosphate per minute.

is centrifuged at 100,000 *g* for 1 hr. To the supernatant, granulated ammonium sulfate is slowly added with stirring to achieve 60% saturation, and the mixture is centrifuged at 26,000 *g* for 30 min. The precipitate is dissolved in a minimal volume of phosphate buffer, and the solution is filtered through Sephadex G-25 equilibrated with 0.01 *M* phosphate buffer, pH 6.8 containing 4 m*M* $MgCl_2$, 1 m*M* EDTA, 1 m*M* glutathione, and 0.05 *M* NaCl. A part of the gel eluate containing 280 mg protein is applied to a column (1.8 × 30 cm) of DEAE Sephadex A-50 equilibrated with the same buffer containing 4 m*M* $MgCl_2$, 1 m*M* EDTA, 1 m*M* glutathione, and 0.05 *M* NaCl. The elution is carried out with a linear concentration gradient of NaCl from 0.05 to 0.35 *M* (total 800 ml). The eluted fractions having enzyme activity are pooled and concentrated by ultrafiltration, and the concentrated solution is dialyzed against 0.2 *M* phosphate buffer, pH 6.8 containing 1 m*M* EDTA and 1 m*M* glutathione. The dialyzed solution is applied on a column (1.8 × 30 cm) of hydroxylapatite, prepared according to the method of Bernardi,[3] equilibrated with the same buffer. The column is developed with the same buffer until protein is no longer eluted. During this elution, farnesylpyrophosphate synthetase, the activity of which is much higher than that of geranylgeranylpyrophosphate synthetase, is collected in the early fractions. After about 460 ml elution, the column is developed with 0.35 *M* phosphate buffer, pH 6.8 containing 1 m*M* each of EDTA and glutathione. Geranylgeranylpyrophosphate synthetase is eluted soon after this change of buffer. Figure 1 shows a typical hydroxylapatite chromatography carried out with 112 mg of DEAE-Sephadex purified enzyme protein. The enzyme solution obtained by hydroxylapatite chromatography is concentrated by ultrafiltration, and chromatographed through a Sephadex G-200 column (3.0 × 80 cm) with 0.01 *M* phosphate buffer, pH 6.8. The table summarizes the results of purification up to hydroxylapatite chromatography starting with 200 g of pig liver.

[3] G. Bernardi, this series, Vol. 22, p. 325.

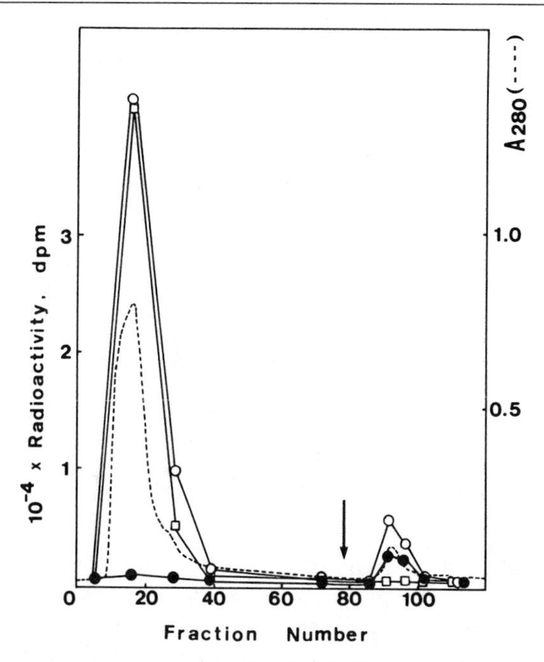

FIG. 1. Hydroxylapatite chromatography. Six milliliter fractions were collected and 0.4-ml aliquots of the fractions were assayed in the presence of [^{14}C]isopentenyl pyrophosphate with dimethylallyl pyrophosphate (□), geranyl pyrophosphate (○), and farnesyl pyrophosphate (●). The arrow indicates the point of the change of elution buffer.

Properties

Stability. The enzyme purified by hydroxylapatite chromatography can be stored frozen at −20° for at least 1 month without loss of activity. The time course of the enzyme reaction at 37° is linear for at least 2 hr.

Activators and Inhibitors. The enzyme requires Mn^{2+} ions for maximum activity. Mg^{2+} and Co^{2+} ions are less effective in activating the enzyme. The activity is more than 90% inhibited by 0.01 *M* inorganic pyrophosphate.

Substrate Specificity. The rate of formation of geranylgeranyl pyrophosphate from geranyl pyrophosphate is almost equal to that from farnesyl pyrophosphate, but that from dimethylallyl pyrophosphate is negligible. This specificity is in contrast to that of geranylgeranylpyrophosphate synthetase of pumpkin[4] and *Micrococcus luteus*.[5] Farnesyl pyrophosphate at concentrations over 25 μM is inhibitory.

[4] K. Ogura, T. Shinka, and S. Seto, *J. Biochem.* (*Tokyo*) **72,** 1101 (1972).

[5] H. Sagami and K. Ogura, *J. Biochem.* (*Tokyo*) **89,** 1573 (1981).

Dependence on Enzyme Concentration. Linear dependence of product formation on the concentration of enzyme is maintained up to 30 μg protein/ml.

Molecular Weight. The molecular weight is estimated to be about 300,000 by Sephadex G-200 filtration.

pH Optimum. The enzyme exhibits maximal activity at pH 7.0 in phosphate buffer.

[22] Geranylpyrophosphate Synthetase–Geranylgeranylpyrophosphate Synthetase from *Micrococcus luteus*

By HIROSHI SAGAMI and KYOZO OGURA

A partially purified geranylpyrophosphate synthetase preparation obtained from *Micrococcus luteus* catalyzes the synthesis of geranyl pyrophosphate from dimethylallyl pyrophosphate and isopentenyl pyrophosphate and of all-*trans*-geranylgeranyl pyrophosphate from farnesyl pyrophosphate and isopentenyl pyrophosphate, but it does not catalyze the synthesis of farnesyl pyrophosphate from geranyl pyrophosphate and isopentenyl pyrophosphate.[1] It is uncertain whether this preparation is a mixture of two enzymes, geranylpyrophosphate synthetase catalyzing the reaction of $C_5 \rightarrow C_{10}$ and geranylgeranylpyrophosphate synthetase catalyzing the reaction of $C_{15} \rightarrow C_{20}$ or a single enzyme catalyzing distinct reactions.

Geranylgeranylpyrophosphate synthetase which catalyzes the three sequential reactions $C_5 \rightarrow C_{10} \rightarrow C_{15} \rightarrow C_{20}$ is also found in *M. luteus*[2,3] as well as in higher plants,[4] but this chapter is concerned with the geranylgeranylpyrophosphate synthetase which lacks geranyl-transferring activity.

Reactions Catalyzed

$$\text{Dimethylallyl-PP} + \text{isopentenyl-PP} \xrightarrow{Mg^{2+}} \text{geranyl-PP} + PP_i$$

$$\text{Farnesyl-PP} + \text{isopentenyl-PP} \xrightarrow{Mg^{2+}} \text{geranylgeranyl-PP} + PP_i$$

[1] H. Sagami and K. Ogura, *J. Biochem. (Tokyo)* **89,** 1573 (1981).
[2] A. A. Kandutsch, H. Paulus, E. Levin, and K. Bloch, *J. Biol. Chem.* **239,** 2507 (1964).
[3] H. Sagami, K. Ogura, S. Seto, and T. Kurokawa, *Biochem. Biophys. Res. Commun.* **85,** 572 (1978).
[4] K. Ogura, T. Nishino, T. Shinka, and S. Seto, this volume [19].

ISBN 0-12-182010-6

Assay Method

The principle of the assay method is the same as described in preceding chapters.

Reagents

Tris–HCl buffer, 1 *M*, pH 7.7
$MgCl_2$, 0.01 *M*
[^{14}C]Isopentenyl pyrophosphate (sp. act., 1 Ci/mol), 0.5 m*M*
Dimethylallyl-PP, 1 m*M*
Farnesyl pyrophosphate, 1 m*M*
HCl, 1 *M*, 6 *M*
NaOH, 1 *M*
Hexane

Procedure. In a 10-ml capped test tube are placed 0.1 ml of Tris buffer, 0.3 ml of $MgCl_2$, 0.1 ml of dimethylallyl pyrophosphate (or 0.05 ml of farnesyl pyrophosphate), and 0.05 ml of [^{14}C]isopentenyl pyrophosphate. Water and enzyme are added to give a final volume of 1.0 ml. The mixture is incubated for 75 min. The reaction of isopentenyl pyrophosphate with dimethylallyl pyrophosphate is stopped by the addition of 0.05 ml of 1 *M* HCl. For the reaction with farnesyl pyrophosphate, 0.1 ml of 6 *M* HCl is added to stop the reaction. After the acidified mixture is kept at 37° for 30 min, the mixture is made alkaline by the addition of NaOH and the mixture is extracted with 3 ml of hexane. The hexane layer is washed with water, and a 1-ml aliquot is assayed for radioactivity in a toluene scintillator. The enzyme activity is proportional to the total radioactivity in the hexane extract. One unit of enzyme activity is defined as the amount required to convert 1 nmol of isopentenyl pyrophosphate to product per minute.

Purification of Enzyme

All steps are carried out at 0–4° unless otherwise stated.

Extraction. Eighty grams of spray-dried cells of *M. luteus,* purchased from Worthington Biochemical, is suspended in 1.6 liter of 0.05 *M* Tris–HCl buffer, pH 7.7 containing 1 m*M* EDTA at room temperature. To the mixture is added 400 mg of lysozyme (Boehringer Mannheim GmbH), and the mixture is manually stirred with a glass rod until a thick gel is formed. Seventy milligrams of deoxyribonuclease (Boehringer Mannheim GmbH) is then added to reduce the viscosity, and the mixture is stirred for an additional 10 min. The mixture is centrifuged at 100,000 *g* for 30 min and the supernatant is collected in a chilled beaker.

Ammonium Sulfate Fractionation. Granulated ammonium sulfate is slowly added to the supernatant (50% saturation) with stirring and the solution is allowed to stand for 30 min. The resulting precipitate is collected by centrifugation at 25,000 *g* for 30 min and stored at −20° before use.

First DEAE-Sephadex Chromatography. One-sixth of the collected precipitate is dissolved in a minimal volume of 0.05 *M* potassium phosphate buffer, pH 6.8 containing 0.05 *M* NaCl and 1 m*M* EDTA and the solution is filtered through Sephadex G-25 equilibrated with the same buffer. The filtrate, containing about 1.8 g of protein, is chromatographed on a column (2.6 × 39 cm) of DEAE-Sephadex A-50 equilibrated with the same buffer. Elution is performed with a linear concentration gradient established between 1 liter of 0.05 *M* NaCl in 0.05 *M* phosphate buffer, pH 6.8 containing 1 m*M* EDTA and 1 liter of 0.85 *M* NaCl in the same buffer containing 1 m*M* EDTA. Fractions containing 250 drops are collected and assayed for enzyme activity. The geranylgeranylpyrophosphate synthetase is eluted at a NaCl concentration of about 0.2 *M*, being separated from nonaprenylpyrophosphate synthetase (see later) eluted at about 0.4 *M* NaCl. Fractions containing geranylgeranylpyrophosphate synthetase are pooled.

Second DEAE-Sephadex Chromatography. The pooled fraction from the first chromatography is concentrated by ultrafiltration using a Diaflo apparatus with a PM-30 membrane (Amicon) and rechromatographed on a DEAE-Sephadex A-50 column (2.6 × 39 cm) equilibrated with 0.05 *M* phosphate buffer, pH 6.8 containing 1 m*M* EDTA and 0.05 m*M* NaCl. Elution is carried out with a linear gradient of concentration from 0.05 to 0.65 *M* of NaCl in the same buffer containing 1 m*M* EDTA (total volume 2 liters). The effluent is collected in 250-drop portions and assayed for enzyme activity.

Hydroxylapatite Chromatography. The active fraction (161 ml) pooled from the preceding step is concentrated to 1.75 mg protein/ml by ultrafiltration with a PM-30 membrane, and the concentrate is brought to 50% saturation of ammonium sulfate. The resulting precipitate is collected by centrifugation at 25,000 *g* for 30 min. One-third of the precipitate is dissolved in a minimum volume of 1 m*M* phosphate buffer, pH 6.4 and the solution is filtered through Sephadex G-25 equilibrated with the same buffer. The gel filtrate is applied to a column (1.6 × 9 cm) of hydroxylapatite equilibrated with the same buffer. Elution is carried out with a linear concentration gradient of 1 to 100 m*M* phosphate buffer pH 6.4 (total volume 480 ml). The effluent is collected in 80-drop portions. The enzyme is eluted in the region with a conductivity of 150 $\mu\Omega$.

PURIFICATION OF GERANYL-PP SYNTHETASE OF *M. luteus*

Step	Protein (mg)	Total activity (units)	Specific activity (units/mg)	Yield (%)	Purity (fold)
100,000 *g* supernatant	56,318	2,100	0.037	100	1
Ammonium sulfate fraction	10,766	2,132	0.198	102	5
First DEAE-Sephadex	936	1,097	1.17	52	32
Second DEAE-Sephadex	125	331	2.65	16	72
Hydroxylapatite	31.6	215	6.80	10	184
Gel filtration	6.0	109	18.2	5	492

Sephadex G-100 Gel Filtration. The pooled fraction (256 ml) from the hydroxylapatite chromatography is concentrated to 1.2 mg protein/ml by ultrafiltration with a PM-30 membrane. To the concentrate is slowly added granulated ammonium sulfate with stirring to achieve 50% saturation, and the mixture is allowed to stand for 30 min. The resulting precipitate is collected by centrifugation at 25,000 *g* for 30 min and dissolved in 5 ml of 0.05 *M* phosphate buffer, pH 6.8 containing 0.05 *M* NaCl, 1 m*M* EDTA, and 1 m*M* dithiothreitol. The resulting solution is subjected to gel filtration through a Sephadex G-100 column (2.2 × 88 cm) equilibrated with the same buffer solution. Elution is carried out at a flow rate of 0.2 ml/min. Fractions of 60 drops are collected and assayed for enzyme activity. The enzyme fractions are combined and concentrated by ultrafiltration and the concentrate is desalted by gel filtration through a column of Sephadex G-25 equilibrated with 0.05 *M* Tris–HCl buffer, pH 7.7. The resulting enzyme solution (about 1 unit/ml) can be kept at −20° for at least 1 week without loss of activity.

The table shows the results of a purification procedure.

Properties

Activators. The enzyme requires Mg^{2+} ions. Dimethylallyl-transferring and farnesyl-transferring activities are maximum in the presence of 3 and 1 m*M* Mg^{2+} ions, respectively, and both activities decline as the Mg^{2+} concentration is increased. Mn^{2+} ions are slightly less effective in activating the enzyme activity.

Dimethylallyl-transferring and farnesyl-transferring activities are affected in different ways by some compounds. Triton X-100 stimulates the farnesyl-transferring activity but inhibits the dimethylallyl-transferring activity. The latter activity is 60% inhibited by 0.02 *M* iodoacetamide, whereas the former activity is not affected at all. Both activities are almost completely inhibited by 0.01 *M* inorganic pyrophosphate, which is one of the products of these reactions.

Effect of Substrate Concentration. The K_m values in the dimethylallyl-transferring reaction are estimated to be 6.2×10^{-5} and 8.0×10^{-6} *M* for dimethylallyl pyrophosphate and for isopentenyl pyrophosphate, respectively. The farnesyl-transferring activity is inhibited by farnesyl pyrophosphate at concentrations over 5×10^{-5} *M*.

Molecular Weight. The molecular weight of the enzyme is about 70,000 as estimated from gel filtration on Sephadex G-100.

pH Optimum. Both dimethylallyl-transferring and farnesyl-transferring activities are maximal at pH 7.7.

Substrate Specificity. E-3-Methyl-2-pentenyl pyrophosphate is active as a substrate with isopentenyl pyrophosphate at a rate of 0.32 relative to that of dimethylallyl pyrophosphate, but neither *E*-3-methyl-2-hexenyl nor *E*-3-methyl-2-heptenyl pyrophosphate is active. *E*-3-Methyl-2-undecenyl and *E*-methyl-2-dodecenyl pyrophosphates react with isopentenyl pyrophosphate in rates of 0.53 and 0.45 relative to that of farnesyl pyrophosphate, respectively.

[23] Hexaprenylpyrophosphate Synthetase of *Micrococcus luteus* B-P 26

By HIROSHI FUJII, TANETOSHI KOYAMA, and KYOZO OGURA

Hexaprenylpyrophosphate synthetase obtained from *Micrococcus luteus* B-P 26 catalyzes the synthesis of all-*trans*-hexaprenyl pyrophosphate by successive condensation of three molecules of isopentenyl pyrophosphate with farnesyl pyrophosphate as a primer.[1] It does not catalyze the reaction between isopentenyl pyrophosphate and either dimethylallyl or geranyl pyrophosphate. This enzyme consists of two essential protein components, designated component A and component B, which can be separated from each other by chromatography. Either has no catalytic activity. The hexaprenylpyrophosphate synthetase activity is restored when they are combined.[1]

Reaction Catalyzed

$$\text{Farnesyl-PP} + 3\ \text{isopentenyl-PP} \xrightarrow{Mg^{2+}} \text{all-}trans\text{-hexaprenyl-PP} + 3\ PP_i$$

[1] H. Fujii, T. Koyama, and K. Ogura, *J. Biol. Chem.* **257,** 14610 (1982).

METHODS IN ENZYMOLOGY, VOL. 110

ISBN 0-12-182010-6

Stereochemistry. The enzyme catalyzes a condensation to yield a trans product with elimination of the *pro-R* hydrogen from C-2 of isopentenyl pyrophosphate.

Assay Method

Principle. The principle of assay method is similar to that for farnesylpyrophosphate synthetase and geranylgeranylpyrophosphate synthetase,[2] but the conditions for acid hydrolysis of longer chain prenyl pyrophosphates is more drastic than that for shorter chain products. When chromatographic fractions are assayed, it should be noted that two protein components which are separable by chromatography are necessary for the hexaprenylpyrophosphate synthetase activity.

Reagents

Tris–HCl buffer, 1 *M*, pH 7.4
$MgCl_2$, 0.05 *M*
[1-^{14}C]Isopentenyl pyrophosphate (sp. act., 1 Ci/mol), 0.5 m*M*
Farnesyl pyrophosphate, 1 m*M*
HCl, 6 *M*
NaOH, 6 *M*
Hexane

Procedure. In a 10 ml capped test tube are placed 0.1 ml of Tris buffer, 0.1 ml of $MgCl_2$, 0.05 ml of [^{14}C]isopentenyl pyrophosphate, 0.05 ml of farnesyl pyrophosphate, an enzyme solution, and water to give a final volume of 1.0 ml. After incubation at 37° for 1 hr, 0.2 ml of HCl is added, and the mixture is heated at 65° for 20 min. The mixture is made alkaline by the addition of 0.4 ml of NaOH and is extracted with 3.0 ml of hexane. The hexane layer is washed with water and an aliquot of the extract is counted for radioactivity in toluene scintillator. The enzyme activity is calculated from the radioactivity in the hexane extract.

Cell Culture

Cells of *M. luteus* B-P 26[3] are grown at 30° in a nutrient medium composed of 25 g of nutrient broth No. 2 (Oxoid) and 1 liter of distilled water. Three-liter shaking flasks containing 1 liter of medium are inoculated with 10 ml of a precultured suspension from an overnight shaking culture. After 12 hr of growth with shaking, cells are harvested by centrifugation and washed twice with 50 m*M* Tris–HCl buffer, pH 7.4.

[2] K. Ogura, T. Nishino, T. Shinka, and S. Seto, this volume [19].

[3] L. Jeffries, M. A. Cawthorne, M. Harris, A. T. Diplock, J. Green, and S. A. Price, *Nature (London)* **215**, 257 (1967).

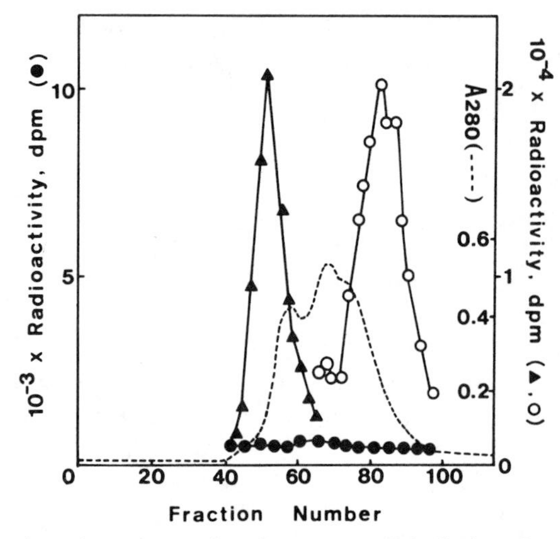

FIG. 1. Separation of component A and component B by hydroxylapatite chromatography. Enzyme activity was assayed in the presence of [^{14}C]isopentenyl pyrophosphate with farnesyl pyrophosphate (●), farnesyl pyrophosphate plus 0.25 ml of fraction 54 (○), and farnesyl pyrophosphate plus 0.25 ml of fraction 82 (▲).

Preparation of Enzyme and Resolution into Essential Components

Cells (wet weight, 20 g) are suspended in 50 ml of 50 m*M* Tris–HCl buffer (pH 7.4) containing 1 m*M* $MgCl_2$ and 1 m*M* 2-mercaptoethanol. The cell suspension is passed six times through a French Press at 1500 kg/cm^2. Subsequent steps are carried out at 0–4°. The mixture is centrifuged at 108,000 *g* for 30 min. To the supernatant, which contains farnesylpyrophosphate synthetase (dimethylallyltransferase, EC 2.5.1.1) and hexaprenylpyrophosphate synthetase, granulated ammonium sulfate is slowly added with stirring to achieve 80% saturation. After centrifugation at 26,000 *g* for 20 min, the precipitate is dissolved in a minimum volume of 10 m*M* potassium phosphate buffer, pH 7.0 containing 10 m*M* 2-mercaptoethanol, 0.5 m*M* EDTA, and 1% (v/v) polyethylene glycol-200 (buffer I) and is applied to a hydroxylapatite column (1.5 × 42 cm) equilibrated with buffer I. Elution is with a linear gradient of 10 to 250 m*M* potassium phosphate buffer, pH 7.0 (total volume 400 ml). The effluent is collected in 3 ml portions. During this chromatography the hexaprenylpyrophosphate synthetase is resolved into two components each of which has no catalytic activity. Therefore, these components cannot be located unless the eluted fractions are assayed for the synthetase activity with a supplement of a fraction containing the counterpart component. Figure 1 shows the chromatographic separation of these two components. As shown by

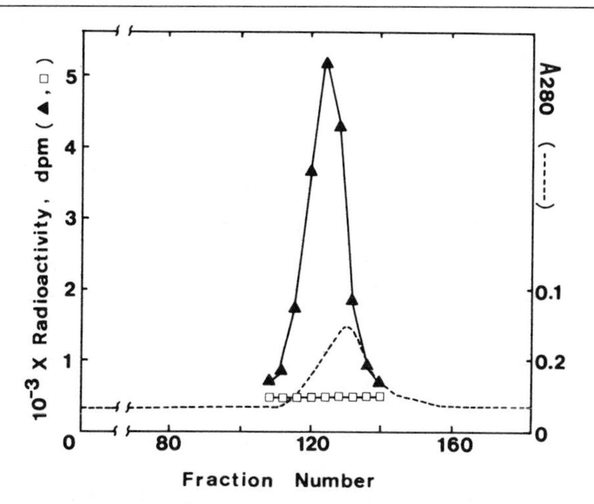

FIG. 2. Sephadex G-200 filtration of component A. Hexaprenylpyrophosphate synthetase activity (▲) was assayed with [^{14}C]isopentenyl pyrophosphate and farnesyl pyrophosphate in the presence of 0.1 ml of fraction 82 from the hydroxylapatite chromatography (Fig. 1). Farnesylpyrophosphate synthetase activity (□) was assayed with [^{14}C]isopentenyl pyrophosphate and geranyl pyrophosphate.

closed circles, enzyme activity is not detected at any fraction when each fraction is assayed with [^{14}C]isopentenyl pyrophosphate and farnesyl pyrophosphate, but the hexaprenylpyrophosphate synthetase activity is detected, as shown by closed triangles, at fractions 46–60, and also, as shown by open circles, at fractions 76–90 when the assay is carried out in the presence of fraction 82 containing component B and fraction 54 containing component A, respectively (Fig. 1). Farnesylpyrophosphate synthetase is eluted at fractions 51–63, overlapping with component A (data not shown in Fig. 1).

Fractions 46–56, which contain component A and farnesylpyrophosphate synthetase, are combined and concentrated by ultrafiltration through a PM-10 membrane (Amicon) with an exclusion limit of M_r 10,000 under a nitrogen atmosphere. The concentrated solution is subjected to gel filtration through a Sephadex G-200 column (2.7 × 86 cm) with buffer I, and a major part of the active fractions is similarly concentrated. The concentrate is rechromatographed on the same column of Sephadex G-200 with buffer I to yield a preparation of component A free of farnesylpyrophosphate synthetase (Fig. 2).

Fractions 76–90, which contain component B, obtained from the hydroxylapatite chromatography (Fig. 1) are combined, concentrated, and rechromatographed on a hydroxylapatite column (1.5 × 40 cm) and fur-

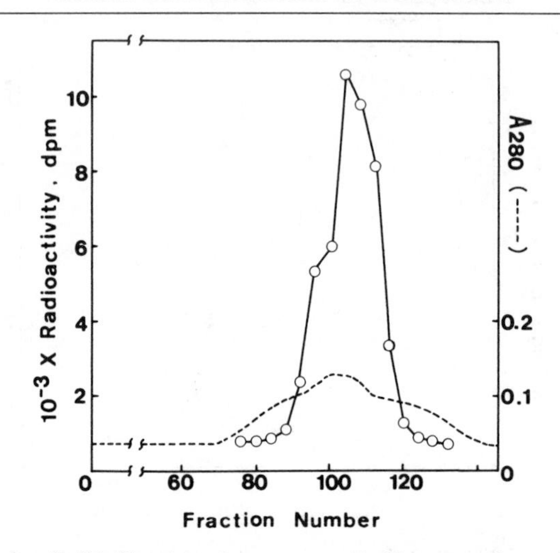

FIG. 3. Sephadex G-200 filtration of component B. Hexaprenylpyrophosphate synthetase activity was assayed in the presence of 0.15 ml of fraction 54 from the hydroxylapatite chromatography (Fig. 1).

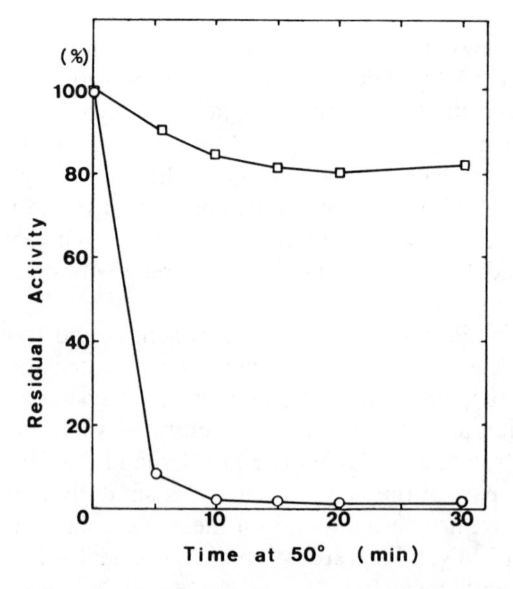

FIG. 4. Heat stability of component A and component B. After component A (□) and component B (○) purified by Sephadex G-200 filtration (Figs. 2 and 3) were heated at 50° for the indicated periods, the activity was assayed with a supplement of their counterpart component.

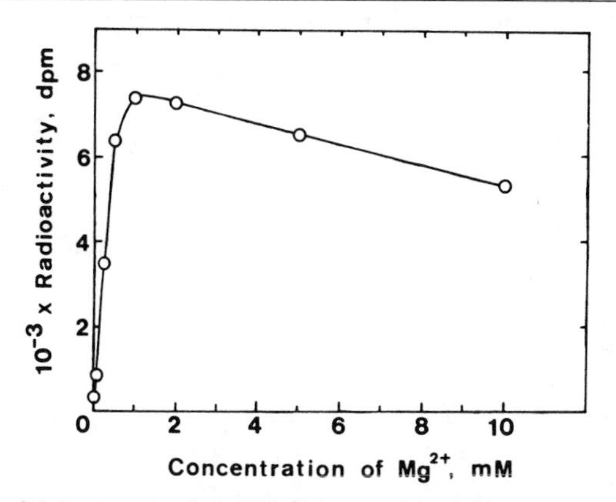

FIG. 5. Effect of the concentration of Mg^{2+} ion on the enzyme activity. Component A (87 μg) and component B (16 μg) were incubated with [^{14}C]isopentenyl pyrophosphate and farnesyl pyrophosphate in the presence of various concentrations of Mg^{2+} ion.

ther chromatographed on a Sephadex G-200 column (2.7 × 86 cm) with buffer I (Fig. 3).

Properties

Stability. Component A is much more stable to heat treatment than component B (Fig. 4). Both components can be stored frozen at −20° for several months.

Activator. The enzyme absolutely requires Mg^{2+} ions for its activity.

Substrate Specificity. Neither dimethylallyl pyrophosphate nor geranyl pyrophosphate is accepted as a primer by this enzyme, but all-*trans*-farnesyl and all-*trans*-geranylgeranyl pyrophosphates are accepted.

Affinity of Substrates. The K_m values obtained from Lineweaver–Burk plots are 25.0, 1.4, and 4.0 μM for isopentenyl, all-*trans*-farnesyl, and all-*trans*-geranylgeranyl pyrophosphate, respectively.

Time Course. The amount of the product increases linearly with the incubation time up to at least 3 hr.

Effect of Mg^{2+} Concentration. Figure 5 shows the dependence of the enzyme activity on the concentration of Mg^{2+} ions. The distribution of the polyprenyl pyrophosphates synthesized by this enzyme varies, depending on the concentration of Mg^{2+} in the reaction medium similarly to the case of nonaprenylpyrophosphate synthetase.[4] The amount of C_{30} product relative to that of C_{25} product increases as Mg^{2+} concentration is elevated.

[4] H. Fujii, H. Sagami, T. Koyama, K. Ogura, S. Seto, T. Baba, and C. M. Allen, *Biochem. Biophys. Res. Commun.* **96,** 1648 (1980).

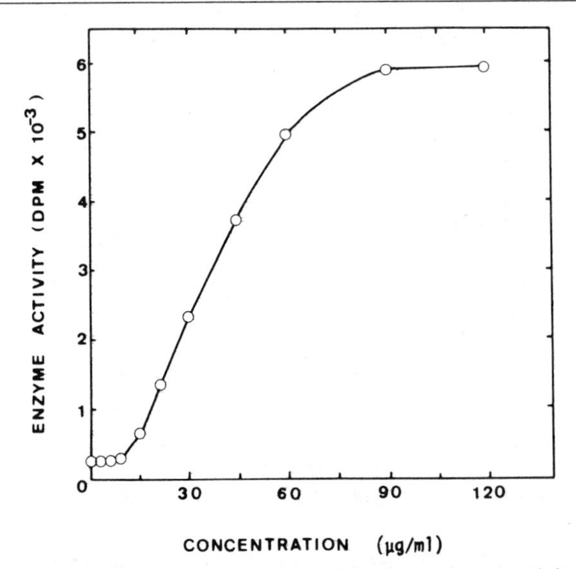

FIG. 6. Effect of the concentration of component A on the enzyme activity. Hexaprenylpyrophosphate synthetase activity was assayed using 16 μg of component B and in the presence of various concentrations of component A as indicated.

For example, at 0.5 mM Mg^{2+}, C_{30} and C_{25} compounds are produced in a ratio of 1 : 0.6, but at 5 mM Mg^{2+}, the ratio is 1 : 0.2. At 20 mM Mg^{2+}, more than 90% of the product is the C_{30} compound.

Effect of the Concentration of Component A on the Enzyme Activity. The dependency of hexaprenylpyrophosphate synthetase activity on the concentration of component A is shown in Fig. 6.

Molecular Weight. From gel filtration on Sephadex G-200, the molecular weights of component A and component B are estimated to be approximately 20,000 and 60,000, respectively. Filtration on Sephadex G-100 of a mixture containing these two components results in the separation of the components, indicating that no tight association occurs between these two components under these conditions.

pH Optimum. The reaction rate is optimal over the pH range 6–7 in phosphate buffer or Tris–HCl buffer.

Physiological Function. The physiological significance of this enzyme is probably to produce the isoprenoid precursor for the biosynthesis of menaquinone-6 contained in *M. luteus* B-P 26.[3]

[24] Heptaprenylpyrophosphate Synthetase from *Bacillus subtilis*

By IKUKO SAGAMI, HIROSHI FUJII, TANETOSHI KOYAMA, and KYOZO OGURA

Heptaprenylpyrophosphate synthetase is obtained from *Bacillus subtilis,* which produces menaquinone-7. This enzyme catalyzes the synthesis of all-*trans*-heptaprenyl pyrophosphate by successive condensation of four molecules of isopentenyl pyrophosphate with farnesyl pyrophosphate as a primer.[1] all-*trans*-Geranylgeranyl pyrophosphate can also be a primer, but neither dimethylallyl nor geranyl pyrophosphate is accepted as a primer.[1] This enzyme consists of two essential protein components which are separated from each other by ion exchange chromatography.[2] Either of the components itself has no catalytic activity but restores the enzyme activity to synthesize heptaprenyl pyrophosphate when combined with each other.

Reaction Catalyzed

$$\text{Farnesyl-PP} + 4\ \text{isopentenyl-PP} \xrightarrow{Mg^{2+}} \text{all-}trans\text{-heptaprenyl-PP} + 4\ PP_i$$

Stereochemistry. The *pro-R* hydrogen is eliminated from the 2-position of isopentenyl pyrophosphate during the condensation reaction forming trans double bonds.

Cell Culture

Cells of *Bacillus subtilis* PCI-219 are grown at 27° in a nutrient medium composed of 25 g of nutrient broth (Oxoid) and 1 liter of distilled water. Five-liter flasks containing 1.5 liters of medium are inoculated with 20 ml of a precultured suspension from a 1-day-old shaking culture. After 22 hr of growth with shaking, cells are harvested by centrifugation, washed twice with 0.05 *M* Tris–HCl buffer, pH 7.5 and used for enzyme preparation.

Assay Methods

Principle. The principle of the assay method is the same as described earlier for hexaprenylpyrophosphate synthetase. Attention must be paid

[1] I. Takahashi, K. Ogura, and S. Seto, *J. Biol. Chem.* **255,** 4539 (1980).
[2] H. Fujii, T. Koyama, and K. Ogura, *FEBS Lett.* **161,** 257 (1983).

ISBN 0-12-182010-6

to the fact the enzyme is resolved by chromatography into two essential components each of which has no activity unless combined with each other.

Reagents

Tris–HCl buffer, 1 *M*, pH 7.5
$MgCl_2$, 0.1 *M*
L-Cysteine, 0.05 *M*
2-Mercaptoethanol, 0.1 *M*
[1-^{14}C]Isopentenyl pyrophosphate (sp. act., 1 Ci/mol), 0.5 m*M*
Geranyl pyrophosphate, 0.5 m*M*
Farnesyl pyrophosphate, 0.5 m*M*
HCl, 6 *M*
NaOH, 6 *M*
Hexane

Procedure. In a 10 ml capped test tube are placed 0.1 ml of Tris–HCl buffer, 0.05 ml of $MgCl_2$, 0.1 ml of L-cysteine, 0.1 ml of 2-mercaptoethanol, and 0.05 ml each of [^{14}C]isopentenyl pyrophosphate and farnesyl pyrophosphate. Water and enzyme are added to give a final volume of 1.0 ml. After incubation at 37° for 1 hr, 0.2 ml of 6 *M* HCl is added and the mixture is heated at 65° for 30 min to hydrolyze the products. The mixture is made alkaline by the addition of 0.4 ml of 6 *M* NaOH and is extracted with 3.0 ml of hexane. The hexane layer is washed with water, and an aliquot of the hexane solution is counted in toluene scintillator for radioactivity. The enzyme activity is expressed as the radioactivity in the hexane extract.

The method of product analysis is similar to that described earlier for hexaprenylpyrophosphate synthetase.[3]

Preparation of Enzyme

Cells (wet weight, 20 g) are suspended in 200 ml of 0.05 *M* Tris–HCl buffer, pH 7.5 containing 1 m*M* 2-mercaptoethanol and 1 m*M* $MgCl_2$. Lysozyme (300 mg, Boehringer Mannheim GmbH) is added to the cell suspension, and the mixture is stirred for 1 hr at room temperature. Subsequent steps are carried out at 0–4°. To the resulting thick gel is added 20 mg of deoxyribonuclease (Boehringer Mannheim GmbH), and the mixture is stirred for 30 min. The mixture is centrifuged at 108,000 *g* for 30 min, and the supernatant is fractionated with ammonium sulfate. The

[3] H. Fujii, T. Koyama, and K. Ogura, this volume [23].

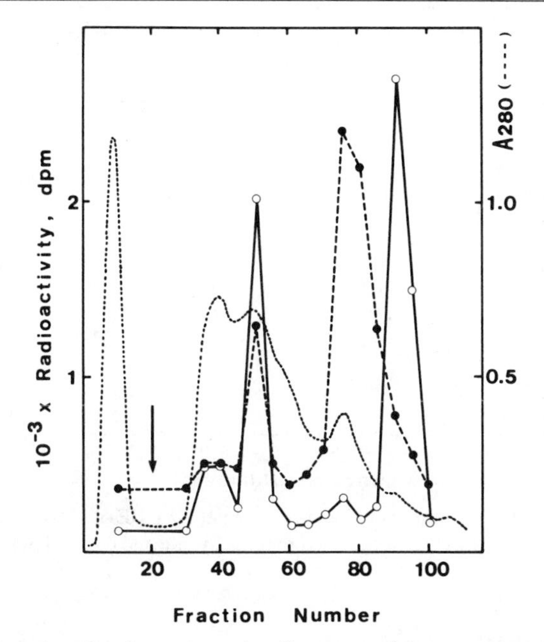

FIG. 1. Hydroxylapatite chromatography. Enzyme activity was assayed using as allylic substrates farnesyl pyrophosphate (●) and dimethylallyl pyrophosphate (○). The arrow indicates the starting point of gradient elution.

protein fraction precipitating between 30 and 80% of saturation with ammonium sulfate is subjected to gel filtration on a Sephadex G-100 column (2.5 × 102 cm) equilibrated with 0.01 *M* phosphate buffer, pH 6.8 containing 0.01 *M* 2-mercaptoethanol. The eluted fractions containing prenyltransferase activities are combined and the resulting solution (~200 mg protein in 6.6 ml) is applied to a hydroxylapatite column (1.2 × 22 cm) equilibrated with the same buffer. After elution with 100 ml of the same buffer, the column is developed with a linear concentration gradient of 0.01 to 0.30 *M* phosphate buffer (total volume 400 ml). Effluent is collected in 4.8 ml portions and assayed for prenyltransferase activity (Fig. 1). By this chromatography heptaprenylpyrophosphate synthetase is separated from farnesylpyrophosphate synthetase,[4] geranylgeranylpyrophosphate synthetase,[5] and undecaprenylpyrophosphate synthetase[5] (data not shown in Fig. 1). The heptaprenylpyrophosphate synthetase purified through hydroxylapatite chromatography is so unstable that the activity is lost in a week at 0°.

[4] I. Takahashi and K. Ogura, *J. Biochem.* (*Tokyo*) **89,** 1581 (1981).

[5] I. Takahashi and K. Ogura, *J. Biochem.* (*Tokyo*) **92,** 1527 (1982).

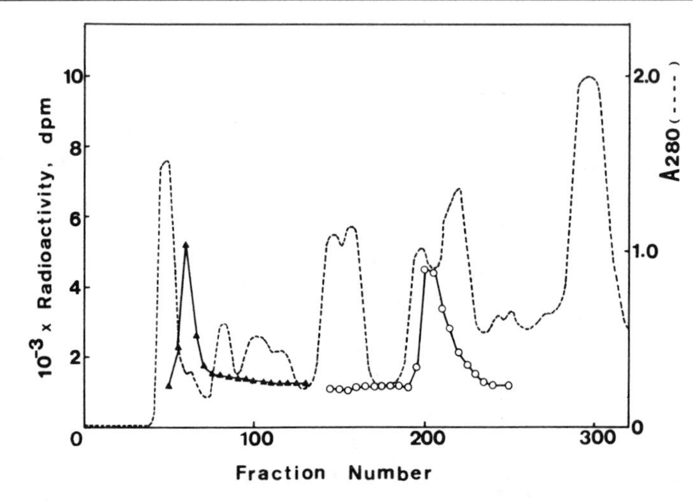

FIG. 2. Separation of component I and component II by DE-52 chromatography. Enzyme activity was assayed with [^{14}C]isopentenyl pyrophosphate and farnesyl pyrophosphate in the presence of fraction 60 (○) and fraction 205 (▲).

Separation of Component I and Component II

Cells (wet weight, 23 g) are suspended in 140 ml of 0.05 *M* potassium phosphate buffer pH 7.0 containing 0.01 *M* 2-mercaptoethanol and 0.05 *M* KCl. To the cell suspension are added 240 mg of lysozyme (Boehringer Mannheim GmbH), and the mixture is stirred at room temperature for 50 min. Twenty-two milligrams of deoxyribonuclease (Boehringer Mannheim GmbH) is then added and stirring is continued for an additional 10 min. Subsequent steps are carried out at 0–4°. The resulting mixture with reduced viscosity is centrifuged at 30,900 *g* for 90 min, and the supernatant is chromatographed on a column (4.5 × 69 cm) of DE-52 with a linear concentration gradient established between 2 liters of 0.05 *M* KCl in 0.05 *M* phosphate buffer, pH 7.0 containing 0.01 *M* 2-mercaptoethanol and 2 liters of 0.85 *M* KCl in the same buffer. Sixteen-milliliter fractions are collected, and aliquots of the fractions are assayed for heptaprenylpyrophosphate synthetase activity with a supplement of a fraction containing component I or component II. In the chromatography shown in Fig. 2, 0.3 ml of fraction 60 and of fraction 205 are supplemented. Farnesylpyrophosphate synthetase is eluted in fractions 164–198 (data not given). The essential components, designated component I and component II in the order of elution, are further purified on Sephadex G-100. Fractions 50–70 and fractions 190–220 are pooled and chromatographed on a column (2.8 × 107 cm) of Sephadex G-100 with 0.05 *M* phosphate buffer,

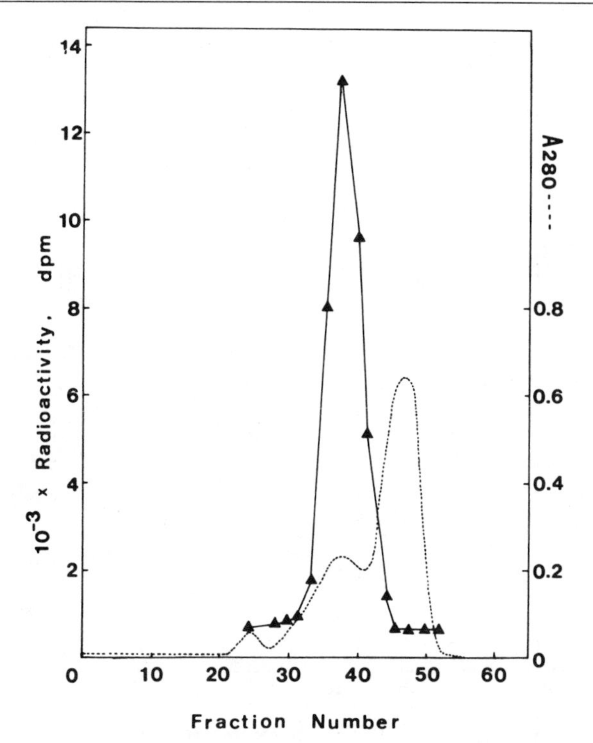

FIG. 3. Sephadex G-100 filtration of component I (fractions 44–70 in Fig. 2). The activity was assayed in the presence of 0.3 ml of a component II solution containing fractions 192–215 in Fig. 2.

pH 7.0 containing 0.01 *M* 2-mercaptoethanol and 0.05 *M* KCl. Effluent is collected in 10 ml portions (Figs. 3 and 4).

Properties

Stability. Component II is inactivated almost completely when heated at 50° for 5 min, whereas component I does not lose its activity even when heated at 50° for 30 min.

Molecular Weight. The molecular weights of component I and component II are estimated to be both about 30,000 from Sephadex G-100 filtration.

Activator. Mg^{2+} is an absolute requirement for the enzymatic reaction. The effectiveness of Mn^{2+} is one-third that of Mg^{2+}.

Substrate Specificity and Kinetic Properties. Neither dimethylallyl pyrophosphate nor geranyl pyrophosphate is active as a primer, but all-

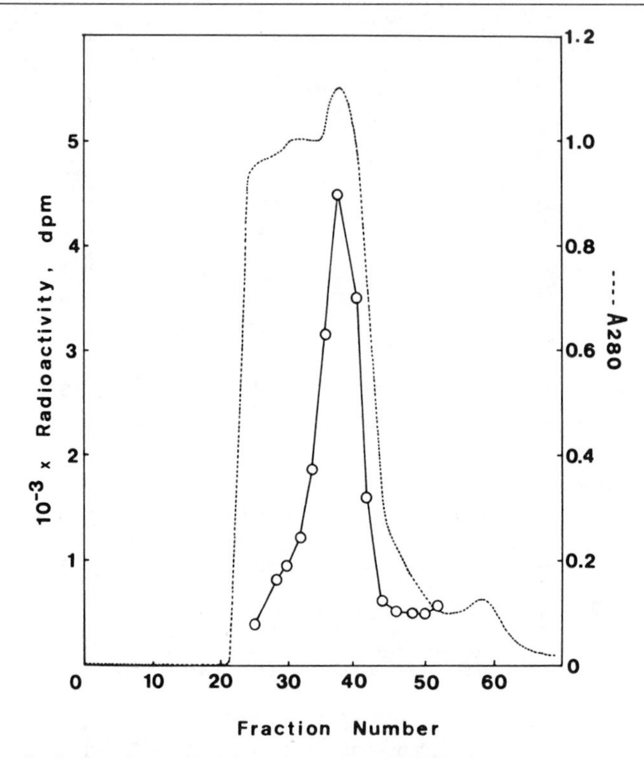

FIG. 4. Sephadex G-100 filtration of component II (fractions 192–215 in Fig. 2). The activity was assayed in the presence of 0.3 ml of a component I solution containing fractions 44–70 in Fig. 2.

trans-farnesyl pyrophosphate and all-*trans*-geranylgeranyl pyrophosphate are active. The K_m values obtained with a hydroxylapatite-purified enzyme are 12.8, 13.3, and 8.3 μM for isopentenyl, all-*trans*-farnesyl, and all-*trans*-geranylgeranyl pyrophosphate, respectively.

Effect of Incubation Time and Protein Concentration. The formation of the product increases linearly with time of incubation at least up to 90 min and with protein concentration up to 2 mg/ml.

pH Optimum. The enzyme activity is maximal at pH 6.5–7.0 in phosphate buffer.

Effect of the Concentration of Component I. The heptaprenylpyrophosphate synthetase activity is absolutely dependent on component I and component II. Figure 5 shows the effect of the concentration of component I on the enzyme activity.

Physiological Significance. Of the four prenyltransferases separated from extracts of *B. subtilis*, undecaprenylpyrophosphate synthetase re-

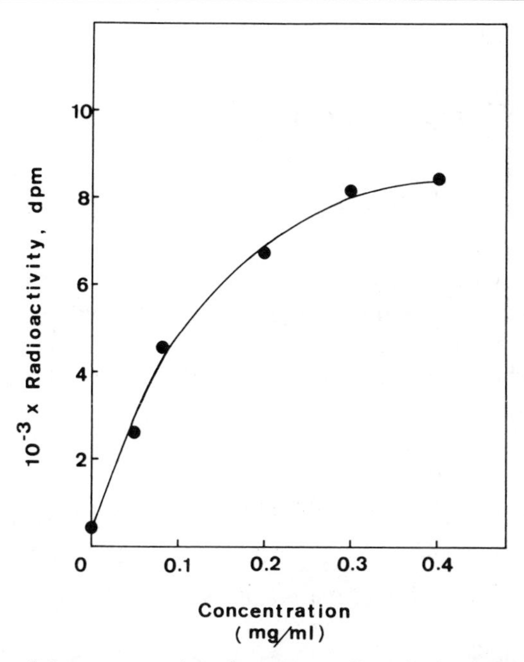

FIG. 5. Effect of the concentration of component I on the heptaprenylpyrophosphate synthetase activity. Component II (0.8 mg) was incubated with [^{14}C]isopentenyl pyrophosphate and farnesyl pyrophosphate in the presence of various concentrations of component I as indicated.

sembles farnesylpyrophosphate synthetase in that both enzymes are greatly stimulated by monovalent cations such as K^+ and NH_4^+ and also by detergents such as Triton X-100 and Tween 80.[4,5] On the other hand, heptaprenylpyrophosphate synthetase resembles geranylgeranylpyrophosphate synthetase in that neither enzyme is susceptible to monovalent cations or detergents. These properties suggest that the former two operate in intracellular environments closely related to each other, probably being peripherally associated with membrane so that undecaprenyl pyrophosphate of cis, trans-mixed stereochemistry can be efficiently synthesized, and that the latter two cooperate in the cytosol to produce all-*trans*-C_{35}-prenyl pyrophosphate which is the precursor for the side chain of menaquinone-7 occurring in *B. subtilis*.

[25] Nonaprenylpyrophosphate Synthetase from *Micrococcus luteus*

By HIROSHI SAGAMI and KYOZO OGURA

A nonaprenylpyrophosphate synthetase fraction obtained from *Micrococcus luteus* catalyzes the condensation of isopentenyl pyrophosphate with geranyl pyrophosphate to yield all-*trans*-prenyl pyrophosphates ranging in carbon number up to C_{45}.[1] Dimethylallyl pyrophosphate does not act as a primer.

Reaction Catalyzed

$$\text{Geranyl-PP} + 7\ \text{isopentenyl-PP} \xrightarrow{Mg^{2+}} \text{all-}trans\text{-nonaprenyl-PP} + 7\ PP_i$$

Stereochemistry. The enzyme catalyzes a condensation to yield trans products with elimination of the *pro-R* hydrogen from C-2 of isopentenyl pyrophosphate.

Assay Methods

Principle. The enzyme activity is assayed as usual for polyprenylpyrophosphate synthetase by determining the amount of incorporation of [^{14}C]isopentenyl pyrophosphate into acid-labile allylic pyrophosphate.

Reagents

Tris–HCl buffer, 1 *M*, pH 7.4
$MgCl_2$, 0.1 *M*
[1-^{14}C]Isopentenyl pyrophosphate (sp. act. 1 Ci/mol), 0.5 m*M*
Geranyl pyrophosphate, 0.5 m*M*
HCl, 6 *M*
NaOH, 6 *M*
Hexane

Procedure. In a 10-ml capped test tube are placed 0.1 ml of Tris–HCl buffer, 0.1 ml of $MgCl_2$, 0.1 ml of [^{14}C]isopentenyl pyrophosphate, and 0.04 ml of geranyl pyrophosphate. Water and enzyme are added to give a final volume of 1.0 ml. After incubation at 37° for 3 hr, 0.2 ml of 6 *M* HCl is added and the mixture is kept at 65° for 30 min. The mixture is then made alkaline by adding 0.4 ml of 6 *M* NaOH and extracted with 3 ml of hexane.

[1] H. Sagami, K. Ogura, and S. Seto, *Biochemistry* **16,** 4616 (1977).

ISBN 0-12-182010-6

The hexane layer is washed with water and an aliquot of the hexane solution is counted in toluene scintillator for radioactivity. The enzyme activity is expressed by the total radioactivity in the hexane extract.

Product Analysis. The polyprenyl pyrophosphates formed by enzymatic reaction are hydrolyzed with acid phosphatase by the method of Fujii *et al.*,[2] and the polyprenols thus liberated are extracted and analyzed by both normal phase and reversed phase thin layer chromatography as described earlier.[3] For mass spectrometric identification, polyprenols are purified by high-pressure liquid chromatography on a Hitachi porous polymer gel No. 3011 column (0.8 × 50 cm) with methanol–hexane (4 : 1, v/v). The elution of prenol is monitored by the absorption at 218 nm and/or radioactivity. The polyprenol thus purified is introduced on a probe directly into the ion source of a mass spectrometer. The temperature of the ion source is 40 ~ 80° for polyprenols of C_{25} ~ C_{45}. The potential of the ionizing electron beam is usually 70 eV.

Purification of Enzyme

All operations are at 0–4° unless otherwise stated.

Extraction. The preparation of extract of *Micrococcus luteus* is based on the work by Allen *et al.*[4] Spray-dried cells (20 g) obtained from Worthington Biochemical are suspended in 400 ml of 0.05 *M* Tris buffer, pH 7.4. A hundred milligrams of lysozyme (Boehringer Mannheim GmbH) is added to the suspension, and the mixture is stirred manually at room temperature for about 30 min until a thick gel forms. Twenty milligrams of deoxyribonuclease (Boehringer Mannheim GmbH) is then added and stirring is continued for an additional 10 min. The resulting mixture with reduced viscosity is centrifuged at 60,000 *g* for 1 hr, and the supernatant is transferred to a beaker.

Ammonium Sulfate Fractionation. To the supernatant granulated ammonium sulfate is added with stirring to achieve 30% saturation, and the mixture is centrifuged at 25,000 *g* for 30 min. To the resulting supernatant ammonium sulfate is similarly added to achieve 50% saturation, and the precipitate which is collected by centrifugation is dissolved in a minimum volume of 0.05 *M* Tris–HCl buffer, pH 7.4 and is filtered through Sephadex G-25 equilibrated with 0.05 *M* potassium phosphate buffer, pH 6.8 containing 0.05 *M* NaCl. The filtrate which contains about 150 mg of protein is applied to a column (1.5 × 20 cm) of DEAE-Sephadex A-50 equilibrated with the same buffer. Elution is performed with a linear

[2] H. Fujii, T. Koyama, and K. Ogura, *Biochim. Biophys. Acta* **712,** 716 (1982); this volume [17].

[3] K. Ogura, T. Nishino, T. Shinka, and S. Seto, this volume [19].

[4] C. M. Allen, W. Alworth, A. Macrae, and K. Bloch, *J. Biol. Chem.* **242,** 1895 (1967).

PURIFICATION OF NONAPRENYLPYROPHOSPHATE SYNTHETASE FROM *M. luteus*

Step	Protein (mg)	Specific activity (unit[a]/mg)	Purification (fold)
60,000 *g* supernatant	20,350	42	1
30–50% ammonium sulfate	3,344	133	3.1
DEAE-Sephadex	384	586	13.9
Ultrogel AcA 44	45	787	18.7

[a] One unit of activity is defined as the amount of enzyme which converts 1 pmol of isopentenyl pyrophosphate into the products per minute.

concentration gradient established between 250 ml of 0.05 *M* NaCl in 0.05 *M* phosphate buffer, pH 6.8 and 250 ml of 1.05 *M* NaCl in the same buffer. The effluent is collected in 6 ml portions and assayed for enzyme activity. Nonaprenylpyrophosphate synthetase is eluted after geranylgeranylpyrophosphate synthetase.[5] Fractions containing the major part of the nonaprenylpyrophosphate synthetase activity are combined, concentrated, and filtered through Ultrogel AcA 44 with 0.05 *M* Tris–HCl buffer, pH 7.4. The table summarizes the results of a purification procedure. Further chromatography on hydroxylapatite results in a decrease of specific activity.

Properties

Stability. The partially purified enzyme can be stored frozen at −20° for at least 2 weeks without loss of activity.

Activators. Mg^{2+} ions are an absolute requirement for the enzymatic reaction, and Mn^{2+} ions are less effective. About 2-fold stimulation is observed when the reaction is carried out in the presence of 0.05–0.1% Tween 80 or 0.03–0.06% bacitracin.

Substrate Specificity and Kinetic Parameters. Geranyl, all-*trans*-farnesyl, and all-*trans*-geranylgeranyl pyrophosphate are active as substrate. Dimethylallyl pyrophosphate and *cis,trans,trans*-geranylgeranyl pyrophosphate are inactive. The apparent K_m values for geranyl pyrophosphate and farnesyl pyrophosphate are 0.75 and 25 μM, respectively.

pH Optimum. The pH optimum of the enzyme is 7.0–8.0 in Tris–HCl buffer.

Molecular Weight. The molecular weight of the enzyme is estimated to be 78,000 by gel filtration on Sephadex G-100.

[5] H. Sagami and K. Ogura, *J. Biochem. (Tokyo)* **89,** 1573 (1983).

Product Variability. The chain length of the synthesized polyprenyl pyrophosphate is dramatically affected by the concentration of Mg^{2+} in the medium.[6] At 0.1 m*M* Mg^{2+}, C_{25}, C_{30}, and C_{35} compounds are produced with C_{25} predominating; at 0.5 m*M* Mg^{2+}, C_{25}, C_{30}, C_{35}, and C_{40} compounds are produced with C_{35} and C_{40} predominating; and at 20 m*M* Mg^{2+}, C_{40} and C_{45} are produced without accumulation of shorter compounds.

Physiological Significance. Since *M. luteus* produces menaquinones having prenyl chains of C_{30}, C_{35}, C_{40}, and C_{45} with a predominance of C_{40}, the role of this enzyme is probably to provide the isoprenoid precursors for these menaquinones, though it is uncertain whether or not this enzyme preparation is a mixture of more than one enzyme protein. Geranyl pyrophosphate, which is needed as a substrate to prime the chain elongation by nonaprenylpyrophosphate synthetase, will be provided by geranylpyrophosphate synthetase which has also been isolated from this bacteria.[5]

[6] H. Fujii, H. Sagami, T. Koyama, K. Ogura, S. Seto, T. Baba, and C. M. Allen, *Biochem. Biophys. Res. Commun.* **96,** 1648 (1980).

[26] Enzymatic Synthesis of Phytoene

By BENJAMIN L. JONES and JOHN W. PORTER[1]

Enzyme systems in the chromoplasts of tomato fruits[1a] and in extracts of spinach leaves[2] convert isopentenyl pyrophosphate to phytoene by the reactions shown in Fig. 1. The enzyme system from tomato fruits (which has been studied more extensively than the spinach system) behaves as a complex on BioGel filtration.[3] However, during subsequent DEAE-cellulose chromatography phytoene-synthesizing activity is lost but activities for the formation of acid-labile compounds are retained.[1a,4] Subsequent purification steps result in the separation of activities for isopentenylpy-

[1] Deceased, June 27, 1984.

[1a] F. B. Jungalwala and J. W. Porter, *Arch. Biochem. Biophys.* **119,** 209 (1967).

[2] C. Subbarayan, S. Kushwaha, G. Suzue, and J. W. Porter, *Arch. Biochem. Biophys.* **137,** 547 (1970).

[3] B. Maudinas, M. L. Bucholtz, C. Papastephanou, S. S. Katiyar, A. V. Briedis, and J. W. Porter, *Arch. Biochem. Biophys.* **180,** 354 (1977).

[4] M. A. Islam, S. A. Lyrene, E. M. Miller, Jr., and J. W. Porter, *J. Biol. Chem.* **252,** 1523 (1977).

ISBN 0-12-182010-6

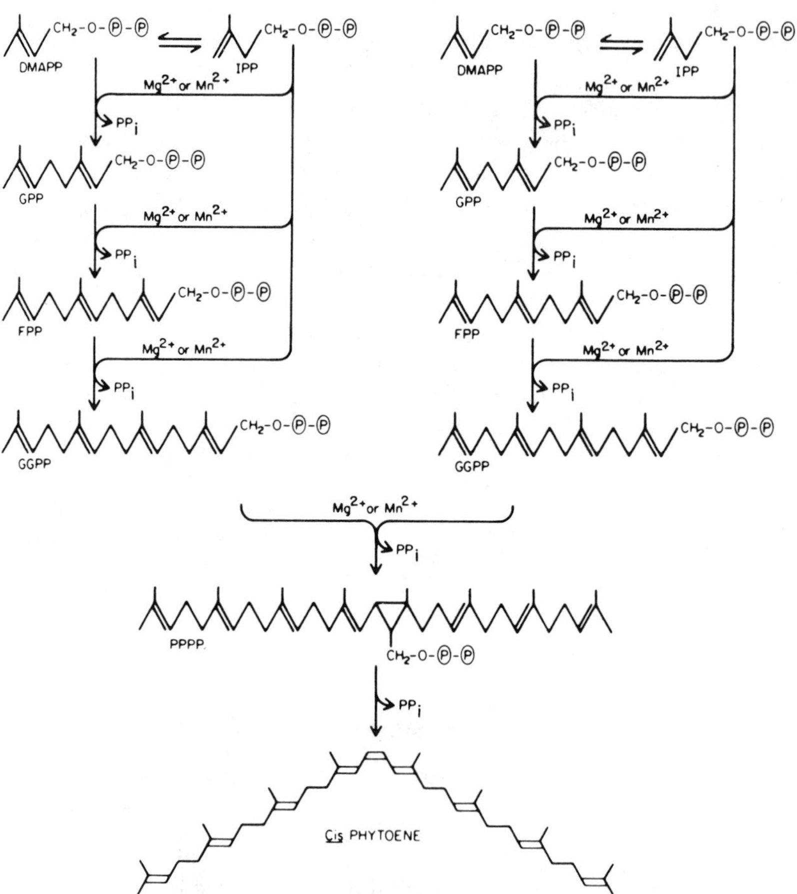
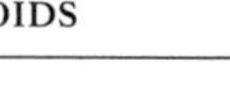

FIG. 1. The enzymatic reactions effected by the enzyme system that converts isopentenyl pyrophosphate to phytoene. The following abbreviations are used: IPP, isopentenyl pyrophosphate; DMAPP, dimethylallyl pyrophosphate; GPP, geranyl pyrophosphate; FPP, farnesyl pyrophosphate; GGPP, geranylgeranyl pyrophosphate; and PPPP, prephytoene pyrophosphate. [Reproduced with permission from J. W. Porter, S. L. Spurgeon, and N. Sathyamoorthy, Biosynthesis of Carotenoids, *in* "Isopentenoids in Plants" (W. D. Nes, G. Fuller, and L.-S. Tsai, eds.), p. 161. Marcel Dekker, New York, 1984.]

rophosphate isomerase and prenyltransferase.[5] Other purification steps result in the separation of the enzymes which convert geranylgeranyl pyrophosphate to phytoene from the isomerase and prenyltransferase.[6]

[5] S. L. Spurgeon, N. Sathyamoorthy, and J. W. Porter, *Arch. Biochem. Biophys.* **230,** 446 (1984).

[6] B. L. Jones and J. W. Porter, unpublished data (1983).

The separation and properties of the enzyme system for phytoene synthesis that exists in tomatoes are the concern of this chapter.

Assay Method—Tomato Enzymes

General Procedure. The enzymatic synthesis of phytoene is determined by assay for the incorporation of radioactivity of [1-^{14}C]isopentenyl pyrophosphate or [1-^{3}H]geranylgeranyl pyrophosphate into this product. The biosynthesized phytoene is extracted from the incubation mixture with petroleum ether and then separated from other compounds by high-pressure liquid chromatography (HPLC) or alumina chromatography. The incorporation of radioactivity into phytoene is confirmed by coelution with authentic phytoene on HPLC and alumina column chromatography.

Tomato System

Reagents

Borate buffer, 500 m*M*, pH 8.2
$MgCl_2$, 375 m*M*
$MnCl_2$, 100 m*M*
Dithiothreitol, 100 m*M*
ATP, 8 m*M*
$NADP^+$, 40 m*M*
Tween 80, 100 mg/ml
[1-^{14}C]Isopentenyl pyrophosphate, 0.2 m*M* or
[1-^{3}H]Geranylgeranyl pyrophosphate, 0.5 m*M*

Unlabeled isopentenyl pyrophosphate is synthesized chemically,[7,8] and purified by HPLC on an anion exchange column with a linear gradient of 0.5 to 4% ammonium bicarbonate. Substrate isopentenyl pyrophosphate (6000 dpm/nmol) is prepared by diluting the labeled compound, purchased from Amersham, with the synthetic compound. [1-^{3}H]Geranylgeraniol is prepared from geranylgeraniol by oxidation with manganese dioxide,[9] and reduction with ^{3}H-labeled sodium borohydride.[10] Phosphorylation of both labeled and unlabeled geranylgeraniol was by the same procedure as for the preparation of isopentenyl pyrophosphate.[7,8] Purification of labeled and unlabeled geranylgeranyl pyrophosphate is carried out on a silica gel column with a convex gradient from 12 : 1,

[7] R. H. Cornforth and G. Popják, this series, Vol. 15, p. 359.
[8] Y. Gafni and I. Shechter, *Anal. Biochem.* **92,** 248 (1979).
[9] L. J. Altman, R. C. Kowerski, and D. R. Laungani, *J. Am. Chem. Soc.* **100,** 6174 (1978).
[10] W. Rudiger, J. Benz, and C. Guthoff, *Eur. J. Biochem.* **109,** 193 (1980).

n-propanol : ammonium hydroxide to 6 : 3 : 1, *n*-propanol : ammonium hydroxide : water.[8] Substrate geranylgeranyl pyrophosphate (30,000 dpm/nmol) is prepared by diluting the labeled compound with unlabeled compound.

Incubation System

The incubation system for the biosynthesis of phytoene, which is a slight modification of previous systems,[1a,11] contains the following: $MnCl_2$, 1 μmol; $MgCl_2$, 7.5 μmol; dithiothreitol, 5 μmol; $NADP^+$, 2 μmol; ATP, 0.4 μmol; Tween 80, 5 mg; borate buffer, pH 8.2, 50 μmol; enzyme protein, 0.5 to 2 mg, and either [^{14}C]isopentenyl pyrophosphate, 20 μmol and 120,000 dpm or [1-^{3}H]geranylgeranyl pyrophosphate, 25 nmol and 750,000 dpm, in a final volume of 0.5 ml. Incubation of this mixture under nitrogen is carried out for 2 hr at 22°.

Isolation of Phytoene

The enzyme reaction is stopped by the addition of four volumes of absolute ethanol and nonpolar compounds are then extracted with two 4.0-ml aliquots each of silica gel purified petroleum ether (bp 40 to 60°). The extract is dried over anhydrous sodium sulfate, concentrated by a stream of nitrogen, and then chromatographed on alumina with increasing concentrations of diethyl ether in petroleum ether.[1a]

More routinely, the extract is concentrated under a stream of nitrogen to near dryness, resuspended in acetonitrile : tetrahydrofuran, 1 : 1, 66%, and methanol : water, 3 : 1, 34%, and analyzed on a C_{18} reverse phase column by HPLC with the same solvent mixture.[12] Phytoene is detected by absorbance at 280 nm. Eluate fractions are collected and aliquots are taken for liquid scintillation counting for radioactivity.

Enzyme Activity

Total enzyme activity is expressed as nanomoles of substrate incorporated into phytoene under the described assay conditions.

Assay for Isopentenyl Pyrophosphate Isomerase

General Procedure. The conversion of isopentenyl pyrophosphate to dimethylallyl pyrophosphate is catalyzed by isomerase. Isopentenyl pyro-

[11] D. V. Shah, D. H. Feldbruegge, A. R. Houser, and J. W. Porter, *Arch. Biochem. Biophys.* **127,** 124 (1968).

[12] P. Beyer, K. Kreuz, and H. Klenig, *Planta* **150,** 435 (1980).

phosphate is stable to acid while dimethylallyl pyrophosphate is not, and the latter is converted to the alcohol form in the presence of acid. This alcohol is extracted with petroleum ether, and an aliquot is taken for liquid scintillation counting for radioactivity.

Reagents

TES buffer, 1 *M*, pH 7.5
$MgCl_2$, 373 m*M*
$MnCl_2$, 100 m*M*
Dithiothreitol, 100 m*M*
Hydrochloric acid, 10.0 *N*
[1-^{14}C]Isopentenyl pyrophosphate, 0.2 m*M*

Incubation System

The incubation system, modified from a previous report,[5] for the assay of isomerase contains $MnCl_2$, 1 μmol; $MgCl_2$, 7.5 μmol; dithiothreitol, 5 μmol; TES buffer, pH 7.5, 100 μmol; [1-^{14}C]isopentenyl pyrophosphate, 20 nmol and 120,000 dpm, and enzyme protein, 0.02 to 0.5 mg in a final volume of 0.5 ml. Incubation is carried out at 37° for 30 min.

Assay of Product

The reaction is terminated by the addition of four volumes of absolute ethanol. The product is hydrolyzed by the addition of 100 μl of 10.0 *N* hydrochloric acid and incubation at 37° for 20 min. Acid-labile product is extracted with 4 ml of petroleum ether (bp 40 to 60°). A 1-ml aliquot of the extract is taken for liquid scintillation counting for radioactivity.

Activity

Total enzyme activity is expressed as nanomoles of substrate converted to acid-labile product.

Prenyltransferase

General Procedure. The condensation of isopentenyl pyrophosphate with dimethylallyl pyrophosphate to form geranyl pyrophosphate and subsequent condensation with isopentenyl pyrophosphate to form farnesyl pyrophosphate and geranylgeranyl pyrophosphate is catalyzed by prenyl transferase. The reaction is assayed by the incorporation of [1-^{14}C]isopentenyl pyrophosphate into acid-labile products.

Reagents

Potassium phosphate buffer, 0.1 *M*, pH 7.0
Tween 80, 50 mg/ml
$MgCl_2$, 375 m*M*
$MnCl_2$, 100 m*M*
Iodoacetamide, 25 m*M*
Dimethylallyl pyrophosphate, 1.3 m*M*
[1-^{14}C]Isopentenyl pyrophosphate, 0.2 m*M*
Hydrochloric acid, 10.0 *M*

Incubation System

The assay for prenyltransferase activity is essentially the same as reported previously.[5] This assay consists of two incubations: the first to destroy isomerase, and the second for prenyltransferase activity. The assay system contains the following: $MnCl_2$, 1 μmol; $MgCl_2$, 7.5 μmol; Tween 80, 5 mg; iodoacetamide, 0.5 μmol; potassium phosphate, pH 7.0, 10 μmol; and enzyme protein, 0.02 to 0.5 mg. The incubation is for 30 min at room temperature. At the end of the incubation, [1-^{14}C]isopentenyl pyrophosphate, 20 nmol and 120,000 cpm, and dimethylallyl pyrophosphate, 33 nmol, are added and the volume of the incubation mixture is adjusted to 0.5 ml. Blank assays are handled in the same manner except that dimethylallyl pyrophosphate is omitted. Incubation is carried out for one hour at 37°.

Acid Hydrolysis

The reaction is terminated and the products are extracted and assayed for radioactivity as described for isomerase. Activity is expressed as nmoles [1-^{14}C]isopentenyl pyrophosphate incorporated into product.

Enzymatic Hydrolysis for Product Analysis

Reagents. The reagents are the same as described above for prenyltransferase with the addition of the following:

EDTA, pH 7.0, 250 m*M*
Ammonium hydroxide, 10.0 *M*
Sodium acetate buffer, pH 4.7, 1 *M*
Triton X-100, 100 mg/ml
Potato acid phosphatase in 10 m*M* sodium acetate, pH 7.5, 2 mg protein/ml (Sigma)
Sodium hydroxide, 6 *N*

The prenyltransferase reaction is stopped by the addition of 25 μmol of EDTA and 1 mmol of ammonium hydroxide. The incubation mixture is then passed through a C_{18} Sep-pak that has been preconditioned according to the manufacturer's instructions. The Sep-pak is washed with 4 ml of water, and the reaction products are eluted with two 2-ml aliquots of methanol. The methanol is blown off under a stream of nitrogen to dryness and the residue is resuspended in 10 m*M* Tris, pH 8.5, to a suitable volume (less than 1 ml). Polyprenyl pyrophosphates are then hydrolyzed with acid phosphatase.[13] The incubation mixture contains sodium acetate buffer, pH 4.7, 0.1 mmol; Triton X-100, 3 mg; acid phosphatase, 3.8 units; an aliquot of the products to be hydrolyzed, and methanol to 60% in a volume of 2.8 ml. The mixture is incubated at 37° overnight, stopped with 0.6 mmol sodium hydroxide, and extracted three times, each with 4 ml of hexane. The extracts are pooled, carrier farnesol, geraniol, and geranylgeraniol are added, and the mixture is blown to dryness under a stream of nitrogen. The sample is resuspended in tetrahydrofuran : acetonitrile, 1 : 1, 60%, and methanol : water, 2 : 1, 40%, and analyzed by HPLC on a C_{18} reverse phase column in the same solvent mixture.[12] Terpenyl alcohols are detected by absorbance at 218 nm or by a change in refractive index. Fractions are collected and aliquots are taken for liquid scintillation counting for radioactivity.

Product analysis of the alcohols may also be performed on reverse phase thin layer chromatography (TLC) plates.[5] Silica gel plates are impregnated with paraffin oil by dipping the plates in a solution of 5% paraffin oil in petroleum ether. The developing system is methanol : water (80 : 20). Geraniol, farnesol, and geranylgeraniol are spotted as standards. The positions of compounds are determined by iodine vapors and radioactive products are determined by cutting or scraping the plate into scintillation vials, adding cocktail and counting for radioactivity.

The pyrophosphate form of the product is analyzed by TLC.[14] The plates are silica gel H buffered with 0.1 M ammonium phosphate, pH 6.5, and they are activated for 20 min at 110° prior to use. The solvent system is chloroform : methanol : water (60 : 40 : 9). Standards of farnesyl pyrophosphate, geranyl pyrophosphate, and geranylgeranyl pyrophosphate are spotted. After development spots are detected by iodine vapor or by spraying for phosphate with Rosenburg's reagent.[15] Radioactive products are determined by scraping the gel into a scintillation vial, adding cocktail, and counting for radioactivity.

[13] H. Fujii, T. Koyama, and K. Ogura, *Biochim. Biophys. Acta* **712,** 716 (1982).

[14] H. Rilling, *J. Lipid Res.* **10,** 183 (1969).

[15] H. Rosenburg, *J. Chromatogr.* **2,** 487 (1959).

Preparation of Enzyme System

Partial Purification. All operations are carried out at 4° unless otherwise stated.

Preparation of Plastids. Tomato fruit plastids are prepared essentially as described previously.[1a] Tomatoes are cut in half, most of the seeds are removed, and the residual tomato sections are placed in a commercial Waring blender with a buffer consisting of 0.2 *M* Tris, pH 8.2, 1 m*M* EDTA, and 1 m*M* 2-mercaptoethanol in a ratio of 800 g tomato tissue to 800 ml of buffer. The mixture is homogenized by three 5-sec bursts of the blender. The slurry is filtered first through a double layer of cheesecloth and then a double layer of cheesecloth with a glass wool mat in it. This mixture is centrifuged at 13,000 *g* for 30 min, the supernatant is poured off, the centrifuge tube refilled with filtrate, and centrifuged again. When the filtrate is exhausted, the pellet is suspended in a minimal volume of 0.1 *M* potassium phosphate buffer, pH 6.5, and 5 m*M* dithiothreitol. The slurry is poured into 20 volumes of −20° acetone and stirred for 20 min at −20°. The acetone is decanted off, fresh acetone added, and the process repeated until the acetone is almost clear. The plastids are then extracted in an analogous fashion with diethyl ether. After extraction plastids are collected on Whatman #1 filter paper and washed with cold diethyl ether. The plastids are placed under vacuum over silica gel overnight, then collected and stored at −20°.

Extraction and Partial Purification of Enzymes of Tomato Fruits

Phytoene Synthetase. Plastids are extracted as described previously.[1a,3] Plastids, usually about 5 g, are extracted in buffer (0.1 *M* potassium phosphate, pH 7.0, 1% Tween 80, 2 m*M* dithiothreitol) at a ratio of 1 g of plastids to 20 ml of buffer. Homogenization is carried out in a tissue homogenizer with a Teflon pestle. The resulting crude extract is centrifuged at 12,000 *g* for 20 min and the resulting supernatant is fractionated with ammonium sulfate. The fraction precipitating at 20 to 60% saturation is collected by centrifugation as described above. The precipitate is dissolved in a minimum volume of 0.1 *M* potassium phosphate buffer, pH 7.0, containing 30% glycerol and 2 m*M* dithiothreitol. The crude enzyme fraction is applied to a column of BioGel A 1.5m (5 × 75 cm), equilibrated in 100 m*M* potassium phosphate, pH 7.0, 2 m*M* dithiothreitol and 30% glycerol. Isomerase, prenyltransferase, and phytoene synthetase activities coelute from this column. Fractions containing activity are pooled and stored at −20°. This preparation synthesizes phytoene from isopentenyl pyrophosphate.

An alternative procedure involves the preparation of the enzyme solution as described for isomerase and prenyltransferase. The majority of the enzyme activity which converts geranylgeranyl pyrophosphate to phytoene is separated from the isomerase and prenyltransferase by ammonium sulfate precipitation between 20 and 40% saturation. The phytoene-synthesizing activity which remains with isomerase and prenyltransferase is eluted from a DEAE-cellulose column by 0.25 *M* potassium phosphate.[6]

Prenyltransferase and Isomerase. This procedure is a modification of a previously published method.[5] Plastids are extracted as described for phytoene synthetase and the protein precipitating between 40 and 80% of saturation with ammonium sulfate is taken. The proteins are resuspended in a minimum volume of 0.1 *M* potassium phosphate buffer, pH 7.0, containing 30% glycerol and 2 m*M* dithiothreitol. The crude enzyme fraction is applied to a Sephadex G-25 column (2.5 × 37 cm) which has been equilibrated with 20 m*M* potassium phosphate, pH 7.0, containing 2 m*M* dithiothreitol. The fractions containing protein are collected and concentrated to less than 20 ml on a PM-10 membrane. The enzyme solution is made 30% with respect to glycerol and then stored at −20°. This is the last stage in which [1-^{14}C]isopentenyl pyrophosphate can be used as substrate for a phytoene synthetase assay without the addition of isomerase and prenyltransferase. The enzyme solution is loaded onto a DEAE-cellulose column (4 × 10 cm) which has been equilibrated with 20 m*M* potassium phosphate, pH 7.0, containing 2 m*M* dithiothreitol and 30% glycerol. The column is washed with the same buffer system containing 50 m*M* potassium phosphate, pH 7.0, then the same buffer system containing 100 m*M* potassium phosphate, pH 7.0. At this buffer concentration isomerase and prenyltransferase coelute.

The fractions containing isomerase and prenyltransferase activity are pooled and concentrated to about 2 to 4 ml on a PM-10 membrane. The enzyme mixture is loaded onto a Sephadex G-100 column (2.6 × 86 cm), equilibrated with 0.1 *M* potassium phosphate buffer, pH 7.0, and 2 m*M* dithiothreitol. The eluate fractions containing isomerase and those containing prenyltransferase activity are separately pooled, concentrated on a PM-10 membrane to about 2 ml, and then made 30% with glycerol. Each fraction is stored at −20° (see the table).

Preliminary results indicate that isomerase may be further purified by passage over hydroxylapatite. A suitable amount of enzyme is loaded onto a Sephadex G-25 column (1 × 45 cm) which has been equilibrated with 10 m*M* potassium phosphate, pH 7.0, 30% glycerol, and 2 m*M* dithiothreitol. The fractions containing protein are pooled and loaded onto a hydroxylapatite column (1.5 × 3 cm) which has been equilibrated with the same buffer. Isomerase elutes in the void volume. Fractions

PURIFICATION OF ISOPENTENYLPYROPHOSPHATE ISOMERASE OF TOMATO CHROMOPLASTS

	Volume (ml)	Protein (mg/ml)	Total protein (mg)	Total isomerase activity (nmol/hr)	Specific activity (nmol/hr/mg)	Yield	Fold purification
Whole homogenate	100.0	25.5	2550	9380	3.7	100	1.0
12,000 *g* Supernatant	82.0	16.6	1361	8315	6.1	89	1.6
40–80% ammonium sulfate precipitation after Sephadex G-25 filtration	31.0	6.4	198	3398	17.1	36	4.6
PM-10 concentrate after DEAE-cellulose chromatography	10.0	2.0	20	1946	96.4	21	26.1
PM-10 concentrate after Sephadex G-100 chromatography	2.8	0.73	2	546	276	6	74.6

containing isomerase activity are pooled, concentrated on a PM-10 membrane, and stored at −20°. Only one or two contaminating proteins are present with isomerase after this procedure is carried out. It should be noted that the G-100 purified isomerase is more stable to storage than the hydroxylapatite-purified enzyme.

Properties of Phytoene Synthetase

Cofactor and Substrate Requirements. Mn^{2+} and Mg^{2+} are required for the synthesis of phytoene from isopentenyl pyrophosphate; however, only Mg^{+} is required for the synthesis of phytoene from geranylgeranyl pyrophosphate.[11] Dithiothreitol is required for maximum phytoene synthetase activity.

Activators and Inhibitors. Dithiothreitol, as noted above, increases activity. In addition, ATP, although not a substrate, markedly stimulates phytoene synthesis.[16] The addition of $NADP^{+}$ stimulates activity up to 3-fold when isopentenyl pyrophosphate is the substrate,[3] but it has little effect when geranylgeranyl pyrophosphate is used.[11] Sulfhydryl reagents

[16] B. Maudinas, M. L. Bucholtz, C. Papastephanou, S. S. Katiyar, A. V. Briedis, and J. W. Porter, *Biochem. Biophys. Res. Commun.* **66**, 430 (1975).

such as *p*-hydroxymercuribenzoate, *N*-ethylmaleimide, and iodoacetamide[1a] markedly inhibit phytoene synthesis.

Substrate Affinity. The apparent K_m for isopentenyl pyrophosphate is 1.0×10^{-5} *M*.[3] Preliminary results indicate that the apparent K_m for geranylgeranyl pyrophosphate is about 2.6×10^{-5} *M*.[6]

Molecular Weight. As determined by gel filtration the phytoene synthetase complex has a molecular weight of about 200,000.[3]

Properties of Isomerase

Cofactor and Substrate Requirements. Only Mn^{2+} and Mg^{2+} are required for the conversion of [1-^{14}C]isopentenyl pyrophosphate to dimethylallyl pyrophosphate.[5] However, dithiothreitol or another thiol reagent is required for maximum enzyme activity.

Optimum pH. The enzyme has an optimum pH of 7.5 in TES buffer.[6]

Activators and Inhibitors. Dithiothreitol increases activity while sulfhydryl reagents such as iodoacetamide, inhibit the enzyme.[5] The enzyme is also inhibited by dimethylallyl and geranyl pyrophosphates and by pyrophosphate, but there appears to be little inhibition by farnesyl pyrophosphate.[5]

Substrate Affinity. The K_m for isopentenyl pyrophosphate is 5.7×10^{-6} *M*.[5]

Molecular Weight. As determined by gel filtration, the molecular weight of this enzyme is 34,000.[5]

Properties of Prenyl Transferase

Cofactor and Substrate Requirements. Mn^{2+} and Mg^{2+} are required for the synthesis of terpenyl pyrophosphates. [1-^{14}C]Isopentenyl pyrophosphate and a "starting" terpenyl pyrophosphate, usually dimethylallyl pyrophosphate, are required. However, geranyl pyrophosphate or farnesyl pyrophosphate may also be used in place of dimethylallyl pyrophosphate.[5]

Products of the Reaction. Analysis of the products of prenyltransferase indicate the production of farnesyl and geranylgeranyl pyrophosphates. Presumably, geranyl pyrophosphate is produced but it is not detected.

Optimum pH. The pH curve is rather broad with an optimum of about 7.0 in potassium phosphate buffer.[5]

Activators and Inhibitors. The enzyme activity is maximal in the presence of Tween 80,[5,6] but in the presence of $NADP^+$ and pyrophosphate, enzyme activity is inhibited.[5]

Substrate Affinity. The K_m for dimethylallyl pyrophosphate is 9.3×10^{-6} *M*, while the apparent K_m for isopentenyl pyrophosphate is 5.7×10^{-6} *M*.[6]

Molecular Weight. The molecular weight of the prenyl transferase, as determined by gel filtration, is 64,000.[5]

Acknowledgment

This work was supported in part by Research Grant HL 16364 from the National Heart and Lung Institute, National Institutes of Health, United States Public Health Service, and by the Medical Research Service of the Veterans Administration.

[27] Carotene Mutants of *Phycomyces*

By ENRIQUE CERDÁ-OLMEDO

Color of *Phycomyces*

The young mycelia and sporangiophores of the Zygomycete fungus *Phycomyces* owe their yellow color to the accumulation of β-carotene. The same pigment is present throughout the life cycle (Fig. 1) but overshadowed, at times, by greenish, brown, or black melanins and sporopollenins. In fact in 1817 the fungus was misclassified as an alga, *Ulva nitens;* the definitive genus name, *Phycomyces,* meaning "algal fungus," is a reminder of this confusion.

Research has concentrated on the species *Phycomyces blakesleeanus*. Color and carotene content differ appreciably from one wild strain to another. Most of the physiological, biochemical, and genetic work has been carried out with strain NRRL1555, belonging to the (−) mating type. Strains largely isogenic with NRRL1555, but belonging to the (+) mating type, have been constructed by repeated backcrosses.[1] All future work should be concentrated on strains isogenic with NRRL1555 if possible. Otherwise the influence of variations in genetic background must be cautiously kept in mind.

Detection of Carotene Mutants

All available carotene mutants were originally detected by simple visual inspection of agar cultures. White, red, and deep-yellow mutants con-

[1] M. I. Álvarez and A. P. Eslava, *Genetics* **105,** 873 (1983).

ISBN 0-12-182010-6

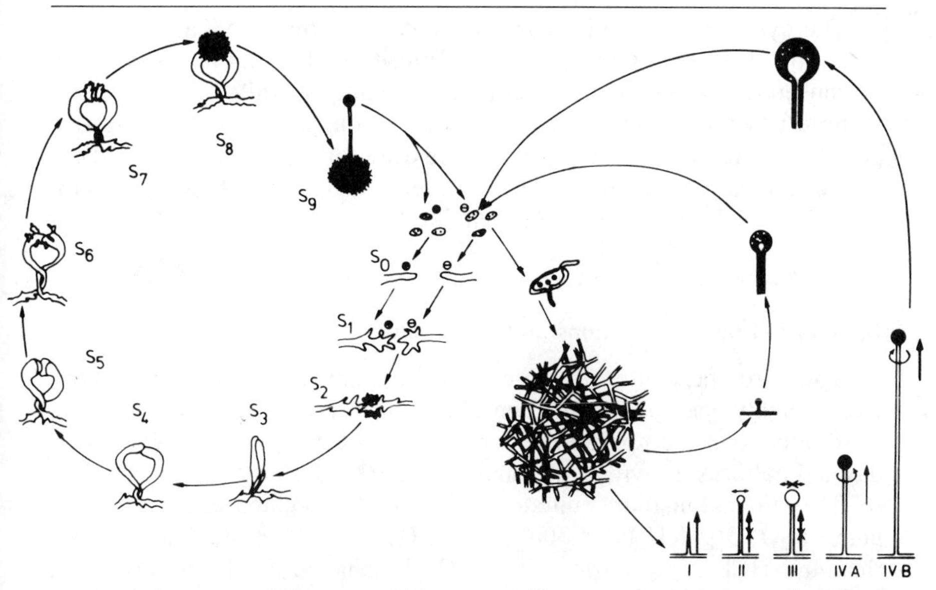

FIG. 1. Anatomy and life history of *Phycomyces*. The spores (center), once activated and provided with an adequate environment, germinate and produce a mycelial mat of branched hyphae. The sporangiophores, or vegetative fruiting structures (lower right) occur as either macrophores (several cm or even dm long) or microphores (about 1 mm long). In either case they carry a sporangium (upper right) containing spores. There are two sexual types; when mycelia of opposite type meet, there is a complex succession of biochemical and morphological events (left), leading to the formation of zygospores (upper left). After a long dormancy, the zygospore germinates into a germ sporangiophore, which forms a germ sporangium. The germ spores are the products of genetic recombination. They are very similar to vegetative spores and able to enter the vegetative cycle.

trast strongly against the light-yellow background of the wild type: in reconstruction experiments, single color mutants are spotted in standard Petri dishes containing over 30,000 colonies.[2] When searching for white mutants in wild-type strains, cultures should be incubated under brilliant white or blue light; photoinduction of carotene synthesis would then enhance the contrast between mutants and wild type. Conversely, red and deep-yellow mutants should be looked for against the light-yellow background of cultures grown in the dark.

The color contrast between some interesting mutants and the wild type would not be very striking. It would then be convenient to search for them against a different background. Red and deep-yellow strains have served well for this purpose. The leaky *carB* mutant S86 and others were isolated in this way.[3]

[2] S. Torres-Martínez, F. J. Murillo, and E. Cerdá-Olmedo, *Genet. Res.* **36,** 299 (1980).
[3] A. P. Eslava and E. Cerdá-Olmedo, *Plant Sci. Lett.* **2,** 9 (1974).

Our eyes are so sensitive to yellow color variations, that a mere duplication of carotene content is clearly noticed. Such variations may easily be caused by environmental factors. Environmental uniformity should be assured when searching for the less-conspicuous mutant types. It is also possible to confuse changes in colony texture (due, for example, to slight mutational alterations in branching pattern) with small changes in color intensity. The quantitative chemical analysis is required to dispel any doubts.

Standard Growth Conditions and Variations

The carotene content is influenced by many environmental factors, most notably age, growth medium, light, and certain chemicals. Culture conditions must be normalized for the sake of comparisons. We use 4-day-old cultures grown on minimal agar at 22–23° in the dark.

The minimal medium contains, per liter: 20 g D-glucose, 2 g L-asparagine · H_2O, 5 g KH_2PO_4, 500 mg $MgSO_4 \cdot 7H_2O$, 28 mg $CaCl_2$, 1 mg thiamin · HCl, 2 mg citric acid · H_2O, 1.5 mg $Fe(NO_3)_3 \cdot 9H_2O$, 1 mg $ZnSO_4 \cdot 7H_2O$, 300 μg $MnSO_4 \cdot H_2O$, 50 mg $CuSO_4 \cdot 5H_2O$, 50 mg $Na_2MoO_4 \cdot 2H_2O$, 15 g agar. Glucose is autoclaved separately. The mineral salts may be prepared as a 100-fold concentrate and kept sterile for a long time in the presence of a few drops of chloroform. This medium is a variation of that proposed by Sutter,[4] following his own experiments and those of Ødegård.[5]

Under these circumstances wild-type cultures consist of thick, velvety mycelial pads, which have already started to produce macrophores and contain about 40 ppm all-*trans*-β-carotene (ppm = μg/g dry weight). Precursors and isomers of β-carotene are present as traces, if at all.

Other media are sometimes used in the laboratory manipulation of *Phycomyces*. Good sexual development is obtained on minimal agar containing 1 g/liter monosodium glutamate, instead of asparagine[4] or on potato dextrose agar. For the microscopic observation of the sexual stages, a thin (1 mm) layer of medium is recommended. Potato dextrose agar, appropriate for the production of the stout macrophores used in studies of sensory physiology, may be purchased or prepared by boiling 200 g fresh, diced potatoes in water for 1 hr, filtering the extract, adding 20 g D-glucose, 1 mg thiamin · HCl, and 15 g agar, completing to 1 liter, and autoclaving. Nutrient media are obtained by adding various amounts of yeast extract or casein hydrolysate to other media.

[4] R. P. Sutter, *Proc. Natl. Acad. Sci. U.S.A.* **72,** 127 (1975).

[5] K. Ødegård, *Physiol. Plant.* **5,** 583 (1952).

Discrete colonial growth, essential in the handling of mutants and in genetic analyses, is obtained by acidifying the media to pH 3.3 with HCl after autoclaving.

The best media for carotene production[6] are much richer than any of the media normally used in genetic studies.

Color Mutants

The Appendix lists the published color mutants of *Phycomyces blakesleeanus*. The first *Phycomyces* color mutants were isolated by M. Heisenberg.[7,8] Here is a short guide to the mutants now available. They are white, red, greenish, light yellow, or intense yellow, depending on the accumulation of β-carotene and its metabolic precursors.

Some white mutants contain large amounts of phytoene (over 1000 ppm), and practically no other carotenoids. They all occur[9] at gene *carB*, located[10] near the centromere of one of the chromosomes. This gene determines phytoene dehydrogenase; four copies of this enzyme are responsible for the four dehydrogenations which convert phytoene successively into phytofluene, ζ-carotene, neurosporene, and lycopene.[11] Leaky *carB* mutants have a greenish-yellowish tinge and accumulate phytoene, phytofluene, etc., in decreasing concentrations. Strain S442 (from our laboratory, unpublished) is interesting because it accumulates important amounts of phytofluene and ζ-carotene, apart from phytoene.

Other white mutants contain small amounts of β-carotene (in the order of 1 ppm), but become very yellow and accumulate large amounts of β-carotene when grown in the presence of retinol and other chemicals.[12] These mutants belong to a single complementation group, originally identified as gene *carA*, but later[2] shown to be a section of a bifunctional gene *carRA*, which maps very close to gene *carB* in the genetic map.[10] The product of this gene section governs substrate transfer[13] in the biosynthesis of β-carotene.

The *carR* section of gene *carRA* determines lycopene cyclase. Two copies of this enzyme, assembled into an aggregate, act successively on

[6] A. Ciegler, *Adv. Appl. Microbiol.* **7,** 1 (1965).

[7] G. Meissner and M. Delbrück, *Plant Physiol.* **43,** 1279 (1968).

[8] M. Heisenberg and E. Cerdá-Olmedo, *Mol. Gen. Genet.* **102,** 187 (1968).

[9] T. Ootaki, A. C. Lighty, M. Delbrück, and W.-J. Hsu, *Mol. Gen. Genet.* **121,** 57 (1973).

[10] M. I. G. Roncero and E. Cerdá-Olmedo, *Curr. Genet.* **5,** 5 (1982).

[11] C. M. G. Aragón, F. J. Murillo, M. D. de la Guardia, and E. Cerdá-Olmedo, *Eur. J. Biochem.* **63,** 71 (1976).

[12] A. P. Eslava, M. I. Álvarez, and E. Cerdá-Olmedo, *Eur. J. Biochem.* **48,** 617 (1974).

[13] F. J. Murillo, S. Torres-Martínez, C. M. G. González-Aragón, and E. Cerdá-Olmedo, *Eur. J. Biochem.* **119,** 511 (1981).

the lycopene substrate molecule and convert it first to γ-carotene and then to β-carotene.[14] The *carR* mutants are red and accumulate large amounts of lycopene (over 1000 ppm). Some of them are leaky and allow the production of small amounts of β-carotene.

Mutants having lost the two functions of gene *carRA* are white, contain small amounts of lycopene, and do not make β-carotene in the presence of retinol.[2,9,12]

Superproduction of β-carotene, resulting in deep-yellow colors, has been attributed to recessive mutations[15] in gene *carS*, located[10] in the same chromosome as the other *car* genes mentioned so far, but only weakly linked to them: the recombination frequency has been estimated as 36%. The β-carotene content of the cell is regulated by a feedback mechanism, which shuts the pathway off when enough β-carotene has been produced. This explains the increased operation of the pathway in *carB* and *carR* mutants or in the wild type exposed to chemical inhibitors of the dehydrogenase[16] or the cyclase.[13] Feedback regulation depends on the *carS* gene: the total production of β-carotene and its precursors in *carS* mutants is not increased by blocking the pathway either genetically[15] or chemically.[17,18] The *carA* mutants may have a lower β-carotene threshold than the wild type for feedback regulation.[19] The best *carS* strains contain about 5000 ppm β-carotene under standard conditions.

The extracts of *carS* mutants have the same carotenogenic potential *in vitro* as wild-type extracts,[20] indicating that they contain about the same amounts of carotenogenic enzymes.

The induction of additional mutations in *carS* strains leads to the isolation of new strains with even higher carotene contents, for example[15] S106 and others now under study.

A third regulatory gene,[21] unlinked to all the others, is *carC*. This gene is defined by whitish mutants, which accumulate about 13 ppm β-carotene in the dark. The double mutants *carA carC*, *carB carC*, *carR carC*, and *carS carC* accumulate the same carotenoids, and in the same concentra-

[14] M. D. de la Guardia, C. M. G. Aragón, F. J. Murillo, and E. Cerdá-Olmedo, *Proc. Natl. Acad. Sci. U.S.A.* **68,** 2012 (1971).

[15] F. J. Murillo and E. Cerdá-Olmedo, *Mol. Gen. Genet.* **148,** 19 (1976).

[16] J. A. Olson and H. Knizley, *Arch. Biochem. Biophys.* **97,** 138 (1962).

[17] M. Elahi, T. H. Lee, K. L. Simpson, and C. O. Chichester, *Phytochemistry* **12,** 1633 (1973).

[18] M. Elahi, R. W. Glass, T.-C. Lee, C. O. Chichester, and K. L. Simpson, *Phytochemistry* **14,** 133 (1975).

[19] F. J. Murillo, *Plant Sci. Lett.* **17,** 201 (1980).

[20] A. de la Concha, F. J. Murillo, E. J. Skone, and P. M. Bramley, *Phytochemistry* **22,** 441 (1983).

[21] J. L. Revuelta and A. P. Eslava, *Mol. Gen. Genet.* **192,** 225 (1983).

tions, as if they contained no *carC* mutations at all. This means that the *carC* gene product influences the regulatory interplay between the *carA* and *carS* gene products and β-carotene, but has no effect if any one of these is missing. The precise function of gene *carC* is unknown.

Light stimulates carotenogenesis[22,23]; 1 W m^{-2} blue light increases the β-carotene content about 10-fold over the dark controls. Mutants in genes *picA* and *picB* are partially defective for photoinduced carotenogenesis, but seem to be normal otherwise.[24,25] Light has little or no effect on carotenogenesis by *carC* and *carA* mutants and *carA carS* double mutants.[21,24] Of the *mad* mutants, grossly defective for phototropism, only those in genes *madA* and *madB* are partially defective in photocarotenogenesis.[22,24,26]

Retinol greatly estimulates β-carotene production[12] while 2-(4-chlorophenyl)thiotriethylamine · HCl inhibits lycopene cyclase.[13,17,27] Mutations blocking simultaneously both effects occur at either a new gene, *carI*, independent of the others, or at the *carRA* gene.[10] These chemoinsensitive mutants make rather normal β-carotene levels, but they are only slightly affected by the two chemicals mentioned above. The action of these chemicals thus requires cellular function(s), different from the enzymes of the pathway.

It is particularly surprising that none of the available mutants blocks the pathway before the formation of phytoene. This may reflect the intervention of the same enzymes in a pathway essential for the cell, most likely in the biosynthesis of sterols.

The existing mutants do not throw any light on the mechanism of isomerization. Phytoene from a *carB* mutant or from the wild type grown with diphenylamine is nearly all in the 15-cis configuration,[28] but β-carotene is overwhelmingly all-trans. There is no report of mutants of the hypothetical isomerase, which would convert 15-*cis*- into all-*trans*-phytoene.[29] Perhaps isomerization is carried out by phytoene dehydrogenase itself, and not by a separate enzyme.

The sequence of the dehydrogenations and cyclizations is not absolutely fixed, since minor amounts of 7,8,11,12-tetrahydrolycopene, 7′,8′,11′,12′-tetrahydro-γ-carotene, and β-zeacarotene have been de-

[22] K. Bergman, A. P. Eslava, and E. Cerdá-Olmedo, *Mol. Gen. Genet.* **123,** 1 (1973).

[23] G. Sandmann and W. Hilgenberg, *Biochem. Physiol. Planzen* **172,** 401 (1978).

[24] I. López-Díaz and E. Cerdá-Olmedo, *Planta* **15,** 134 (1980).

[25] M. Jayaram, D. Presti, and M. Delbrück, *Exp. Mycol.* **3,** 42 (1979).

[26] M. Jayaram, L. Leutwiler, and M. Delbrück, *Photochem. Photobiol.* **32,** 241 (1980).

[27] C. W. Coggins, Jr., G. L. Henning, and H. Yokoyama, *Science* **168,** 1589 (1970).

[28] A. Than, P. M. Bramley, B. H. Davies, and A. F. Rees, *Phytochemistry* **11,** 3187 (1972).

[29] B. H. Davies, *Pure Appl. Chem.* **35,** 1 (1973).

tected in diphenylamine-inhibited cultures.[29–31] β-Zeacarotene is also present in some heterokaryons.[11] The phenotypes of the existing mutants and their heterokaryons suggest, however, that these minor components are not made by specific enzymes, but result from an altered operation of the *carB-carR* enzyme aggregate.

An extranuclear genetic factor, *carE*, is involved in carotenogenesis in a most intriguing way.[32] The double mutants *carS carE* grown in the dark are red, because they contain large amounts of a β-carotene-protein complex with maximal absorption at 538 nm. In the light they are yellow and contain only large amounts of free β-carotene.

Mutagenesis

Spontaneous color mutants have never been reported in *Phycomyces*. Their frequency is probably high enough for the easy detection of conspicuous phenotypes due to dominant mutations and, with substantially more work, of some recessive mutations. Spontaneous mutants would be welcome because they would not be suspected of harboring multiple mutations.

Several mutagens have been used to increase the frequency of color mutants. All of them are potential health hazards. Appropriate safety precautions[33] should be followed.

Mutagenesis with N-Methyl-N′-nitro-N-nitrosoguanidine

For an extensive review on the mutagenicity of *N*-methyl-*N*′-nitro-*N*-nitrosoguanidine see Gichner and Velemínský.[34]

The drug is a potent carcinogen; fortunately the low volatility and the high chemical reactivity of the compound diminish the risks of accidental exposure and of long-term environmental contamination. Contaminated glassware and other materials should be exposed to 2% $Na_2S_2O_3 \cdot 10H_2O$ for 1 hr for the complete destruction of the drug.

Working under a safety hood, a small amount of the chemical (a few mg) is poured into a preweighed tube. After weighing the tube, distilled water is added to obtain a concentration of 1 mg/ml. The tightly closed tube is shaken to accelerate solution. Aliquots of about 0.5 ml are distributed to small tubes and kept frozen in the dark until used. Unfrozen

[30] R. J. H. Williams, B. H. Davies, and T. W. Goodwin, *Phytochemistry* **4**, 759 (1965).

[31] B. H. Davies and A. F. Rees, *Phytochemistry* **12**, 2745 (1973).

[32] A. de la Concha and F. J. Murillo, *Planta* **161**, 233 (1984).

[33] L. Ehrenberg and C. A. Wachtmeister, *in* "Handbook of Mutagenicity Test Procedures" (B. J. Kilbey, M. S. Legator, and C. Ramel, eds.), p. 411. Elsevier, Amsterdam, 1977.

[34] T. Gichner and J. Velemínský, *Mutat. Res.* **99**, 129 (1982).

samples should not be reused. To check for accidental unfreezing and refreezing, a small thumbtack may be dropped into a similar tube containing ice. The yellow color of the solution serves as a quality guarantee; if any doubt arises, the absorbance at 400 nm of a 1 mg/ml solution through a 1-cm light path should be 1.1.

N-Methyl-*N'*-nitro-*N*-nitrosoguanidine has been successfully used to induce mutations in *Phycomyces* spores and mycelia following many diverse protocols. A recent systematic study[35] led to the following conclusions and recommendations.

The effects of *N*-methyl-*N'*-nitro-*N*-nitrosoguanidine on *Phycomyces* depend on the dose, defined as the product concentration of the mutagen and the exposure time, measured in M · s (mol second liter^{-1}). The two factors may be inversely altered within a wide interval. The buffer in which spores and mutagen are brought together is not very critical.

Newly activated spores, suspended in buffer (about 10^6 spores per ml 0.05 *M* Tris-maleate buffer, pH 7.5) are exposed to *N*-methyl-*N'*-nitro-*N*-nitrosoguanidine at room temperature. The spores are gently and continuously shaken during exposure to prevent sedimentation, then washed three times by centrifugation. To maximize the number of mutants in the initial spore population, the dose should be about 0.1–0.15 M · s, which is conveniently achieved by a 15-min exposure to 20 μg/ml *N*-methyl-*N'*-nitro-*N*-nitrosoguanidine. The frequency of mutants among the survivors increases monotonously with the dose, but very heavy doses have clear drawbacks, including the induction of several mutations in the same genome. For the attainment of a high mutant frequency among the survivors, the authors recommended 0.4 M · s, or a 10-min exposure to 100 μg/ml *N*-methyl-*N'*-nitro-*N*-nitrosoguanidine, which leads to the survival of about 5% of the cells.

The exposed spores are then plated out and the resulting colonies screened for color mutants.

Mutagenesis with 4-Nitroquinoline 1-Oxide

4-Nitroquinoline 1-oxide (Pharmaceuticals Inc., New York) is dissolved in acetone at 10 mg/ml and diluted to 1 μg/ml with phosphate buffer (0.1 *M*, pH 7). Spores are exposed to the mutagen solution for 60 min at 22°; the suspension should contain about 10^6 spores per ml and be shaken occasionally. The exposure is stopped by adding an equal volume of sodium thiosulfate solution (50 μg/ml). The spores are washed three times by centrifugation and resuspension in water. About 5% of the spores survive this treatment.[21]

[35] M. I. G. Roncero, C. Zabala, and E. Cerdá-Olmedo, *Mutat. Res.* **125,** 195 (1984).

Mutagenesis with ICR-170

This compound, 2-methoxy-6-chloro-9-[3-(ethyl-2-chloroethyl)amino-propylamino]acridine · 2HCl (Terochem Labs. Ltd., Edmonton, Alberta, Canada) induces frameshift mutations, which usually lead to total loss of gene function. In fact, the different kinds of color mutants isolated after exposures to ICR-170 and *N*-methyl-*N'*-nitro-*N*-nitrosoguanidine may contribute to the understanding of gene structure and function.[2]

A suspension of about 10^6 spores per ml is exposed at 22° to a freshly prepared solution of the mutagen. Exposures of 1 hr to 40 μg/ml ICR-170 in 0.1 phosphate buffer at pH 7.0 and 4 hr to 20 μg/ml ICR-170 in water have been successfully used by Álvarez *et al.*[36,37] and Torres-Martínez *et al.*[2] respectively.

The reaction may be stopped by adding a 5-fold volume of 1% NaCl. The mutagen is destroyed by reaction with 0.2 *M* KOH in methanol.

Isolation of Recessive Mutations in Multinucleate Cells

Phycomyces remains mutlinucleate throughout the entire cell cycle. The mycelia are large coenocytes, containing millions of nuclei, and the spores are multinucleate cells, containing several nuclei. The distribution of nuclei per spore depends on the strain. Only 0.3% of the spores of strain NRRL1555 are uninucleate.[8] Insignificant improvements of the uninucleate frequency were obtained through cultural changes. Fractions containing up to 80% uninucleate spores may be obtained by sedimentation,[38] but the procedure is more cumbersome than productive.

The presence of several nuclei in *Phycomyces* spores does not prevent the isolation of dominant mutations.[35] The trouble is that mutants are isolated in heterokaryotic form, that is, the dominant mutant nuclei constitute only a fraction of the nuclei in the organism. Repetition of the vegetative life cycle eventually leads to the isolation of homokaryotic mutants. These are recognized because they breed true: their spore progeny show an uniform mutant phenotype.

Recessive mutant hunts have relied so far in the death of some of the nuclei by a chemical or a physical agent, which may well be the same used to induce a high mutation frequency. A fraction of the survivors becomes functionally uninucleate and able to express recessive mutations. X Rays are excellent nuclear killers, but they induce so many chromosome alterations that their use is not recommended. About one-third of the survivors

[36] M. I. Álvarez, M. I. Peláez, and A. P. Eslava, *Mol. Gen. Genet.* **179,** 447 (1980).
[37] M. I. Álvarez, T. Ootaki, and A. P. Eslava, *Mol. Gen. Genet.* **191,** 507 (1983).
[38] P. Reau, *Planta* **108,** 153 (1972).

of the usual exposures to *N*-methyl-*N'*-nitro-*N*-nitrosoguanidine are uninucleated.[39] Most of the available mutants have been isolated in this way.

Selection of survivors containing a single functional nucleus, following a genetic method[35] saves more than half the screening work needed to isolate recessive, unselectable mutants. To this end, a heterokaryotic mycelium, carrying a selectable recessive marker in 10–20% of the nuclei, is allowed to sporulate. The spores are exposed to 0.4 M · s. *N*-Methyl-*N'*-nitro-*N*-nitrosoguanidine as above and plated, at about 10^5 survivors per plate, on media selecting the marker. After a whole vegetative cycle, spore stocks are separately collected from each plate, lyophilized, and kept indefinitely for mutant hunts. Convenient recessive markers are *dar* (conferring resistance to 1 μg/ml 5-carbon-5-deazariboflavin)[40] and *gal* (conferring resistance to 100 μg/ml 2-deoxygalactose in the presence of 20 g/liter sorbitol as carbon source).[41]

Genetic Analyses

Preparation of Heterokaryons

The attribution of mutations to genes is based on complementation analysis. Nuclei from two different strains are brought together in the same cell, which is called a heterokaryon. The method most often used for heterokaryon formation was developed by Ootaki[42] and is based on the grafting of young macrophores (before the differentiation of the sporangium).

One young macrophore from each of the two strains to be grafted is picked up by the base with fine tweezers and placed on an agar cube (about 5 mm/side) so that the tip (about 5 mm) protrudes from the agar block. The tips (0.5–2 mm) are cut off with microdissecting scissors (whose blades are 5 mm long), dipped first in alcohol, then in sterile water, while checking under a dissecting microscope that the cell walls are not squashed together. One of the two tips is cut right across, the other oblique to the long axis. The two agar blocks are maneuvered to bring the cut surfaces into contact and to insert one slightly into the other. The agar blocks glide more easily if water drops are placed at their edge. The grafts are incubated in a moist chamber. A standard Petri dish may accommodate three grafted pairs. Quick operation is essential to avoid desiccation. After 1 hr a drop of water exudate may have formed at the

[39] E. Cerdá-Olmedo and P. Reau, *Mutat. Res.* **9,** 369 (1970).

[40] M. Delbrück and T. Ootaki, *Genetics* **92,** 27 (1979).

[41] F. Rivero, personal communication (1983).

[42] T. Ootaki, *Mol. Gen. Genet.* **121,** 49 (1973).

junction; it is removed with the closed tips of fine tweezers, without disturbing the graft. After 10 to 18 hr the grafts are tested by gently pulling one of the agar blocks; those coming apart are rejected. Many of the graft junctions develop a small sporangiophore. Its sporangium is picked up in a sterile water drop between tweezer tips, crashed, and streaked on agar medium. Colonies from successful grafts are not uniform: some are identical to each of the grafted parent and the rest correspond to the desired heterokaryon. The heterokaryosity of any colony is tested by restreaking its spores and checking for diversity of progeny.

A much simpler method for heterokaryon formation[43] is restricted to mutants of opposite sex able to interact sexually. This method is only seldom useful for *car* mutants, since most of these are sexually defective.

Qualitative Complementation

When heterokaryons containing both mutant and wild-type nuclei show essentially the mutant phenotype, the mutation is called dominant. If they show the wild phenotype, the mutation is called recessive. Two recessive mutations altering the same phenotypic feature are said to complement if the heterokaryon containing the two mutations resembles the wild type. Thus two white mutants complement if they make up a yellow heterokaryon. The β-carotene content of the heterokaryon need not be as high as that of the wild type, and usually it is not.

Experimenters with good graft success rates sometimes establish complementation by making at least 30 grafts between each pair of mutants. If they never obtain phenotypes similar to the wild type, they conclude that the two mutations do not complement.[9] Another approach is to have complementing markers in the two mutants to be tested. They may be, for example, auxotroph for different nutrients. The attainment of the desired heterokaryon is then confirmed by complementation of the markers (phototrophy, in our example) and the carotene phenotype of the heterokaryon is then registered.

A gene is defined by a set of noncomplementing mutants. Two complementing mutations occur usually at different genes, but there are many cases in the fungal literature of intragenic complementation, that is, of pairs of complementing mutations, both unable to complement with many other mutations. Intragenic complementation in *Phycomyces* was found at the *carRA* gene[2]; the gene is defined by the set of mutants (white with traces of lycopene) which do not complement each other or any other mutants of the same gene. Other mutants of this gene are able, however, to complement in certain combinations (the red and the white with a little β-carotene).

[43] W. Gauger, M. I. Peláez, M. I. Álvarez, and A. P. Eslava, *Exp. Mycol.* **4,** 56 (1980).

Quantitative Complementation

Complementation seldom reproduces the wild phenotype faithfully. Dominance, recessivity, and complementation are rather quantitative concepts, best described by a function relating the phenotype of the heterokaryon to the proportions of the component nuclei. *Phycomyces* lends itself particularly well for the investigation of these "genophenotypic functions," as they have been called by Medina.[44]

The heterokaryons between isogenic strains of *Phycomyces* are remarkably uniform and stable. The nuclear proportion keeps constant during vegetative growth and during repeated mycelial transfers to fresh medium.[8] It is thus easy to obtain any desired amount of a heterokaryotic mycelium with a fixed nuclear proportion.

Heterokaryons make three kinds of spores: homokaryons for one or the other component and heterokaryons. The proportion of the spores homokaryotic for component A is

$$S_A = \sum_{n=1}^{n=\infty} p^n f(n) \tag{1}$$

where p is the proportion of component A among the nuclei of the heterokaryon, and $f(n)$ the proportion of the spores having n nuclei. For strain NRRL1555, $f(1) = 0.003, f(2) = 0.090, f(3) = 0.420, f(4) = 0.410, f(5) = 0.074$, and $f(6) = 0.003$.

The proportion of spores homokaryotic for the other component is

$$S_B = \sum_{n=1}^{n=\infty} (1 - p)^n f(n) \tag{2}$$

The remaining spores ($S_H = 1 - S_A - S_B$) are heterokaryotic.[8]

When the two homokaryons and the heterokaryon are easily distinguished, the values of S_A and S_B are readily obtained after culturing the spores of the heterokaryon on acid medium, which allows each spore to form a discrete, limited colony. Colonies must be marked as they appear (for example, by marking the underside of the Petri plate after 36, 48, and 60 hr incubation) and their phenotypes scored when feasible (for colors, after a 4-day incubation). About 50 colonies can be scored in a 10-cm plate and each analysis should be based on several hundred colonies. If the three kinds of spores grow into colonies with very different rates one would be well advised to follow the more complicated method described by Heisenberg and Cerdá-Olmedo.[8]

The nuclear proportion p is derived from Eqs. (1) and (2). A micro-

[44] J. R. Medina, *Rev. Biomath.* **67**, 39 (1979).

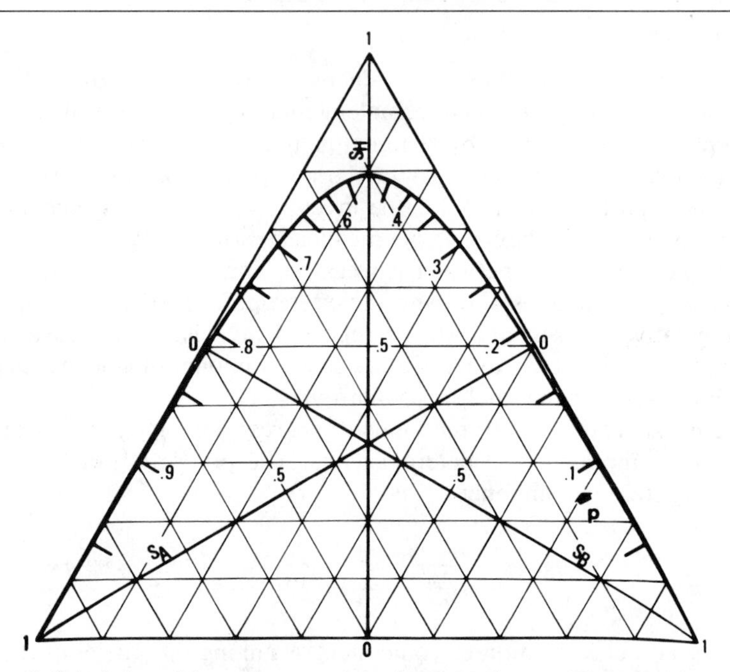

FIG. 2. Calculation of the nuclear proportion in a heterokaryon. The proportions S_A and S_B of spores homokaryotic for each nuclear component and the proportion S_H of heterokaryotic spores add up to one and define a point within the triangle whose distances to each of the three sides equal the three S values. The proportion p of component A among the nuclei of the heterokaryon is read along the theoretical line, based on the equations of Heisenberg and Cerdá-Olmedo[8] and redrawn after E. Cerdá-Olmedo, *in* "Handbook of Genetics" (R. C. King, ed.), Vol. 1, p. 343. Plenum, New York, 1974.

computer may be programmed to obtain the best value of p with criteria of maximum likelihood or minimum squares. A simple graphical solution for p was given by Heisenberg and Cerdá-Olmedo.[8] A modified version is presented in Fig. 2. Though theoretically possible, it is unadvisable to calculate the nuclear proportion from only S_A or S_B or S_H, because the errors tend to be very high, at least for some intervals of possible nuclear proportions.

Each heterokaryotic mycelium is subject to a phenotypic analysis (for example, content of β-carotene or other carotenoid) and to a genetic analysis to determine the nuclear proportion. The resulting genophenotypic functions may be interpreted in terms of the structure and function of the gene products affected by the mutations. This method was introduced by de la Guardia *et al.*,[14] and has been applied repeatedly to carotene biosynthesis.[11,13,15,21]

Recombination

The sexual cycle of *Phycomyces* (Fig. 1, left) occurs when mycelia of opposite sex meet each other under appropriate conditions. In each interaction leading to the formation of a zygospore, a single nucleus from each mated strain fuse to form a diploid which undergoes meiosis; the four meiotic products divide to produce the usually uninucleate germspore primordia and then continue dividing to produce the germspores. The germspores of a single germsporangium thus constitute an expanded, unordered tetrad. However, the germspores in a germsporangium derive sometimes from more than one diploid nucleus.[45–47]

There is no systematic study of the best conditions for the completion of the sexual cycle. The method given here follows essentially that given in Ref. 45.

Spores or mycelial bits of the parent strains to be crossed are inoculated on opposite sides of a plate containing potato dextrose agar or glutamate minimal medium and incubated at 17° under constant dim illumination. After 2–8 weeks the zygospores are scraped from the plates with a spatula, cleaned of adherent mycelia, individually placed on moist filter paper kept wet by dipping one or two sides in water, and incubated at room temperature in a translucent box. Water agar (10 g highly purified agar in water, without salts or nutrients) or silicate gels may be used instead of the filter paper. Zygospores from the isogenic strains[1] have a dormancy of 2–2.5 months.

Individual germ sporangia are collected in a drop of sterile water between tweezer tips, crushed, and placed in a tube containing 1 ml sterile water. For tetrad analyses only one germ sporangium is placed in each tube. For mass analyses, many germ sporangia are pooled in a single tube. The contents of each tube are thermally activated and plated on acid agar at different dilutions. Individual colonies (at least 40 for each tetrad, at least 200 for each mass analysis) are then classified phenotypically as to the markers involved in the cross. This classification may be based on visual observation, on chemical, biophysical, or genetic analyses, or on the behavior of subcultures. For example, to determine the sexual type, (+) or (−), a subculture of each colony is confronted against two tester strains of known sex on potato dextrose or glutamate minimal medium and incubated for 4–6 days at 17°. The typical morphological structures of the sexual cycle should appear between the unknown and one of the tester strains.

[45] E. Cerdá-Olmedo, *Genet. Res.* **25,** 285 (1975).

[46] A. P. Eslava, M. I. Álvarez, P. V. Burke, and M. Delbrück, *Genetics* **80,** 445 (1975).

[47] A. P. Eslava, M. I. Álvarez, and M. Delbrück, *Proc. Natl. Acad. Sci. U.S.A.* **72,** 4076 (1975).

Tetrad analysis and mass analysis are interpreted as in other genetic systems. Linkage of the mutations with one another and, to a certain extent, with chromosomal centromeres, may be thus investigated. Most of the markers investigated turn out to be independent (recombination frequency 50%). A sizable recombination frequency between two mutations (at least a few percent) is a good sign that the mutations occur at different genes. Intragenic recombination is relatively frequent in *Phycomyces*. Recombinants between two *nicA* mutations were found in a pool of only 175 germ sporangia.[46]

When the mated strains are not isogenic, the results of crosses are rather erratic, i.e., certain genotypes may be missing from some germ sporangia (incomplete tetrads) or there are different counts of genotypes expected to occur with equal frequencies. Regularity and reliability of the results is greatly improved by use of the new isogenic strains.[1]

Most of the *car* mutants are unable to complete the sexual cycle. A heterokaryon trick has been used to circumvent this difficulty.[10,45] Complementing heterokaryons containing *car* mutant nuclei are fertile. If strain *a* carries a *car* mutation totally blocking the sexual cycle, an appropriate cross would be $(a * b) \times c$, where *b* and *c* are strains of opposite sex which share an allele for auxotrophy. The offspring results from $a \times c$ and $b \times c$ crosses. In mass analysis, the germ spores are plated on minimal medium, where the offspring of the second cross and half the offspring of the first cross cannot grow. Genetic analysis is then conducted on half the offspring of the $a \times c$ cross. Tetrad analysis is restricted to germ sporangia containing genetic markers from strain *a* and the other germ sporangia are disregarded.

Conservation of Mutants

Each mutant is subcultivated two to three times before it is added to the collection; each subculture is initiated from spores of a single sporangium of the preceding culture and checked for the desired phenotype. To ensure that the mutant nuclei occur in homokaryosis, the subculturing procedure must be restarted if the colonies of any subculture are not uniform.

Phycomyces spores keep well for several months, sometimes years, in the refrigerator (about 4°), either in water suspension or air-dried. Well sporulated mycelia may be kept frozen at −20° or less for several years.[48] Lyophilized spore suspensions in cow plasma or other protein solution

[48] J. W. Carmichael, *Mycologia* **54,** 432 (1962).

[49] K. B. Raper and D. F. Alexander, *Mycologia* **37,** 499 (1945).

keep well at room temperature or in the refrigerator for many years, as long as the ampoules remain intact.[49,50]

Mutant collections should be subcultured as seldom as possible, to minimize evolution in the laboratory. Master stocks and working stocks should be maintained. It is recommended that working stocks be kept and used for 1 or 2 years, before deriving new ones from the master stocks.

Application of the Mutants

β-Carotene is not an essential component of *Phycomyces*. The *car* mutants grow normally in the laboratory, even those containing no β-carotene. Strains C173 and C174, for example, carry *carB* and *carR* mutations which block all the last six reactions in the pathway and contain no detectable β-carotene, according to analyses which would have detected 0.01 ppm of this pigment.[51]

Lack of β-carotene does not affect phototropism,[22,51] implying that the photoreceptor contains no β-carotene or carotene derivatives. On the contrary, the photoreceptor contains a flavin.[52] The white, blind mutants obtained from the wild type after a single mutagen exposure carry separate *car* and *mad* mutations.[22,53]

The white mutants are convenient in the investigation of phototropism because they eliminate an unnecessary pigment absorbing the same wavelengths as the photoreceptor. Many examples of this application are referred to in the Appendix. The *car* mutants offer a variety of optical densities and have been used to study the effects of shading pigments in sensory transduction.[54]

Strains with very high β-carotene and lycopene contents have been obtained by putting together different *car* mutations and other genetic alterations.[55,56]

β-Carotene is essential to carry out the sexual cycle. Most of the *car* mutants form no zygospores at all. The *carA* mutants, which make small amounts of β-carotene, form fewer zygospores than the wild type. The

[50] J. J. Ellis and J. A. Roberson, *Mycologia* **60,** 399 (1968).

[51] D. Presti, W.-J. Hsu, and M. Delbrück, *Photochem. Photobiol.* **26,** 403 (1977).

[52] M. K. Otto, M. Jayaram, R. M. Hamilton, and M. Delbrück, *Proc. Natl. Acad. Sci. U.S.A.* **78,** 266 (1981).

[53] M. I. G. Roncero, Tesis Doctoral, Universidad de Sevilla, Spain (1982).

[54] K. Bergman, P. V. Burke, E. Cerdá-Olmedo, C. N. David, M. Delbrück, K. W. Foster, E. W. Goodell, M. Heisenberg, G. Meissner, M. Zalokar, D. S. Dennison, and W. Shropshire, Jr., *Bacteriol. Rev.* **33,** 99 (1969).

[55] F. J. Murillo, I. L. Calderón, I. López-Díaz, and E. Cerdá-Olmedo, *Appl. Environ. Microbiol.* **36,** 639 (1978).

[56] F. J. Murillo, I. L. Calderón, I. López-Díaz, and E. Cerdá-Olmedo, U.S. Patent 4,318,987 (1982).

carC mutants, which contain detectable β-carotene, also make zygospores.

Mutants lacking β-carotene fail to sexually stimulate their partners, but are stimulated by them to produce zygophores, the first differentiated structures of the sexual cycle.[4] This is understandable since sexual stimulation is mediated by sex-specific compounds derived from β-carotene.[57–59]

The *carS* mutants, with more than enough β-carotene, form pairing zygophores, but these do not develop further.[4] A second step in the sexual cycle is thus blocked in the *carS* mutants. The block probably results from excessive β-carotene, but not from the *carS* mutation itself, since strain S283, carrying both *carA* and *carS* mutations and making moderate amounts of β-carotene, completes the sexual cycle satisfactorily.[10]

The need for adequate regulation of carotenogenesis in the sexual cycle is seen in the chemoinsensitive mutants, such as S119 and S144, which produce rather normal amounts of β-carotene, but no zygospores.

β-Carotene is a component of sporopollenins, structural polymers found in zygospores, sporangiophores, and vegetative spores of *Phycomyces*.[60] Metabolism of β-carotene into sporopollenins, sexual hormones, or other compounds is small compared to the existing content of β-carotene.[13]

β-Carotene is required for photoinduction of the pathway, since *carB* and *carR* mutants accumulate the same amounts of phytoene and lycopene, respectively, when grown in the dark.[3,61] This does not necessarily mean that the pigment is part of the corresponding photoreceptor.

Mutants with little or no β-carotene differ from the wild type in the development of sporangiophores, both macrophores[62,63] and microphores,[64] and have abnormally high alcohol dehydrogenase activities.[65] The reasons for these differences are obscure.

Many of the references given in the Appendix have used the *car* mutants as convenient markers in basic and applied genetic experiments.

[57] D. J. Austin, J. D. Bu'lock, and D. Drake, *Experientia* **26,** 348 (1970).

[58] J. D. Bu'lock, B. E. Jones, D. Taylor, N. Winskill, and S. A. Quarrie, *J. Gen. Microbiol.* **80,** 301 (1974).

[59] R. P. Sutter, *in* "Eukaryotic Microbes as Model Developmental Systems (D. H. O'Day and P. A. Horgen, eds.), Microbiol. Ser., Vol. 2, p. 251. Dekker, New York, 1977.

[60] B. Furch and G. W. Gooday, *Trans. Br. Mycol. Soc.* **70,** 307 (1978).

[61] E. R. Bejarano, personal communication (1983).

[62] P. Galland and V. E. A. Russo, *Photochem. Photobiol.* **29,** 1009 (1979).

[63] P. Galland and V. E. A. Russo, *Planta* **146,** 257 (1979).

[64] F. Gutiérrez, Tesis Doctoral, Instituto Politécnico Nacional, México (1983).

[65] R. Garcés, M. Tortolero, and J. R. Medina, *Exp. Mycol.* (in press).

Appendix: Genetic Nomenclature and Strain List

Genetic nomenclature in *Phycomyces* follows the recommendations of Demerec *et al.*[66] for bacteria and was established by agreement of the members of Max Delbrück's group at the California Institute of Technology in 1968.

1. All mutations producing the same gross phenotypic change or affecting the same biochemical pathway are designated by the same three letter, lower-case, italicized code.
2. Different genes, any one of which may mutate to produce the same gross phenotypic change or affect the same biochemical pathway, are distinguished from each other by a gene symbol, that is, an italicized capital letter added immediately after the three-letter code.
3. An allele is designated by placing an italicized isolation number after the gene symbol. If the gene is unknown, a hyphen is placed between the three letter code and the isolation number, instead of the gene symbol. Each isolation number is used only once for each three-letter code. Independently isolated mutants are assumed to carry different alleles, even if they are phenotypically similar.
4. Each strain is designated by a collection code, consisting of one or more capital letters, and a serial number. This designation is not italicized and does not contain any dashes or spaces. Independently isolated mutants and recombinants constitute new strains.
5. The genotype of a strain is a list of all the alleles known to be present in the strain and the symbol (+) or (−) for the sexual type.
6. Phenotypes should be described in words or by the use of abbreviations defined each time and not resembling genetic designations.
7. Heterokaryons are indicated by placing an asterisk between the strain designations or the genotypes of the components.
8. Crosses are indicated by placing the symbol × between the strain designations or the genotypes of the parents. The (+) parent is written first.

In the following table[67–99] *car* designates genes involved in carotene biosynthesis and regulation; *dar*, riboflavin permease (resistance to 5-carbon-5-deazariboflavin); *fur*, resistance to 5-fluorouracil; *geo*, abnornal geotropism (very fast in strain C5); *mad*, abnormal phototropism; *nic*, synthesis of nicotinic acid (auxotrophy); *pic*, photoinduced carotenogenesis; *pil*, bulging sporangiophore growing zone (piloboloides); *pur*, synthesis of purines (auxotrophy for adenine or hypoxanthine); *tri*, synthesis of trisporic acids (sexual hormones).

[66] M. Demerec, E. A. Adelberg, A. J. Clark, and P. E. Hartman, *Genetics* **54,** 61 (1966).

Obsolete designations, used in some old papers, are given between quotation marks.

The origin of mutants is indicated by writing the parent strain and the mutagenesis procedure: (NG), *N*-methyl-*N'*-nitro-*N*-nitrosoguanidine; (NQO), 4-nitroquinoline 1-oxide; (ICR), ICR-170; (Spon), spontaneous. The two parents of crosses are separated by ×. Several cycles of mutant isolation and crosses are indicated by several levels of brackets and parentheses. For complicated cases, please consult the references.

Strain NRRL1555 is the standard (−) wild type. Strain A56 is a (+) wild type, largely isogenic with the standard.[1] Strains NRRL1554, UBC21, and UBC24 are other wild types, with genetic backgrounds quite different from the standard.

The published color mutants belong to the following collections:

A, Departamento de Genética, Facultad de Biología, Universidad de Salamanca, Spain (Prof. A. P. Eslava).

B, Max-Planck-Institut für Molekulare Genetik, Berlin, Federal Republic of Germany (Dr. V. E. A. Russo).

C, Division of Biology, California Institute of Technology, Pasadena, California (the late Prof. M. Delbrück).

H, University of California, Santa Cruz, California (Dr. P. V. Burke).

L, Department of Physics, Syracuse University, Syracuse, NY (Prof. E. D. Lipson).

M, Department of Biology, University of West Virginia, Morgantown, WV (Prof. R. P. Sutter).

S, Departamento de Genética, Facultad de Biología, Universidad de Sevilla, Sevilla, Spain (Prof. E. Cerdá-Olmedo).

Y, Faculty of Education, Yamagata University, Yamagata, Japan (Prof. T. Ootaki).

The strains in this list and many others may be obtained from the addresses above. The most complete collections are kept at Yamagata and Sevilla.

Genetic Nomenclature and Strain List for *Phycomyces*

Strain	Genotype	Origin	References
A98	*carC652* (−)	NRRL1555 (NQO)	21
A220	*carA5 carC652* (−)	Several mutagenesis and crosses	21
A227	*carB10 carC652 nicA101* (+)	Several mutagenesis and crosses	21
A229	*carC652 carR21 nicA101* (−)	Several mutagenesis and crosses	21
A344	*carC660 furA401* (−)	NRRL1555 (Spont) (NQO)	21
A346	*carC662* (−)	NRRL1555 (NQO)	21

GENETIC NOMENCLATURE AND STRAIN LIST FOR *Phycomyces* (*continued*)

Strain	Genotype	Origin	References
A348	*carC664* (−)	NRRL1555 (NQO)	21
A350	*carC652 carS42* (+)	Several mutagenesis and crosses	21
A354	*carC668* (−)	NRRL1555 (NQO)	21
B41	*carA5 purA51* (+)	Several mutagenesis and crosses	36
B80	*carA5 madA7* (+)	Several mutagenesis and crosses	43
B401	*car* (−)	NRRL1555 (NG)	62
B402	*car* (−)	NRRL1555 (NG)	62
C1	*carB1* (−) ["Alb. 1"]	NRRL1555 (NG)	7, 9, 12, 22
C2	*carA5* (−) ["Alb. 5"]	NRRL1555 (NG)	2, 4, 7–15, 19–22, 24, 25, 32, 35, 39, 40, 42, 45, 47, 54, 62, 63, 65–85
C3	*carA5* (+)	NRRL1554 (NG) × C2	4, 12, 15, 40
C5	*carB10 geo-10* (−) ["Alb. 10"]	NRRL1555 (NG)	4, 7, 9, 11, 12, 15, 17, 20–22, 28, 29, 39, 40, 42, 45, 54, 62, 63, 65, 78, 79, 83, 86–89
C6	*carRA12 madF48* (−) ["Alb. 12"]	NRRL1555 (NG)	2, 4, 7, 9, 11, 12, 22, 39, 62, 63, 72, 76, 79
C7	*carA13* (−)	NRRL1555 (NG)	4, 9
C8	*carA14* (−)	NRRL1555 (NG)	4, 9, 12
C9	*carR21* (−) ["R1"]	NRRL1555 (NG)	2, 4, 7–15, 17, 18, 20, 22, 29, 39, 40, 42, 45, 54, 55, 62, 63, 65, 68, 69, 75, 78, 80, 83, 84, 86, 88–90, 92
C10	*carR22* (−)	NRRL1555 (NG)	4, 9, 12, 22, 39, 68
C11	*carR23*	NRRL1555 (NG)	4, 9, 12
C12	*carR24* (−)	NRRL1555 (NG)	4, 9, 12
C13	*car-25* (−)	NRRL1555 (NG)	9, 68
C14	*carR26* (−)	NRRL1555 (NG)	9

(*continued*)

Genetic Nomenclature and Strain List for *Phycomyces* (*continued*)

Strain	Genotype	Origin	References
C115	*carS42 mad-107* (−) ["Ph. 107"]	NRRL1555 (NG)	2, 4, 7, 10, 11, 13, 15, 17, 18, 20, 24, 29, 31, 32, 35, 79, 83, 86, 89, 91–95
C141	*carA5 madC51* (−)	NRRL1554 (NG) × C2	22, 74, 96, 97
C142	*carA5 madC51* (+)	NRRL1554 (NG) × C2	39
C143	*carRA2 mad-49* (−)	NRRL1555 (NG)	4, 9, 12, 22
C144	*carRA3 mad-98* (−)	NRRL1555 (NG)	4, 9, 12, 22
C148	*carA5 madC119* (−)	C2 (NG)	9, 22, 43, 62, 63, 72, 76–78, 97
C152	*carA15 mad-123* (−)	NRRL1555 (NG)	4, 9, 12, 19
C153	*carB10 mad-124* (−)	C5 (NG)	9
C158	*car-41* (−)	NRRL1555 (NG)	69
C167	*carA5 mad-125* (−)	C2 (NG)	9
C169	*carA5* (+)	UBC21 × C2	4, 12, 21, 36, 46
C170	*carA5* (+)	UBC21 × C2	4, 12, 45–47
C171	*carA30 carR21* (−)	C9 (NG)	2, 4, 9, 12, 62, 63, 65
C172	*carB31 carR21 mad-126* (−)	C9 (NG)	9
C173	*carB32 carR21* (−)	C9 (NG)	4, 9, 12, 51, 62, 63
C174	*carB33 carR21* (−)	C9 (NG)	4, 9, 12, 51
C175	*carA34 carR21* (−)	C9 (NG)	4, 9, 12
C176	*carA34 carR21* (−)	C9 (NG)	9, 12, 84, 85
C177	*carA36 carR21 mad-127* (−)	C9 (NG)	4, 9
C178	*carA37 carR21 mad-128* (−)	C9 (NG)	4, 9
C179	*carA38 carR21 mad-129* (−)	C9 (NG)	4, 9, 12
C180	*carA39 carR21 mad-130* (−)	C9 (NG)	4, 9
C242	*carA5 nicA101* (+)	C170 × NRRL1555 (NG)	2, 10, 35, 47
C312	*carA5 pilA7* (−)	C2 (NG)	84, 85
C313	*carA5 pilA8* (−)	C2 (NG)	84, 85
H7	*carR51* (+)	UBC21 (NG)	55
L5	*madA7 carA701* (−)	NRRL1555 (NG) (NG)	78
L6	*madB103 carA702* (−)	NRRL1555 (NG) (NG)	78
L7	*madD59 carB703* (−)	NRRL1555 (NG) (NG)	78
L8	*madF48 carA704* (−)	Several mutagenesis crosses	78
L9	*madG131 carRA705* (−)	Several mutagenesis and crosses	78
L23	*madE102 carRA718*	NRRL1555 (NG) (NG)	78
L33	*madA7 carB733* (−)	NRRL1555 (NG) (NG)	78
L89	*carA5* (+)	A56 × C2	98

GENETIC NOMENCLATURE AND STRAIN LIST FOR *Phycomyces* (*continued*)

Strain	Genotype	Origin	References
M1	*carS43 tri-751* (+)	NRRL1554 (NG)	4, 15, 20, 55, 83
S5	*carA51 madC202* (−)	UBC24 (NG)	4, 9, 12, 19, 22, 67, 97
S13	*carR95* (−)	UBC24 (NG)	12
S14	*carA53 madC205* (−)	UBC24 (NG)	4, 9, 12, 22, 43, 67, 74, 75, 97
S16	*carA55 mad-207* (−)	UBC24 (NG)	4, 9, 12, 22
S18	*carA57 madC209* (−)	UBC24 (NG)	9, 12, 19, 22, 45, 67, 74, 97, 99
S19	*carA58 mad-210* (−)	UBC24 (NG)	9, 12, 22
S20	*carB59* (−)	UBC24 (NG)	3, 4, 9, 12, 22
S21	*carB60* (−)	UBC24 (NG)	3, 4, 9, 12, 22
S22	*carB61* (−)	UBC24 (NG)	3, 4, 9, 12, 22
S23	*carA62 mad-211* (−)	UBC24 (NG)	4, 9, 12, 22
S25	*car-64 mad-213* (−)	UBC24 (NG)	12
S28	*carA67 mad-216* (−)	UBC24 (NG)	4, 12
S29	*carA68 mad-217* (−)	UBC24 (NG)	12
S37	*carA76 mad-225* (−)	UBC24 (NG)	9, 12, 22, 67
S39	*carA78 mad-227* (−)	UBC24 (NG)	12
S40	*carA79* (−)	UBC24 (NG)	4, 9, 12, 19, 22
S45	*carA84* (−)	UBC24 (NG)	4, 9, 12, 22
S47	*carA86 mad-232* (−)	UBC24 (NG)	12
S48	*carA87 mad-233* (−)	UBC24 (NG)	4, 9, 12, 22
S49	*carA89 mad-234* (−)	UBC24 (NG)	4, 9, 12, 22
S57	*carA5 madC242* (−)	C2 (NG)	75
S80	*aux carR21* (−)	C9 (NG)	3, 45, 79, 99
S86	*aux carB96 carR21* (−)	S80 (NG)	3
S100	*carS98* (−)	NRRL1555 (NG)	15
S101	*carS42 mad-107* (+)	UBC21 × (C115*NRRL1555)	15
S103	*carS42 carB99 mad-107* (−)	C115 (NG)	15
S104	*carRA115 carS42 mad-107* (−)	C115 (NG)	2, 15
S105	*carA101 carS42 mad-107* (−)	C115 (NG)	15
S106	*carS42 car-102 mad-107* (−)	C115 (NG)	15, 55
S113	*carA109 carS42 mad-107* (−)	C115 (NG)	15
S114	*carA110 carS42 mad-107* (−)	C115 (NG)	15
S115	*carA111 carS42 mad-107* (−)	C115 (NG)	15
S116	*carA5* (+)	C170 × NRRL1555	45
S117	*carA112 carS42 mad-107* (−)	C115 (NG)	15
S119	*carA113* (−)	NRRL1555 (NG)	10
S124	*carS42 car-119 mad-107* (−)	C115 (ICR)	2
S125	*carS42 car120 mad-107* (−)	C115 (ICR)	2

(*continued*)

GENETIC NOMENCLATURE AND STRAIN LIST FOR *Phycomyces* (*continued*)

Strain	Genotype	Origin	References
S131	*carA158 madA7* (−)	NRRL1555 (NG) (NG)	75, 77
S132	*carA159 madB104* (−)	NRRL1555 (NG) (NG)	75, 77
S133	*carA160 madD59* (−)	NRRL1555 (NG) (NG)	75
S134	*carA161 madE102* (−)	NRRL1555 (NG) (NG)	75, 77
S136	*carR127* (+)	NRRL1554 (NG)	55
S144	*carI131* (−)	NRRL1555 (NG)	10, 79
S160	*carRA115 carS42 car-136 mad-107* (−)	S104 (NG)	2
S161	*carRA115 carS42 car-137 mad-107* (−)	S104 (NG)	2
S165	*carRA115 carS42 car-141 mad-107* (−)	S104 (NG)	2
S240	*carR21 nicA101* (−)	Several mutagenesis and crosses	10
S241	*carR21* (+)	Several mutagenesis and crosses	10
S252	*picB1* (−)	NRRL1555 (NG)	24, 25
S253	*picA2* (−)	NRRL1555 (NG)	24, 25
S265	*carB10 dar-54* (−)	C5 (Spont.)	10
S273	*carS42 nicA101* (+)	C242 × [C115*(NRRL1555 [NG])]	21
S274	*carS42 nicA101* (−)	C242 × [C115*(NRRL1555 [NG])]	32
S276	*carS42* (+)	C242 × [C115*(NRRL1555 [NG])]	65
S283	*carA5 carS42 nicA101* (−)	C242 × [C115*(NRRL1555 [NG])]	10, 24
S290	*madB103 picB1* (−)	Several mutagenesis and crosses	24
S303	*carS179 dar-52* (−)	NRRL1555 (Spont.) (NG)	32
S323	*carS180 dar-52* (−)	NRRL1555 (Spont.) (NG)	32
S324	*carS181 dar-52* (−)	NRRL1555 (Spont.) (NG)	32
Y12	*carR21 pilA905* (−)	C9 (NG)	84, 85
Y13	*carR21 pilA903* (−)	C9 (NG)	84, 85
Y15	*carR21 pilA904* (−)	C9 (NG)	84, 85
Y16	*carR21 pilA905* (−)	C9 (NG)	84, 85
Y17	*carR21 pilA906* (−)	C9 (NG)	84, 85
Y18	*carR21 pilA907* (−)	C9 (NG)	84, 85
Y19	*carR21 pilA908* (−)	C9 (NG)	84, 85
Y26	*carR21 pilA910* (−)	C9 (ICR)	84, 85
Y30	*carR21 pilA914* (−)	C9 (ICR)	84, 85
Y31	*carA5 pilA915* (−)	C2 (ICR)	84, 85
Y32	*carA5 pil-916* (−)	C2 (ICR)	84, 85
Y35	*carR21 pilD919* (−)	C9 (NQO)	84, 85
Y37	*carA5 pil-921* (−)	C2 (NQO)	84, 85
Y38	*carA5 pil-924* (−)	C2 (NQO)	84, 85

[67] K. W. Foster and E. D. Lipson, *J. Gen. Physiol.* **62,** 590 (1973).
[68] W.-J. Hsu, D. C. Ailion, and M. Delbrück, *Phytochemistry* **13,** 1463 (1974).
[69] A. J. Jesaitis, *J. Gen. Physiol.* **63,** 1 (1974).
[70] E. D. Lipson, *Biophys. J.* **15,** 989 (1975).

[71] E. D. Lipson, *Biophys. J.* **15,** 1013 (1975).
[72] E. D. Lipson, *Biophys. J.* **15,** 1033 (1975).
[73] M. Delbrück, A. Katzir, and D. Presti, *Proc. Natl. Acad. Sci. U.S.A.* **73,** 1969 (1976).
[74] A. P. Eslava, M. I. Álvarez, E. D. Lipson, D. Presti, and K. Kong, *Mol. Gen. Genet.* **147,** 235 (1976).
[75] J. R. Medina and E. Cerdá-Olmedo, *Exp. Mycol.* **1,** 286 (1977).
[76] V. E. A. Russo, *J. Bacteriol.* **130,** 548 (1977).
[77] E. D. Lipson and D. Presti, *Photochem. Photobiol.* **25,** 203 (1977).
[78] E. D. Lipson, D. T. Terasaka, and P. S. Silverstein, *Mol. Gen. Genet.* **179,** 155 (1980).
[79] J. R. Medina and J. L. Micol, *Microbios Lett.* **16,** 69 (1981).
[80] C. Pueyo and E. Cerdá-Olmedo, *Exp. Mycol.* **5,** 112 (1981).
[81] I. López-Díaz and E. D. Lipson, *Mol. Gen. Genet.* **190,** 318 (1983).
[82] E. D. Lipson and S. M. Block, *J. Gen. Physiol.* **81,** 845 (1983).
[83] I. E. Clarke, A. de la Concha, F.J. Murillo, G. Sandmann, E. J. Skone, and P. M. Bramley, *Phytochemistry* **22,** 435 (1983).
[84] K. Koga and T. Ootaki, *Exp. Mycol.* **7,** 148 (1983).
[85] K. Koga and T. Ootaki, *Exp. Mycol.* **7,** 161 (1983).
[86] T.-C. Lee, D. B. Rodríguez, I. Karasawa, T. H. Lee, K. L. Simpson, and C. O. Chichester, *Appl. Microbiol.* **30,** 988 (1975).
[87] E. D. Lipson, I. López-Díaz, and J. A. Pollock, *Exp. Mycol.* **7,** 241 (1983).
[88] T.-C. Lee, T. H. Lee, and C. O. Chichester, *Phytochemistry* **11,** 681 (1972).
[89] P. M. Bramley, A. Than, and B. H. Davies, *Phytochemistry* **16,** 235 (1977).
[90] T.-C. Lee and C. O. Chichester, *Phytochemistry* **8,** 603 (1969).
[91] P. M. Bramley and B. H. Davies, *Phytochemistry* **14,** 463 (1975).
[92] M. Elahi, C. O. Chichester, and K. L. Simpson, *Phytochemistry* **12,** 1627 (1973).
[93] G. J. P. Riley and P. M. Bramley, *Cytobios* **34,** 97 (1982).
[94] P. M. Bramley and B. H. Davies, *Phytochemistry* **15,** 1913 (1976).
[95] G. J. P. Riley and P. M. Bramley, *Biochim. Biophys. Acta* **450,** 429 (1976).
[96] I. López-Díaz and E. Cerdá-Olmedo, *Curr. Genet.* **3,** 23 (1981).
[97] T. Ootaki, E. P. Fischer, and P. Lockhart, *Mol. Gen. Genet.* **131,** 233 (1974).
[98] I. López-Díaz and E. D. Lipson, *Curr. Genet.* **7,** 313 (1983).
[99] J. R. Medina, *Genet. Res.* **30,** 211 (1977).

[28] Carotene Synthesis in *Capsicum* Chromoplasts

By BILAL CAMARA

The ripening of *Capsicum* fruit is accompanied by a chloroplast to chromoplast transformation. The latter is the site of an active terpenoid synthesis, especially carotenoids. Chromoplasts isolated from this system efficiently incorporate terpene precursors into carotenes while xanthophylls are poorly labeled. This article summarizes some of the properties of this cell-free system.

Preparation of Chromoplasts

Intact chromoplasts are prepared from orange pericarp tissue (2 kg) of *Capsicum annuum* L. fruits. The procedure is described in Fig. 1.

After the purification procedure outlined in Fig. 1, the intact chromoplasts settle at the 0.84 to 1.45 *M* interface. The fraction recovered at 0.45 to 0.84 *M* interface is enriched with broken chromoplasts and membranes. Attention must be drawn on the following points: (1) The use of firm mature fruits greatly improves the yield of intact chromoplasts. (2) As an osmotic agent for the homogenizing medium, sucrose can be replaced by sorbitol. (3) The presence of Mg^{2+} or other polyvalent cations induces agglutination of heterogenous membranes which sometimes have the same sedimentation characteristics as the chromoplasts.[1] Therefore $MgCl_2$ must be excluded during the purification procedure.

Preparation of Chromoplast Subfractions

The method developed by Mackender and Leech[2] to rupture the plastid envelope by osmotic shock was used. To adapt their procedure, the chromoplast band taken at the 0.84–1.45 *M* of the sucrose gradient was diluted (1.5 ml of suspension + 1 ml of Tris 50 m*M* pH 7.6). This mixture was centrifuged at 4000 *g* for 10 min. At the completion, the chromoplast pellet, free of sucrose, is resuspended in 50 ml of Tris 50 m*M* pH 7.6 and left for 10 min at 0–4° before centrifugation at 100,000 *g* for 3 hr in a 42.1 fixed angle rotor, Beckman. The supernatant containing the stromal proteins is carefully pipetted and, if desired, the protein content is concen-

[1] B. Camara, F. Bardat, O. Dogbo, J. Brangeon, and R. Moneger, *Plant Physiol.* **73,** 94 (1983).

[2] R. O. Mackender and R. M. Leech, *Nature (London)* **228,** 1347 (1970).

ISBN 0-12-182010-6

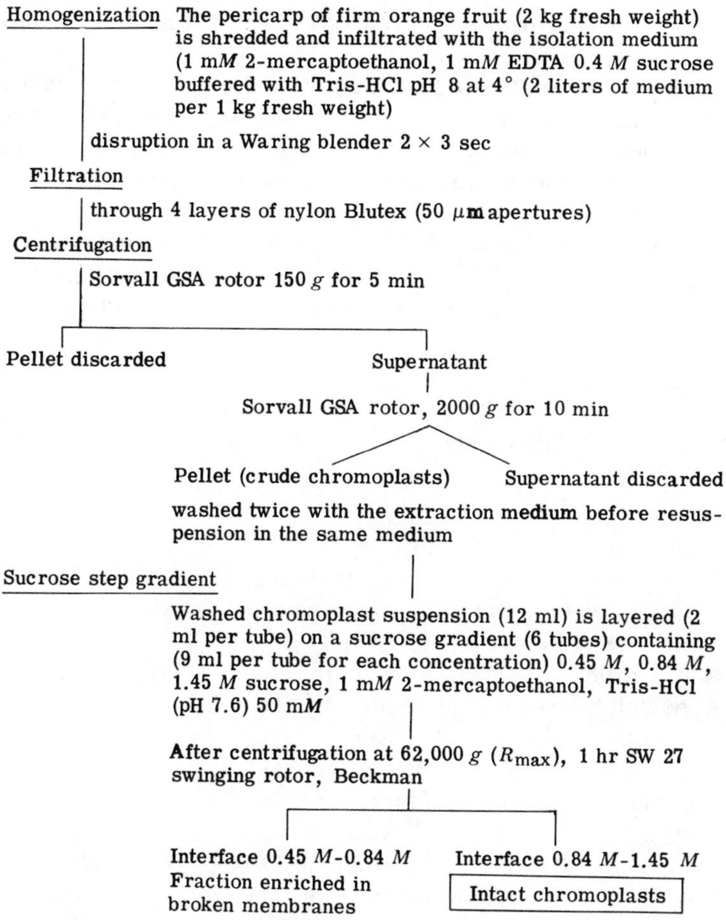

Homogenization The pericarp of firm orange fruit (2 kg fresh weight) is shredded and infiltrated with the isolation medium (1 m*M* 2-mercaptoethanol, 1 m*M* EDTA 0.4 *M* sucrose buffered with Tris-HCl pH 8 at 4° (2 liters of medium per 1 kg fresh weight)

disruption in a Waring blender 2 × 3 sec

Filtration

through 4 layers of nylon Blutex (50 μm apertures)

Centrifugation

Sorvall GSA rotor 150 *g* for 5 min

Pellet discarded

Supernatant

Sorvall GSA rotor, 2000 *g* for 10 min

Pellet (crude chromoplasts)

Supernatant discarded

washed twice with the extraction medium before resuspension in the same medium

Sucrose step gradient

Washed chromoplast suspension (12 ml) is layered (2 ml per tube) on a sucrose gradient (6 tubes) containing (9 ml per tube for each concentration) 0.45 *M*, 0.84 *M*, 1.45 *M* sucrose, 1 m*M* 2-mercaptoethanol, Tris-HCl (pH 7.6) 50 m*M*

After centrifugation at 62,000 *g* (R_{max}), 1 hr SW 27 swinging rotor, Beckman

Interface 0.45 *M*-0.84 *M* Fraction enriched in broken membranes

Interface 0.84 *M*-1.45 *M* Intact chromoplasts

FIG. 1. Method for the isolation of intact *Capsicum chromoplasts* (the working temperature is 0–4°).

trated using an Amicon cell equipped with a PM 10 filter. The pellet containing achlorophyll lamellae and envelope membranes is washed with Tris 50 m*M* pH 7.6 before use.

Criteria of Purity

The purity of the chromoplasts prepared as described above is verified by several points.

Phosphatidylethanolamine which is considered as an extraplastidial membrane lipid is barely detectable.[1]

The NADPH cytochrome *c* reductase[1] and succinate cytochrome *c* reductase activities[1] are negligible.

Thin layer chromatography (TLC) on Silica gel 60 precoated plates developed with light petroleum, bp 40–60°–diethyl ether (10 : 1, v/v)[3] followed by high-performance liquid chromography (HPLC) on μBondapak C_{18} (Waters) developed with methanol–tetrahydrofuran (80 : 20, v/v) reveal any ubiquinones.

The analysis of the subfractions reveals that the stromal fraction, which peaks at 280 nm, is devoid of carotenoids and galactosyltransferase.[4] The latter is a positive marker for membrane fractions.

Carotene Composition

Reagents and Equipment

NaCl 0.9%
Chloroform–methanol (2 : 1, v/v)
Light petroleum, bp 60–40°
Light petroleum, bp 60–40°–diethyl ether (98 : 2, v/v)
Light petroleum, bp 60–40°–benzene (9 : 1, v/v)
Diethyl ether–acetone (1 : 1, v/v)
Acetonitrile–ethyl acetate–chloroform (70 : 20 : 10, v/v) for HPLC
Activated Silica gel G plate for TLC
Magnesium oxide : kieselguhr G (1 : 1) plate for TLC
C_{18} μBondapak column (Waters) for HPLC
Chromoplast suspension (approximately 2–4 mg protein)
Vortex mixer

Experimental Method. The chromoplast suspension (2 ml) is homogenized with 9 ml of chloroform–methanol (2 : 1, v/v) for 3 min in a screw capped tube shaken by a vortex mixer; after which, 3 ml of chloroform is added and shaking is continued for 2 min, then 3 ml of 0.9% sodium chloride is added. The mixture is homogenized again and the solution is cleared by centrifugation. The chloroform layer is withdrawn and concentrated to dryness *in vacuo* or by a stream of nitrogen. The lipid residue is spotted on a silica gel G plate under a continuous stream of nitrogen. The plate is developed with light petroleum, bp 40–60°–diethyl ether (98 : 2, v/v) in a dim light atmosphere at 0–4°. The carotene fraction (R_f 0.5–0.7) detected by visible and UV light is clearly separated from the xanthophyll fraction. The carotene fraction is scraped from the plate and eluted with acetone before rechromatography on a magnesium oxide : kieselguhr G

[3] H. K. Lichtenthaler, *in* "Lipids and Lipid Polymers in Higher Plants" (M. Tevini and H. K. Lichtenthaler, eds.), p. 231. Springer-Verlag, Berlin, 1977.

[4] B. Camara, F. Bardat, and R. Moneger, *Eur. J. Biochem.* **127,** 255 (1982).

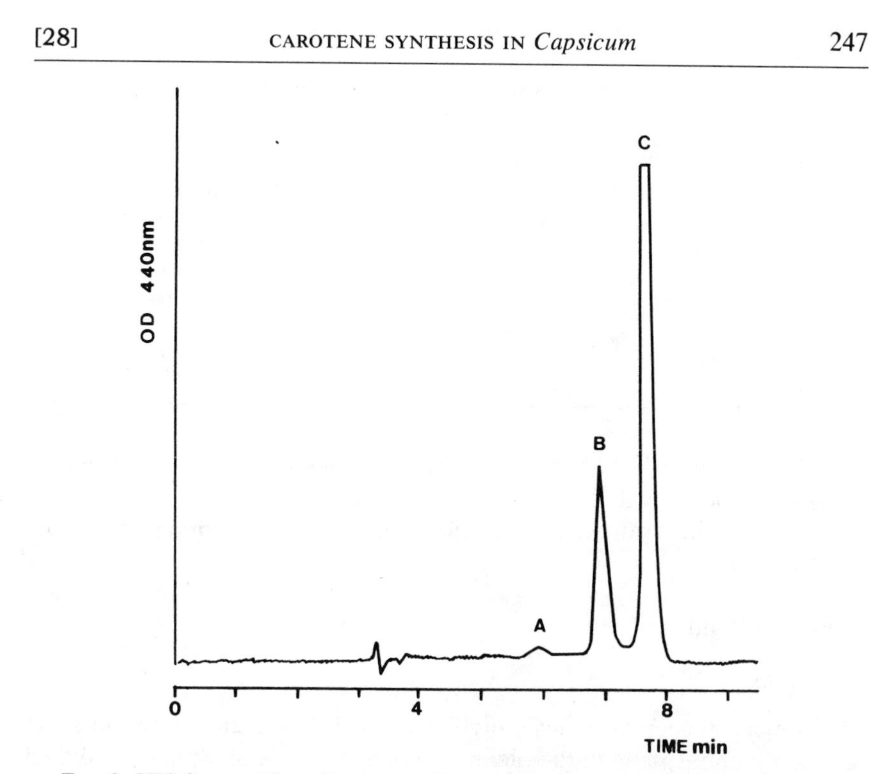

FIG. 2. HPLC separation of lycopene, ζ-carotene and β-carotene recovered after the incubation procedure described in the text. The extract was analyzed using a Waters system (Pump model 510, variable wavelength detector M450, data module M730). Conditions of analysis (column: C_{18} μBondapak 30 cm; solvent: acetonitrile–ethyl acetate–chloroform (70 : 20 : 10, v/v); flow rate 1 ml/min; detection at 440 nm, AUFS 0.04). (A) Lycopene; (B) ζ-carotene; (C) β-carotene. The position of lycopene was verified by an authentic standard.

(1 : 1) plate developed with light petroleum–benzene (9 : 1, v/v).[5,6] In this system phytoene moves with the solvent front and is well separated from *cis*-phytofluene (R_f 0.81), *trans*-phytofluene (R_f 0.72), α-carotene (R_f 0.64), and β-carotene + ζ-carotene + lycopene (R_f 0–0.4). The latter fraction is eluted with diethyl ether–acetone (1 : 1, v/v) and chromatographed on silica gel G plate developed with light petroleum. Under these conditions β-carotene (R_f 0.45), ζ-carotene (R_f 0.30), and lycopene (R_f ~0) are separated. A more rapid procedure for the separation of β-carotene + ζ-carotene + lycopene is conveniently achieved in less than 10 min by HPLC on C_{18} Bondapak column developed with acetonitrile–ethyl acetate–chloroform (70 : 20 : 10, v/v) (see Fig. 2).

[5] G. Britton and T. W. Goodwin, this series, Vol. 18C, p. 654.

[6] B. Camara, C. Payan, A. Escoffier, and R. Moneger, *C. R. Hebd. Seances Acad. Sci., Ser. D* **291,** 303 (1980).

CAROTENE COMPOSITION OF *Capsicum* CHROMOPLASTS

Carotene	Composition	Configuration	
		cis (%)	trans (%)
Phytoene	22	99	Traces
Phytofluene	14	86	14
ζ-Carotene	11		100
Lycopene	Traces		
α-Carotene	1		100
β-Carotene	52		100

The identification of the different carotenes is based on procedures previously described.[7]

Results. The carotene composition of *Capsicum* chromoplasts is given in the table.

Carotene Synthesis

Principle

The carotene biosynthesis involves primarily the condensation of an allylic pyrophosphate to the olefinic isopentenyl pyrophosphate mediated by a prenyltransferase:

Isopentenyl-PP + dimethylallyl-PP ⟶ geranyl-PP
Geranyl-PP + isopentenyl-PP ⟶ farnesyl-PP
Farnesyl-PP + isopentenyl-PP ⟶ geranylgeranyl-PP

The dimerization of geranylgeranyl-PP yields phytoene. The latter is desaturated, isomerized, and cyclized to yield α- or β-carotene.

Experimental Procedure

Preparation of Substrates. The commonly used substrate, i.e., [^{14}C]isopentenyl pyrophosphate is commercially available. The other intermediates involved in the reactions specially geranylgeraniol[8] can also be prepared satisfactorily.

Step 1: Oxidation and Reduction of Terpenoids. Three to 4 g of activated manganese dioxide[9,10] is added to 2 mmol of the alcohol dissolved in

[7] B. H. Davies *in* "Chemistry and Biochemistry of Plant Pigments" (T. W. Goodwin, ed.), 2nd ed., Vol. 2, p. 38. Academic Press, New York, 1976.

[8] O. P. Vig, J. C. Kapur, J. Singh, and B. Vig, *Indian J. Chem.* **7,** 574 (1969).

[9] J. Attenburrow, A. F. B. Cameron, J. H. Chapman, R. M. Evans, B. A. Hems, A. B. A. Jansen, and T. Walker, *J. Chem. Soc.* 1094 (1952).

[10] E. J. Corey, N. W. Gilman, and B. E. Gamen, *J. Am. Chem. Soc.* **90,** 5616 (1968).

light petroleum. The flask is flushed with nitrogen and maintained at 0° under constant agitation. The time course of the reaction is detected by TLC on silica gel plates developed with dichloromethane and revealed by 0.04% rhodamine in acetone or by spraying the plate with 95% H_2SO_4 and heating at 105°. Two spots are detected, the aldehyde having the higher R_f value. At the end of the reaction (1–1.5 hr) the suspension is filtered and the residue is rinsed with diethyl ether.

The resulting aldehyde is taken by 10 ml of tetrahydrofuran containing 100 μl of 0.5 *N* sodium hydroxide and added to a known quantity of NaB^3H_4 dissolved in 100 μl of 0.5 *N* sodium hydroxide. After 1 hr at room temperature, the solvent is evaporated and the residue is dissolved in a biphasic system, diethyl ether–water (1 : 1, v/v). The labeled alcohol is recovered after evaporation of the diethyl ether phase and purified by column chromatography on Silica gel 60 developed by sequential elution with 100% light petroleum, to 50% diethyl ether in light petroleum. This yields a [1-^{3}H]terpenol.

Step 2: Pyrophosphorylation and Purification of Terpenols. The labeled terpenol (2 mmol) is dissolved in 20 mmol of trichloroacetonitrile and stirred for 30 min under nitrogen at room temperature. Then 10 mmol of ditriethylamine phosphate, prepared as described previously[11] and dissolved in 60 ml of acetonitrile, is added dropwise under constant stirring. Then the mixture is left at room temperature, under nitrogen, for 15 hr, after which the solution is evaporated to dryness. The purification procedure depends on the length of the carbon chain.

For C_5 and C_{15} alcohol the residue is dissolved with 5 ml of 0.14 *N* ammonium hydroxide. The aqueous solution is extracted twice with diethyl ether (for C_5 and C_{10} alcohols) or light petroleum (for C_{15} alcohol). The organic phase is discarded and the aqueous layer evaporated to dryness. The residue is dissolved with 2 ml of *n*-propanol–ammonium hydroxide–water (60 : 30 : 10, v/v)[12] and applied to a silica gel column (15 × 1.5 cm, 18 g of adsorbent) previously equilibrated and developed with the same solvent system, the flow being aided by aspirating the effluent with a peristaltic pump. Fractions of 1 ml are collected and tested on silica gel 60 plate developed with the above solvent mixture. The spots are detected after drying the plates and pulverization of Zindzade reagent.[13] The monophosphates are light blue while pyrophosphates and polyphosphates are dark blue.

For C_{20} alcohol or higher homologs, the dry residue is dissolved with 10 ml of chloroform–methanol (2 : 1, v/v). To this solution, 3 ml of water is added, and the mixture is homogenized with a Vortex mixer. The lower

[11] P. V. Bhat, L. M. de Luca, and M. L. Wind, *Anal. Biochem.* **102,** 243 (1980).
[12] R. H. Cornforth and G. Popják, this series, Vol. 15, p. 359.
[13] V. E. Vaskovsky and V. I. Svetashev, *J. Chromatogr.* **65,** 451 (1972).

chloroform phase is withdrawn and concentrated to dryness in a rotating evaporator. The residue is dissolved in 2 ml of chloroform–methanol (2 : 1, v/v) before application to a diethylaminoethyl(DEAE)-Sephadex LH-20 column (13 × 1 cm, 6 g of adsorbent) prepared according to Peterson and Sober[14] and previously equilibrated by successive elution with 50 ml of 1 *M* formic acid–0.5 *M* ammonium hydroxide in chloroform–methanol (1 : 1, v/v).[15] The free alcohols are eluted by 75 ml of chloroform–methanol (2 : 1, v/v) while the monophosphate and pyrophosphate derivatives are recovered by sequential elution with 35 ml of 0.04 *M* and 50 ml of 0.4 *M* ammonium formate dissolved in chloroform–methanol (2 : 1, v/v).[15,16] Fractions of 5 ml are collected and homogenized with 1 ml of 0.015 *M* ammonium hydroxide. After centrifugation the chloroform layer containing the monophoshate and the pyrophosphate derivatives is withdrawn. The remainder of the procedure is the same as that described above.

The structures of the different alcohol pyrophosphates are verified, first by phosphorus determination according to the method described by Chen,[17] and second by cleavage of the phosphate group via alkaline phosphatase and high-performance liquid chromatography using the system described below (Fig. 3). Finally the specific radioactivity of the labeled substrate is based on the phosphorus content and radioactivity determination by liquid scintillation counting. The labeled substrate is dissolved in Tris 50 m*M* pH 7.6 and stored at −20° until use.

Assay Method

Reagents and Equipment. In addition to those described under "Carotene Composition," the following materials described below are used.

Tris–HCl buffer, 50 m*M*, pH 7.6
$MgCl_2$, 1 *M*
$MnCl_2$, 1 *M*
Dithiothreitol (DTT), 1 *M*
NADP, 0.5 *M*
FAD, 100 μM
KF, 1 *M*
[1-^{14}C]Isopentenyl pyrophosphate, 53 mCi/mmol
[1-^{3}H]Geranylgeranyl pyrophosphate, 20 mCi/mmol
Chromoplast suspension in 0.84 *M* sucrose buffered with 50 m*M*

[14] E. A. Peterson and H. A. Sober, *Biochem. Prep.* **8,** 39 (1961).
[15] J. C. Dittmer, *J. Chromatogr.* **43,** 512 (1969).
[16] O. Samuel, Z. El Hachimi, and R. Azerad, *Biochimie* **56,** 1279 (1974).
[17] P. S. Chen, T. Y. Toribara, and H. Warner, *Anal. Chem.* **28,** 1756 (1956).

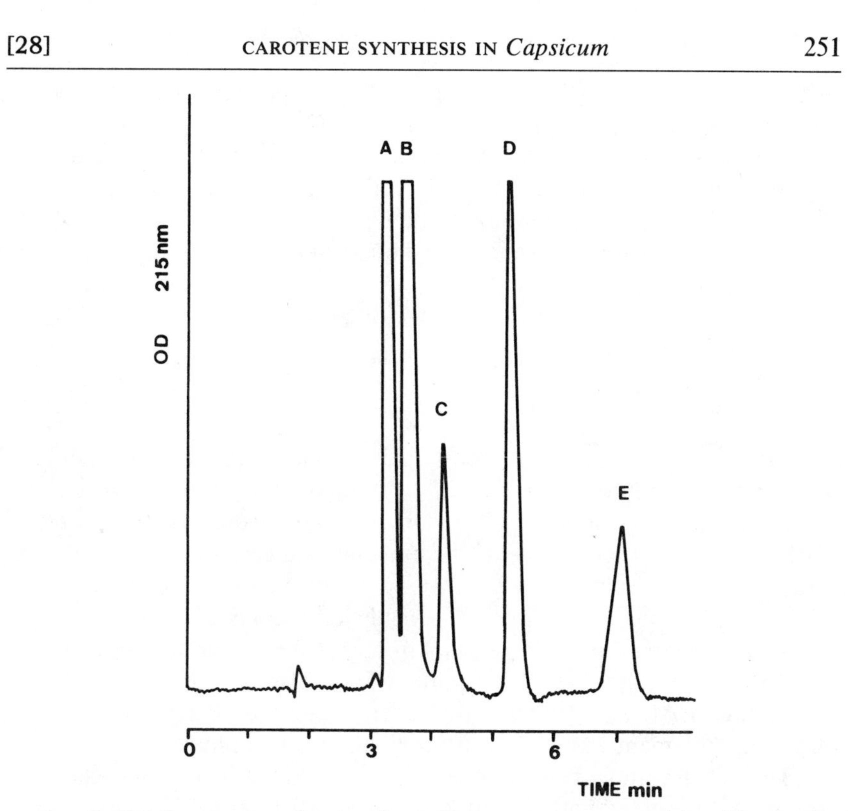

FIG. 3. HPLC separation of terpenols. A Waters system was used (see Fig. 2). The conditions of analysis are described in Fig. 2, except the solvent [methanol–water (97 : 3, v/v)] and the detection at 215 nm. (A) Dimethylallyl alcohol + isopentenol; (B) geraniol; (C) farnesol; (D) geranylgeraniol; (E) phytol.

Tris–HCl (pH 7.6) withdrawn from the centrifugation tube and homogenized in a tight fitting Potter-Elvejhem homogenizer (approximatively 2.5 mg protein/ml)

Chromoplast membranes suspended in 0.4 *M* sucrose buffered with 50 m*M* Tris–HCl (pH 7.6) (approximatively 2 mg protein/ml)

Chromoplast stroma in 0.25 *M* sucrose buffered with 50 m*M* Tris–HCl (pH 7.6) (approximatively 200 μg protein/ml)

Liquid scintillation cocktail (Permafluor Packard)

Incubation Medium. The enzymatic reaction is carried out for various times at 25° in a medium buffered with 50 m*M* Tris · HCl (pH 7.6) (2 ml final volume) containing 10 m*M* $MgCl_2$, 5 m*M* $MnCl_2$, 2 m*M* DTT, 1.25 μM FAD, 1.25 m*M* $NADP^+$, 1 m*M* KF, 0.4 *M* sucrose, [1-^{14}C]isopentenyl pyrophosphate 0.2 μCi, or [1-^{3}H]geranylgeranyl pyrophosphate 0.1 μCi, plastid suspension (chromoplast 2 mg protein, chromoplast membranes 2 mg, or chromoplast stroma 10 μg to 200 μg protein). The reaction is stopped by 4.5 ml of chloroform–methanol (2 : 1, v/v).

Extraction and Purification. The method described under "Carotene Composition" gives radiochemically pure carotenes for liquid scintillation counting. The unmetabolized and labeled isopentenyl pyrophosphate remains in the water phase during the extraction procedure. Prenols or prenyl pyrophosphate formed during the incubation remains at the origin of the silica gel G chromatoplate developed with light petroleum : diethyl ether (98 : 2, v/v).

Results

Chromoplasts. Ruptured chromoplasts convert efficiently isopentenyl pyrophosphate into carotenes (25% conversion can be observed). Geranylgeranyl pyrophosphate is converted to a lesser extent (10% conversion). *cis*-Phytoene is by far the most labeled carotene, and its desaturation up to α-carotene and β-carotene via *cis*-phytofluene, *trans*-phytofluene, ζ-carotene, and lycopene takes place.[4] Geranylgeraniol is detected during the incubation while geraniol or farnesol is apparently not formed.

Mn^{2+}, Mg^{2+}, and DTT are required for maximal synthesis.

Though FAD and $NADP^+$ are added to the incubation medium their direct involvement in the desaturation steps is not established.

The desaturation of *cis*-phytoene is drastically reduced when the chromoplasts are treated with Triton X-100 at ratio Triton X-100 (mg)/chromoplast protein (mg), 1 : 9.[4] Tween 80, which only slightly solubilizes the membrane protein, has a lower effect. When a chlorophyllide preparation emulsified in Tween 80 is added to the incubation medium, the radioactivity incorporated is channeled toward chlorophyllide esterification.[18]

Chromoplast Subfractions. With the same precursors, the chromoplast stroma synthesizes only *cis*-phytoene.[4] The same requirement for Mn^{2+}, Mg^{2+}, and DTT is observed.

The reaction is stimulated (up to 5-fold) by neutral detergents at ratio Triton X-100 (mg)/Stromal protein (mg): 1 : 5; Tween 80 (mg)/Stromal protein (mg): 1 : 10; and liposomes prepared from digalactosyldiacylglycerol (DGDG) at ratio DGDG (mg)/Stromal protein (mg): 1 : 4. Polyethylene glycol 1500 (10%) stimulates the reaction. The reaction is drastically inhibited by inorganic pyrophosphate (1 m*M*) and ethylenediaminetetraacetic acid (EDTA) (20 m*M*). The apparent Michaelis constant K_m for isopentenyl pyrophosphate is approximatively 15 μM.

Under the same conditions the chromoplast membranes are unable to metabolize isopentenyl pyrophosphate. However when the stromal pro-

[18] O. Dogbo, F. Bardat, and B. Camara, *Physiol. Veg.* **22,** 75 (1984).

teins are added to the chromoplast membrane fraction, *cis*-phytoene is desaturated, isomerized, and cyclized to yield colored carotenes ζ-carotene, lycopene, α-carotene, and β-carotene. This shows that the prenyl transfer reactions up to geranylgeranyl pyrophosphate and phytoene synthetase are soluble enzymes associated with the chromoplast stroma, while the enzymes involved in the isomerization, desaturation, and cyclization reactions are localized in the membrane fraction.

Acknowledgments

I thank Prof. Moneger for his interest during the work. I appreciate the technical assistance given by C. Agnes, F. Bardat, and A. Nargeot.

[29] Carotenoid Synthesis in *Neurospora crassa*

By WERNER RAU and URSULA MITZKA-SCHNABEL

Introduction

Since the earliest descriptions by Went[1,2] and, in particular, since the pioneer work of Haxo[3,4] and Zalokar[5,6] it has been known that the orange-red color of *Neurospora crassa* is due to carotenoids, and that the production of pigment is light-dependent. In conidiospores carotenoids are synthesized independently of light, while in mycelia only small amounts of pigments are formed in complete darkness and bulk production starts only after an illumination period. Up to now many investigations, not restricted to *Neurospora,* have contributed to the elucidation of the mechanism of photoregulation and of biosynthesis; however, we are still far from a complete understanding (for references, see 7,8).

[1] F. A. F. C. Went, *Zentralbl. Bakteriol. Parasitenkd.* **7,** 544 (1901).
[2] F. A. F. C. Went, *Recl. Trav. Bot. Neerl.* **1–4,** 106 (1904).
[3] F. Haxo, *Arch. Biochem. Biophys.* **20,** 400 (1949).
[4] F. Haxo, *Fortschr. Chem. Org. Naturst.* **12,** 169 (1955).
[5] M. Zalokar, *Arch. Biochem. Biophys.* **50,** 71 (1954).
[6] M. Zalokar, *Arch. Biochem. Biophys.* **56,** 318 (1955).
[7] W. Rau, *in* "The Blue Light Syndrome" (H. Senger, ed.), p. 283. Springer-Verlag, Berlin and New York, 1980.
[8] W. Rau, *in* "Biosynthesis of Isoprenoid Compounds" (J. W. Porter and S. L. Spurgeon, eds.), Vol. 2, p. 124. Wiley, New York, 1983.

METHODS IN ENZYMOLOGY, VOL. 110

ISBN 0-12-182010-6

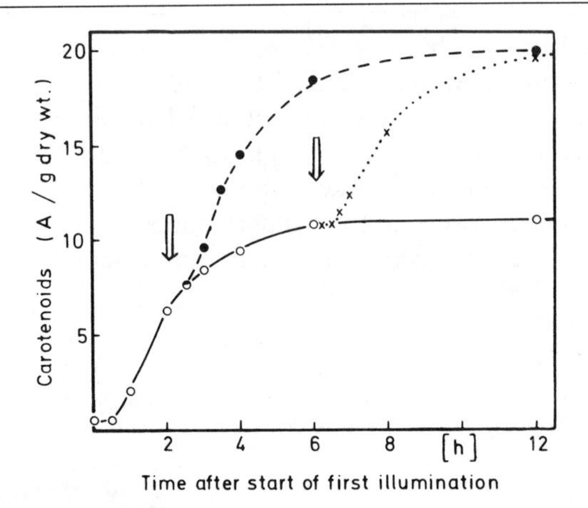

FIG. 1. Time course of carotenoid accumulation in *Neurospora crassa* after a first and second illumination (10 min, white light). (○) First illumination only; (●) second illumination after 2 hr; (×) second illumination after 6 hr. Arrows indicate time of second illumination. Reprinted from Rau and Rau-Hund.[18]

The kinetics of photoinduced accumulation of carotenoids in the mycelium following a brief, i.e., a few minutes, illumination (Fig. 1) show that after a lag period the amount of pigment increases rapidly for a period of time after which carotenoid synthesis ceases. From the nature of these kinetics it has been concluded (1) that carotenogenic enzymes are present in dark-grown mycelia only to a low extent, (2) that the enzymes are synthesized *de novo* as a consequence of photoinduction, and (3) that the enzymes are active only for a few hours.

Therefore, successful isolation and characterization of carotenogenic enzymes are highly dependent on photoinductive conditions and one may gather indispensable hints from *in vivo* results. Consequently, we will first summarize briefly the characteristics of photoregulation elaborated for *N. crassa,* then compile the *in vivo* data on pigment synthesis and finally describe methods for isolation and characterization of the carotenogenic enzymes.

Photoregulation

From the data gathered from experiments using brief periods of illumination it is obvious that in photoinduced carotenogenesis in *N. crassa*—as in other organisms (for references see 7,8)—photoreactions and subsequent dark reactions are involved. The mechanism exhibits all the features of a "classical" induction, including a very early "point of no return" in the sequence of events triggered by light.

Photoinduction

The shapes of the action spectra for carotenoid synthesis in *Neurospora*[6,9] resemble those obtained for other photoresponses of the blue/UV-light or "cryptochromal" reactions (for references see 8,10). These spectra show that only irradiation with wavelengths of 340–500 nm and around 280–300 nm is effective.

Earlier investigations on the dose–response relationship[6,11,12] indicated a saturation of the photoreaction with low doses of light, i.e., illumination for a few minutes.

More recently a biphasic response has been found[13,14] indicating additional photoinduction by prolonged illumination periods. For a brief exposure the Bunsen-Roscoe law of reciprocity is valid.[9,11,13] The photoreaction has been found to be independent of temperature,[4,12] thereby indicating that the primary event is, as in other organisms, a photochemical reaction. For optimal photoinduction the presence of oxygen is necessary although minor induction takes place under anaerobic conditions.[5,15]

Level of Regulation

Separation of photoinduction and subsequent responses could be demonstrated by "memory" phenomenon, i.e., temporal and reversible inhibition of pigment synthesis by low temperatures or anaerobic conditions.[6,16]

The existence of a lag-phase in pigment accumulation triggered by illumination points to a *de novo* synthesis of carotenogenic enzymes. Further evidence for this comes from the following results. Photoinduced carotenoid accumulation is completely blocked by cycloheximide, when applied prior to or immediately after illumination. Inhibition is reduced with increasing time between illumination and addition of the inhibitor.[11,12,17] Application of the inhibitor at a time after illumination, when the

[9] E. C. DeFabo, R. W. Harding, and W. Shropshire, Jr., *Plant Physiol.* **57,** 440 (1976).
[10] J. Gressel and W. Rau, *Encycl. Plant Physiol., New Ser.* **16,** 603 (1983).
[11] W. Rau, I. Lindemann, and A. Rau-Hund, *Planta* **80,** 309 (1968).
[12] R. W. Harding, *Plant Physiol.* **54,** 142 (1974).
[13] E. L. Schrott, *Planta* **150,** 174 (1980).
[14] E. L. Schrott, *Planta* **151,** 371 (1981).
[15] W. Rau, *Planta* **84,** 30 (1969).
[16] W. Rau, *Planta* **101,** 251 (1971).
[17] R. W. Harding and H. K. Mitchell, *Arch. Biochem. Biophys.* **128,** 814 (1968).

rate of pigment accumulation is highest, proved to be nearly ineffective in blocking carotene synthesis, indicating that the inhibitor does not interfere with the catalytic activity of the enzymes. Furthermore, from the gradual loss of effectiveness of the inhibitor, it has been concluded that the photoinduced machinery for the synthesis of carotenogenic enzymes is active only for a limited time. Only a second illumination leads to a renewed *de novo* synthesis of enzymes, again after a lag phase[18] (Fig. 1). Inhibition by cycloheximide probably via inhibition of protein synthesis is partly reversible by rinsing the mycelium followed by further incubation in an inhibitor-free medium.[16,17] These experiments again reveal the "memory" phenomenon.

The problem of whether photoinduction regulates the synthesis of carotenogenic enzymes at the transcription or at the translation level has been approached using a few fungal and bacterial species.[7,10] In *N. crassa,* actinomycin D inhibits pigment synthesis to about 80% when applied prior to illumination but not when applied 15 min after illumination.[19] From induced and uninduced mycelia, poly(A)$^+$-RNA was extracted which shows good activity in an *in vitro* translational system obtained from rabbit reticulocytes. The differences in the pattern of the translated products[20] provide evidence for the photoinduction of specific mRNAs; however further proof is needed.

In Vivo Synthesis

Culture Conditions

In some investigations mycelia were grown for several days in standing cultures and were used as pads for the experiments.[6,12] As an attempt to achieve more standardized conditions and in order to get a younger, metabolically active mycelia, the following culture conditions, modified from earlier investigations,[13] are used: *N. crassa,* strain ATCC 10816 (C.B.S., Baarn, The Netherlands) is grown for conidia production on agar medium ("conidiating medium"), for 3 days at 32° and for additional 2 days at room temperature. Submerged mycelia are grown in 1.8-liter Fernbach flasks, containing 400 ml (total inoculum 2.5×10^8 conidia) "Fries minimal medium" plus 0.5% Difco yeast extract and 1% sucrose. They are shaken aerobically in the dark at 27° for 20 hr, at which time the cells are at the end of the log phase.

[18] W. Rau and A. Rau-Hund, *Planta* **136,** 49 (1977).

[19] R. E. Subden and G. Bobowski, *Experientia* **29,** 965 (1973).

[20] U. Mitzka-Schnabel, E. Warm, and W. Rau, *in* "Blue Light Effects in Biological Systems" (H. Senger, ed.), p. 264. Springer-Verlag, Berlin and New York, 1984.

Illumination Procedure and Incubation Conditions

Mycelia are harvested under dim red safety light (Philips TL 40 W/15 and red Plexiglas filter Röhm und Haas Nr. 501, 3 mm) by suction filtration and then suspended in incubation medium [16 m*M* KH_2PO_4; 2% (w/v) glucose] in gas washing bottles. Two hour incubation at 20° in the dark to these "step down" conditions is favorable for optimal photoinduction of carotenogenesis. Samples are illuminated with 2 banks of white light fluorescent tubes (Osram 65 W/19) from both sides, providing together 40 W m^{-2} within the spectral range of 400–520 nm as a convenient source for blue light. Constant temperature is maintained with a water bath. If not indicated otherwise, temperature during illumination and subsequent dark incubation is 20°. During illumination, a stream of oxygen (4 liter/hr) is passed through the bottles. Oxygen is replaced by air during dark incubation.

It is well established that in contrast to photoinduction, the biosynthesis of carotenoids has an absolute requirement for oxygen.[3,5,16,21] *N. crassa* shows an unusual temperature dependence of photoinduced carotenogenesis. In mycelia grown for several days in standing cultures and used as pads for the experiments,[6,12] optimum carotenoid production was found at 6°. Mycelia grown in liquid cultures formed maximal amounts at 10°.[13]

Mevalonate is the only substrate which has been shown by radioactive labeling to be incorporated into carotenoids during *in vivo* experiments.[22,23]

Carotenoid Content

Harvesting of mycelia, extraction and determination of carotenoids, as well as the amounts of the different carotenoids present in *N. crassa* have been reported by several investigators.[3,5,11,21–27] The following carotenoids have been identified: phytoene, phytofluene, ζ-carotene, neurosporene, lycopene, 3,4-dehydrolycopene, β-zeacarotene, γ-carotene, β-carotene, torulene, neurosporaxanthin.

[21] E. C. Grob, *Biosynth. Terpenes Sterols, Ciba Found. Symp., 1958,* p. 267 (1959).

[22] R. W. Harding, P. C. Huang, and H. K. Mitchell, *Arch. Biochem. Biophys.* **129,** 696 (1969).

[23] U. Mitzka, Ph.D. Thesis, Universität München (1978).

[24] S. Liaaen-Jensen, *Phytochemistry* **4,** 925 (1965).

[25] B. H. Davies, *in* "Carotenoids other than Vitamin A," Vol. III, p. 1. Butterworth, London, 1973.

[26] J. C. Lansbergen, R. L. Renaud, and R. E. Subden, *Can. J. Bot.* **54,** 2445 (1976).

[27] U. Mitzka and W. Rau, *Arch. Microbiol.* **111,** 261 (1977).

Neurosporaxanthin, an acidic pigment, was first isolated and characterized by Zalokar[28] and later the chemical structure was elucidated.[29]

Comparison of the results[30] clearly demonstrates that in the same species pigment levels may vary quite considerably; most striking are differences in the ratio of phytoene to colored pigments and in the ratio of neutral pigments to neurosporaxanthin. The differences may be due to different growth conditions, to the use of different strains, and to variations of the light regime used. The conidiospores contain much larger quantities of pigments on a dry weight basis than the mycelium itself. The relative amounts of several carotenoids are quite different from those of the mycelium; most remarkable is the prominent increase in lycopene and 3,4-dehydrolycopene.[5]

As in other microorganisms (for references, see 31,32) diphenylamine inhibits dehydrogenation of carotenoids thus leading to accumulation of saturated carotenes; nicotine decreases drastically the amounts of cyclized compounds.[33]

Biosynthetic Step(s) under Photocontrol

Experiments for elucidation of the photocontrolled reaction are hampered by the fact that dark-grown cultures synthesize small amounts of all pigments.

Reports on the content of different carotenoids in dark-grown mycelia of *N. crassa*[6,12,26] agree that phytoene is the dominant component; however, differences in the presence and amounts of the other carotenoids are found.

The kinetics of accumulation following illumination[6,12] reveal that, with some simplification, carotenoids are synthesized in a sequence according to the assumed biosynthetic pathway. However, the substrates for cyclization and the possible end-products are not definitely established.

Investigations using a mutant strain lacking phytoene dehydrogenation have shown that in this organism formation of phytoene is enhanced by photoinduction indicating that phytoene synthesis is partly photoregu-

[28] M. Zalokar, *Arch. Biochem. Biophys.* **70,** 568 (1957).

[29] A. J. Aasen and S. Liaaen-Jensen, *Acta Chem. Scand.* **19,** 1843 (1965).

[30] W. Rau, *Pure Appl. Chem.* **47,** 237 (1976).

[31] T. W. Goodwin, "The Biochemistry of the Carotenoids," 2nd ed., Vol. 1, Chapman & Hall, London, 1980.

[32] S. L. Spurgeon and J. W. Porter, *in* "Biosynthesis of Isoprenoid Compounds" (J. W. Porter and S. L. Spurgeon, eds.), Vol. 2, p. 1. Wiley, New York, 1983.

[33] W. Rau and U. Mitzka-Schnabel, unpublished results.

lated.[26] Further results derived from cell-free systems of this mutant will be discussed below.

In Vitro Synthesis

Synthesis of Carotenoids by Crude Enzyme Systems

Different cell-free systems that convert mevalonate[34–36] or isopentenyl pyrophosphate[37] to phytoene and more unsaturated carotenoids[36] have been obtained from *N. crassa*.

Preparation of the System.[36] Mycelia are grown and illuminated (30 min) as described above. Based on *in vivo*[11,27] and *in vitro*[36] studies, preparation of the homogenate seems to be optimal from mycelia at 2 hr after the start of photoinduction (Fig. 1).

Mycelia are harvested by suction filtration and rinsed carefully with deionized H_2O and homogenization buffer. Mycelial pads are suspended in ice-cold homogenization buffer containing 0.2 *M* Tris–HCl, pH 8.0, 0.2 *M* nicotinamide and 2 m*M* DTT (1 ml/g wet wt) and ground with sand (5 g/20 g wet wt) by mortar and pestle for 8 min. The homogenate is filtered through 2 layers of gauze and then centrifuged at 700 *g* for 20 min, thus removing cell debris and spores. The resulting supernatant is used as a crude enzyme extract. Protein concentration is determined by the method of Lowry *et al.*[38] with bovine serum albumin as a standard. All procedures are at 0–4°.

Incubation System.[36] In the incubation mixture (final vol. 7.5 ml) concentrations of components are 0.2 *M* Tris–HCl (pH 8.0), 0.2 *M* nicotinamide, 14.5 m*M* DTT, 5 m*M* ATP, 18.8 m*M* $MgCl_2$, 1.3 m*M* $MnCl_2$, 0.37 m*M* of each NAD, NADP, NADH, NADPH, 0.2 m*M* FAD, and 0.03 m*M* (2.5 μCi) [2-^{14}C]MVA (mevalonic acid). To initiate the incubation homogenate (21 mg protein) is added. DL-[2-^{14}C]MVA lactone is converted prior to incubation to its K-salt with aqueous KOH.[39] The mixture is incubated for 3 hr in 18 ml gas washing flasks at 25° in a H_2O bath under slow aeration and in dim daylight. Reactions are stopped by addition of ethanol (5 ml) and the protein precipitated with acetone (5 ml).

[34] U. Mitzka-Schnabel and W. Rau, *Phytochemistry* **19,** 1409 (1980).

[35] G. S. Bobowski, W. G. Barker, and R. E. Subden, *Can. J. Bot.* **55,** 2137 (1977).

[36] U. Mitzka-Schnabel and W. Rau, *Phytochemistry* **20,** 63 (1981).

[37] S. L. Spurgeon, R. V. Turner, and R. W. Harding, *Arch. Biochem. Biophys.* **195,** 23 (1979).

[38] O. H. Lowry, N. J. Rosebrough, A. L. Farr, and R. J. Randall, *J. Biol. Chem.* **193,** 265 (1951).

[39] G. Britton and T. W. Goodwin, this series, Vol. 18, Part C, p. 654.

Isolation and Identification of Products. Samples are saponified with 40% (w/v) methanolic-KOH (final conc. 9.6%) overnight at room temperature under N_2. For extraction of unsaponifiable lipids,[40] protein is precipitated with 40 ml acetone and centrifuged for 5 min at 500 *g*; from the precipitate lipids are extracted three times with 6 ml acetone–methanol (3 : 1). The residual precipitate is free from radioactivity. To the combined acetone extracts unlabeled carotenoids (~5 μg), and 300 ml 5% $(NH_4)_2SO_4$ solution are added and the lipids extracted three times with 15 ml petroleum ether. This extract is washed, dried (Na_2SO_4), evaporated to dryness, and then dissolved in petroleum ether (20 ml). Radioactivity is determined in an aliquot (0.5 ml).

Unsaponifiable lipids are separated on a column (14 × 0.5 cm) of Al_2O_3 (Woelm neutral, Brockmann activity grade III) using increasing concentrations (v/v) of diethyl ether in petroleum ether (40–60°) slightly modified from Goad and Goodwin.[41] The substances eluted are 50 ml 0% (squalene, phytoene and more saturated carotenes); 50 ml 2% (more unsaturated carotenoids): 80 ml 12% (prenols, sterols); 50 ml 100% ether (ergosterol). The different lipids are identified by their chromatographic behavior as compared with authentic compounds and by their absorption spectra. The identity of phytoene[42] is confirmed, in addition, by chromatography on columns of Al_2O_3 (Brockmann activity grade I-II and I). The lipid fractions eluted by column chromatography are further analyzed by using various TLC systems.

System 1: Activated Si gel 60 F 254 (Merck) developed with benzene–petroleum ether (1 : 24) using nonequilibrated chambers to separate phytoene (R_f 0.28), lycopersene (R_f 0.45), and squalene (R_f 0.5–0.6) from more unsaturated carotenoids. Marker spots of authentic samples are used for comparison. Substances are located under both visible and UV light and by exposure to J_2 vapor.

System 2: Paraffin-impregnated Si gel layers [obtained by dipping plates in a 4% solution of paraffin oil in petroleum ether (40–60°)] are developed with acetone–H_2O (13 : 7) to separate prenols from sterols. Sterols remaining at the plate origin are eluted and rechromatographed on activated Si gel 60, F 254 with ethylacetate–benzene (1 : 5).

When labeled terpenols have been identified, and their purity is established, the unsaturated hydrocarbons may be separated only by TLC using System 3: Si gel 60 F 254 developed with toluene–petroleum ether (2 : 98, 5 : 95, 10 : 90) using nonequilibrated chambers. Different solvent combinations are used depending on the desired separation of phytoene

[40] J. E. Graebe, *Phytochemistry* **7,** 2003 (1968).

[41] L. J. Goad and T. W. Goodwin, *Biochem. J.* **99,** 735 (1966).

[42] A. Than, P. M. Bramley, B. H. Davies, and A. F. Rees, *Phytochemistry* **11,** 3187 (1972).

and more unsaturated carotenoids. With toluene–petroleum ether (10 : 90) R_f values are approximately 3,4-dehydrolycopene (0.15), lycopene (0.2), neurosporene (0.25), γ-carotene (0.35), ζ-carotene (0.4), β-carotene, phytoene and phytofluene (0.5–0.55), squalene (0.7), and kaurene (0.8). Prenols, sterols, and neurosporaxanthin remain at the origin. Radioactivity on the TLC plates is detected by scanning. For radioassay by liquid scintillation spectrometry it is most convenient to scrape the silica gel regions to be assayed directly into a toluene-butyl-PBD scintillation mixture (counting efficiency 96%).

Synthesizing Activity. The crude enzyme system described incorporates up to 22% (~25 nmol) of the active isomer of mevalonic acid into the unsaponifiable lipids during a 3 hr incubation. The enzymes involved are active for 2–3 hr. However no linearity of incorporation is found during the incubation period. Most of the radioactivity is found in squalene and sterols (80%), ~12% in phytoene, and only low levels (0.9%) in the bulk of more unsaturated carotenoids. Therefore one cannot analyze the formation of individual carotenoids. Moreover, a low level of radioactivity is found in prenols, e.g., farnesol and geranylgeraniol (GG). The condensation of 2 molecules of geranylgeranyl pyrophosphate (GGPP) leads to the formation of 15-*cis*-phytoene,[36,37,42] which is accompanied *in vivo* by small amounts of the all-trans isomer.[42] Consistent with *in vivo* analysis no label is found in prephytoene pyrophosphate or lycopersene, the latter is in contrast to early reports with *N. crassa*.[43]

The conversion of MVA or isopentenyl pyrophosphate (IPP) to phytoene (and more unsaturated carotenoids) is significantly affected by different parameters, i.e., the age of mycelia, conditions of light treatment such as irradiation time or temperature[36,37] and the time interval between illumination and preparation of cell-free extracts.[36] With enzyme preparations additional evidence for light-induced *de novo* synthesis of carotenogenic enzymes in *N. crassa*[36,37] was obtained, thus confirming conclusions from *in vivo* inhibitor experiments (for review, see 8,30). Enzyme activity for synthesis of phytoene[36,37] and more unsaturated carotenoids[36] is low or absent[37] in preparations from dark-grown cells and is significantly increased (8- to 12-fold) in preparations from photoinduced cultures. The inducing or enhancing effect of illumination can be blocked[36,37] by an *in vivo* treatment with inhibitors of protein synthesis. Photoinduction of carotenogenesis *in vitro* has not been possible. Results obtained using a cell-free system show that the first step(s) in the biosynthetic pathway of carotenoids which are photoinduced are among those leading to the formation of the first C_{40} compound, i.e., phytoene,[36,37] while in some mu-

[43] E. C. Grob, K. Kirschner, and F. Lynen, *Chimia* **15**, 308 (1961).

tants the conversion of farnesyl pyrophosphate (FPP) to GGPP seems also to be under photocontrol.[44] The "white collar" mutant appears to be a regulatory mutant.[44]

Localization of Carotenogenic Activity

Carotenoid biosynthesis is a multistep reaction. Investigation of single steps requires both isolation of the enzyme systems involved and *in vitro* assays. The steps of the biosynthetic pathway are compartmented. The enzyme(s) converting MVA to GGPP [prenyltransferase(s)] are soluble, while carotenogenic enzymes [as phytoene synthase, isomerase, dehydrogenase(s), and cyclase(s)] are membrane-bound.[36] Corresponding results have been obtained from mutants of *N. crassa* which are blocked in early steps of the biosynthetic pathway.[44]

By the following *procedure* carotenogenic enzyme activities in *N. crassa* can be localized.[36] The crude extract is subfractionated by differential centrifugation: 2300 *g*, 10 min; 13,000 *g*, 12 min; 21,000 *g*, 30 min; 115,000 *g* 2.5 hr. The final supernatant is referred to as soluble fraction (115 KS) and has a lipid layer on top. The pellets are washed and resuspended by hand in a homogenizing vessel (5 mg/ml). The resulting enzyme preparations are incubated, as soon as possible, for 3 hr with [^{14}C]MVA in the incubation mixture as described. The incorporation of radioactivity from [^{14}C]MVA into labeled metabolites is measured by radioassay after extraction and chromatographic separation of the unsaponifiable lipids.

The data in Table I show that preparation procedures do not cause severe damage to the enzymes as indicated by nearly no decrease of incorporating activity in a recombined system consisting of aliquots of all the cell fractions as compared with the crude extract.

With single or combined particulate fractions no incorporation of radioactivity is observed, while both the soluble fraction and the lipid layer show reduced incorporation into carotenoids including phytoene. The conclusion that carotenogenic enzymes are located in the particulate fractions, while a GGPP-synthesizing system is soluble, can be tested by the following procedure. Particulate fractions (1.5 mg protein) characterized by marker enzyme activities[34] are incubated together with the soluble protein fraction (16.6 mg) and [^{14}C]MVA for 3 hr. Potassium fluoride (13.3 m*M*) is added to inhibit phosphatases. Since during the relatively long incubation there might be an insufficient supply of substrate, the ratio of soluble to particulate protein should be as high as possible and the assay for carotenogenic activity should be not longer than 10 min. Moreover,

[44] R. W. Harding and R. V. Turner, *Plant Physiol.* **68,** 745 (1981).

TABLE I
INCORPORATION OF RADIOACTIVITY FROM [2-^{14}C]MVA INTO SEVERAL METABOLITES BY DIFFERENT CELL FRACTIONS[a]

		Specific incorporation (cpm/mg protein)				
Fraction	Total incorporation (cpm)	Phytoene	Carot-enoids	GG	Farnesol	Presqualene, squalene, and sterols
Crude extract	219,941	1324	167	128	214	8310
Recombined pellets	604	—	—	—	—	—
115 KS	185,261	303	—	847	615	7990
Lipid layer	165,697	252	—	313	454	8850
Recombined system	210,467	2431	100	66	372	6930

[a] Three hour incubation; ~20 mg protein.

GGPP may be available by preincubation of the soluble fraction alone; however, GGPP may not accumulate to high amounts.

To overcome a possible deficiency of [^{14}C]GGPP (which is not commercially available) a supplementation with a GGPP-synthesizing *in vitro* system provided by the semiliquid endosperm of maturing seeds of pumpkin (*Cucurbita pepo*) may be used.[36,40,45]

Fruits of pumpkins are harvested when not fully matured. The tips are cut from the seeds and the semiliquid endosperm is squeezed out and frozen immediately at −60°. The endosperm can be stored for up 2 years. The thawed endosperm is homogenized gently in a Potter-Elvehjem glass homogenizer and centrifuged at 35,000 *g* for 20 min. The supernatant (35 KS) is used as an enzyme source (3 ml; ~30 mg protein) for synthesis of [^{14}C]GGPP in the presence of [^{14}C]MVA. Cofactors added are as described for carotenoid synthesis (see "Incubation System"), although these conditions are not optimal for GGPP-formation. In addition, incubation mixture contains an ATP-regenerating system (0.13 m*M* PEP; 100 μg pyrophosphate kinase), as well as KF (13.3 m*M*) and 2-isopropyl-4-dimethylamino-5-methylphenyl-1-piperidine carboxylate methyl chloride (AMO 1618) to a final concentration of 1 μ*M*. AMO 1618, an inhibitor of cyclases,[46] inhibits the conversion of GGPP to kaurene to some extent.[36]

[45] M. O. Oster and C. A. West, *Arch. Biochem. Biophys.* **127,** 112 (1968).
[46] D. T. Dennis, C. D. Upper, and C. A. West, *Plant Physiol.* **40,** 948 (1965).

To generate some GGPP the enzyme preparation from pumpkin endosperm is preincubated for 2.5 hr with substrate. Subsequently, carotenoid synthesis is initiated by addition of particular isolated cell fractions (1 mg) prepared from illuminated mycelia of *N. crassa*. Incubation for 10 min proved to be optimal for incorporation into carotenoids.

The synthetic capacity of enzymes from *Cucurbita pepo* endosperm for isoprenoids is very low, except for GGPP and kaurene particularly when calculated on a 1 mg protein basis. Data listed in Table II show that maximum carotenogenic activity (~80%) is located in two membrane fractions which contain plasma membranes and in particular that which contains membranes of the endoplasmic reticulum (ER).[34,36] Studies on storage of carotenoids have shown that these two fractions contain most of the particulate bound carotenoids whereas the lipid layer on top of the supernatant contains the bulk of pigments.[34]

Comments. If incubation mixtures without AMO 1618 are used, incorporation of radioactivity into unsaturated carotenoids is increased (3-fold) while label of phytoene decreases. Furthermore, extensive labeling of kaurene is found (up to 80% of unsaponifiable lipids).

Properties of the Membrane-Bound System

To optimize the *in vitro* assay for carotenoid synthesis variation of the following conditions would be useful[47,48]: For photoinduction of carotenogenic enzymes the duration of *in vivo* illumination should be extended to 128 min followed by immediate preparation of homogenates.[11,13,14,27,47] Also after removing a mitochondrial enriched fraction (13 KP; 15 min) a combined membrane fraction is gained by subsequent centrifugation at 115,000 *g*, for 2.0 hr (115 KP). Membranes are suspended (15–20 mg protein/ml), in 0.2 *M* Tris–HCl, pH 8.0 containing 2 m*M* DTT and 30% (w/v) glycerol. The new incubation system contains the following components: 3 ml 35 KS from the *Cucurbita endosperm* homogenate (pH 6.5); 3.3 ml homogenization buffer (0.2 *M* Tris–HCl, pH 8.0; 0.2 *M* nicotinamide; 2 m*M* DTT); 0.5 ml Tris–HCl (2 *M*, pH 7.8); 0.05 ml DTT (0.94 *M*); components to final concentrations: 5 m*M* ATP, 18.8 m*M* $MgCl_2$, 1.3 m*M* $MnCl_2$, 0.37 m*M* of each NAD, NADP, NADH, 1.0 m*M* NADPH, 0.1 m*M* FAD, 0.13 m*M* PEP, 13.3 m*M* KF; finally 100 μg pyruvate kinase and 500 nmol MVA. The final volume is 7.5 ml and the pH is 7.8. This system is preincubated for 2.5 hr; 0.5 mg membrane protein is added and incubation is continued for 10 min at 27°. Using [^{14}C]MVA and a GGPP-

[47] U. Mitzka-Schnabel and W. Rau, *Carotenoid Chem. Biochem., Proc. Int. Symp. Carotenoids, 6th, 1981* Abstracts (1982).

[48] U. Mitzka-Schnabel, I. Stark, and W. Eiberger, *Tag. Dtsch. Bot. Ges., 1982* Abstracts (1982).

TABLE II

DISTRIBUTION OF SOME MARKER ENZYME ACTIVITIES, CAROTENOGENIC ENZYME ACTIVITY, AND OF CAROTENOIDS IN CELL FRACTIONS OF *Neurospora crassa*

	Marker enzyme activity (%)[a]				Carotenogenic activity (%)[b]		Distribution of carotenoids (%)[c]
	Cytochrome *c* oxidase	Succinate–cytochrome *c* reductase	NADH–cytochrome *c* reductase	NADPH–cytochrome *c* reductase			
Fraction	Mitochondria		Mitoch. + ER	ER	Phytoene	Unsaturated carotenoids	
2.3 KP	10.7	1.8	12.4	1.4	1.4	1.0	4.5
13 KP	63.4	86.9	23.3	6.9	14.6	16.8	4.2
21 KP	12.7	5.3	9.8	20.4	23.9	31.5	9.0
115 KP	7.3	6.0	28.1	25.6	58.1	49.2	17.9
115 KS	4.8	0.9	23.9	35.5	2.0 (115 KS + Lipid layer)	1.5 (115 KS + Lipid layer)	6.6
Lipid layer	1.1	0	2.5	9.2			57.8

[a,b] Percentage activity represents percentage of overall total enzyme activity (*a*) respectively of specific incorporation of radioactivity (*b*).

[c] Percentage of the amount of carotenoids formed *in vivo*.

TABLE III
OPTIMIZATION OF THE MEMBRANE-BOUND CAROTENOID-SYNTHESIZING SYSTEM[a]

		MVA incorporated (nmol/mg/min) into		Products formed (nmol/mg/min)	
System	Fraction	Phytoene	Unsaturated carotenoids	Phytoene	Unsaturated carotenoids
System used previously	21 KP	0.93	0.02	0.116	0.003
	115 KP	0.85	0.01	0.106	0.001
Optimized system	Membrane fraction (21 KP + 115 KP)	1.78	0.23	0.223	0.029

[a] Incubation of 0.5 mg membrane protein for 10 min at 27°. Incubation system as described. Substrate source: [^{14}C]MVA + GGPP-synthesizing system from *Cucurbita pepo* endosperm (2.5 hr preincubation).

synthesizing system from *Cucurbita* endosperm as substrate source, saturation for MVA is achieved with 400–500 nmol. Data obtained under these optimized conditions are summarized in Table III. The conversion of [^{14}C]MVA to phytoene is up to 1.9, to the more unsaturated carotenoids up to 0.25 nmol/mg protein/min. Incorporation proved to be nearly linear to 0.75 mg protein and 30 min incubation time. Analysis of neurosporaxanthin, the end-product formed, is limited by the chromatographic procedure.

Using chemically prepared [2-^{14}C]geranylgeranyl pyrophosphate as substrate characterizes the membrane-bound enzyme system. The conversion of GGPP to phytoene and more unsaturated carotenoids required no additional cofactors (e.g., iron ions, pyridine nucleotides). The formation of phytoene, however, is dramatically increased by Mg^{2+} (conc. > 10 m*M*).[48] In contrast, phytoene formation from IPP using a 100,000 *g* supernatant, requires Mg^{2+}, while Mn^{2+} has also an inhibitory effect.[37]

Membrane-bound enzymes of *N. crassa* (15–20 mg/ml) can be stored frozen (−60°) for at least 2 weeks without appreciable loss of carotenogenic activity.

Solubilization of Carotenogenic Enzymes

Attempts to solubilize carotenogenic enzymes by treatment (30 min at 4° with gentle homogenization) with detergents such as Tween 20 (1%), Tween 80 (0.5%; 1%) Triton-X-100 (1%), Na-cholate (1%), Na-deoxycholate (1%) yields partial solubilization of carotenogenic activity but also decrease of carotenoid-synthesizing activity in the solubilized protein fraction (115 KS; due to centrifugation at 115,000 *g* for 2.5 hr). Subsequent dialysis (15 hr; against 0.2 *M* Tris–HCl, pH 8.0) could not restore carotenogenic activity.[47]

Using Na-cholate (1 mg/10 mg protein), up to 20% protein is recovered in the soluble protein fraction. Resulting decrease of carotenogenic activity in the solubilized enzyme fraction can be partially restored by addition of lipids extracted from membranes of *N. crassa* after *in vivo* illumination. Further attempts to purify solubilized carotenogenic enzymes of *N. crassa* are in progress.

[30] Carotene Synthesis in Spinach (*Spinacia oleracea* L.) Chloroplasts and Daffodil (*Narcissus pseudonarcissus* L.) Chromoplasts

By HANS KLEINIG and PETER BEYER

In plant cells the site of carotenoid synthesis from isopentenyl pyrophosphate and of carotenoid location is in plastids (chloroplasts, etioplasts, chromoplasts).

In chloroplasts, several other prenyl lipid pathways compete with carotenoid formation for the substrates isopentenyl pyrophosphate/geranylgeranyl pyrophosphate (C_{20} side chains of tocopherols, quinones, chlorophylls; C_{45} side chain of plastoquinone). In isolated spinach cloroplasts the esterification of chlorophyllide *a* with geranylgeranyl pyrophosphate by chlorophyll synthase dominates over the head to head condensation of geranylgeranyl pyrophosphates in the carotenogenic pathway. Therefore, substantial *in vitro* synthesis of carotenes is only achieved after subfractionation of these organelles (see below).

In etioplasts from cereals examined so far, study of *in vitro* carotene formation is severely hampered by very active phosphatases.

Chromoplasts in most cases are very fragile organelles, especially the globulous-type chromoplasts (for classification of chromoplasts types, see ref. 1), and can hardly be isolated. Only a few exceptions are known, where intact organelles can be isolated in sufficient amounts: The tubulous-type chromoplast from *Capsicum* fruits[2] and the membranous-type

[1] P. Sitte, H. Falk, and B. Liedvogel, *in* "Pigments in Plants" (F. C. Czygan, ed.), p. 117. Fischer, Stuttgart, 1980.

[2] B. Camara, this volume [28].

ISBN 0-12-182010-6

chromoplast from daffodil (*Narcissus pseudonarcissus*) petals (see below). These chromoplasts have proven to be very well suited for investigations on carotene biosynthesis as they exhibit high activities of carotenogenic enzymes.

Spinach Chloroplasts

Isolation Procedure

Spinach leaves (500 g) are homogenized in 1500 ml isolation medium (sucrose, 0.3 *M*; $MgCl_2$, 4 m*M*; Tris–HCl, 50 m*M* pH 7.6) using a razor-blade-equipped food mixer. The homogenate is filtered through three layers of nylon gauze (40 μm mesh) and centrifuged at 150 *g* for 1 min. The supernatant is centrifuged at 1200 *g* for 1 min and the chloroplast pellet resuspended in isolation medium in a total volume of 24 ml. The suspension is layered on top of sucrose step-gradients (sucrose, 1.8, 1.0, and 0.75 *M*; Tris–HCl, 10 m*M*, pH 7.6; $MgCl_2$, 4 m*M*) and centrifuged at 2000 *g* for 20 min. Intact chloroplasts (about 75% intactness) are obtained from the 1.8/1.0 *M* sucrose boundary, diluted with isolation medium and pelleted at 3400 *g* for 2 min. For isolation of envelope membranes according to Douce and Joyard[3] these chloroplasts are suspended in "swelling medium" ($MgCl_2$, 4 m*M*; Tris–HCl, 10 m*M*, pH 7.6) with the aid of a Pasteur pipet for disintegration. This suspension is layered on sucrose step-gradients (sucrose, 0.93 and 0.6 *M*; $MgCl_2$, 4 m*M*; Tris–HCl, 10 m*M*, pH 7.6) and centrifuged in a swinging bucket rotor at 60,000 *g* for 1.5 hr. Envelope membranes are found in the 0.93/0.6 *M* sucrose interphase, thylakoids in the pellet. Envelope membranes are removed, diluted with "swelling medium," and pelleted at 120,000 *g* for 45 min.

Chloroplasts with a yield of more than 90% intact are obtained by the Percoll gradient method. Spinach leaves are homogenized in isolation medium (sorbitol, 0.33 *M*; EDTA, 2 m*M*; $MgCl_2$, 1 m*M*; $MnCl_2$, 1 m*M*; Tricin–NaOH, 50 m*M*, pH 7.8), passed through nylon filters, and centrifuged as described above. The chloroplasts are layered on linear Percoll gradients (Percoll, 5% to 85% v/v in isolation medium) and centrifuged at 1200 *g* for 10 min in a swinging bucket rotor. Intact chloroplasts are recovered from the gradient at a density of about 1.11 g cm^{-3}, diluted with a 14-fold volume of isolation medium, and pelleted at 1200 *g* for 1 min.

Incubations

Chloroplasts (intact, or disintegrated by use of a French Pressure Cell at 5000 psi), envelope membranes, or thylakoids are suspended in a

[3] R. Douce and J. Joyard, *in* "Plant Organelles" (E. Reid, ed.), p. 47. Ellis Horwood Ltd., Chichester, 1979.

final protein concentration in the range of 1–2 mg ml^{-1} in incubation buffer ($MgCl_2$, 5 m*M*; $MnCl_2$, 1 m*M*; dithioerythritol, 2 m*M*; Tris–HCl, 100 m*M*, pH 7.2) containing NADP (2 m*M*), FAD (0.02 m*M*) in a total volume of 1 ml. In the case of incubations with intact chloroplasts, sorbitol (0.3 *M*) is added to adjust isotonic conditions. As radioactive precursor [1-^{14}C]isopentenyl pyrophosphate (about 10 μM, corresponding to approximately 20 kBq, depending on the commercially available specific activity used) or [^{14}C]geranylgeranyl pyrophosphate, formed from [1-^{14}C]isopentenyl pyrophosphate by an isolated prenyltransferase (see below) is used. Incubations are carried out at 25° for 3 hr in the dark.

Preparation of [^{14}C]Geranylgeranyl Pyrophosphate

For the biosynthetic formation of geranylgeranyl pyrophosphate from isopentenyl pyrophosphate any prenyltransferase preparation with chain length specificity of C_{20} can be used (e.g., ref. 4). We use a purified enzyme derived from the thermophilic bacterium *Chloroflexus aurantiacus,* grown under anaerobic conditions.[5] The cells (ten 250 ml cultures at the end of logarithmic growth) are suspended in buffer ($MgCl_2$, 10 m*M*; dithioerythritol, 2 m*M*; Tris–HCl, 100 m*M*, pH 7.4), homogenized in a French Pressure Cell at 10,000 psi, and centrifuged at 140,000 *g* for 2 hr. The clear supernatant is further centrifuged at 180,000 *g* for 14 hr. The resulting pellet, containing the prenyltransferase activity, is resuspended in buffer and applied in a volume of about 4 ml and a protein concentration of approximately 35 mg ml^{-1} to a gel filtration column (AcA 22, LKB) with a gel bed of 25 × 3 cm which is developed with the same buffer at a flow rate of 84 ml hr^{-1}. Under these conditions the active fraction elutes with the second peak. This fraction is dialyzed against 20% $(NH_4)_2SO_4$ in buffer for enhancement of the hydrophobic effect and subsequently applied to hydrophobic interaction chromatography on phenyl-Sepharose (Pharmacia, gel bed 7 × 2 cm; flow rate, 73.2 ml hr^{-1}). Binding to this column material also takes place in the absence of $(NH_4)_2SO_4$; the presence of this salt, however, has a great influence on the resolution of this separation which is performed stepwise, after removal of nonbinding protein with 20% $(NH_4)_2SO_4$ in buffer, with three solvents: ethylene glycol in buffer, 10% (30 ml), 35% (30 ml), and 50%, until complete elution of the protein peak, which contains the prenyltransferase activity, has taken place. After dialysis against buffer the sample is applied to a Blue Sepharose column (Blue Sepharose CL-6B, Pharmacia, gel bed 5 × 2 cm; flow rate 57 ml hr^{-1}, at 45°). For removal of nonbinding protein the column is washed with buffer; the prenyltransferase is then eluted with

[4] G. Sandmann, W. Hilgenberg, and P. Böger, *Z. Naturforsch., C: Biosci.* **35C,** 927 (1980).
[5] B. K. Pierson and R. W. Castenholz, *Arch. Microbiol.* **100,** 5 (1974).

0.5 *M* KCl in buffer (protein concentration of the active peak about 0.2 mg ml^{-1} in a volume of approximately 20 ml), and dialyzed against buffer. Concentration, where necessary, is performed by dialysis using polyethylene glycol 20,000, 20% in buffer.

The enzyme obtained in this way is apparently homogeneous (as judged by isoelectric focusing), showing a pH optimum at pH 7.5 in Tris buffer, a dependency on Mn^{2+} and $NADP^+$, and a temperature optimum at 50°. Upon incubation (see below) up to 70% of the applied radioactive isopentenyl pyrophosphate is incorporated into geranylgeranyl pyrophosphate. For rapid isolation of the enzyme with lesser purity, the Phenyl Sepharose step can be omitted.

For use in incubation experiments with chloroplasts and chloroplast subfractions, the enzyme solution (0.5 ml) is supplemented with $MnCl_2$ (1 m*M*), $NADP^+$ (2 m*M*), FAD (0.02 m*M*), preincubated at 50° for 1 hr in the presence of [1-^{14}C]isopentenyl pyrophosphate (about 10 μM, see above), cooled down to 25°, and postincubated in the presence of chloroplast protein (see above).

Analysis of Products of the in Vitro Assays

After incubation the samples are extracted with chloroform/methanol as usual.[6] The chloroform phase containing chlorophylls, carotenes, and polyprenols is used directly for separations on thin layer chromatography and HPLC.

The compounds are separated on precoated silica gel thin layer plates with the solvent system petroleum ether/diethyl ether/acetone (40 : 10 : 5) into chlorophyll, polyprenol, and carotene fractions (with decreasing polarity). Radioactive zones can be detected with a radio scanner. The carotene zone is eluted with acetone and further separated by HPLC [Nucleosil 10 C_{18} column SS 300/6/4, Macherey and Nagel; solvent system for isocratic run, methanol/water (3 : 1), 34%, tetrahydrofuran/acetonitrile (1 : 1), 66%, flow rate 1.5 ml min^{-1}].[7] The carotenes elute from the column in the sequence tetradehydrolycopene, lycopene, β-carotene, phytofluene, phytoene and are monitored by a radio column chromatography monitor (Berthold) with a glass scintillator cell and by absorbance detectors (for nonlabeled standards).

The chloroform/methanol extract can also be applied directly to HPLC analysis using the same column with the gradient system methanol/water (3 : 1) and tetrahydrofuran/acetonitrile (1 : 1), 100/0 to 30/70 within 15 min, with a flow rate of 1.5 ml min^{-1}.[7] The labeled compounds elute

[6] E. S. Bligh and W. J. Dyer, *Can. J. Biochem. Physiol.* **37,** 911 (1959).

[7] P. Beyer, K. Kreuz, and H. Kleinig, *Planta* **150,** 435 (1980).

TABLE I
SPINACH CHLOROPLASTS AND CHLOROPLAST SUBFRACTIONS: INCORPORATION OF $[1\text{-}^{14}C]$ISOPENTENYL PYROPHOSPHATE INTO PRENYL LIPIDS AND PRODUCT PATTERN

Incubation conditions	Incorporated of the radioactivity applied (%)	Product pattern (total = 100%)				
		Polyprenols	Chl a_{GG}	Phytoene + phytofluene	Lycopene	Tetradehydrolycopene
Intact chloroplasts[a]	44	18	82			
Disintegrated chloroplasts	35	20	80			
Chloroplast envelopes[b]	—					
Chloroplast envelopes + prenyltransferase	23	34	31	1	22	12
Chloroplast envelopes + chromoplast stroma	39	54	13	8	16	9
Chloroplast thylakoids[c]	—					
Chloroplast stroma[d]	8	100				

[a] About 2 mg protein per ml.
[b] About 0.3 mg protein per ml.
[c] About 0.3 mg protein per ml.
[d] About 2 mg protein per ml.

here in the sequence, geranylgeraniol, chlorophyll a esterified with geranylgeraniol (Chl a_{GG}), tetradehydrolycopene, lycopene, β-carotene, phytofluene, phytoene. In more complex compound mixtures, however, serious overlapping occurs.

Results

Table I shows the results of incubation experiments. With intact or disintegrated spinach chloroplasts no *in vitro* carotene synthesis is obtained, only polyprenols (mainly geranylgeraniol) and Chl a_{GG} from chlorophyllid a plus geranylgeranyl pyrophosphate, the esterification catalyzed by chlorophyll synthase, are formed. The chloroplast envelope per se is not able to incorporate isopentenyl pyrophosphate. When supplemented with a geranylgeranyl pyrophosphate-forming system (prenyltransferase from *Chloroflexus aurantiacus*), however, phytoene, lycopene, and tetradehydrolycopene are formed. A similar pattern is obtained when coincubated with daffodil chromoplast stroma, containing a soluble phytoene synthase complex (see below).

Thylakoids exhibit no synthetic activities with isopentenyl pyrophosphate, while the chloroplast stroma contains only prenyltransferase(s).

These results demonstrate that, under the *in vitro* conditions used, the only site of carotenoid formation is the envelope fraction.[8] These membranes, however, have to be supplemented at least with a geranylgeranyl pyrophosphate-forming system, which means that prenyltransferase activity is not present. Furthermore, a cyclase activity, forming β-carotene from lycopene, cannot be demonstrated in this preparation, rather, only the phytoene dehydrogenase(s) are active in the membrane.

Daffodil Chromoplasts

Isolation Procedure

Daffodil flowers (50 inner coronae) are homogenized in 250 ml isolation medium (sucrose, 0.74 *M*; $MgCl_2$, 5 m*M*; polyvinylpyrrolidone K 90, 0.2% w/v; phosphate buffer, 67 m*M*, pH 7.5) using a razor-blade-equipped food mixer. After filtration through three layers of nylon cloth (40 μm mesh) and centrifugation at 1400 *g* for 5 min, a crude chromoplast pellet is obtained from the supernatant by a further centrifugation at 16,500 *g* for 20 min. The pellet is resuspended in 50% (w/v) sucrose in buffer ($MgCl_2$, 5 m*M*; phosphate buffer, 67 m*M*, pH 7.5), transferred to 34 ml centrifugation tubes, and overlayered with equal volumes of 40, 30, and 15% (w/v) sucrose in the same buffer and centrifuged at 50,000 *g* for 60 min. The chromoplasts, banded by flotation at the 40%/30% and 30%/15% interphases are removed, diluted with buffer to a final sucrose concentration of 15% (w/v), collected by centrifugation at 16,500 *g* for 20 min, and resuspended in incubation buffer ($MgCl_2$, 10 m*M*; dithioerythritol, 2 m*M*; Tris–HCl, 100 m*M*, pH 7.2).[9] Chromoplasts are routinely homogenized before incubation in a French Pressure Cell at 5000 psi. This homogenate is centrifuged at 210,000 *g* for 1 hr for separation of chromoplast membranes from the chromoplast stroma.

Incubations

Incubations are performed in incubation buffer ($MgCl_2$, 10 m*M*; dithioerythritol, 2 m*M*; Tris–HCl, 100 m*M*, pH 7.2), supplemented with $MnCl_2$ (1 m*M*), $NADP^+$ (2 m*M*), FAD (0.02 m*M*), [1-^{14}C]isopentenyl pyrophosphate (about 10 μM, corresponding to about 20 kBq) in 1 ml at 25° for 3 hr. Protein concentrations are indicated in Table II.

Analysis of Products of the in Vitro Assays

The samples are extracted and analyzed as described above for spinach chloroplasts. For HPLC separation the gradient system is used.

[8] F. Lütke-Brinkhaus, B. Liedvogel, K. Kreuz, and H. Kleinig, *Planta* **156,** 176 (1982).

[9] B. Liedvogel, P. Sitte, and H. Falk, *Cytobiologie* **12,** 155 (1976).

TABLE II
DAFFODIL (*Narcissus pseudonarcissus*) CHROMOPLASTS AND CHROMOPLAST SUBFRACTIONS: INCORPORATION OF [1-^{14}C]ISOPENTENYL PYROPHOSPHATE INTO PRENYL LIPIDS AND PRODUCT PATTERN

Incubation conditions	Incorporated of the radioactivity applied (%)	Product pattern (total = 100%)			
		Geranylgeraniol	Prephytoene alcohol[a]	Phytoene	β-Carotene
Chromoplast homogenate[b]	50	13	13	28	46
Chromoplast membranes[c]	—				
Chromoplast stroma[d]	18	30	22	48	—

[a] Tentatively identified.
[b] About 1 mg protein per ml.
[c] About 0.8 mg protein per ml.
[d] About 0.2 mg protein per ml.

Results

Table II shows the results of the *in vitro* assays. *In vitro* β-carotene synthesis is very high in daffodil chromoplasts. Phytoene and dephosphorylated intermediates (geranylgeraniol and prephytoene alcohol) are also formed to some extent. Isolated membranes show no synthetic activities under the conditions used, while the membrane-free stroma contains the complete system for phytoene synthesis. Addition of liposomes (prepared from isolated lipids by sonication) to the stroma increases the total incorporation of [1-^{14}C]isopentenyl pyrophosphate as well as the yield of phytoene.[10] From the results it can be inferred that in these chromoplasts (in contrast to the chloroplasts, see above) a phytoene synthase complex (isopentenylpyrophosphate isomerase, prenyltransferase, prephytoene pyrophosphate synthase, and phytoene synthase activities) is obtained in soluble form, which can be enriched as such by protein separation methods (unpublished). The phytoene dehydrogenase and lycopene cyclase activities, on the other hand, are tightly membrane-bound. Upon combining stroma and membranes the full β-carotene forming pathway can be restored.

Acknowledgment

This work was supported by Deutsche Forschungsgemeinschaft.

[10] K. Kreuz, P. Beyer, and H. Kleinig, *Planta* **154,** 66 (1982).

[31] Prenylation of Chlorophyllide *a* in *Capsicum* Plastids

By BILAL CAMARA

The later steps in chlorophyll *a* synthesis involve the prenylation of chlorophyllide (Chlide) *a* and *b* with geranylgeranyl pyrophosphate[1-3] after which the prenyl chain is sequentially reduced to phytol in the presence of NADPH. This article deals with the characteristics of the prenylation of chlorophyllide a by *Capsicum* chloroplasts, chromoplasts, and related subfractions.

Preparation of *Capsicum* Plastids and Plastid Subfractions

Intact chloroplasts are prepared from green tissue (2 kg) of *Capsicum annuum* L. fruits. The procedure is given in Fig. 1.

The method for chromoplast preparation has been given previously in this volume.[4]

The method described previously in this volume[4] is used for the preparation of plastid subfractions.

The chloroplast preparations are devoid of cytoplasmic contaminations as shown by the absence of phosphatidylethanolamine and negligible NADPH cytochrome *c* reductase and succinate cytochrome *c* reductase activities.[5]

Chlorophyll Composition of Plastids

Reagents and equipment

Chloroform–methanol (2 : 1, v/v)
Light petroleum–diethyl ether (9 : 2, v/v)
Light petroleum–diethyl ether (8 : 2, v/v)
Acetone–diethyl ether (1 : 1, v/v)
NaCl 0.9%
SepPak, silica cartridge (Waters)
C_{18} μBondapak column for HPLC (Waters)
Methanol–acetone–water (90 : 17 : 3, v/v) for HPLC
Vortex mixer

[1] C. Costes, *Phytochemistry* **5,** 311 (1966).
[2] A. R. Wellburn, K. J. Stone, and F. W. Hemming, *Biochem. J.* **100,** 23 (1966).
[3] W. Rüdiger, J. Benz, and C. Guthoff, *Eur. J. Biochem.* **109,** 193 (1980).
[4] B. Camara, this volume [28].
[5] O. Dogbo, B. Camara, and R. Moneger, *C.R. Hebd. Seances Acad. Sci.* **295**(III), 477 (1982).

ISBN 0-12-182010-6

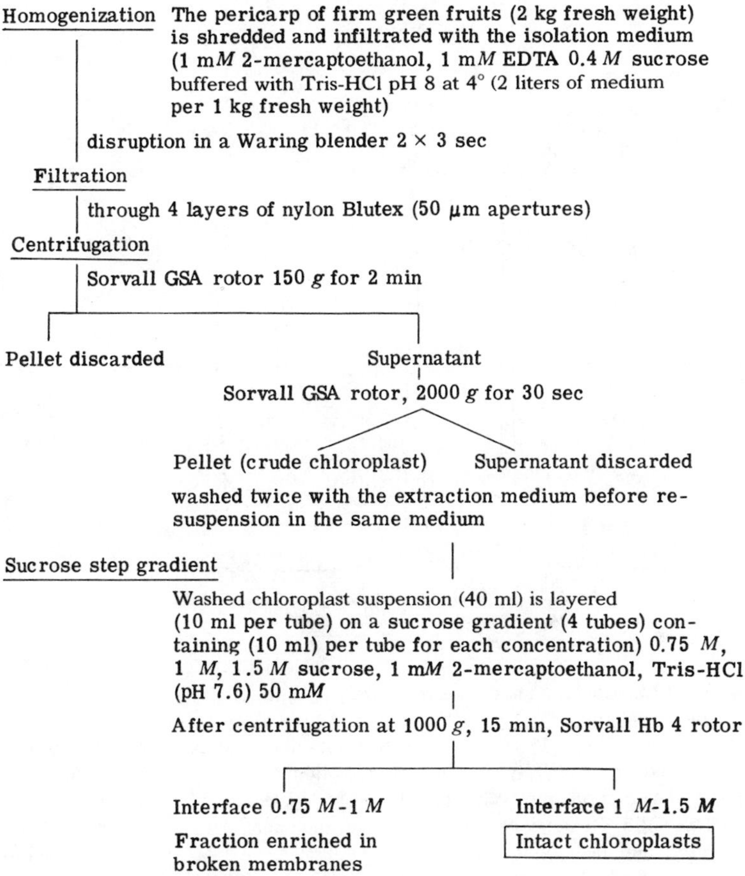

Homogenization — The pericarp of firm green fruits (2 kg fresh weight) is shredded and infiltrated with the isolation medium (1 m*M* 2-mercaptoethanol, 1 m*M* EDTA 0.4 *M* sucrose buffered with Tris-HCl pH 8 at 4° (2 liters of medium per 1 kg fresh weight)

disruption in a Waring blender 2 × 3 sec

Filtration

through 4 layers of nylon Blutex (50 μm apertures)

Centrifugation

Sorvall GSA rotor 150 *g* for 2 min

Pellet discarded | Supernatant

Sorvall GSA rotor, 2000 *g* for 30 sec

Pellet (crude chloroplast) | Supernatant discarded

washed twice with the extraction medium before re-suspension in the same medium

Sucrose step gradient

Washed chloroplast suspension (40 ml) is layered (10 ml per tube) on a sucrose gradient (4 tubes) containing (10 ml) per tube for each concentration) 0.75 *M*, 1 *M*, 1.5 *M* sucrose, 1 m*M* 2-mercaptoethanol, Tris-HCl (pH 7.6) 50 m*M*

After centrifugation at 1000 *g*, 15 min, Sorvall Hb 4 rotor

Interface 0.75 *M*-1 *M* — Fraction enriched in broken membranes

Interface 1 *M*-1.5 *M* — Intact chloroplasts

FIG. 1. Method for the isolation of intact *Capsicum* chloroplasts (the working temperature is 0–4°).

Experimental Procedure. The chloroplasts (2 ml) or chromoplast membranes (2 ml) suspension is homogenized with 9 ml of chloroform–methanol (2:1, v/v) for 3 min in a screw capped tube shaken by a vortex mixer, after which, 3 ml of chloroform is added and shaking is continued for 2 min, then 3 ml of 0.9% sodium chloride is added. The mixture is again homogenized and the solution is cleared by centrifugation. The chloroform layer is withdrawn and concentrated to dryness *in vacuo* or by a stream of nitrogen. The lipid residue, dissolved in light petroleum–diethyl ether (9:2, v/v), is applied to a SepPak cartridge, after which, the latter is eluted by 15 ml of light petroleum–diethyl ether (8:2, v/v). The

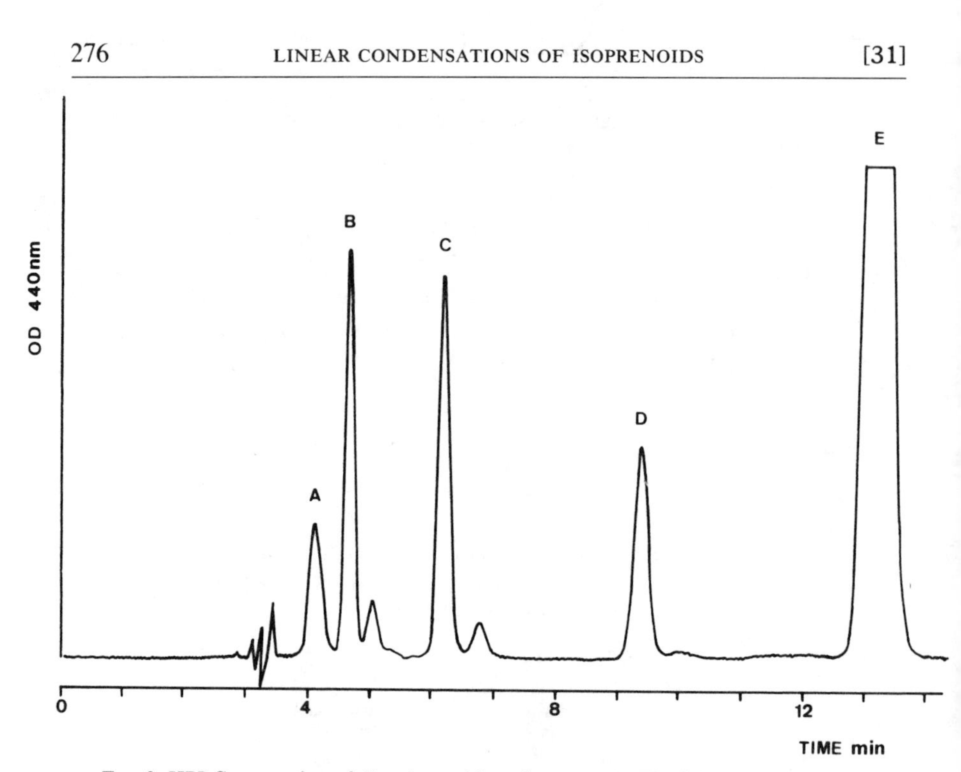

FIG. 2. HPLC separation of *Capsicum* chloroplast extract. The lipid residue dissolved in light petroleum–diethyl ether (98 : 2, v/v) was applied to a SepPak silica cartridge (Waters), successively eluted with 15 ml of light petroleum–diethyl ether (8 : 2, v/v) and 10 ml of acetone–diethyl ether (1 : 1, v/v). The latter fraction was analyzed on a Waters system (Pump model 510, variable wavelength detector M 450, data module M 730). Conditions of analysis [column C_{18} μBondapak 30 cm; solvent: methanol–acetone–water (90 : 17 : 3, v/v) flow rate: 1 ml/min; detection at 440 nm, AUFS 0.04]. (A) Neoxanthin; (B) violaxanthin; (C) lutein; (D) chlorophyllide *b* prenylated with phytol; (E) chlorophyllide *a* prenylated with phytol.

fraction containing chlorophylls *a* and *b* is eluted with (10 ml) acetone–diethyl ether (1 : 1, v/v) and subjected to HPLC on C_{18} μBondapak column isocratically developed with methanol–acetone–water (90 : 17 : 3, v/v).

Results. The chlorophyll composition (shown in Fig. 2) reveals that only chlorophyll *a* and chlorophyll *b* esterified with phytol are detected in mature chloroplasts. In chromoplasts the latter are absent.

Chlorophyllide Prenylation

Principle

The prenylation is followed by measuring the rates of incorporation of radioactive isopentenyl pyrophosphate or geranylgeranyl pyrophosphate into chlorophyll *a*:

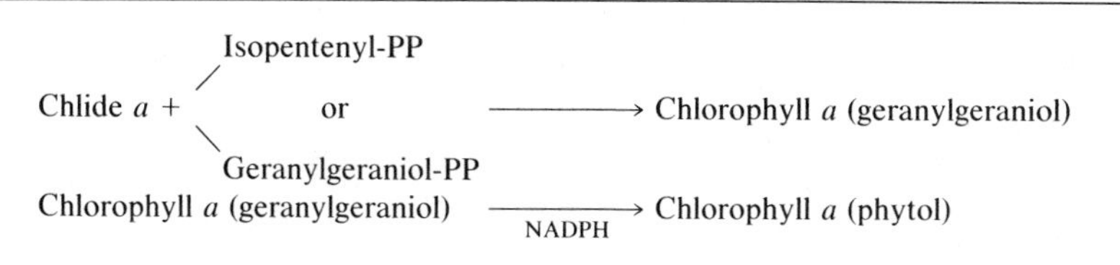

Experimental Procedure

Preparation of Substrates

Reagents and Equipment

Geranylgeraniol[6]
Fresh spinach leaves
Acetone
Light petroleum, bp 40–60°
Diethyl ether
Light petroleum–acetone (80 : 20, v/v)
Cellulose plate for TLC
NaCl

Step 1: Labeling and pyrophosphorylation of geranylgeraniol. The method for [1-^{3}H]geranylgeranyl pyrophosphate preparation has been described previously in this volume.[4]

Step 2: Chlorophyllide preparation from fresh spinach leaves: the method described by Holden[7] is used. Twenty grams of fresh deveined spinach leaves are homogenized in 200 ml of acetone–H_2O (1 : 1, v/v). The mixture is kept in the dark at room temperature for 18 hr before filtration. The filtrate is extracted with light petroleum in order to eliminate the remaining chlorophylls. Then the acetone phase is extracted with diethyl ether. The ether phase is dried over NaCl and dried in a rotating evaporator. Small portions of the extract are purified on cellulose thin layer plates (which allows quantitative recovery of the polar chlorophyllide) developed with light petroleum–acetone (80 : 20, v/v). The chlorophyllides *a* and *b* move in the region (R_f 0–0.2) and are eluted together with diethyl ether. The chlorophyllide content is determined by the method of Comar and Zscheile[8]:

$$\text{chlorophyllides } (a + b) \text{ in mg/ml} = 7.12 \text{ OD}_{660} + 16.8 \text{ OD}_{642,5}$$

The purified chlorophyllides are stored in peroxide-free diethyl ether at −20°.

[6] O. P. Vig, J. C. Kapur, J. Singh, and B. Vig, *Indian J. Chem.* **7**, 574 (1969).
[7] M. Holden, *Biochem. J.* **78,** 359 (1961).
[8] C. L. Comar and F. P. Zscheile, *Plant Physiol.* **17,** 198 (1942).

Assay Method

Reagents and Equipment

Barley seedlings
Silica gel plate for TLC, adjusted to pH 7 with KOH
Diethyl ether
Light petroleum–acetone (6 : 4, v/v)
Light petroleum–diethyl ether (98 : 2, v/v)
Light petroleum–diethyl ether (8 : 2, v/v)
Acetone–diethyl ether (1 : 1, v/v)
Chloroform–methanol (2 : 1, v/v)
SepPak, silica cartridge (Waters)
C_{18} μBondapak column for HPLC (Waters)
Methanol–acetone–water (90 : 17 : 3) for HPLC
Tris–HCl, 50 m*M* (pH 7.6)
$MgCl_2$, 1 *M*
DTT, 0.5 *M*
KF, 1 *M*
NADPH, 0.2 *M*
Chlorophyllide (*a* + *b*)
Tween 80
[1-^{14}C]Isopentenyl pyrophosphate, 53 mCi/mmol
[1-^{3}H]Geranylgeranyl pyrophosphate 20 mCi/mmol

Authentic standards: For good detection of the reaction products, the different chlorophylls *a* (chlorophyllides *a* prenylated with geranylgeraniol, dihydrogeranylgeraniol, tetrahydrogeranylgeraniol and phytol) must be added just before the purification procedure. For this, chlorophyllide esterified with phytol is isolated from 9-day light grown barley leaves cultivated at 25°, while the other intermediates are isolated from 9-day dark grown barley leaves cultivated at 25° and illuminated for 1 hr.[9] In both cases, the leaves are extracted with acetone in the dark, in the presence of magnesium carbonate and sand. The extract is filtered and transferred into a separatory funnel containing the biphasic system diethyl ether–water (1 : 1, v/v). After repeated washing with water, the ether phase is evaporated to dryness and streaked on silica gel G chromatoplate and developed with light petroleum–acetone (60 : 40, v/v). The different chlorophylls *a* have the same R_f(0.6) in this system and are well separated from chlorophyll *b* (R_f 0.46). The chlorophyll *a* band is scraped eluted with acetone–diethyl ether (1 : 1, v/v) and rechromato-

[9] W. Rudiger, J. Benz, U. Lempert, S. Schoch, and D. Steffens, *Z. Pflanzenphysiol.* **80,** 131 (1976).

graphed on the same system. The nature of the prenol is determined as described previously.[9] A fraction containing 200 μg of total chlorophylls *a* is used as unlabeled carrier. The method of Arnon[10] is used for chlorophyll determination.

Incubation Medium. The enzymatic reaction is carried out for various times at 25° in a medium buffered with 50 m*M* Tris–HCl (pH 7.6) (2 ml final volume) containing 10 m*M* $MgCl_2$, 1 m*M* DTT, 1 m*M* KF, 2 m*M* NADPH, 100 μg chlorophyllide *a* + *b* emulsified in 100 μg Tween 80, [1-^{14}C]isopentenyl pyrophosphate 0.1 μCi or [1-^{3}H]geranylgeranyl pyrophosphate 0.1 μCi, plastid suspension (chloroplast 1.5 mg protein or chromoplast 2.5 mg protein), chloroplast membranes (1 mg protein), chromoplast membranes (2 mg protein). The reaction is stopped by 4.5 ml of chloroform–methanol (2 : 1, v/v).

Extraction and Purification. To the plastid suspension previously treated with 4.5 ml of chloroform–methanol (2 : 1, v/v) is added 4.5 ml of the same solvent mixture containing 200 μg of unlabeled chlorophylls *a* esterified with different prenyl chains (see above, "Reagents and Equipment Used for Assay Method"). After shaking for 5 min with a Vortex mixer, 2 ml of 0.9% sodium chloride is added. The mixture is homogenized again and the solution is cleared by centrifugation. The chloroform layer is withdrawn and concentrated to dryness *in vacuo* or by a stream of nitrogen. For chloroplast preparation only, a preliminary chromatography on silica gel G plate developed with light petroleum–acetone (60 : 40, v/v) is performed (this step is necessary since the endogenous chlorophyllide *b* esterified with phytol may interfere with chlorophyllide *a* esterified with geranylgeraniol) and the chlorophylls a fraction (R_f 0.61) is eluted with acetone–diethyl ether (1 : 1, v/v). The residue is dissolved in light petroleum–diethyl ether (98 : 2, v/v) like the total chromoplast extract and is applied to a SepPak cartridge. The latter is eluted by 15 ml of light petroleum–diethyl ether (8 : 2, v/v) in order to eliminate any labeled geranylgeraniol formed during the incubation. The different chlorophylls *a* are eluted with acetone–diethyl ether (1 : 1, v/v) and fractionated by HPLC on C_{18} μBondapak column isocratically developed with methanol–acetone–water (90 : 17 : 3, v/v).

Results

Capsicum chloroplasts as well as chromoplasts (provided that exogenous chlorophyllide *a* is added) incorporate isopentenyl pyrophosphate in the lateral prenyl chain of chlorophyll *a*.[5] The analysis of the prenyl chain after saponification shows that geranylgeraniol pyrophosphate is primar-

[10] D. I. Arnon, *Plant Physiol.* **24,** 1 (1949).

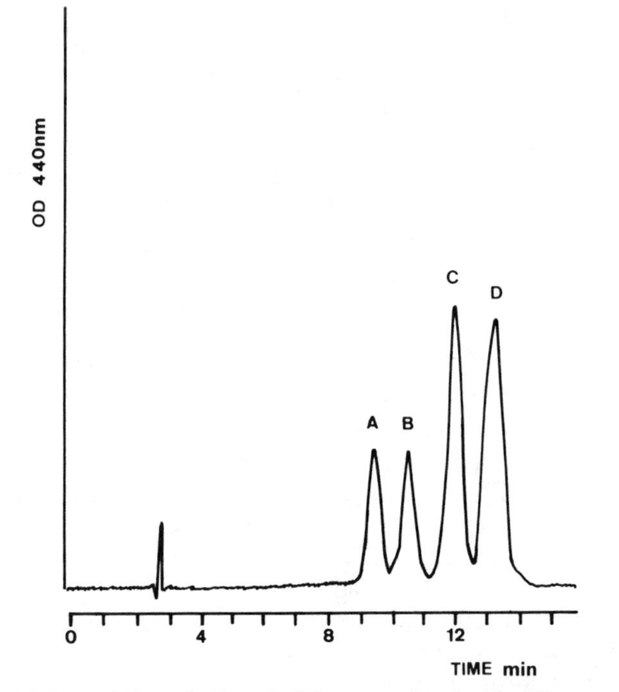

FIG. 3. HPLC separation of chlorophyllide *a* prenylated with different prenols. The purified chlorophyll *a* fraction recovered after chloroplast incubation (see the text for details) was analyzed as described in Fig. 2. (A, B, C, D) Chlorophyllide *a* prenylated respectively with geranylgeraniol, dihydrogeranylgeraniol, tetrahydrogeranylgeraniol, and phytol. For the chromoplast lipid extract obtained as described in the text, additional peaks (retention time below 8 min) corresponding to carotenoids are detected.

ily engaged in the prenylation[5] of chlorophyllide *a*. This trend is confirmed when geranylgeranyl pyrophosphate is used as a substrate.[11]

When isopentenyl pyrophosphate is used as a substrate Mn^{2+} and Mg^{2+} are required for maximum activity. Inorganic pyrophosphate (1.25 m*M*) and ethylenediaminetetraacetic acid (20 m*M*) strongly inhibit the reaction.[11] Also, the reaction is inhibited by Triton X-100 at ratio Triton X-100 (mg)/Plastid suspension protein (mg) = 10.[11]

The stepwise reduction up to phytol is dependent on the availability of NADPH in the reaction medium.[5] The following intermediates chlorophyll *a* (geranylgeraniol), chlorophyll *a* (dihydrogeranylgeraniol), chlorophyll *a* (tetrahydrogeranylgeraniol), and chlorophyll *a* (phytol) are detected (Fig. 3).[11]

[11] O. Dogbo, F. Bardat, and B. Camara, *Physiol. Veg.* **22,** 75 (1984).

The chloroplast thylakoids and the chromoplast achlorophyll lamellae are the sites of chlorophyllide prenylation. In both types of plastid these membranes are unable to metabolize isopentenyl pyrophosphate in contrast to the stroma which is only concerned in the synthesis of geranylgeranyl pyrophosphate.[11]

Acknowledgments

I thank Prof. Moneger for his interest during the work. I appreciate the technical assistance given by C. Agnes, F. Bardat, and A. Nargeot.

[32] Purification and Characterization of Undecaprenylpyrophosphate Synthetase

By CHARLES M. ALLEN

Undecaprenylpyrophosphate ($C_{55}PP$) synthetase catalyzes the biosynthesis of the long-chain polyprenyl pyrophosphate required as a carbohydrate carrier in the biosynthesis of a variety of bacterial cell envelope components.[1,2] It has been described in several bacteria including *Salmonella newington*,[3] *Micrococcus luteus*,[4,5] and *Bacillus subtilis*,[6] but it has been studied most extensively in *Lactobacillus plantarum*.[7-10] These enzymes differ in two major respects from the better studied prenyltransferases which catalyze the formation of short and intermediate chain length all-*trans*-polyprenyl pyrophosphates. The undecaprenylpyrophosphate synthetases catalyze the addition of only *cis*-isoprene residues instead of trans residues during product formation and require detergent or phospholipid for *in vitro* activity.

[1] I. W. Sutherland, *in* "Surface Carbohydrates of the Prokaryotic Cell" (I. W. Sutherland, ed.), p. 27. Academic Press, London, 1977.
[2] H. Hussey and J. Baddiley, *in* "The Enzymes of Biological Membranes (A. Martonosi, ed.), Vol. 2, p. 227. Plenum, New York, 1976.
[3] J. G. Christenson, S. K. Gross, and P. W. Robbins, *J. Biol. Chem.* **244,** 5436 (1969).
[4] T. Kurokawa, K. Ogura, and S. Seto, *Biochem. Biophys. Res. Commun.* **45,** 251 (1971).
[5] T. Baba and C. M. Allen, *Arch. Biochem. Biophys.* **200,** 474 (1980).
[6] I. Takahashi and K. Ogura, *J. Biochem.* (*Tokyo*) **92,** 1527 (1982).
[7] M. V. Keenan and C. M. Allen, Jr., *Arch. Biochem. Biophys.* **161,** 375 (1974).
[8] C. M. Allen, M. V. Keenan, and J. Sack, *Arch. Biochem. Biophys.* **175,** 236 (1976).
[9] C. M. Allen and J. D. Muth, *Biochemistry* **16,** 2908 (1977).
[10] T. Baba and C. M. Allen, *Biochemistry* **17,** 5598 (1978).

ISBN 0-12-182010-6

Assay Method

Principle. Undecaprenylpyrophosphate synthetase activity is determined by measuring the amount of an acid-labile polyprenyl pyrophosphate produced from the reaction of $^3\Delta$-[^{14}C]isopentenyl pyrophosphate (IPP) with the allylic isoprenyl pyrophosphate, farnesyl pyrophosphate (FPP).

$$\text{FPP} + 8\ \text{IPP} \rightarrow \text{C}_{55}\text{PP}$$

Substrates. Isopentenyl pyrophosphate. [1-^{14}C]IPP (40–60 mCi/mmol) may be purchased from Amersham Corp. or New England Nuclear. Alternatively, it may be synthesized by the phosphorylation of 3-methyl-3-butenol. [1-^{14}C]-3-Methyl-3-butenol is synthesized by the method of Yuan and Bloch[11] and phosphorylated by the general method of Popják *et al.*[12] [1-^{14}C]IPP is separated from the unphosphorylated alcohol and the polyphosphorylated products by DEAE-cellulose chromatography as described by Dugan *et al.*[13] [^{14}C]IPP with a specific activity of approximately 0.5 mCi/mmol is adequate to assay C_{55}PP synthetase activity, therefore, commercially obtained [^{14}C]IPP can be diluted with unlabeled IPP, which is prepared as described for the radiolabeled substrate. If the concentration of IPP is measured from the amount of acid labile phosphate, it is particularly important to establish that the purified synthetic IPP is free of isopentenyl monophosphate, isopentenyl triphosphate, and inorganic pyrophosphate, which are frequent contaminants when the published purification procedures are followed. Stereospecifically labeled IPPs, 2*R*-[2-^{3}H]IPP and 2*S*-[2-^{3}H]IPP are prepared[8,10] enzymatically from (4*S*)-[4-^{3}H]mevalonic acid and (4*R*)-[4-^{3}H]mevalonic acid, respectively, by the method of Tchen[14] with a crude enzyme preparation from yeast. The tritiated IPP is separated from other mevalonate metabolites and purified by DEAE-cellulose chromatography[15] and descending paper chromatography.[10]

Allylic isoprenyl pyrophosphates. These must be synthesized since they are not commercially available. Geraniol is purified from a commercial preparation, whereas, *cis,trans*-farnesol, *trans,trans*-farnesol, *cis, trans,trans*-geranylgeraniol, and *trans,trans,trans*-geranylgeraniol are synthesized from geraniol as previously described.[10] The stereochemical purity of the alcohols is established from NMR analysis. The [1-^{3}H]allylic

[11] C. Yuan and K. Bloch, *J. Biol. Chem.* **234,** 2605 (1959).

[12] G. Popják, J. W. Cornforth, R. H. Cornforth, R. Ryhage, and D. S. Goodman, *J. Biol. Chem.* **237,** 56 (1962).

[13] R. E. Dugan, E. Rasson, and J. W. Porter, *Anal. Biochem.* **22,** 249 (1968).

[14] T. T. Tchen, this series, Vol. 6, p. 505.

[15] D. N. Skilleter, and R. G. O. Kekwick, *Anal. Biochem.* **20,** 171 (1967).

isoprenols are prepared by reduction of the ethyl farnesoates and geranyl geranates with $LiAl^3H_4$. The allylic isoprenyl pyrophosphates are prepared by phosphorylation of the allylic isoprenols as described by Popják *et al.*[12] and purified by Amberlite XAD-2 and DEAE-cellulose chromatography by adapting the procedures of Holloway and Popják.[16]

Enzyme Assay Procedure. The assay solutions contain in a final volume of 0.5 ml: enzyme, 100 m*M* Tris–HCl buffer (pH 7.5), 0.5% Triton X-100, 200 μM $MgCl_2$, 5 μM *trans,trans*-FPP, and 30 μM IPP (0.5 mCi/mmol, 15,000 dpm). The enzyme is added last to start the reaction. This also results in more reproducible assays, since the enzyme is stabilized by detergent in dilute protein solutions. The enzyme may be assayed satisfactorily with 8–60 μM IPP and 5–50 μM *trans,trans*-FPP. The higher concentrations of substrate were used in the earlier reported experiments but recently the lower concentrations have been shown to be sufficient. Frequently, mixed stereochemistry (*trans,trans* and *cis,trans*) FPP, which is prepared from commercially available farnesol, is used for reasons of economy to monitor enzymatic activity in routine enzyme purifications, where a precise measurement of the absolute enzyme specific activity is not necessary. The mixed stereochemistry FPP has a reactivity with the enzyme similar to *trans,trans*-FPP because of the similar reactivities of *trans,trans*-FPP and *cis,trans*-FPP.

The reaction mixtures are incubated for 30 min at 35° in 16 × 125-mm screw top culture tubes closed with a Teflon lined cap. The reaction is terminated by the addition of 0.5 ml of 50% trichloroacetic acid (w/v). This mixture is heated at 100° for 30 min to hydrolyze completely the allylic pyrophosphate to the primary and tertiary allylic isoprenols and hydrocarbons. After cooling, the acid hydrolyzate is neutralized with 0.5 ml of 5 *M* NaOH, then 2.5 ml of 2 *M* KCl is added. The KCl promotes solubilization of the isoprenoid hydrolysis products into the organic phase during the subsequent solvent extraction. The neutralized hydrolysis mixture is extracted two times with 5-ml aliquots of petroleum ether. The emulsion, which results from each extraction, is dispersed by the addition of several drops of ethanol and the phases separated by centrifugation in a bench top clinical centrifuge. Both petroleum ether extracts are combined and backwashed with 2 ml of water. A 5-ml aliquot of the resulting clear petroleum ether layer is evaporated to dryness in a scintillation vial placed on a hot plate at low setting. Toluene based scintillation fluid is added for subsequent determination of the radioactivity.

This assay is dependent on the acid stability of IPP and its insolubility in petroleum ether. The conditions for acid hydrolysis of the long-chain

[16] P. W. Holloway and G. Popják, *Biochem. J.* **104,** 57 (1967).

allylic isoprenyl pyrophosphates, $C_{55}PP$[7] and dehydrodolichyl-PP[17] are considerably more stringent than those described for the shorter chain (C_{10}, C_{15}, and C_{20}) pyrophosphates.[18,19]

Isolation and Analysis of Polyprenyl Phosphates

Polyprenyl Pyrophosphate Isolation. When the pyrophosphorylated products are required for subsequent chromatographic analysis, they are removed from larger scale reaction mixtures (3 ml) by extraction three times with 5 ml of water-saturated 1-butanol or $CHCl_3$–CH_3OH (2 : 1, v/v). The combined organic extracts are then washed with 2–3 ml of water. The butanol is removed from the products by rotary evaporation under vacuum (oil pump) at 30–40°. Control experiments, in which protein–[^{14}C]IPP mixtures without FPP were extracted in the same way, showed that essentially no radioactive IPP was extracted into the organic phase.

The ^{14}C-labeled isoprenoid products also associate with protein and detergent, therefore they can be separated from the [^{14}C]IPP by Sephadex G-25 column chromatography.[7]

Preparation of Polyprenyl Monophosphate and Free Polyprenols. The reaction product, polyprenyl pyrophosphate is not readily hydrolyzed by *E. coli* alkaline phosphatase at pH 9.2,[4,19] even in the presence of the detergent Triton X-100. The polyprenyl pyrophosphate may be hydrolyzed, however, to the polyprenyl monophosphate with the membrane associated polyprenyl pyrophosphatase from *M. luteus*.[20] The sample of polyprenyl pyrophosphate (dried in the bottom of the assay tube) is incubated with 2 mg of *M. luteus* membrane protein, 95 μmol Tris–HCl buffer pH 7.5, 7 μmol EDTA, 1% Triton X-100, and 70 μmol $MgCl_2$ in a 0.6 ml reaction volume for 5 hr at 37°. The polyprenyl monophosphate is isolated by extraction with an equal volume of water-saturated 1-butanol, and the butanol extract is washed with 0.5 volume of butanol-saturated water. When the free polyprenol is desired it is not necessary to first isolate the polyprenol monophosphate, but a second hydrolysis[7] is carried out by adding a yeast homogenate, 120 μmol of additional $MgCl_2$ and incubating for an additional 5 hr at 37°.

The isoprenoid monophosphates may also be prepared by treating the pyrophosphate with a wheat germ phosphatase at pH 6.2 for 4 hr by the method of Carson and Lennarz.[21] This hydrolysis can be accomplished

[17] D. K. Grange and W. J. Adair, Jr., *Biochem. Biophys. Res. Commun.* **79,** 734 (1977).
[18] G. Popják, J. L. Rabinowitz, and J. M. Baron, *Biochem. J.* **113,** 861 (1969).
[19] C. M. Allen, W. Alworth, A. Macrae, and K. Bloch, *J. Biol. Chem.* **242,** 1895 (1967).
[20] R. Goldman and J. L. Strominger, *J. Biol. Chem.* **247,** 5116 (1972).
[21] D. D. Carson and W. J. Lennarz, *J. Biol. Chem.* **256,** 4679 (1981).

either in the presence or absence of octyl glucoside. The ability to hydrolyze the product in the absence of octyl glucoside is probably due to the presence of small quantities of Triton X-100 in the isoprenoid pyrophosphate preparation. Little free isoprenol was obtained under these conditions in this laboratory.

The free polyprenols may also be prepared by the treatment of the polyprenyl pyrophosphates with potato acid phosphatase at pH 5.0 in 60% methanol for 3 hr by the method of Fujii *et al.*[22] When this enzyme is used at pH 6.5, it gives good yields of the polyprenol monophosphates. A yeast supernatant[4,7] or a microsomal membrane preparation from rat or bovine liver[8] has also been used to prepare the free polyprenol, but each gives highly variable and generally low yields of free alcohol.

The alcohols are extracted in each case with either water-saturated 1-butanol or petroleum ether. These extracts are evaporated to dryness, and the residues redissolved in $CHCl_3$ and passed through a column of silicic acid that was previously equilibrated with $CHCl_3$. The free polyprenols are eluted with $CHCl_3$.[23]

Determination of Chain Length. The product chain length may be determined by several criteria including (1) the mobility of the free polyprenols on reverse phase TLC in comparison with known standards,[4,7,8,10,19] (2) double labeling experiments with ^3H-labeled allylic pyrophosphate and ^{14}C-labeled IPP,[5,10] (3) the elution times of free polyprenols on gas–liquid chromatography[24] or high pressure liquid chromatography[6,25] (see also this volume [33]) in comparison with known standards, and (4) mass spectral analysis.[6,25–29] The first two have been employed most extensively in the author's laboratory. The mass spectral method requires a quantity of pure product which is not generally available from the biosynthetic procedures described.

The chain length of the long-chain isoprenoid products is not readily assessable by TLC analysis of either the *in vitro* products (the polyprenyl pyrophosphates) or their monophosphates. The free polyprenols of different chain length, may however be characterized by TLC in the appropri-

[22] H. Fujii, T. Koyama, and K. Ogura, *Biochim. Biophys. Acta* **712,** 716 (1982).

[23] J. R. Kalin and C. M. Allen, Jr., *Biochim. Biophys. Acta* **574,** 112 (1979).

[24] K. J. Stone and J. L. Strominger, *J. Biol. Chem.* **247,** 5107 (1972).

[25] P. L. Donnahey and F. W. Hemming, *Biochem. Soc. Trans.* **3,** 775 (1975).

[26] K. J. I. Thorne, *Biochem. J.* **135,** 567 (1973).

[27] D. P. Gough, A. L. Kirby, J. B. Richards, and F. W. Hemming, *Biochem. J.* **118,** 167 (1970).

[28] Y. Higashi, J. L. Strominger, and C. C. Sweeley, *Proc. Natl. Acad. Sci. U.S.A.* **57,** 1878 (1967).

[29] A. Wright, M. Dankert, P. Fennessey, and P. W. Robbins, *Proc. Natl. Acad. Sci. U.S.A.* **57,** 1798 (1967).

ate solvent systems. It is sometimes convenient, however, to assess the relative amounts of a given chain length polyprenol and its phosphorylated derivatives. This is accomplished readily by TLC on silica gel 60 coated plastic sheets with diisobutyl ketone–acetic acid–water (8 : 5 : 1, v/v/v), Solvent A. The R_f values of the $C_{55}PP$, $C_{55}P$, and $C_{55}OH$ are 0.47, 0.60, and 0.94, respectively. A convenient TLC system for analysis of the polyprenols is reverse phase chromatography on paraffin oil coated kieselguhr G plates using acetone–water (23 : 2, v/v), Solvent B. The plates are coated with oil by immersion in a 5% (w/v) solution of paraffin oil in petroleum ether, bp 30–60°. The R_f of the radioactive free polyprenol is compared with a naturally occurring polyprenol such as $C_{55}OH$, which may be isolated from *Lactobacillus plantarum* as previously described.[7] The migration of the polyprenol is dependent on the chain length as well as the number and stereochemistry of the double bonds, so it is important to consider this when drawing conclusions about chain length from comparisons of the migration of the radioactive components with the migration of the standards. More than one standard is obviously desirable but this is sometimes not possible, so alternative methods are necessary for an accurate determination of the product chain length.

The chain length determination by the method of evaluating isotopic ratios is the most reliable. One method requires the utilization of [1-^{14}C]IPP and [1-^{3}H]*trans,trans*-FPP as substrates. The specific activity of both the substrates must be known accurately. The double-labeled free polyprenols are prepared from the phosphorylated enzyme product as described above and partially purified by reverse phase TLC. The double-labeled products localized with a radiochromatogram scanner are extracted from the kieselguhr G with two 1-ml aliquots of $CHCl_3$ and three 1-ml aliquots of acetone. The pooled eluates are then analyzed for ^{3}H and ^{14}C. The quantity of each isotopically labeled substrate, which is incorporated into the products, may be determined from the measured radioactivity and the known specific activities of the two substrates. Since only one FPP molecule is used in the synthesis of each product, the number of isoprene units added to FPP during polymerization is easily assessed.

An alternative method is an analysis of the isotopic ratios in a polyprenol ester prepared from [^{3}H]acetic anhydride and the ^{14}C-labeled polyprenol, which is prepared from unlabeled FPP and [^{14}C]IPP of known specific activity. The ^{14}C-labeled polyprenol is acetylated by the addition of 50 μl of a pyridine solution containing an excess (48 μmol) of [^{3}H]acetic anhydride of known specific activity and incubating at room temperature for 150 min with occasional shaking. Absolute ethanol (0.1 ml) is added to the reaction solution and this mixture is allowed to stand for 30 min at room temperature. The solvent and volatile components are removed by evaporation *in vacuo*. The resulting residue was dissolved in benzene and the acetylated polyprenols purified by chromatography on silica gel 60 (300 mg) in benzene–ethyl acetate (10 : 1, v/v). These double-labeled polyprenyl acetates are then purified by reverse phase TLC as described above in acetone–water (92 : 8, v/v). The chain length may be determined from the [^{3}H]/[^{14}C] ratio and the known specific activities of [^{3}H]acetic anhydride and the [^{14}C]IPP, which was used to prepare the [^{14}C]polyprenol.

Separation and Identification of C_{55}PP Synthetase in Mixtures of Prenyltransferases

When it is desirable to study the C_{55}PP synthetase in microorganisms which contain more than one long-chain prenyltransferase, it is necessary to separate the different prenyltransferases and identify the desired enzyme. DEAE-cellulose chromatography has been effective in accomplishing this in two cases we have studied. Figure 1 illustrates the elution profile of protein and prenyltransferase activity from DEAE-cellulose columns to which high-speed supernatant fractions from *L. plantarum* (Fig. 1A), *E. coli* (Fig. 1B), and *M. luteus* (Fig. 1C) have been applied. Column fractions are assayed by the method described above for enzymes which catalyze the addition of [1-^{14}C]IPP to FPP. The prenyltransferases, GPP and FPP synthetase are not measured in this assay since the allylic pyrophosphate substrate is the C_{15} isoprenoid FPP. The inclusion of an appropriate nonionic detergent in the assays mixtures also is necessary to detect the C_{55}PP synthetase activity. Other prenyltransferases are not markedly affected by these detergents.

The C_{55}PP synthetase (Peak I) is the only detectable prenyltransferase measured in this way from *L. plantarum,* but both *M. luteus* and *E. coli* exhibit a prenyltransferase (Peaks II) which catalyzes the formation of an intermediate chain length polyprenyl pyrophosphate with all-*trans* isoprene residues. *M. luteus* also has a third prenyltransferase (Peak III) which catalyzes the synthesis of geranylgeranyl pyrophosphate.

The enzymes are identified by characterization of their products. A direct way of identifying the enzyme is to isolate each product and determine its chain length by one of the procedures described above. This process, although eventually necessary is tedious and lengthy. An alternative method assumes that the C_{55}PP is the only polyprenyl product with *cis*-isoprene residues. The procedure for this method employs either the stereospecifically labeled 2*R*- or 2*S*-[2-^{3}H]IPP as substrate. The ^{3}H label

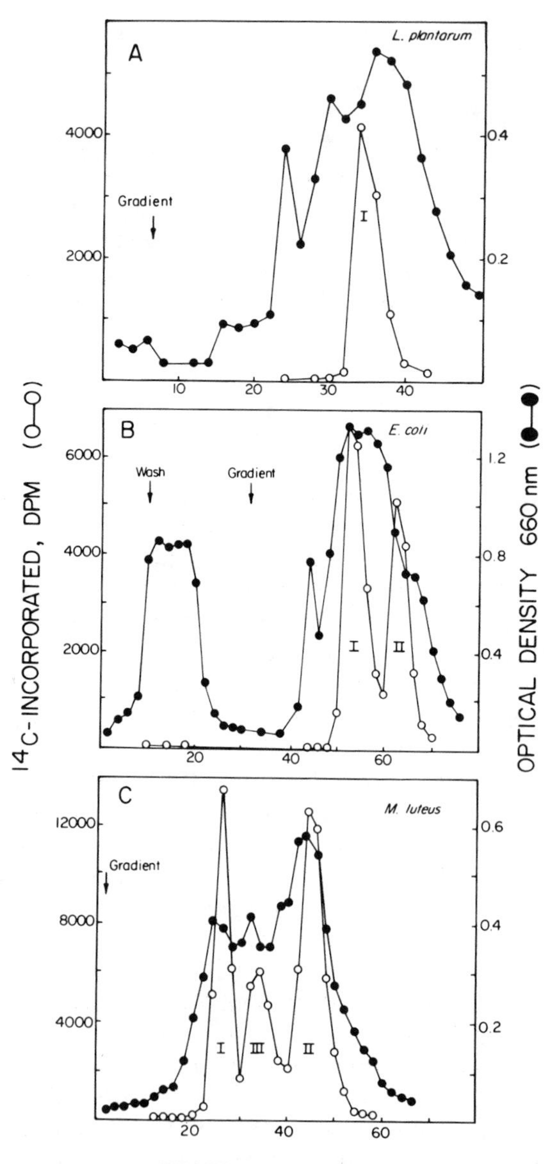

FIG. 1. DEAE-cellulose chromatography of long-chain prenyltransferases. High-speed centrifugal supernatant fractions were prepared from cell lysates and applied to the columns in 10 m*M* Tris–HCl buffer pH 7.5. The columns were washed with 10 m*M* Tris buffer and the enzyme was eluted with a linear salt gradient in 10 m*M* Tris buffer as indicated below. (A) *L. plantarum* (0–0.5 *M* NaCl gradient), (B) *E. coli* (0–0.6 *M* NaCl gradient), (C) *M. luteus* (0–0.6 *M* KCl gradient). Protein was determined by the Lowry method and is represented by optical density (660 nm). I, II, and III represent the C_{55}PP synthetase, intermediate all-*trans*-polyprenylpyrophosphate synthetase and geranylgeranylpyrophosphate synthetase, respectively.

from 2*R*-[2-^{3}H]IPP is retained during the formation of the *cis*-isoprene residue and lost during the formation of the *trans*-isoprene residue.[30] Alternatively, the ^{3}H label from 2*S*-[2-^{3}H]IPP is retained during the formation of the *trans*-isoprene residue and lost during the formation of the cis residue.[31] Unfortunately this method is not currently of general use because the appropriately labeled metabolic precursors, 4*R*- and 4*S*-[4-^{3}H]mevalonic acid are not now commercially available.

Detailed Purification of *L. plantarum* C_{55}PP Synthetase[32]

Bacterial Growth. L. plantarum (ATCC 8014) is grown at 30° in Skeggs medium[33] as previously described[7] except that Tween 80 is omitted from the medium. The cells are harvested from late log cultures and washed with 100 m*M* Tris–HCl buffer pH 7.5. The cells may be kept frozen for 6 months at −15° without any noticeable effect on enzyme activity. Longer storage causes the cells to be more resistant to disruption and there is also a decrease in the specific activity of the isolated enzyme.

Cell Lysis and Ammonium Sulfate Precipitation. Thirty grams of cells (wet weight) are treated for 30 min at 35° with 67 mg of lysozyme in 270 ml of a solution containing 30 m*M* Tris–HCl buffer pH 8.0, 1 m*M* glutathione, 2 m*M* phenylmethylsulfonyl fluoride, and 10 m*M* iodoacetamide. The resulting lysate is treated for 5 min with 0.5 mg of deoxyribonuclease to reduce the viscosity. Subsequent steps are carried out at 4°. This suspension is passed 3–5 times through a French pressure cell at 18,000 psi and centrifuged at 30,000 *g* for 30 min. A good indication of effective cell breakage is the appearance of a white membrane layer on top of the centrifugal pellet. The resulting supernatant is recentrifuged for 1 hr at 100,000 *g* to obtain a supernatant which contained up to 80% of the total activity of the cell homogenate.

The enzyme in the 100,000 *g* supernatant is precipitated with ammonium sulfate (0–65%). This precipitate may be stored frozen for up to 6 days without loss of activity. If two of these precipitates are combined, a larger protein load can be applied to chromatographic columns and higher recovery of enzymatic activity results after column chromatography and dialysis. The eluted enzyme is more stable in concentrated protein solutions.

[30] J. W. Cornforth, R. H. Cornforth, C. Donninger, and G. Popják, *Proc. R. Soc. London, Ser. B* **163,** 492 (1966).

[31] B. L. Archer, D. Barnard, E. G. Cockbain, J. W. Cornforth, R. H. Cornforth, and G. Popják, *Proc. R. Soc. London, Ser. B* **163,** 519 (1966).

[32] J. D. Muth and C. M. Allen, *Arch. Biochem. Biophys.* **230,** 49 (1984).

[33] H. R. Skeggs, L. D. Wright, E. L. Cresson, G. D. E. MacRae, C. H. Hoffman, D. E. Wolf, and K. Folkers, *J. Bacteriol.* **72,** 519 (1956).

DEAE-Cellulose Step. The ammonium sulfate precipitated enzyme is desalted by dialysis in 10 m*M* Tris–HCl buffer pH 7.5, diluted to 100 ml with the Tris buffer, and up to 2.5 g of protein is applied to a 2.5 × 40 cm column of DEAE-52 cellulose preequilibrated with the Tris buffer. The column is washed with 100 ml of 10 m*M* Tris buffer, then protein is eluted with a linear gradient (800 ml) of NaCl (0–0.5 *M*) in the Tris buffer and protein was determined by the method of Lowry *et al.*[34] The enzyme elutes at about 60% of completion of the gradient. The column fractions with the highest enzymatic activity are pooled, dialyzed against 10 m*M* KPO_4 buffer pH 7.5 and applied the same day to a hydroxylapatite column. Alternatively, the enzyme pool may be precipitated with ammonium sulfate (80%) before dialysis, stored frozen overnight, and then dialyzed the next day against 10 m*M* KPO_4 buffer pH 7.5. If the pooled DEAE-cellulose chromatographic fractions are frozen without concentration then poor recovery of activity is observed.

Hydroxylapatite Step. This dialyzed pool (50 ml) containing about 200 mg of protein is applied to a 2.5 × 20 cm hydroxylapatite (Hypatite C, Clarkson) column equilibrated with 10 m*M* KPO_4 buffer pH 7.5. The column is washed with 50 ml of 10 m*M* KPO_4 buffer and the enzyme is eluted with a linear gradient (300 ml) of KPO_4 buffer pH 7.5 (50–150 m*M*). The enzyme elutes at about 40% of the gradient with the primary protein band. The fractions with the greatest specific activity are pooled and can be kept frozen for over 2 years without loss of enzymatic activity. There is only a small increase in enzyme specific activity after this step but far greater enzyme stability is achieved.

Sephadex Step. Several hydroxylapatite pools are concentrated by ultrafiltration with an Amicon XM-50 filter without noticeable loss of activity. The concentrated sample is applied to a 5 × 47 cm Sephadex G-100 column preequilibrated with 100 m*M* KPO_4 buffer pH 7.5. Protein is eluted with 100 m*M* KPO_4 buffer pH 7.5. The enzyme activity is found following the major protein peak. The column fractions with enzymatic activity may be kept frozen for over 2 years without loss of activity, when the protein concentration is greater than 0.5 mg/ml.

The protein and enzyme elution profiles from these various chromatographies have been previously described.[7,8]

Notes on Enzyme Solubilization. The general method now used for the preparation of the $C_{55}PP$ synthetase from *L. plantarum* has changed remarkably from the earlier descriptions.[8,35] We had reported[8] that the yield

[34] O. H. Lowry, N. J. Rosebrough, A. L. Farr, and R. J. Randall, *J. Biol. Chem.* **193,** 265 (1951).

[35] I. F. Durr and M. Z. Habbal, *Biochem. J.* **127,** 345 (1972).

of soluble protein and active $C_{55}PP$ synthetase after lysis of the bacterium with lysozyme was at best 30% of the whole cell lysate. The addition of Triton X-100 had also been shown to assist the release of enzyme into the high speed supernatant and to stabilize the enzyme during DEAE-cellulose chromatography.[7] Furthermore, Triton X-100 had been used in every step of earlier reports for the purification of the enzyme.

The method described above is an improved procedure for the partial purification of the $C_{55}PP$ synthetase in the complete absence of Triton X-100. Treatment of cells by osmotic shock following lysozyme treatment was apparently insufficient to permit optimal liberation of the enzyme into the soluble fraction unless Triton X-100 was added.[8] The use of the French pressure cell in the current procedure results in the release of 50–80% of the total enzymatic activity into the high-speed supernatant without the aid of detergent. Furthermore, the utilization of bacteria grown to the late log phase (OD = 1.0) instead of stationary phase (OD = 1.6) is also more productive, since late log cells are easier to disrupt and the recovery of total activity into the high-speed supernatant fraction is greater.

The use of a cell mass greater than 40 g/270 ml of lysing buffer leads to a poor yield of enzymatic activity in the supernatant obtained after high-speed centrifugation. When a greater mass of cells is required, successive batchs of cells are processed through ammonium sulfate precipitation after the high-speed centrifugation step. These protein precipitates are stored until several precipitates are accumulated for a large scale DEAE-cellulose chromatography.

If cell disruption during the French pressure cell step is not very effective, the 30,000 *g* pellet may contain up to 50% of the activity. The observation that the enzyme is assayable in the pellet indicates that the substrate is able to permeate the partially disrupted cells and be metabolized. The recovery of activity from this pellet is poor, if it has not been released after five passages through the French pressure cell. The enzyme in the pellet is not solubilized by treatment with Triton X-100. Previous results have shown that a variety of agents such as KCl, EDTA, and deoxycholate are ineffective in increasing the solubilization of the enzyme.[8] There is only a trace of activity found in the white membrane layer covering the 30,000 *g* centrifugal pellet or in the 100,000 *g* pellet. Furthermore, recombination of the centrifugal pellets and supernatant fractions gives no greater activity than the sum of the different fractions.

The current use of highly concentrated protein solutions and rapid processing has apparently substituted for the earlier requirement for Triton X-100 in the chromatography steps. Now the entire enzyme preparation through polyacrylamide gel electrophoresis can be done in the ab-

TABLE I
PURIFICATION OF *L. plantarum* C_{55}PP SYNTHETASE

Step[a]	Total protein (mg)	Specific activity (nmol IPP consumed/mg protein)	Purification	Total activity (nmol IPP consumed)	Recovery
1. Crude lysate	6075	11.6	1	70240	100%
2. 100,000 *g* supernatant	3314	15.7	1.3	53735	76%
3. 0–65% ammonium sulfate precipitate	2343	22.3	1.9	52360	74%
4. DEAE-cellulose pool	405	113	9.8	45800	65%
5. Hydroxylapatite pool	219	174	15.1	38106	54%
6. Sephadex G-100 (Superfine) pool	11	1763	152	17657	25%

[a] The data given for steps 1 and 2 are the averages of the specific activities and sums of total protein and total activity obtained from processing two 30 g batches of cells. The ammonium sulfate precipitates from step 2 were pooled for subsequent purification.

sence of Triton X-100 with the same resultant purification as when Triton X-100 is present. Purification in the absence of Triton X-100 makes enzyme concentration procedures and subsequent purification steps easier, since Triton X-100 is poorly removed during the concentrating procedures and gives highly viscous and less manageable solutions.

Table I illustrates a purification scheme through the Sephadex G-100 step showing good recovery and high specific activity of the enzyme starting from 60 g of cells. The enzyme has therefore been prepared in at least 10 times the previously reported yields.

Polyacrylamide Gel Electrophoresis. The enzyme may be purified further by preparative slab gel electrophoresis. Even though a large degree of protein purification is achieved in this electrophoresis step there is only a 2- to 6-fold increase in specific activity from the previous step since enzyme denaturation is large.

Analytical nondenaturing discontinuous gel electrophoresis is performed at 4° on 1.5 mm slab gels of 22.5% (w/v) acrylamide and 0.15% (w/v) bisacrylamide by the Canalco method using 25 m*M* Tris–HCl buffer pH 8.3 as the running buffer.[36] The stacking of the protein is done at a

[36] L. Shuster, this series, Vol. 22, p. 412.

constant voltage (100 V) for 1 hr and the separation is achieved at 300–370 V for 18 hr or until an azobovine serum albumin marker migrates about two-thirds of the length of the gel. After electrophoresis, 2 mm horizontal cuts are made in a small vertical section of the gel leaving the slices attached to the main portion of the slab. A part of each slice is assayed for enzymatic activity, while the other part, remaining attached to the slab, is stained with Coomassie Blue R-250. Comparison of the assayed slices with the stained slices permits the localization of the enzyme with great precision.

A larger quantity of enzyme is needed for preparative gel electrophoresis. The required amount (15–30 mg) is obtained by processing 90 g of cells through the Sephadex G-100 step (Table I). The enzyme is then concentrated by ultrafiltration with an Amicon XM-50 filter. Preparative gel electrophoresis is carried out at 4° on a 3 mm slab gel of 14% acrylamide and 0.74% bisacrylamide. The proteins are stacked at 80 V for 1 hr and the separation achieved in 3–5 hr at 400 V or until an azobovine serum albumin marker migrates about two-thirds of the length of the gel.

The enzyme is localized by staining a small vertical section of the gel with Coomassie Blue and comparing the stained bands with the pattern of enzyme migration previously established in an analytical gel. Three or four 3 mm horizontal sections, corresponding to the estimated position of the enzyme migration, are cut out along the entire width of the gel, extracted and analyzed for enzyme activity. Cutting several narrow slices instead of a wide slice minimizes contamination of the enzyme with closely migrating proteins. Each of these gel sections is passed through a 1 ml hypodermic syringe. The enzyme is then extracted by suspending each pulverized gel twice in 2.5 ml of 50 mM KPO_4 buffer, pH 7.5 for 30 min at 4° with frequent vortexing. The gel is removed from the extract by vacuum filtration. The gel extract with highest activity is used for subsequent experiments. The first gel extract has the highest enzymatic activity therefore this is the most useful for subsequent work. Pooling of the first extract with subsequent extracts is not advised since substantial losses of enzymatic activity are observed during the process of concentrating dilute enzyme solutions. The electrophoretic separation and the enzyme extraction are carried out on the same day to minimize denaturation of the enzyme. The activity of the extracted enzyme is lost in 2–3 days, even when it is kept frozen.

Affinity Chromatography. Alkyl Agarose. A promising method for purification of prenyltransferases is hydrophobic affinity chromatography. Crude or hydroxylapatite purified $C_{55}PP$ synthetase is retained on columns of octyl(C_8)- and decyl(C_{10})-agarose (Miles Chemical Corp.).[37]

[37] J. D. Muth, T. Baba, and C. M. Allen, *Biochim. Biophys. Acta* **575**, 305 (1979).

The enzyme is not eluted with 2 *M* NaCl. The enzyme may, however, be eluted from the C_{10}-agarose column with 0.5% Triton X-100 and from the C_8-agarose column with 0.1% Triton X-100. The C_6-agarose column only slightly retards the chromatographic elution of the enzyme. Its elution may be facilitated with 0.02% Triton X-100. The major portion of the protein in all cases is unadsorbed to the alkyl agarose columns without Triton X-100 addition. Furthermore, the prenyltransferases C_{55}PP synthetase and C_{40}PP synthetase from *M. luteus* also bind to C_{10}-agarose columns and are eluted with 0.5% Triton X-100. However, it was also shown that although geranylgeranyl-PP synthetase from *M. luteus* and FPP synthetase from Baker's yeast both adsorbed to C_{10}-agarose, only 10–15% of the activity can be eluted with 1% Triton X-100.

Blue Agarose. An alternate method of affinity chromatography uses Blue-agarose columns. An hydroxylapatite purified enzyme (12 mg protein) in 24 ml of 50 m*M* KPO_4 buffer pH 7.5 is applied to a 1.5 ml column of Blue-agarose, which has been preequilibrated with 10 m*M* KPO_4 buffer pH 7.5. The column is successively washed with 20 ml of 10 m*M* KPO_4 and 60 ml of 100 m*M* KPO_4, then the enzyme is eluted with 100 ml of 225 m*M* KPO_4 buffer while 5 ml fractions are collected. Seventy-five percent of the applied protein is unadsorbed to the column. Furthermore, no protein is detected by the method of Lowry *et al.*[34] in the later eluting fractions containing the enzyme. A large volume of elutant is required to displace the enzyme from the small column, but 50% of the activity is recovered in the first 30 ml of eluate. Concentration of this eluate for subsequent electrophoretic analysis by a variety of methods, including ultrafiltration, membrane dialysis, or lyophilization invariably results in the loss of 60–100% of the activity. When the Blue-agarose affinity column is charged with C_{55}PP synthetase and subjected to elution with 5 μM FPP only 10% of the activity is recovered in the eluate. ATP (100 m*M*) does not elute any enzyme from the column. The enzyme is also adsorbed tightly to a Red-agarose column, but it is not eluted with 500 m*M* KPO_4 buffer.

Properties

Molecular Weight Estimation. Studies using Sephadex G-100 show that the enzyme chromatographed as a protein of 53,000–60,000. A second molecular weight estimation of 52,000–58,000 has been shown by sucrose gradient centrifugation.

SDS Gel Electrophoresis.[32] The molecular weight and polypeptide constitution of the enzyme is evaluated by SDS gel electrophoresis. Aliquots of Sephadex G-100 chromatography fractions containing enzymatic activity are subjected to nondenaturing electrophoresis in a 22.5% polyacrylamide gel. There is only one band of enzymatic activity detected in these gels. The Coomassie Blue stained band in each fraction, which corresponds to the position of migration of the enzyme, is cut out separately from the adjoining bands and subjected to SDS-polyacrylamide electrophoresis.[38] It has been shown that a 30,000 ± 1000 MW component is the major band observed and that the degree of Coomassie Blue and silver staining[39] of the 30,000 MW component from different column fractions parallels the enzymatic activity in each fraction. This estimation of molecular weight indicates that the $C_{55}PP$ synthetase is a dimer of 60,000 ± 2000. This falls within the range of 52,000–60,000 MW obtained by Sephadex chromatography and sucrose gradient centrifugation.

Although the subunit character of this prenyltransferase is described by these techniques, frequently preparative electrophoresis fractions are obtained which contain the 30,000 MW component with either some accompanying 52,000 or 21,000 MW polypeptide. It is clear, however, that neither the 52,000 nor 21,000 MW component is necessary for enzymatic activity, since enzymatic activity is still observed in some gel fractions, where either the 50,000 or 21,000 MW component is absent.

Isoelectric Focusing. The polyacrylamide purified enzyme is subjected to isoelectric focusing in polyacrylamide gels having a pH gradient from 4 to 10. There is a single peak of enzymatic activity associated with a protein of p*I* 5.1.

Substrate Specificity. $C_{55}PP$ synthetases partially purified from *L. plantarum, M. luteus,* and *B. subtilis* have been examined for their preference for various allylic prenyl pyrophosphate substrates.[5,6,10] Table II summarizes the reactivities of several of these substrates relative to *trans*, trans-FPP. The reactivity of the allylic substrate increases with the chain length, C_5, C_{10}, C_{15}, C_{20}. The all-*trans*-solanesyl (C_{45}) PP is unreactive. Moreover, the enzyme is not only active with all-trans derivatives but also with those substrates having both cis and trans stereochemistry. A consideration of the K_m values also shows that the larger the number of *trans*-isoprene residues in the substrates, the lower the K_m values.[10]

Farnesol, geraniol, geranyl monophosphate, and citronellyl pyrophosphate have no stimulatory or inhibitory effect on the *L. plantarum* enzyme when it is assayed with *trans,trans*-FPP as a substrate. Binding of the allylic pyrophosphate substrate to the enzyme is therefore apparently dependent on the presence of an α-unsaturated isoprene unit and a pyrophosphate moiety.

[38] H. Jäckle, *Anal. Biochem.* **98,** 81 (1979).

[39] W. Wray, T. Boulikas, V. Wray, and R. Hancock, *Anal. Biochem.* **118,** 197 (1981).

TABLE II
COMPARISON OF SUBSTRATE SPECIFICITY OF *L. plantarum*, *M. luteus*, AND *B. subtilis* C_{55}PP SYNTHETASES FOR ALLYLIC PYROPHOSPHATE SUBSTRATES

Allylic pyrophosphate substrate	Prenyltransferase activity relative to *t,t*-FPP[a]		
	L. plantarum[b]	*M. luteus*[c]	*B. subtilis*[d]
C_5PP (DMAPP)	0	0	0
t-C_{10}PP (GPP)	0.1	0.1	0.1
c-C_{10}PP (NPP)	0.1	0.4	0.3
t,t-C_{15}PP (FPP)	1.0	1.0	1.0
c,t-C_{15}PP (FPP)	1.0	1.0	0.6
t,t,t-C_{20}PP (GGPP)	0.7	1.1	0.9
c,t,t-C_{20}PP (GNPP)	1.3	1.8	1.9
t,c,t,t-C_{25}PP	—	—	0.1[e]
c,c,t,t-C_{25}PP	—	—	1.3[e]
all-*trans*-C_{45}PP (SolPP)	0	0	—

[a] Each enzyme was assayed at saturating levels of substrate except as indicated. The enzymatic activity with each substrate was compared relative to the activity found when *t,t*-FPP was used as a substrate.
[b] Data taken from T. Baba and C. M. Allen, *Biochemistry* **17,** 5598 (1978).
[c] Data taken from T. Baba and C. M. Allen, *Arch. Biochem. Biophys.* **200,** 474 (1980).
[d] Data taken from I. Takahashi and K. Ogura, *J. Biochem.* (*Tokyo*) **92,** 1527 (1982).
[e] Estimates made from the reaction rates and K_m values cited.

Product Chain Length and Stereochemistry. Each of the enzymes studied has the ability to synthesize long-chain products without the accumulation of any intermediate chain length products. The *L. plantarum* enzyme[10] gives almost exclusively the C_{55} product when either *trans,trans*-FPP or *cis,trans,trans*-GGPP are used as a substrate, whereas the *M. luteus* and *B. subtilis* enzymes give a mixture of C_{55} (major) and C_{50} (minor) products. The stereochemistry of the newly added isoprene units has been shown[5,6,10] to be cis in each case as assessed by using 2*R*-[2-^{3}H]IPP and 2*S*-[2-^{3}H]IPP as substrates.

Kinetic Properties. The K_m of the partially purified *L. plantarum, M. luteus*, and *B. subtilis* C_{55}PP synthetases for *trans,trans*-FPP were determined to be 0.13, 8, and 9.1 μM, respectively.[5,6,10] The K_m of the *L. plantarum* enzyme for IPP is 1.92 μM. These K_m values for the *L. plantarum* enzyme are much different from the respective values of 130 and 14 μM for FPP and IPP, which were reported earlier[10] for the enzyme prepared with Triton X-100 through the purification scheme. The reason for this difference is not apparent, but it is not due to the use of Triton X-100 in the preparation of the enzyme since the same K_m values were obtained with enzymes prepared from the same batch of cells with or without Triton X-100.

The ^{14}C-labeled product obtained with the enzyme prepared without Triton X-100 was reconfirmed to be the C_{55} isoprenoid by showing that the mobility of the free prenol on reverse phase TLC was the same as the mobility of authentic *L. plantarum* $C_{55}OH$.

Lipid and Detergent Activation of $C_{55}PP$ Synthetase. The $C_{55}PP$ synthetase from *L. plantarum, M. luteus,* and *B. subtilis* requires detergent or phospholipid for activity. The most extensive studies of these lipid and detergent requirements have been made on the *L. plantarum* enzyme.[9,40] Activating detergents and the concentrations required for half maximum effect are as follows: deoxycholate (330 μM), dodecyl sulfate (75 μM), cetyl sulfate (40 μM), and Triton X-100 (0.2%, 3.4 mM). Brij 35, 56, and 96 and cetyl trimethylammonium bromide are ineffective as activators. *L. plantarum, E. coli,* and bovine cardiolipin, egg phosphatidic acid, and oleate are all good activators of the enzyme in the absence of detergent. C_6–C_{18} saturated fatty acids are only effective activators when they are mixed with a nonionic detergent. It has been shown that a fixed molar ratio (2:1) of fatty acid to detergent (Brij 35) is required for optimal activity. *L. plantarum* phosphatidylglycerol and lysylphosphatidylglycerol, several lecithins, phosphatidylserine, and saturated fatty acids (without detergent) are all ineffective over a wide concentration range.

The *M. luteus* enzyme is less specific for the type of phospholipid which stimulates its activity.[5] Triton X-100, egg lecithin, and a phospholipid extract from *M. luteus* stimulate enzymatic activity, whereas cardiolipin and deoxycholate are far less stimulatory. Furthermore, the replacement of Triton X-100 with a phospholipid extract in the enzyme reaction mixture causes a shift in the product distribution such that the C_{55} product is greatly favored over the C_{50} product when *trans-trans*-FPP is used as a substrate.

The activity of the *Bacillus subtilis* enzyme is stimulated strongly by Triton X-100 and Tween 80.[6] Phosphatidylethanolamine is less effective, whereas egg lecithin and deoxycholate were poor activators.

Innumerable attempts with the *L. plantarum* enzyme have failed to establish that the enzyme is tightly associated with the natural membrane which was isolated by sucrose gradient centrifugation. Furthermore, the partially purified enzyme does not associate with this membrane fraction or micelles of egg phosphatidic acid or cardiolipin as determined by gel

[40] M. V. Keenan and C. M. Allen, *Biochem. Biophys. Res. Commun.* **61,** 338 (1974).

chromatography. The *L. plantarum* membrane also does not substitute for phospholipids, such as cardiolipin, as a stimulator of enzymate activity. The neutral lipid fraction extracted from *L. plantarum* exhibits good stimulatory activity. This function is not associated with the glycolipid fraction.

Assay of Lipid Stimulation. The assay for measuring lipid activation is the same as the standard assay except that 200 m*M* Tris–HCl buffer pH 7.5 is used and the enzyme is incubated with detergent, sonicated phospholipid, or fatty acid at 40 instead of 35°.

The following is a description of the preparation of the phospholipid activator. An organic solution of each phospholipid to be tested is brought to dryness with a nitrogen stream. The residue is then suspended in a solution containing 200 m*M* Tris–HCl buffer, pH 7.5, 200 μM $MgCl_2$, and 50 μM FPP, flushed with nitrogen, and dispersed immediately prior to assay by sonication in a plastic test tube for 5 min at 0° with a Biosonic II sonicator at 150 W. Aliquots are then transferred to the assay tubes. FPP is included in the sonication mixture to maintain small and constant enzyme reaction volumes. Sonicated mixtures of fatty acids are prepared in the same way. Either sonicated or unsonicated cardiolipin gives the same stimulatory response.

Preparation of Detergent-Depleted Enzyme. If the enzyme has been prepared in the presence of Triton X-100, it is necessary to deplete the enzyme solution of detergent before testing the enzyme for stimulation by various lipids. This is accomplished by the addition of the nonionic resin Amberlite XAD-2 (or Biobeads SM-2, Bio-Rad Lab) by the procedure of Holloway.[41] The methanol-washed beads are added in small portions to a stirred solution of enzyme at 0–4° until a final ratio of 100–150 mg of resin per mg of Triton X-100 is obtained. This mixture is stirred for 30 min, then it is filtered under a slight vacuum through a layer of glass wool on a small Hirsh funnel to remove the resin. A Triton X-100 concentration of less than 0.1% remains after this treatment as determined from the detergent absorbance at 275 nm. Dilution of the enzyme into assay mixtures after this procedure gives levels of Triton X-100 which are ineffective in stimulating the enzyme.

Divalent Cation Requirement. The C_{55}PP synthetases require divalent cation for activity. The *L. plantarum* enzyme[8] is equally stimulated by 200 μM Mg^{2+}, 200 μM Co^{2+}, or 100 μM Mn^{2+}. Ca^{2+} and Cu^{2+} are ineffective. The *M. luteus* enzyme[5] is optimally stimulated by 100 μM Mg^{2+}, 50 μM Mn^{2+}, or 200 μM Co^{2+}. Ba^{2+}, Cu^{2+}, Ca^{2+}, and Zn^{2+} failed to stimulate this enzyme. The *B. subtilis* enzyme[6] is reported to use Mg^{2+} best although

[41] P. W. Holloway, *Anal. Biochem.* **53,** 304 (1973).

Mn^{2+} and Zn^{2+} may partially replace Mg^{2+}. This enzyme is also markedly stimulated by the addition of the monocations K^+, Na^+, or NH_4^+ when they are added with Mg^{2+}. The mono cations apparently enhance the V_{max} values but not the K_m values. Polyamines also stimulate enzymate activity.

pH Optima. The pH optima for the different bacterial enzymes are *L. plantarum,* 7.5 and 10.2, double optima, *M. luteus,* 7.5, and *B. subtilis,* 8.5.

Other Properties. Mercaptoethanol (10–100 m*M*) and iodoacetamide (10 m*M*) have no effect on the *L. plantarum* enzyme although iodoacetate (10 m*M*) completely inhibits the enzyme. The enzymate activity is enhanced with increasing ionic strength when the enzyme is tested with up to 0.5 *M* Tris–HCl and up to 0.8 *M* KCl. Phosphate, an inhibitor of the pig liver prenyltransferase,[16] does not inhibit the *L. plantarum, M. luteus,* or *B. subtilis* enzymes. The *L. plantarum* enzyme is inactivated by the arginine specific reagents phenylglyoxal, butanedione, and cyclohexanedione, but this inactivation is not prevented by either of the substrates.[32]

Acknowledgment

This work was supported by NIH Grant GM-23193.

[33] *In Vitro* Synthesis of C_{15}–C_{60} Polyprenols in a Cell-Free System of *Myxococcus fulvus* and Determination of Chain Length by High-Performance Liquid Chromatography

By PETER BEYER and HANS KLEINIG

Prenyl alcohols of varying chain length are commonly found as products in incubation assays for investigations of *in vitro* biosynthesis of prenyl lipids from low-molecular-weight precursors (isopentenyl pyrophosphate, mevalonate). These polyprenols may arise either as biosynthetic end products or, to a small extent, by prenyl transfer onto water in an aberrant reaction of prenyltransferase[1] or as artifacts of unspecific phosphatase action.

[1] H. C. Rilling, *in* "Carotenoids V" (T. W. Goodwin, ed.), p. 597. Pergamon, Oxford, 1979.

ISBN 0-12-182010-6

In order to facilitate the analysis of such alcohols, we describe in the following the biosynthetic formation of ^{14}C-labeled C_{15}–C_{60} polyprenols which may be used as standards, and their separation and chain length determination by HPLC. Phosphorylated polyprenols may also be analyzed in this way when converted to their corresponding alcohols by alkaline phosphatase,[2] hydrolysis,[2] or reduction with $LiAlH_4$,[3] because handling of the free alcohols is much easier than of phosphorylated compounds. For the separation and determination of phosphorylated C_5 and C_6 precursors for prenyl lipid synthesis and of short chain prenyl mono- and pyrophosphates see Beyer *et al.*[4]

In Vitro System

Myxococcus fulvus strain Mxf2 (Myxobacterales) is grown in Erlenmeyer flasks under continuous shaking, or in an aerated fermenter using a medium containing Bacto Casitone (10 g/liter) and $MgSO_4$ (2 g/liter) at 27°. The cells are harvested by centrifugation (10,000 *g*), resuspended in incubation buffer (Tris–HCl, 100 m*M*; $MgCl_2$, 10 m*M*; mercaptoethanol, 2 m*M*; pH 7.6) and centrifuged again. The cell pellet is resuspended in incubation buffer (1 : 5 by vol.) and disintegrated using a French Pressure Cell (16,000 psi). The homogenate is centrifuged (20,000 *g*, 40 min). The supernatant is centrifuged again (140,000 *g*, 2 hr). The slightly colored supernatant of this centrifugation step is the active fraction and is used for incubations with [1-^{14}C]isopentenyl pyrophosphate, delivering all polyprenols ranging from C_{15} to C_{60} in chain length. For incubation 1 ml of the 140,000 *g* supernatant is supplemented with NADP (2 m*M*), $MnCl_2$ (1 m*M*), FAD (20 μM), and [1-^{14}C]isopentenyl pyrophosphate (18.5 kBq, corresponding to about 10 μM). Incubation is carried out at 30° for 4 hr in the dark.

For extraction 2 ml acetone is added and precipitating proteins removed by centrifugation. To the supernatant 3 ml petroleum ether/ether (1 : 4) and an excess of water are added. The polyprenols formed are distributed into the organic upper phase and contain about 30% of the applied radioactivity. Small amounts of labeled hydrocarbons in this extract which may interfere in the HPLC separation described below can easily be removed from the polyprenol fraction by thin layer chromatography on silica gel plates, solvent system petroleum ether/ether/acetone (40 : 10 : 5).

[2] G. R. Daelo and R. Pont Lezica, *FEBS Lett.* **74,** 247 (1977).
[3] D. S. Goodman and G. Popják, *J. Lipid Res.* **1,** 286 (1960).
[4] P. Beyer, K. Kreuz, and H. Kleinig, this series, Vol. 111 [10].

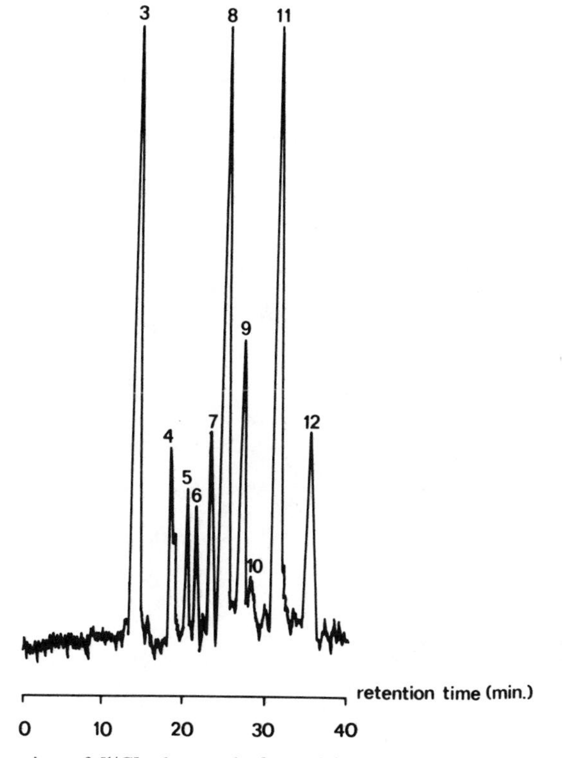

FIG. 1. Separation of [^{14}C]polyprenols formed in the *Myxococcus in vitro* assay on HPLC, using a C_{18} reversed phase column in the gradient system. 3, Farnesol; 4, geranylgeraniol; 5, pentaprenol; 6, hexaprenol; 7, heptaprenol; 8, octaprenol; 9, nonaprenol; 10, decaprenol; 11, undecaprenol; 12, dodecaprenol (the numbers given correspond to the number of prenyl units of the different polyprenols).

Analysis of Products

Standards: Nonradioactive geraniol, farnesol, geranylgeraniol, and undecaprenol. Undecaprenol can be isolated from *Myxococcus fulvus* cells where it occurs in considerable amounts. It has been identified by mass spectrography.

For the separation and chain length determination a HPLC system (Waters) equipped with a C_{18} reversed phase column (Nucleosil 10, C_{18}, SS 300/6/4, Macherey and Nagel) is used. Nonradioactive standards are detected using a differential refractometer, and radioactive polyprenols formed *in vitro* using a Berthold radio column chromatography monitor with a glass scintillator cell.

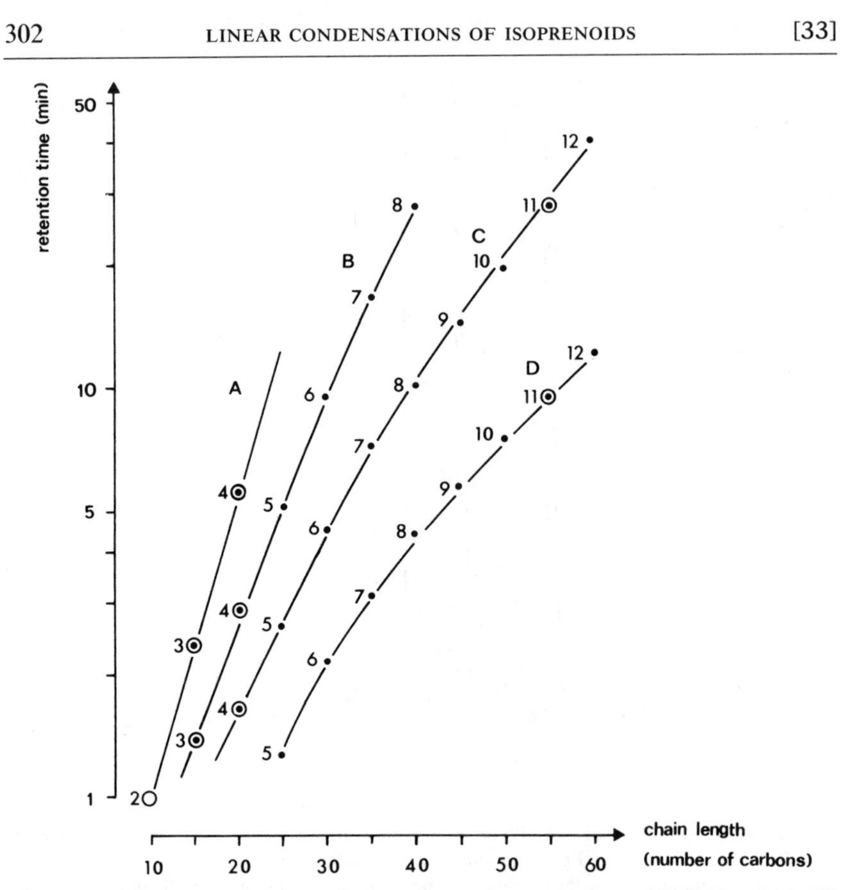

FIG. 2. Semilogarithmic plot of retention times of the *in vitro* formed [^{14}C]polyprenols (●) and standards (○) versus chain length in the four isocratic systems A–D. The numbers given correspond to the number of prenyl units of the different polyprenols. 2, Geraniol; 3, farnesol; 4, geranylgeraniol; 5, pentaprenol; 6, hexaprenol; 7, heptaprenol; 8, octaprenol; 9, nonaprenol; 10, decaprenol; 11, undecaprenol; 12, dodecaprenol.

With the gradient system methanol/H_2O (3 : 1) and tetrahydrofuran/acetonitrile (1 : 1), 100 : 0 to 30 : 70 within 15 min and subsequent isocratic final conditions (flow rate 1.5 ml/min), ten radioactive polyprenols differing in chain length can be resolved from the lipid extract of *Myxococcus fulvus in vitro* assay (Fig. 1).

As homologous polyprenols do not elute logarithmically from the column under gradient conditions, their length determination by extrapolating graphical methods is not possible in this system. This can be obtained under isocratic conditions. In our case, however, several isocratic runs are necessary in order to avoid very long retention times and, thus, large broadening of peaks due to a wide divergency in chain length of the homologues.

In Fig. 2 the retention times of the compounds in the different isocratic runs are plotted in a semilogarithmic scale versus the chain length. The following systems are used:

	Acetonitrile/tetrahydrofuran (%)	Methanol/H_2O (%)
System A, for short chain compounds	60	40
System B, for medium chain compounds	70	30
System C, for medium chain compounds	80	20
System D, for long chain compounds	90	10

A solvent flow of 1 ml/min is applied.

As can be seen in Fig. 2, a number of graphs are obtained, demonstrating the homologous behavior of the compounds. Due to this and the overlapping separation patterns of the different graphs an unequivocal assignment of chain length can be made, although relatively few standards are available. The slight curvature of these graphs is explained by the nonideality of partition on reversed phase columns.

Acknowledgment

This work was supported by Deutsche Forschungsgemeinschaft.

[34] Xanthophyll Cycles

By HARRY Y. YAMAMOTO

Xanthophyll cycles are the light-dependent, reversible interconversions of 5,6-epoxycarotenoids in photosynthetic organisms.[1] Two types of cycles (Fig. 1) are known. The violaxanthin cycle occurs in plants and the

[1] D. Sieffermann-Harms, *in* "Lipids and Lipid Polymers in Higher Plants" (M. Tevini and H. K. Lichtenthaler, eds.), p. 218. Springer-Verlag, Berlin and New York, 1977.

ISBN 0-12-182010-6

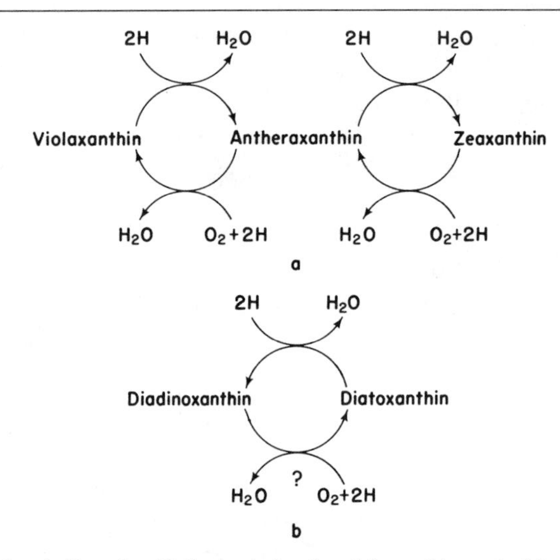

FIG. 1. Xanthophyll cycles. Reductants for the violaxanthin cycle (a) are ascorbate and reduced pyridine nucleotides for deepoxidation and epoxidation reactions respectively. The biochemistry of the diadinoxanthin cycle (b) is unclear.

algal classes, Chlorophyceae, and Phaeophyceae whereas the diadinoxanthin cycle is found in many other algal classes but not in the Cyanophyceae, Cryptopyceae, and Rhodophyceae.[2] The latter group of algae, except for a few Rhodophyceae,[3] do not have epoxy carotenoids.

Xanthophyll-cycle activity is indicated by stoichiometric shifts in pigment concentrations under illumination or other experimental conditions. No net change takes place in total carotenoids.[1] The kinetics of the interconversions are relatively slow and the changes are large enough that potentially any method capable of quantitatively distinguishing the specific carotenoids can be the basis of an assay. Chapters on HPLC and general methods of carotenoid analyses may be found in this volume by M. Ruddat and G. Britton.[3a] Among the various analytical methods which have been used for xanthophyll studies two thin-layer chromatographic methods and one spectrophotometric method have proven useful and are described. Biochemical methods for the violaxanthin cycle in higher plants are also described. Although xanthophyll cycles are widely distributed, biochemical methods have been adequately developed only for the violaxanthin cycle in chloroplasts.

[2] H. Stransky and H. Hager, *Arch. Mikrobiol.* **73,** 315 (1970).

[3] M. S. Aihara and H. Y. Yamamoto, *Phytochemistry* **7,** 497 (1968).

[3a] M. Ruddat and O. H. Will, this series, Vol. 111 [7]; G. Britton, this series, Vol. 111 [5].

Analytical Methods

Chromatographic and spectrophotometric methods are complementary and detailed studies on xanthophyll cycles require application of both techniques. Chromatographic methods, though specific and widely applicable, are time consuming and as a practical matter give limited data points. In contrast, spectrophotometric methods are rapid, continuous, and sensitive but require special instrumentation and are not adaptable to some situations. Chromatographic methods are especially appropriate in earlier stages of study while spectrophotometric methods are advantageous in biochemical investigations.

Chromatographic

Hager–Stransky Alkaline Plates.[4,5] Plates are coated with a 0.20-mm-thick layer of a slurry composed of 30 g $CaCO_3$ (Merck 2066), 6 g MgO (Merck 5865), and 5 g $Ca(OH)_2$ (Merck 2047) in 60 ml 1.7% KOH solution. The plates are activated at 115° for 40 min and used within a few hours after preparation. Pigment extract in acetone is streaked on the plate in a narrow band and the plate is developed in a mixture of 50 ml petroleum ether (bp 100–140°), 50 ml acetone, 40 ml chloroform, and 1 ml methanol. The carotenoid bands are immediately scraped and eluted with ethanol then quantitated using the extinction coefficients in Table I. Cycle activity is indicated by stoichiometric changes in pigment concentrations per unit leaf area, dry weight, or other suitable basis.

This system can separate α- from β-carotene and lutein from zeaxanthin. When resolving higher plant pigments, methanol may be omitted from the developing solvent.[5] The chlorophylls, retained near the origin, are degraded and are not suitable for use as an internal reference.

Egger Reverse-Phase Plates.[6] Plates coated with 0.250-mm-thick Kieselguhr G (Merck) are dipped in a solution of 7% hydrogenated coconut oil in light petroleum ether, impregnating all but about a 3-cm strip of the plate with fat, and dried at room temperature for at least 24 hr. No further activation is required and the plates remain usable for several weeks. Acetone extracts are applied to the fat-free portion and the plates are developed in a mixture of methanol–acetone–water (20 : 4 : 3). The bands are scraped, eluted with light petroleum ether (60–80°) for carotenes, acetone for chlorophylls, and ethanol for xanthophylls then quantitated using the extinction coefficients in Table I. Cycle activity is indi-

[4] A. Hager and H. Stransky, *Arch. Mikrobiol.* **71,** 132 (1970).
[5] A. Hager and T. Meyer-Bertenrath, *Planta* **69,** 198 (1966).
[6] K. Egger, *Planta* **58,** 664 (1962).

TABLE I
SPECTRAL AND R_f VALUES FOR SELECTED XANTHOPHYLLS AND CHLOROPHYLLS

	In ethanol		R_f	
Pigment	λ_{max} (nm)	Extinction coefficient, $E^{1\%}_{1\,cm}$	Hager–Stransky[a] system	Egger[b,c] system
Violaxanthin	418, 441, 471	2500[d]	0.74	0.84
Lutein	422, 446, 474	2550[e]	0.65	0.56
Antheraxanthin	423, 445, 474	2350[d]	0.52	0.66
Neoxanthin	418, 440, 470	2243[e]	0.43	0.95
Diadinoxanthin	(425), 446, 476	2500[g]	0.41	0.70
Zeaxanthin	(428), 451, 478	2540[e]	0.39	0.56
Diatoxanthin	(427), 451, 478	2500[g]	0.28	0.59
Chlorophyll *a*	663	840[d,f]	—	0.13
Chlorophyll *b*	645	518[d,f]	—	0.25

[a] A. Hager and H. Stransky, *Arch. Mikrobiol.* **73,** 77 (1970).
[b] K. Egger, *Planta* **58,** 664 (1962).
[c] K. Egger, H. Nitsche, and H. Kleinig, *Phytochemistry* **8,** 1583 (1969).
[d] A. Hager and T. Meyer-Bertenrath, *Planta* **69,** 198 (1966).
[e] B. H. Davies, *in* "Chemistry and Biochemistry of Plant Pigments" (T. W. Goodwin, ed.), p. 38. Academic Press, London, 1976.
[f] In acetone.
[g] Nominal value.

cated by stoichiometric changes in pigment concentrations and chlorophyll *a* can be used as an internal reference.

Although pigments which have close partition coefficients such as lutein and zeaxanthin are not resolved, this method has several advantages. Samples can be applied rapidly as a moderately broad band because the pigments concentrate at the starting edge of the fat coating before development begins. Since contact with catalytic surfaces and air is minimal, the pigments are relatively stable. Precoated Kieselguhr-G (Merck) plates can be used. Lack of resolution between lutein and zeaxanthin is not a problem for violaxanthin-cycle studies because lutein does not undergo interconversions.

Spectrophotometric

The absorbance spectra of epoxy carotenoids are blue-shifted from their parent olefins by approximately 4 to 6 nm. This shift gives a characteristic difference spectrum which can be used to detect interconversions

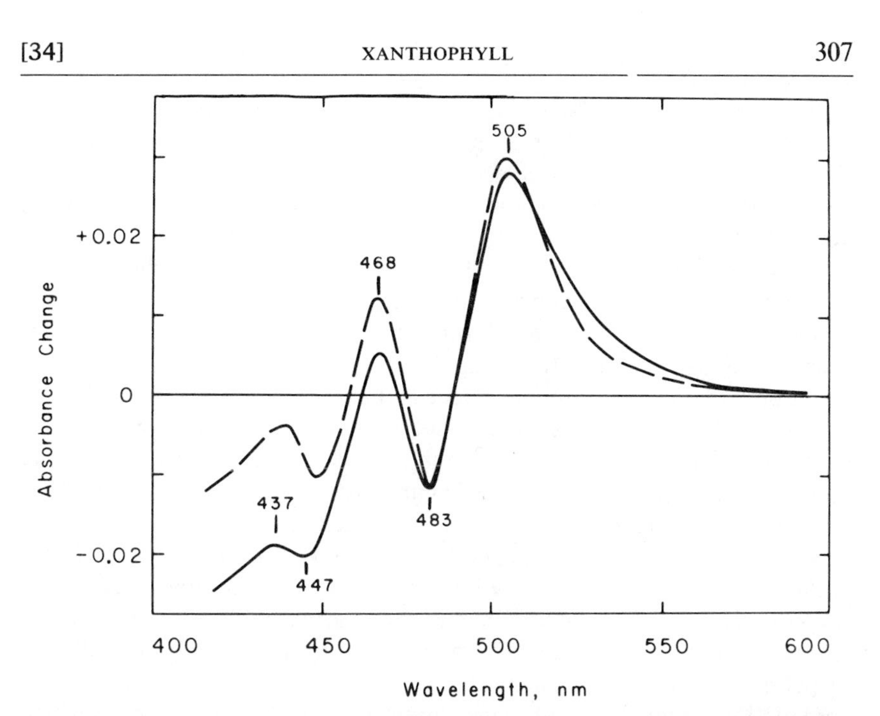

FIG. 2. Difference spectrum of violaxanthin deepoxidation in chloroplasts during (——) and after (-----) continuous illumination. Chloroplasts (41 μg chlorophyll per ml) suspended in SNH solution containing 30 m*M* sodium ascorbate were illuminated with red actinic light (Corning CS2-58 filter) at 2×10^5 ergs cm^{-2} sec^{-1}.

in situ and, when applied as two-wavelength spectrophotometry, provides detailed kinetic data.

Difference spectrophotometry has been employed mainly on the violaxanthin cycle in chloroplasts but in principle could be applicable to other systems, including the diadinoxanthin cycle. A sensitive spectrophotometer with end-on type photomultiplier tube and optics suitable for measurement of light-scattering samples is required. Opal glass, fiber optics, or cuvette stirrers may be used to further reduce nonspecific light-scattering changes.

Difference Spectrum for the Violaxanthin Cycle. Figure 2 shows the difference spectrum for violaxanthin deepoxidation in isolated chloroplasts.[7] Cuvettes are filled with active chloroplasts and the baseline is established. Sodium ascorbate is added to both cuvettes, the sample is illuminated with red actinic light, and the spectrum is scanned until no further change is observed. Deepoxidation is irreversible under these conditions. The difference spectrum represents primarily the absorbance

[7] H. Y. Yamamoto, L. Kamite, and Y. Wang, *Plant Physiol.* **49,** 224 (1972).

difference between violaxanthin and zeaxanthin, the contribution by antheraxanthin being small presumably because its conversion to zeaxanthin is rapid.[8] Opal glass is placed directly before and behind the cuvettes and a Corning filter CS-96 is used to shield the photomultiplier tube from actinic light.

It is known that violaxanthin deepoxidation is light independent at pH 5.[9] The difference spectrum for deepoxidation thus can also be obtained by balancing cuvettes with a chloroplast suspension in pH 5 citrate buffer then adding ascorbate to the sample cuvette.

Deepoxidation difference spectra has been observed in intact leaves.[10,11] Because leaf areas are heterogeneous, balancing absorbance by conventional double-beam techniques is difficult. It is more readily done with a computerized single beam instrument.[12] The spectra thus obtained have agreed with those from isolated chloroplasts.

Two-Wavelength Assay for the Violaxanthin Cycle. Kinetics of violaxanthin cycle activities in isolated chloroplasts can be followed readily by two-wavelength spectrophotometry. Measuring and reference wavelengths selected from the difference spectrum are at 505 and 540 nm, respectively. Absorbance increases for deepoxidation and decreases for epoxidation. Opal glass and a photomultiplier filter are used as described above. The difference extinction coefficient for the change at 505 nm as violaxanthin is 27.3 mM^{-1} cm^{-1}.[13] Continuous mechanical stirring of the reaction mixture is essential for baseline stability.

Violaxanthin Cycle in Isolated Chloroplasts

Chloroplast Isolation

Various isolation methods have been used in xanthophyll studies.[9,11,13] While differing in detail, they have common elements. For optimal activity fresh starting material is used, isolations are done in the cold (0–4°) and as rapidly as possible, whole chloroplasts (class 1) are isolated, followed by osmotic shock if envelope-free thylakoids are needed, and ascorbate is used to keep polyphenols reduced. Deepoxidation activity is relatively easily isolated, requiring mainly the preservation of an active

[8] H. Y. Yamamoto and R. M. Higashi, *Arch. Biochem. Biophys.* **190,** 514 (1978).

[9] A. Hager, *Planta* **89,** 224 (1969).

[10] R. J. Strasser and W. L. Butler, *Plant Physiol.* **58,** 371 (1976).

[11] D. Siefermann-Harms, J. Michel, and F. Collard, *Biochim. Biophys. Acta* **587,** 315 (1980).

[12] W. L. Butler, this series, Vol. 24, p. 3.

[13] D. Siefermann and H. Y. Yamamoto, *Biochim. Biophys. Acta* **357,** 144 (1974).

proton pump.[9] Epoxidation activity requires, in addition, avoiding formation of or removing free fatty acids from the reaction mixture.[14]

The following procedure has been used routinely in this laboratory.[13] Isolations are carried out at 0–4° under green light. About 8 g of washed and deribbed fresh lettuce leaves (*Lactuca sativa*) in 50 ml cold SNHA solution (400 m*M* sorbitol, 10 m*M* NaCl, 50 m*M* HEPES, and 16 m*M* sodium ascorbate) are homogenized for 5 sec in a semimicro Waring blendor, filtered through 37-μm-mesh nylon cloth and centrifuged by accelerating the rotor to 1500 *g* and stopping it quickly by hand. The pellet is washed in 50 ml SNHA, centrifuged as before, and finally resuspended in 1–2 ml SNHA. Chloroplasts thus prepared have retained full activity for up to 3 hr.

Reversible Violaxanthin Cycle in Chloroplasts

The violaxanthin cycle is a transmembrane process wherein deepoxidation occurs on the loculus side and epoxidation on the stroma side of the membrane.[1] The complete cycle has been observed in both spinach[15] and lettuce[14] chloroplasts. Isolated lettuce chloroplasts (0.1 ml concentrated suspension) are osmotically shocked for 15 sec in 1 ml water containing 0.1% BSA (defatted), then mixed with 1.9 ml reaction medium to a final concentration of 400 m*M* sorbitol, 50 m*M* HEPES pH 7.2, 16 m*M* sodium ascorbate, 0.03% BSA, 0.5 m*M* NaDPH, and 12 μg chlorophyll. The kinetics of violaxanthin cycle changes are followed by the two-wavelength procedure. Actinic light (670 nm) is at saturating intensity (36 kerg cm^{-2} sec^{-1}).

Figure 3 shows a typical result.[14] Deepoxidation proceeds in light and reverses in dark. The maximal deepoxidation under continuous illumination is the same as after a light–dark cycle. BSA, NaDPH, and O_2 are required. Deepoxidation follows first-order kinetics and hence activities should be based on the corresponding rate constant if the initial substrate concentration varies. The initial substrate concentration available for deepoxidation varies at light intensities below saturation.[13] Initial rates may be used if light is saturating.

Violaxanthin Deepoxidase

Only violaxanthin deepoxidase has been isolated and partially purified.[8,16] Zeaxanthin epoxidase apparently is membrane bound and no re-

[14] D. Siefermann and H. Y. Yamamoto, *Biochem. Biophys. Res. Commun.* **62,** 456 (1975).
[15] A. Hager, *Ber. Dtsch. Bot. Ges.* **88,** 27 (1975).
[16] A. Hager and H. Perz, *Planta* **93,** 314 (1970).

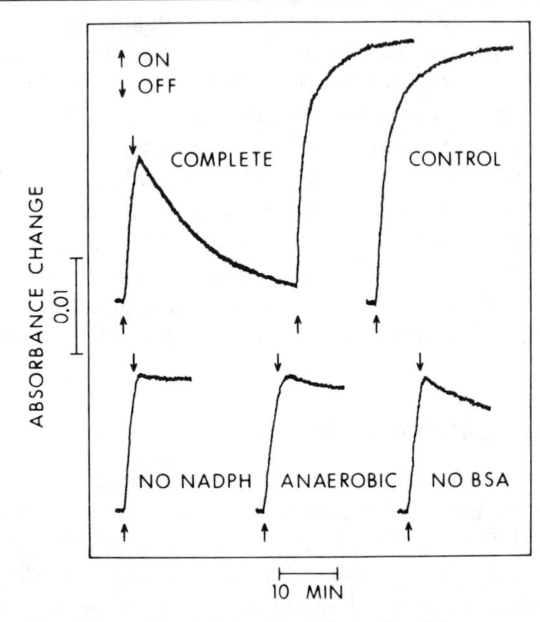

FIG. 3. Reversible violaxanthin cycle in chloroplasts. Conditions were as described in the text. Anaerobiosis was achieved with glucose oxidase-catalase.

ports have appeared on its successful purification. The deepoxidase assay is based on the fact that pure violaxanthin is a poor substrate unless suspended with monogalactosyldiacylglycerol.[17]

Assay[8]

Reagents

10 μM violaxanthin in methanol
270 μM monogalactosyldiacylglycerol (MG) in methanol
0.2 M sodium citrate buffer, pH 5.1
3 M sodium ascorbate

Procedure. The reaction mixture is prepared by pouring 1.5 ml of buffer rapidly into a premix containing 0.10 ml each of violaxanthin and MG, then adding the balance of water needed for a 3 ml final reaction volume. Enzyme (up to 1.2×10^{-3} units in 100 μl) is added, and after the baseline is established, the reaction is initiated by adding 30 μl of ascorbate. The enzyme unit is defined as 1 μmol violaxanthin deepoxidized per

[17] H. Y. Yamamoto, E. E. Chenchin, and D. K. Yamada, *in* "Proceedings of the Third International Congress on Photosynthesis" (M. Avron, ed.), p. 1999. Elsevier, Amsterdam, 1975.

minute. Activity is determined from the initial rate of the absorbance change at 502 nm *minus* 540 nm using an extinction coefficient of 63 mM^{-1} cm^{-1}.

For optimum activity, violaxanthin must be all-trans in the polyene chain. The spectrum of optimally active violaxanthin has nearly equal absorbance at 441 and 471 nm in ethanol, the absorbance ratio being 1.022 or less. Partially isomerized mixtures can be repurified by chromatography on silica gel thin-layer plates using 25% acetone in light petroleum as developing solvent. A 10 μM violaxanthin solution in methanol has 1.5 absorbance at 441 nm. MG is isolated from chloroplasts.[18] Some commercial MG preparations are hydrogenated for stability and are unsuitable for this assay.

Purification

Violaxanthin deepoxidase has been purified from spinach[16] and lettuce[8] by similar strategies. The procedure described for lettuce uses an additional pH 5.2 extraction step that takes advantage of the pH-dependent differential binding of the enzyme to the membrane.

All steps are carried out at 0 to 4°. Chloroplasts are isolated from 400 g of lettuce leaves (*Lactuca sativa*, var. Romaine), washed, and sonicated (Biosonk IV) in citrate buffer pH 5.2 for 4 min, stopping periodically to cool the preparation, and centrifuged at 37,000 *g* for 30 min. The resultant pellet is suspended in 100 ml 0.18 *M* sodium phosphate buffer (pH 7.2) and sonicated as before for 4 min. Membrane debris is removed by centrifugation at 37,000 *g*, soluble pigment proteins removed by passage through a DEAE-Sephadex A-25 column (10 × 2.5 cm) previously equilibrated with 0.18 M phosphate buffer, and the eluate fractionated between 0.5 to 0.7 saturation of ammonium sulfate. The deepoxidase is further purified on a 80 × 2.5-cm column of Sephadex G-100 with 0.1 *M* phosphate buffer, pH 7.2. The deepoxidase elutes at retention volume equivalent to 60,000 Da. Active fractions are located by enzyme assay which is many fold more sensitive than 280 nm absorbance.[8] Fractions containing more than 50% of the peak activity are pooled. Typical preparations yield about 1 unit of deepoxidase at 0.6 units per mg specific activity. Active fractions may be stored for long periods with little activity loss by freezing small aliquots under liquid nitrogen and holding at −60°.

Properties

Violaxanthin deepoxidase has an apparent molecular weight of 5.5 to 6.0×10^4 and a pH 5.1 optimum.[16] Ascorbate is required for activity.

[18] C. F. Allen, this series, Vol. 23, p. 523.

TABLE II
SUBSTRATES FOR VIOLAXANTHIN DEEPOXIDASE[a]

Xanthophyll	Difference spectrum (λ_{max}, nm)	Difference extinction coefficient, E (mM^{-1} cm^{-1})[b]	Relative rate[c]
Violaxanthin	502,468,436	63	1.00
Antheraxanthin	508,471,438	20	5.25
Neoxanthin	499,467,437	25	0.10
Diadinoxanthin	506,475,441	37	2.25
Lutein epoxide	498,466,435	32	1.33
β-Cryptoxanthin epoxide	504,470,438	22	0.92

[a] Adapted from H. Y. Yamamoto and R. M. Higashi, *Arch. Biochem. Biophys.* **190,** 514 (1978).
[b] Difference extinction coefficients are for the longest wavelength peak with reference wavelength at 540 nm.
[c] Rates are relative to violaxanthin, normalized to equivalent epoxide content.

Purified violaxanthin, free of contaminating lipid, is inactive.[17] The presence of MG activates purified violaxanthin which can then be completely converted to zeaxanthin. The deepoxidase itself contains MG which is essential for activity.[8] Also present are plastoquinones but they do not appear to be required for activity.[19]

Violaxanthin deepoxidase is stereospecific for 3-hydroxy, 5,6-epoxy carotenoids in a 3*S*, 5*R*, 6*S* configuration and all-trans in the polyene chain. Purified violaxanthin deepoxidase from chloroplasts can deepoxidize other epoxy carotenoids which conform to the above configurations. Table II is a summary of the activities and difference spectra for various active carotenoids. In case of violaxanthin the difference spectrum in the assay is blue-shifted by 3 nm compared to isolated chloroplasts. The difference extinction coefficient for violaxanthin deepoxidation in the assay is also higher. Accordingly the *in situ* difference spectrum for diadinoxanthin deepoxidation can be expected to be red-shifted and to have a lower extinction coefficient than those given in Table II. Neoxanthin in Table II is the all-trans form and not the naturally occurring 9-*cis*-neoxanthin which is inactive.

[19] T. Fan and H. Y. Yamamoto, unpublished (1980).

[35] Biosynthesis of Plastoquinone

By J. F. PENNOCK

The biosynthesis of plastoquinone has been established by studies with whole plants but in particular with chloroplasts of higher plants.[1] The pathway involves three main biosynthetic precursors. Tyrosine is converted to homogentisate which provides the quinone ring and one nuclear methyl group while *S*-adenosylmethionine is the immediate precursor of the other nuclear methyl group. Nonaprenyl (solanesyl) pyrophosphate is formed by the normal isoprenoid pathway to provide the side chain of plastoquinone. The pathway is shown in Fig. 1 and has been reviewed recently.[2] There is only one main intermediate in the biosynthetic route. Homogentisate decarboxylase–nonaprenyltransferase brings about condensation of homogentisate and nonaprenyl pyrophosphate, with the concomitant loss of CO_2 and pyrophosphate, to yield 2-methyl-6-nonaprenylbenzoquinol. This compound and the related octaprenyl isoprenologue have been found in high concentration in *Iris hollandica* bulbs.[3]

To study plastoquinone biosynthesis it is essential to use radioactive procedures and a variety of radioactivity labeled precursors may be used with different plant tissues. The radioactive precursors used are not absolutely specific and so a careful separation procedure is required to purify the biosynthesized plastoquinone and its precursor. This chapter describes several ways of examining plastoquinone biosynthesis.

Isolation of Marker Quinones and Quinols

Small amounts of plastoquinone and 2-methyl-6-nonaprenylbenzoquinol (this may be referred to more conveniently by the trivial name 2-demethylplastoquinol) as well as the quinone related to the latter compound, i.e., 2-demethylplastoquinone are required as markers for chromatographic separations or as carrier compounds in the isolation procedure. Any green leaf will be a reasonable source of plastoquinone but in general the darker the leaf, i.e., the greater the chlorophyll concentration, then the higher will be the amount of plastoquinone present. In a plant

[1] K. G. Hutson and D. R. Threlfall, *Biochim. Biophys. Acta* **632,** 630 (1980).
[2] J. F. Pennock and D. R. Threlfall, *in* "Biosynthesis of Isoprenoid Compounds" (J. W. Porter and S. L. Spurgeon, eds.), p. 191. Wiley, New York, 1983.
[3] C. Etman-Gervais, C. Tendille, and J. Polonsky, *Nouvo Cimento* **1,** 323 (1977).

METHODS IN ENZYMOLOGY, VOL. 110

ISBN 0-12-182010-6

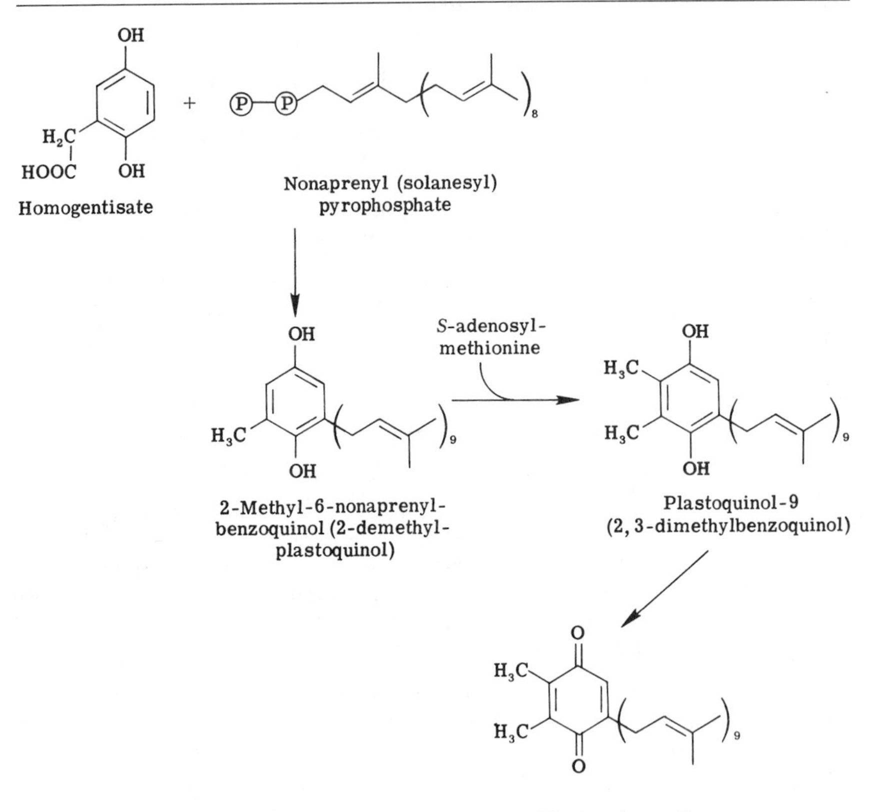

FIG. 1. Biosynthesis of plastoquinone.

such as *Ficus elastica* with dark green leaves, 1 mg of plastoquinone can be obtained from about 10 g of leaves (fresh weight).

2-Demethylplastoquinol can be obtained from *Iris hollandica* bulbs. A survey of other varieties of *Iris* rhizomes and of bulbs from several different plants showed that only the bulbs of *Iris hollandica* contained 2-demethylplastoquinol (and the related quinone).[4] The amount of 2-demethylplastoquinol present is about 50 μg/g fresh weight and there is an approximately equal amount of plastoquinone present.

Extraction of Plant Tissue

About 10 g of chopped tissue (leaf or *Iris* bulb) is macerated in 50 ml of acetone (distilled before use) and filtered through glass wool or a sintered

[4] K. E. Belcher and J. F. Pennock, unpublished observations.

glass funnel into a 250 ml separating funnel. An equal volume of light petroleum ether (bp 40–60° and distilled before use) is added and then water is added carefully to form two layers. The lower aqueous acetone layer is removed and the upper light petroleum ether layer is washed twice with water to remove any remaining acetone. The light petroleum ether is decanted carefully from the separating funnel and evaporated to dryness preferably in a vacuum or in a stream of oxygen-free nitrogen. If the tissue extracted is chlorophyll-containing the extraction should be carried out as quickly as possible, avoiding sunlight or other bright lights to restrict photooxidation of the lipid components.

Purification of Quinones and Quinols

Small amounts of the required compounds can be purified from total lipids by thin-layer chromatography. Although solvent systems and R_f values will be given, it is rare to find such values reproducible in different laboratories and therefore they should be used only as guidelines. The lipid should be examined analytically by TLC and by the use of various stains. Small aliquots of lipid (equivalent to about 0.2 g of tissue) are applied to silica gel GF_{254} plates as small lines (about 0.5 cm). The plates are developed in 6% diethyl ether in light petroleum (the ether is dried over sodium/lead alloy and distilled over "reduced" iron immediately prior to use). In this system both plastoquinone (R_f 0.60) and 2-demethylplastoquinone (R_f 0.43) can be detected readily under UV light (254 nm).[4a] Confirmation of the quinone nature of compounds can be obtained by spraying the chromatogram with leucomethylene blue in ethanol. To prepare the leucomethylene blue, dissolve about 20 mg of methylene blue in 10 ml ethanol and add 50–100 mg of zinc dust and a few drops of 1 *M* HCl. The mixture is stirred and allowed to stand for a few minutes. After spraying the chromatogram (taking care not to disturb the sedimented zinc) quinones appear as bright blue spots and should be marked immediately as the leucomethylene blue quickly oxidizes in air to return to a blue color over the whole chromatogram. In the 6% diethyl ether/light petroleum system 2-demethylplastoquinol will not be separated from polar compounds near the origin.

Similar aliquots of lipid can be chromatographed on silica gel GF_{254} plates using 20% diethyl ether/light petroleum as solvent. In this system plastoquinone has an R_f of 0.85 and 2-demethylplastoquinone 0.75. 2-Demethylplastoquinol (R_f 0.30) is not easily detected under UV light unless present in large amounts because of its relatively low molecular absorbance coefficient but its presence can be shown by use of the

[4a] Phylloquinone (vitamin K_1) is a possible congener (R_f 0.54) which can be purchased.

"Emmerie-Engel reagent" i.e., equal volumes of 0.2% ferric chloride in ethanol and 0.2% 2,2-dipyridyl in ethanol (mixed immediately before use). Reducing compounds such as quinols or chromanols produce a pink to red color with this reagent. In lipid from *Iris hollanidica* bulbs, 2-demethylplastoquinol gives a strong spot with this reagent (R_f 0.30) and two further reducing materials, α-tocopherol (R_f 0.55) and plastochromanol (R_f 0.52) are also present. The chromatogram can be stained further by spraying with a 10% ethanolic solution of phosphomolybdate which shows all lipids as blue spots after heating to 100° for a few minutes.

Once the identity of the various quinones and quinols has been established the remaining lipid can be subjected to preparative TLC using the above systems (10 mg of lipid on a 0.25/0.30-mm-thick, 20 × 20-cm film of silica gel). The various compounds can be eluted from silica gel with diethyl ether. In hexane plastoquinone shows λ_{max} 255 nm (ε mol 18030) and a slightly lower maximum at 263 nm, 2-demethylplastoquinone shows λ_{max} 252.5 nm with a shoulder at 260.5 nm and 2-demethylplastoquinol has a maximum at 290 nm (no quantitative values are available for the last two compounds). 2-Demethylplastoquinol can be converted into 2-demethylplastoquinone by oxidation with silver oxide. The quinol is dissolved in diethyl ether and 50–100 mg silver oxide is added and the mixture is stirred at room temperature for 1 hr. The silver oxide can be removed by filtration or centrifugation and the resulting lipid (after evaporation of the diethyl ether) can be separated using the 6% diethyl ether/light petroleum ether solvent system.

High-Performance Liquid Chromatography

Final purification of the quinones is best achieved on HPLC. Most adsorption systems will give good separations of the various quinones and the system used in the author's laboratory is on an Altex Ultrasphere Silica 5 μm column (4.6 × 250 mm) using 2% methyl *t*-butyl ether in hexane solvent. At a flow rate of 1 ml/min plastoquinone is eluted at 7.8 min and 2-demethylplastoquinone is eluted at 9.7 min.

Incubations of Seedlings with Radioactive Precursors

Plastoquinone is a chloroplast component but not all green photosynthetic tissue is suitable for a study of plastoquinone biosynthesis. In general very young actively growing tissue is preferable for studies, in mature or dark green leaves most isoprenoid synthesis is proceeding at a very low rate. This is probably because isoprenoid components are present in very high levels in mature leaves and there is considerable storage of isoprenoid quinones, such as plastoquinone, in osmiophilic globules within the chloroplast. Cereal shoots such as barley, wheat, and maize used between 5 and 10 days after germination are suitable or other rapidly growing seedlings such as broad beans or French dwarf beans are equally successful but must be older, i.e., about 14 days after germination.

In using whole tissue the shoots are cut (about 20–50 g tissue) and the stems are placed in a small beaker containing an aqueous solution of the radioactive precursor. A flow of air across the seedling is helpful to ensure transpiration and therefore rapid uptake of the precursors. When all the solution has been taken up by the plant more water is added and the incubation can be left for up to 24 hr. Either light or dark conditions can be attempted.

An early approach to plastoquinone biosynthesis involved the use of etiolated seedlings for the experiment which was carried out in the light so that "greening up" occurred during the course of the experiment.[5] This procedure has the advantage of encouraging particularly high incorporation of precursor because of the very large formation of chloroplast components during the "greening up" process. However the experiment utilizes relatively artificial conditions and clearly does not indicate normal levels of plastoquinone formation.

Precursors of the isoprenoid side chain such as [2-^{14}C]mevalonate may be used to study plastoquinone biosynthesis but many isoprenoid compounds are radiolabeled with this precursor and incorporation is relatively low. However the precursor is readily available and if 10–20 μCi of mevalonate is used per incubation then incorporation into plastoquinone is clearly demonstrable. As precursors of the benzoquinone ring of plastoquinone, either radiolabeled tyrosine or homogentisate may be used. Either [U-^{14}C]- or [α-^{14}C]tyrosine is suitable as a precursor and surprisingly DL-tyrosine is a better precursor than L-tyrosine.[6] This is presumably because the D-isomer is degraded to homogentisate in greater proportions than L-tyrosine which will be utilized for protein synthesis. The most specific precursor for plastoquinone is homogentisate which can be prepared from L-[U-^{14}C]tyrosine[7] or [*methylene*-^{3}H]homogentisate which can be obtained from the Radiochemical Centre, Amersham, U.K. as a custom preparation.[1] The main components labeled from either radiolabeled tyrosine or homogentisate are plastoquinone, α- and γ-tocopherol, and α-tocopherolquinone but in some cases there is a general incorporation into all isoprenoids albeit at a lower level. This is because homogentisate can be degraded to acetoacetate and fumarate and the former of these two can

[5] G. R. Whistance and D. R. Threlfall, *Biochem. J.* **109,** 577 (1968).
[6] G. R. Whistance and D. R. Threlfall, *Biochem. J.* **117,** 593 (1970).
[7] D. R. Threlfall and G. R. Whistance, this series, Vol. 18, p. 369.

be used for mevalonate formation. About 20 μCi of one of these precursors is adequate for a study of plastoquinone biosynthesis. Finally biosynthesis can be studied in seedlings using L-[*Me*-^{14}C]methionine.[5] Plastoquinone is highly labeled with this precursor but α- and γ-tocopherols, ubiquinone, phylloquinone and 3β-hydroxysterols are also labeled. Obviously 2-demethylplastoquinone or quinol will not be labeled using this precursor.

Incubations with Chloroplast Preparations

Perhaps the most convenient method of studying plastoquinone biosynthesis is to use chloroplast preparations.[1,8] Both intact chloroplasts[1] and chloroplast subfractions[8] have been used to study plastoquinone biosynthesis and both have been successful. The chloroplast preparation as described by Huston and Threlfall[1] is made by macerating (MSE top drive blender) leaves of lettuce, spinach, or pea seedlings (previously washed with distilled water) with ice-cold suspension buffer (10 g leaf/30 ml) for 2 × 10 sec. The suspension buffer is 30 m*M* HEPES/0.5 *M* sucrose/5 m*M* cysteine/1 m*M* $MgCl_2$/1 m*M* EDTA/0.2% (w/v) bovine serum albumin fraction V adjusted to pH 7.6 with KOH. The homogenate is filtered through cheesecloth and centrifuged at 500 *g* for 90 sec and the supernatant centrifuged at 1500 *g* for 5 min. The pellet is resuspended in suspension medium and resedimented at 1500 *g* for 5 min and finally suspended in suspension buffer (1 mg chlorophyll/ml). Chlorophyll may be assayed by the method of Arnon.[9]

The assay mixture is 2.3 ml suspension buffer 0.3 ml chloroplast preparation; 0.1 ml 10 m*M* trilithium isopentenyl pyrophosphate; 0.1 ml 5 m*M* *S*-adenosylmethionine; 0.1 ml 100 m*M* $MgCl_2$ and 20 μl of ethanolic [*methylene*-^{3}H]homogentisate (2 μCi). The assay is incubated at 30° with shaking for 60 min. In experiments described by Hutson and Threlfall[1] the assays were illuminated for the period of incubation with 500 lux (150 W reflector lamp).

Extraction of lipids at the end of the incubation period and isolation of plastoquinone is essentially as described under *Extraction of plant tissue*. If *S*-adenosylmethionine is omitted from the assay medium there is a greater accumulation of radioactivity in the demethylplastoquinone precursor and less in plastoquinone.

Soll and co-workers[8] used broken chloroplasts or chloroplast subfractions to study plastoquinone biosynthesis. Broken chloroplasts prepared

[8] J. Soll, M. Kemmerling and G. Shultz, *Arch. Biochem. Biophys.* **204,** 544 (1980).

[9] D. I. Arnon, *Plant Physiol.* **24,** 1 (1949).

according to Jensen and Bassham[10] were incubated in 0.05 *M* HEPES; 0.005 *M* $Na_4P_2O_7 \cdot H_2O$; 20 μM [^{3}H]homogentisic acid (tritium exchange service of New England Nuclear); 2.5 m*M* $NaHCO_3$; 80 μM nonaprenyl pyrophosphate (or 80 μM nonaprenol plus 100 μM ATP) and adjusted to pH 7.6. Addition of ATP and nonaprenol brings about a satisfactory synthesis of nonaprenyl pyrophosphate. Using this system synthesis of 2-demethylplastoquinol can be shown.

Chloroplast envelope membrane can be used to show formation of 2-demethylplastoquinol from [^{3}H]homogentisate and nonaprenyl pyrophosphate and also the formation of plastoquinol from 2-demethylplastoquinol and *S*-[*Me*-^{14}C]adenosylmethionine.[8] The envelope membrane can be prepared by fractionation of intact chloroplasts on a biphasic dextran gradient.[11] After osmotic shock envelope membranes are isolated by sucrose gradient centrifugation.[12] The incubation medium is as described for broken chloroplasts but without the $NaHCO_3$. When 2-demethylplastoquinol is used as substrate it is 100 μM added in 25 μl of ethanol. The incubations are kept at 20° in the dark for 80 min. Products can be determined as before.

Experiments have been described as requiring light[1] or not requiring light[8] and it is likely that this discrepancy can be attributed to whether sufficient ATP and NADPH are present for biosynthetic purposes.

Possible Areas for Study

The main aspects of plastoquinone synthesis as a biochemical pathway have been worked out but little is known of turnover, the importance of different pools within the chloroplast or regulation. Although studies of biosynthesis with chloroplast, or chloroplast fragments are the most convenient way of studying the effect of varying conditions on plastoquinone synthesis, regulation of biosynthesis requires an assay system using intact cells. The biosynthesis of plastoquinone takes place in the chloroplast but it is likely that some enzymes concerned with the pathway are of cytosolic ribosomal origin and thus studies with intact leaves may still have major roles to play.

[10] R. G. Jensen and J. A. Bassham, *Proc. Natl. Acad. Sci. U.S.A.* **56,** 1095 (1966).

[11] C. Larson, B. Anderson, and G. Roos, *Plant Sci. Lett.* **8,** 291 (1977).

[12] R. Douce and J. Joyard, *in* "Plant Organelles" (E. Reid, ed.), p. 47. Ellis Harwood, Chichester, 1979.

[36] Aspulvinone Dimethylallyltransferase

By IKUKO SAGAMI, NOBUTOSHI OJIMA, KYOZO OGURA, and SHUICHI SETO

Aspulvinones are yellow pigments isolated from *Aspergillus terreus*.[1-3] They include dihydroxypulvinone (aspulvinone E) and trihydroxypulvinone (aspulvinone G) and their prenylated derivatives. Dimethylallyl pyrophosphate:aspulvinone dimethylallyltransferase has been purified from mycelia of *A. terreus*.[4]

Cell Culture

Aspergillus terreus IAM 2054 (origin NRRL 1960 Raper ATCC 10020) are grown at 27° on Czapek Dox medium in stationary culture. Three-liter stationary culture flasks containing 1 liter of medium are inoculated with 20 ml of a mycelial suspension from a 3-day-old shaking culture. After the culture has grown stationarily for 1 week, mycelia are harvested by filtration, washed with water, and used for the enzyme preparation.

Assay Methods

Principle. The prenylated product synthesized from aspulvinone E and dimethylallyl pyrophosphate, either one of which is radiolabeled, in the presence of enzyme is extracted with ether and assayed for radioactivity. Neither aspulvinone E nor dimethylallyl pyrophosphate is extractable at neutral pH with ether. Aspulvinone E can be replaced by aspulvinone G which is also prenylated under the same conditions.

Reagents

[G-^{14}C]Aspulvinone E, 1 m*M*
Dimethylallyl pyrophosphate, 10 m*M*
Tris–HCl buffer, pH 7.0, 1 *M*
The same buffer containing 2 *M* NaCl

[1] N. Ojima, S. Takenaka, and S. Seto, *Phytochemistry* **12,** 2527 (1973).
[2] N. Ojima, S. Takenaka, and S. Seto, *Phytochemistry* **14,** 573 (1975).
[3] N. Ojima, I. Takahashi, K. Ogura, and S. Seto, *Tetrahedron Lett.* p. 1013 (1976).
[4] I. Takahashi, K. Ogura, and S. Seto, *Biochemistry* **17,** 2696 (1978).

ISBN 0-12-182010-6

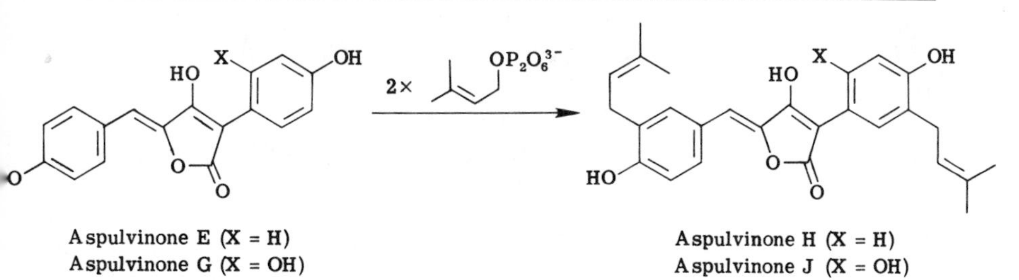

SCHEME I.

Procedure. To a screw-capped test tube are added 0.04 ml of [G-^{14}C]aspulvinone E, 0.02 ml of dimethylallyl pyrophosphate, 0.05 ml of Tris–HCl buffer, 0.10 ml of enzyme solution (usually less than 0.05 mg protein/ml), and sufficient water so that the final volume will be 1.0 ml. After the mixture is incubated at 37° for 1 hr, 1.0 ml of Tris–HCl buffer and 4.0 ml of diethyl ether are added, and the mixture is shaken. The ether layer is taken and washed with 1 ml of the same buffer containing 2 *M* NaCl, and an aliquot of the ether extract is removed for determination of radioactivity. The enzyme activity is expressed in terms of the amount of radioactivity found in the ether extract. An enzyme unit is 1 μmol aspulvinone incorporated into product per minute. For the reaction with crude enzyme, 10 μmol of potassium fluoride is added to the above incubation mixture to inhibit phosphatase activity.

Preparation of [G-^{14}C]Aspulvinone. [G-^{14}C]Aspulvinone E is prepared biosynthetically from [^{14}C]tyrosine according to the following procedure. A 1-ml solution of L-[U-^{14}C]tyrosine (50 μCi/ml, 0.103 μmol) is added to 100 ml of a 3-day-old culture of *A. terreus*. After 3 weeks of stationary culture, the medium is filtered and the filtrate was repeatedly extracted out at pH 7.0 with ether. The resulting aqueous layer is adjusted to pH 2–3 with hydrochloric acid and is extracted with ether until yellow material is no longer transferred into the ether layer. The ether layer is concentrated, and the extract is subjected to preparative silica gel thin-layer chromatography with a solvent system of ether–hexane–acetic acid (40 : 8 : 1, v/v). Radioactive regions of the developed plate are located with a radiochromatoscanner, and the section corresponding to aspulvinone E (R_f 0.28) is scraped and extracted with ether. The radioactive aspulvinone E thus obtained is further purified on a column (8 × 500 mm) packed with porous polymer (Hitachi Gel 3011) with methanol–hexane–acetic acid (180 : 20 : 1, v/v) at a flow rate of 1.5 ml/min. The elution of aspulvinones is monitored by recording the absorbance at 370 nm. [^{14}C]Aspulvinone E, 6.7 mg (specific activity, 5.3 × 10^5 dpm/μmol), is obtained in 11% yield based on [^{14}C]tyrosine added. The radioactive aspulvinone E is dissolved in 0.01 *M* Tris–HCl buffer (pH 7.4) and used as substrate for the enzymatic reaction.

Dimethylally Pyrophosphate. This substrate is prepared as described by Poulter (this volume [15]) or by the method described by Nishino *et al.*[5]

Purification of Enzyme

All steps are carried out at 0–4° unless otherwise stated.

Step 1. Extraction. Mycelia (200 g) of 7-day-old cultures are ground well in a mortar with HCl washed sea sand (200 g) and 800 ml of 0.05 *M* phosphate buffer, pH 6.8, until the mixture becomes a slurry. To remove cell debris and sea sand, the slurry is centrifuged at 26,000 *g* for 30 min, and the supernatant is centrifuged at 108,000 *g* for 30 min. The resulting supernatant is transferred to a beaker chilled on ice.

Step 2. Ammonium Sulfate Fractionation. To the above supernatant (750 ml), granulated ammonium sulfate (132 g) is slowly added with stirring to achieve 30% saturation and the mixture is stirred for 30 min. The resultant precipitate is removed by centrifugation at 26,000 *g* for 30 min, and then the supernatant (800 ml) is brought to 55% saturation by further addition of ammonium sulfate (130 g). After standing for 30 min, the precipitate is collected by centrifugation at 26,000 *g* for 30 min and dissolved in a minimum volume of 0.05 *M* phosphate buffer (pH 6.8).

Step 3. DEAE-Sephadex A-50 Chromatography. The solution from step 2 is filtrated through a Sephadex G-25 column (2.0 × 45 cm) with 0.05 *M* phosphate buffer (pH 6.8). In the case of a large-scale preparation, the solution is dialyzed overnight against 3–5 liters of the same buffer. The resulting protein fraction is referred as "crude enzyme." The crude enzyme is applied to a DEAE-Sephadex A-50 column (1.2 × 25 cm) previously equilibrated with 0.05 *M* phosphate buffer (pH 6.8) containing 0.1 *M* KCl. Elution is carried out with a linear concentration gradient established between 200 ml of buffer containing 0.1 *M* KCl and 200 ml of buffer containing 0.4 *M* KCl, and 5-ml fractions are collected. Under these conditions, the transferase is eluted from the column at a KCl concentration of about 0.22 *M*. The fractions showing the transferase activity are combined (200 ml) and concentrated to about 4 ml by ultrafiltration using a Diaflo apparatus with a UM-10 membrane with an exclusion limit of M_r = 10,000.

Step 4. Gel Filtration with Sephadex G-200. The enzyme solution from the preceding step is applied to a column (2.5 × 38 cm) of Sephadex G-200, equilibrated with 0.05 *M* phosphate buffer, pH 6.8. Elution is

[5] T. Nishino, K. Ogura, and S. Seto, *J. Am. Chem. Soc.* **94,** 6849 (1972).

PURIFICATION SCHEME

Step	Protein (mg)	Activity		Specific activity (unit/mg)	Purity (fold)
		Unit[a]	Recovery (%)		
108,000 *g* supernatant	4700	1140	100	0.24	1
30–50% ammonium sulfate fraction	658	617	54	0.96	4.0
Dialysis	517	911	80	1.85	7.7
DEAE-Sephadex	103	898	79	7.44	30.9
Sephadex G-200	71	978	86	13.8	57.1
Hydroxylapatite	36	1058	93	29.4	122

[a] A unit of activity is defined as the amount of enzyme which converts 1 nmol of aspulvinone E into the prenylated products per minute.

carried out with the same buffer and 5-ml fractions are collected. In the case of a large scale preparation, it is convenient to subject the sample to Ultrogel AcA-22 filtration prior to Sephadex G-200 filtration to remove proteins of a larger molecular weight.

Step 5. Hydroxylapatite Chromatography. The active pool (70 ml) from step 4 is concentrated to about 8 ml using a Diaflo apparatus and applied to a hydroxylapatite column (1.2 × 21 cm) for further purification. The elution is performed with 100 ml of the same buffer and then with a linear concentration gradient established between 120 ml of 0.05 *M* phosphate and 120 ml of 0.30 *M* phosphate at pH 6.8. Six milliliter fractions are collected. The transferase emerges at about 0.18 *M* phosphate; the chromatographic curves for enzyme activity and protein concentration are coincident.

The fractions containing the transferase are combined and concentrated, and the concentrate is subjected to ultrafiltration to exchange the buffer for 0.05 *M* Tris–HCl buffer (pH 7.0) using a Diaflo apparatus with a UM-10 membrane. The enzyme thus obtained can be stored frozen at −20° without loss of activity for at least 1 week, but 50% of the activity is lost when kept at 4° for 3 weeks.

The table summarizes this purification. After the final step of purification the purity of the transferase is satisfactorily high, as judged by gel electrophoresis. The enzyme activity is reversibly inhibited by ammonium sulfate.

Properties

Products. The enzyme catalyzes the transfer of the dimethylallyl moiety from dimethylallyl pyrophosphate to activated positions ortho to the hydroxyl group of the two aromatic rings of aspulvinone E to give aspulvinone I (monoprenyl derivative) and aspulvinone H (diprenyl derivative) both of which are found as metabolites in *A. terreus*.[3] The amount of aspulvinone I relative to aspulvinone H increases with incubation time. The products of a short-term incubation are aspulvinone I and aspulvinone H, but the former is not detected when the incubation is prolonged. Up to three dimethylallyl groups can be introduced, though the third prenylation is sluggish. The enzyme also catalyzes the prenylation of aspulvinone G which is another metabolite of this series of pigments to give the corresponding prenylated product, aspulvinone J, which has not been found in the fungi but is assumed to be a precursor for aspulvinone C, D, and F.[2]

Aspulvinone I, aspulvinone H, and aspulvinone J are chromatographed on a silica gel plate in a system of ether–hexane–acetic acid (40 : 8 : 1, v/v) with R_f values of 0.50, 0.67, and 0.27, respectively.

Substrate Specificity. Geranyl pyrophosphate does not act as substrate, but several analogues of dimethylallyl pyrophosphate, including *E*-3-methylpent-2-enyl, cyclopentylideneethyl, and cyclohexylideneethyl pyrophosphates, act as substrate to prenylate aspulvinone E at 35% the rate of dimethylallyl pyrophosphate. *E*-Butenyl and cycloheptylideneethyl pyrophosphate also prenylate the acceptor, but the rate is only about 10% that of the normal substrate. Dimethylallyl monophosphate is not active.

Effect of Substrate Concentration. The apparent K_m values obtained from Lineweaver–Burk plots are 40.0, 13.7, and 7.7 μM for dimethylallyl pyrophosphate, aspulvinone E, and aspulvinone G, respectively.

Molecular Weight. The molecular weight of the transferase is estimated to be 240,000–270,000 by Sephadex G-200 filtration. SDS–polyacrylamide gel electrophoresis shows a single protein band corresponding to a molecular weight of 45,000, suggesting that the enzyme consists of six subunits of identical molecular weight.

Behavior on Electrophoresis. When the enzyme purified through the hydroxylapatite chromatography step is subjected to polyacrylamide disc gel electrophoresis, two protein bands (fractions I and II) are observed, the major one (fraction I) running faster. These two protein fractions are bound tightly to bromphenol blue used as the tracking dye, and as a result two greenish blue bands can be discerned without usual staining for protein. When the gel is treated with amido black or Coomasie brilliant blue, two bands are stained, being superimposed on the bands colored with bromphenol blue. The electrophoresis of the crude enzyme, 30–50% ammonium sulfate fraction, also shows a band colored with bromphenol blue at the same distance as that for the major band (fraction I) of the purified

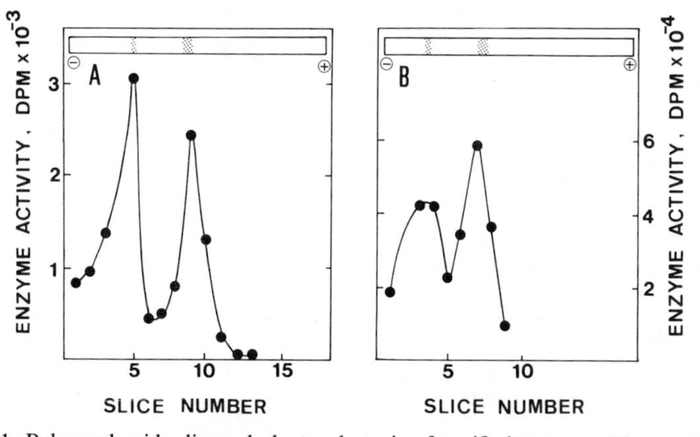

FIG. 1. Polyacrylamide disc gel electrophoresis of purified enzyme. Electrophoresis is performed with 7.5% gels. After electrophoresis, the gel is sliced into 5-mm sections, and the slices are crushed in Tris buffer (pH 7.0) and assayed for the transferase activity with [^{14}C]aspulvinone E and dimethylallyl pyrophosphate (A) and with aspulvinone G and [^{3}H]dimethylallyl pyrophosphate (B).

enzyme, indicating that the coloration with the dye is due to a specific binding to the transferase. In the electrophoresis of the crude enzyme, no band is colored at the region corresponding to fraction II. Both fractions I and II are able to catalyze the prenylation of aspulvinone E and aspulvinone G (Fig. 1). The products obtained by the action of these two enzyme fractions are identical with each other.

The reelectrophoresis of fraction I recovered from the developed gel shows again two protein bands in a similar manner to the first electrophoresis, indicating that fraction II is derived from fraction I. However, the reelectrophoresis of fraction II recovered from the gel shows only one band at the original distance. These fractions appear to be size isomers, as judged by the criteria of Hedrick and Smith,[6] since the log of relative mobility vs gel concentration shows nonparallel lines extrapolating to a point near 0% gel concentration (Fig. 2). Therefore, fraction II may be an aggregated form of fraction I. The molecular weights of fractions I and II are estimated to be approximately 280,000 and 470,000, respectively.

Inhibitory Effect. Bromophenol blue inhibits the enzyme reaction even more strongly than substrate analogs, and the inhibition is of mixed type when aspulvinone E is the varied substrate. The degrees of inhibition by 40, 100, and 200 μM of bromphenol blue are 69, 85, and 95%, respectively. Although bromopyrogallol red is structurally close to bromphenol

[6] J. L. Hedrick and A. J. Smith, *Arch. Biochem. Biophys.* **126,** 155 (1968).

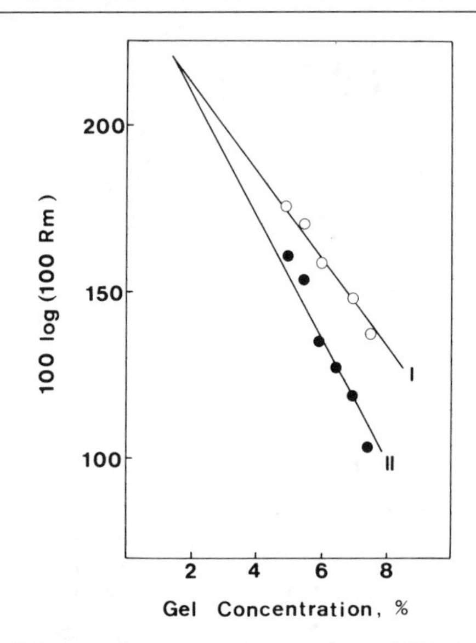

FIG. 2. Effect of different gel concentrations on the mobilities of fractions I and II.

blue, it has little effect. Inorganic pyrophosphate, which is a product of the transferase reaction, shows a strong inhibitory effect. The reciprocal plots of initial rates against the reciprocal concentrations of dimethylallyl pyrophosphate or aspulvinone E shows that the inhibition of inorganic pyrophosphate is of mixed type against dimethylallyl pyrophosphate ($K_i = 17.8\ \mu M$, $K_i = 37.4\ \mu M$) and noncompetitive against aspulvinone E ($K_i = 50\ \mu M$).

Effect of Metal Ions. The enzyme activity is not affected by the addition of Mg^{2+} ion or EDTA, and other divalent metal ions show some inhibitory effect. It is noteworthy that no metal ion is required for enzyme activity, nor is the reaction retarded when EDTA is added to the incubation.

Effect of Detergents. The transferase is activated 2.3 times by 0.025% Tween 80, and 1.5 times by 0.025% Triton X-100, but is 69% inhibited by lecithin of the same concentration.

pH Optimum. The transferase shows the maximum activity at pH 7.0 of Tris–HCl buffer. In phosphate buffer at the same pH, the enzyme shows about 70% of the activity in Tris buffer.

[37] 4-Hydroxybenzoate Polyprenyltransferase from Rat Liver

By ARUN GUPTA and HARRY RUDNEY

4-Hydroxybenzoate polyprenyltransferase is a key enzyme in ubiquinone synthesis because it catalyzes the step which brings together the precursor of the benzoquinone ring and the polyisoprenoid side chain. The substrates are 4-hydroxybenzoate derived from tyrosine metabolism, and polyprenyl pyrophosphates derived from mevalonic acid and the product is 3-polyprenyl-4-hydroxybenzoate (PPHB) and pyrophosphate. The 3-polyprenyl-4-hydroxybenzoate, after hydroxylation, decarboxylation, and methylation, gives rise to ubiquinone.[1,2] The length of polyprenyl side chain may vary from C_{30} to C_{50} depending upon the substrate and the source of the enzyme. This subject has been recently reviewed.[3] In rat liver, a major product of the enzymatic reaction is 3-nonaprenyl-4-hydroxybenzoate, whereas in human and guinea pig, it is 3-decaprenyl-4-hydroxybenzoate.[1-5]

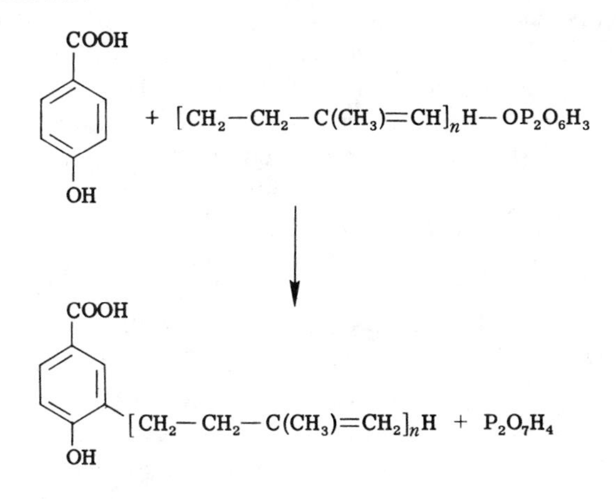

Assay Method

Principle. The enzyme is most easily measured utilizing as labeled substrate 4-hydroxybenzoate and measuring its rate of incorporation into

[1] K. Momose and H. Rudney, *J. Biol. Chem.* **247,** 3930 (1972).
[2] B. L. Trumpower, R. M. Houser, and R. E. Olson, *J. Biol. Chem.* **249,** 3041 (1974).
[3] H. Rudney and R. E. Olson, *Vitam. Horm.* (*N.Y.*) **40,** 1 (1983).
[4] T. Nishino and H. Rudney, *Biochemistry* **16,** 605 (1977).
[5] T. S. Raman, H. Rudney, and N. K. Buzzelli, *Arch. Biochem. Biophys.* **130,** 164 (1969).

ISBN 0-12-182010-6

PPHB. The other substrate, nonaprenyl pyrophosphate (NPP), could also be labeled but is more difficult to obtain. Depending on the absence of PPHB decarboxylase, 4-hydroxybenzoate labeled in the carboxyl position can be used. The reaction mixture is extracted with chloroform leaving unreacted 4-hydroxybenzoate in the aqueous phase.

Reagents

Tricine NaOH buffer 20 m*M*, pH 7.4
Nonaprenyl pyrophosphate (solanesyl pyrophosphate) 10 mg/ml in benzene stock
0.125 *M* EDTA
0.5 *M* Magnesium chloride
4-[*carboxyl*-^{14}C]Hydroxybenzoate (55 mCi/mmol) in distilled water

Procedure. Enzyme activity was assayed by minor modification of the method described earlier.[4] The complete assay mixture contained 10 μl of 0.15% Triton X-100 with or without 6 nmol of nonaprenyl pyrophosphate (20 μl of stock solution was evaporated under a stream of nitrogen and redissolved in 200 μl of 0.15% Triton X-100), 10 μl of 0.125 *M* EDTA, 10 μl of 0.5 *M* magnesium chloride, and 100 μl of inner mitrochondrial membrane fragments in a final volume of 0.5 ml. Tricine buffer (20 m*M*) is used to make up the volume. The order of the addition was as indicated. The tubes were preincubated at 37 or 25° for 5 min. The reaction was started by the addition of 0.25 μCi 4-[*carboxyl*-^{14}C]hydroxybenzoate. After incubation for 30 min, the reaction was stopped by the addition of 2.0 ml of chloroform : methanol (1 : 2 v/v). The tubes were agitated with a Vortex mixer and allowed to stand for 30 min at room temperature. Chloroform (0.5 ml) and 0.5 ml of distilled water were added to each tube. The tubes were agitated again with a Vortex mixer and centrifuged in a clinical centrifuge for 2 min to separate the chloroform layer. This procedure extracts PPHB into the chloroform phase. The chloroform layer was washed two times with 1.0 ml of theoretical upper phase [theoretical upper phase is obtained by mixing 3.75 vol of chloroform : methanol (1 : 2), 1.2 vol of chloroform and 2.2. vol of distilled water]. The chloroform layer was finally transferred to a scintillation vial with one washing of chloroform. After evaporation, either under a stream of nitrogen or by allowing the vial to stand in the fume hood overnight, the radioactivity was measured by scintillation spectrophotometer (Beckman) using scintillation formula 950A with a counting efficiency of approximately 95%.

Assay of Protein. The protein was estimated by the method of Schacterle and Pollock[6] using bovine serum albumin as a standard.

Unit. One unit of 4-hydroxybenzoate polyprenyltransferase is defined

[6] G. R. Schacterle and R. L. Pollock, *Anal. Biochem.* **51,** 6554 (1973).

as pmol of 3-polyprenyl 4-hydroxybenzoate formed per minute under the assay conditions and specific activity is expressed as units per mg of protein.

Isolation of Inner Mitochondrial Membrane Fragments. Adult male Sprague–Dawley-derived rats were killed by decapitation and their livers were rapidly removed and rinsed with isolation buffer (50 m*M* potassium hydrogen phosphate; 100 m*M* sucrose; 50 m*M* potassium chloride; 30 m*M* EDTA; and 3 m*M* dithioerythritol, pH 7.2) and then homogenized with two volumes of the same buffer using a Potter-Elvehjem glass homogenizer and motor driven Teflon pestle. The resulting homogenate was centrifuged at 12,000 *g* for 15 min. The resulting pellets were resuspended in 9 volumes of homogenization buffer (280 m*M* sucrose; 10 m*M* Tris; and 0.1 m*M* EDTA, pH 7.4) and homogenized. Nuclei and cell debris were removed by centrifuging twice at 750 *g* for 10 min. The supernatant was centrifuged at 12,000 *g* for 10 min to sediment mitochondria. The mitochondrial pellets were then homogenized in 5 volumes of homogenization buffer and centrifuged at 12,000 *g* for 10 min. The resulting pellets were suspended in 5 volumes of 20 m*M* tricine–NaOH buffer, pH 7.4, and either used the same day or stored frozen at −20° for preparation of mitochondrial membrane fragments. Inner mitochondrial fragments were prepared by a modification of the method of Parsons and Williams.[7] It was observed that the mitochondrial suspension stored frozen after thawing showed more activity than fresh mitochondria. The frozen mitochondrial suspension was thawed in cold water with gentle mixing and then centrifuged at 9750 *g* for 10 min, the resulting pellets were suspended in one volume of 20 m*M* tricine–NaOH buffer, pH 7.4. This suspension was placed in an ice bath and sonicated twice for 6 min. (Heat Systems—Ultrasonics, Inc., 30% duty cycle, pulsed, output setting of 6 and with a tapered microtip). After sonication, the membrane fragments were centrifuged at 251,000 *g* for 60 min. The pellet was suspended in 2.5 to 3.0 ml of tricine–NaOH buffer, pH 7.4 per liver and sonicated with 20 bursts at the above setting. This mitochondrial membrane fragment preparation was kept on ice until use.

Preparation of Nonaprenyl Pyrophosphate from Solanesol: all-*trans*-Nonaprenyl Alcohol

Nonaprenyl pyrophosphate was prepared from solanesol, obtained from tobacco leaves by the method of Popják *et al.* described earlier.[8]

[7] D. F. Parsons and G. R. Williams, this series, Vol. 10, p. 443.

[8] G. Popják, J. W. Cornforth, R. H. Cornforth, R. Ryhage, and D. S. Goodman, *J. Biol. Chem.* **237,** 56 (1962).

Trichloroacetonitrile, 520 mg (3.6 mmol), was added slowly to 250 mg (0.4 m*M*) of solanesol in a flask fitted with dropping funnel and a stirrer. Diethyl ammonium phosphate, 360 mg (1.25 mmol) dissolved in 50 ml of dry acetonitrile was then added through a dropping funnel over a period of 4 hr. The reaction mixture was kept at room temperature and stirred continuously. After 24 hr, the mixture was concentrated by means of a rotary evaporator. The solid yellow residue was transferred to a centrifuge tube with 10 ml of acetone and to this 0.5–1.0 ml of concentrated ammonium hydroxide was added. The precipitated ammonium salt was separated by centrifugation. The crude preparation of phosphorylated solanesol was dissolved in minimum amount of 1-propanol : ammonia : water (9 : 4 : 1 v/v) and purified on a silica gel column (2.5 × 19 cm) containing 40 g of acid-washed silica gel equilibrated with the above buffer. The column was eluted with the same buffer and 3.0 ml fractions were collected. Fractions 1 to 11 contained unreacted solanesol, fractions 12–52 contained solanesyl monophosphate with a minor contamination of pyrophosphate, and fractions 53–75 contained solanesyl pyrophosphate with a minor contamination of mono- and polyphosphates. Fractions 53–75 were pooled and concentrated by rotary evaporation. Solanesyl pyrophosphate was finally purified on silica gel thin layer plates using chloroform : methanol : water (65 : 25 : 4) as a solvent. The R_f of solanesyl pyrophosphate ranges between 0.10 and 0.13.

In recent years solanesyl pyrophosphate has been made commercially available by the Collection of Polyprenols of the Institute of Biochemistry and Biophysics, Polish Academy of Sciences, Ul. Rakowiecka 36, Warsaw, Poland.

Product Analysis

The lipid extract from the assay reaction mixture was subjected to thin layer chromatography (TLC) on ChromAR 1000 silica gel sheets (Mallinckrodt) with 15% acetone in petroleum ether. The radioactive areas were cut out and extracted with acetone and further analyzed by reverse phase TLC. This was done by impregnating precoated kieselguhr G (Merck) plate with 5% liquid paraffin oil in *n*-hexane. The plates were developed with acetone : water : acetic acid (80 : 20 : 1) saturated with paraffin oil. Radioactive areas on TLC plates were detected with a Packard radiochromatogram scanner, Model 7201.

Properties of the Enzyme

Distribution. 4-Hydroxybenzoate polyprenyltransferase activity is ubiquitous in all cells where ubiquinone is synthesized. In the rat heart,

TABLE I
SUBCELLULAR DISTRIBUTION OF 4-HYDROXYBENZOATE POLYPRENYLTRANSFERASE IN RAT LIVER

Cell fraction	4-Hydroxybenzoate polyprenyltransferase (pmol/min/mg protein)
Homogenate	0.256
Cytosol	0.016
Swollen mitochondria	1.48
Inner mitochondrial membrane fragment	1.264
Outer mitochondria	0.155
Microsomes	0.13

kidney, and liver are rich sources of enzyme with heart containing maximum polyprenyltransferase activity.[9,10] Rat liver preparations were used to locate the subcellular distribution of the enzyme. The enzyme activity was found predominantly in mitochondria (Table I). Less than 5% of the total activity is observed in other fractions which may be due to contamination by the mitochondrial enzyme. Within the mitochondria, the enzyme seems to be tightly bound to inner mitochondrial membrane. Earlier attempts to solubilize the enzyme with detergents were unsuccessful.

Catalytic Properties. The optimum pH for polyprenyltransferase reaction was 7.4. The apparent K_m value for nonaprenyl pyrophosphate with rat liver enzyme was 0.833 μM and K_m for 4-hydroxybenzoate was 1.14 μM. Nonaprenyl monophosphate did not act as substrate for the enzyme and its presence in the assay mixture with nonaprenyl pyrophosphate did not cause any change in enzyme activity.[4]

Effect of Substrate on the Product of Enzyme Reaction. The end product of the enzyme reaction varied depending upon the source of the enzyme and the substrate used to provide polyprenyl side chain.[4] With a rat liver mitochondrial preparation, when IPP was used to provide polyprenyl side chain, the major product of enzymatic reaction was 3-nonaprenyl-4-hydroxybenzoate; however, small amount of decaprenyl derivatives were also observed. Conversely with mitochondria from guinea pig liver, the sole product was 3-decaprenyl-4-hydroxybenzoate. When nonaprenyl pyrophosphate (NPP) was used as a source of polyprenyl side chain in rat liver, 3-nonaprenyl-4-hydroxybenzoate was the major product but there was also minor incorporation into 3-octaprenyl-4-hydroxybenzoate (Table II). The formation of octaprenyl derivative might be due to

[9] N. Schechter, K. Momose, and H. Rudney, *Biochem. Biophys. Res. Commun.* **48,** 833 (1972).

[10] N. Schechter, T. Nishino, and H. Rudney, *Arch. Biochem. Biophys.* **158,** 282 (1973).

TABLE II
EFFECT OF SUBSTRATE ON THE FORMATION OF PRENYLATED 4-HYDROXYBENZOATE IN RAT AND GUINEA PIG LIVER[a]

Tissue	Incubation mixture			Picomoles of PPHB formed		
	PHB	NPP	IPP	Decaprenyl-PHB	Nonaprenyl-PHB	Octaprenyl-PHB
Rat liver	+	+	−	0	19.9	3.6
	+	−	+	15.5	39.5	0.0
	+	+	+	0.7	12.5	1.1
Guinea pig liver	+	+	−	0.0	78.1	11.2
	+	−	+	44.6	0.0	0.0
	+	+	+	35.7	35.7	13.4

[a] Part of data is adopted from Nishino and Rudney.[4]

observed contamination of solanesol which was used for the preparation of nonaprenyl pyrophosphate. In rat liver, the addition of IPP along with NPP resulted in formation of a minor amount of decaprenyl derivative but still nonaprenyl derivative was the major component of the products formed. Whereas in guinea pig liver, the addition of IPP along with NPP resulted in formation of octa, nona, and decaprenyl derivatives all in significant quantities. The addition of Triton X-100 caused a redistribution of products both in rat liver and guinea pig liver preparations.[4]

Effect of Detergents on Enzyme Activity. Detergents, especially Triton X-100, showed a variable effect on enzyme activity in fresh mitochondrial preparations. With aged mitochondria, the effect of various detergents was clear and reproducible. Triton X-100 and Tween 80 were very effective in activating enzyme activity, whereas deoxycholate, Brij W_1, and Nonidet P40 were least effective (Table III). All these detergents were inhibitory at concentrations of 0.5% or more.

Inhibitors and Activators of Enzyme. The addition of magnesium ions greatly enhanced the enzyme activity. Maximum activation was observed at 10 m*M* concentration. The enzyme activity is inhibited by Ca^{2+}, and Mg^{2+} could overcome the inhibition by Ca^{2+}.[9,10] Bacitracin, an antibiotic known to form a complex with polyprenyl pyrophosphate and metal ions,[11] was a strong inhibitor of enzyme. Potassium cyanide, sodium fluoride, and diphenylamine were good inhibitors.[12] *p*-Chloromercuribenzoate and iodoacetamide also inhibited the enzyme and thereby suggesting that

[11] J. K. Stone and J. L. Strominger, *Proc. Natl. Acad. Sci. U.S.A.* **69**, 1287 (1972).
[12] M. J. Winrow and H. Rudney, *Biochem. Biophys. Res. Commun.* **37**, 833 (1969).

TABLE III
EFFECT OF DETERGENTS ON THE 4-HYDROXYBENZOATE 3-POLYPRENYLTRANSFERASE ACTIVITY FROM RAT LIVER[a]

Detergent	Concentration (%)	4-Hydroxybenzoate polyprenyltransferase (pmol/min/mg protein)
None	—	1.99
Triton X-100	0.010	11.15
	0.050	8.47
	0.5	1.18
Tween 80	0.01	10.66
	0.05	9.50
	0.5	6.04
Na deoxycholate	0.01	3.42
	0.05	6.38
	0.50	0.30
Nonidet P-40	0.05	5.60
Brij W_1	0.05	4.78

[a] Part of data is adopted from Nishino and Rudney.[4]

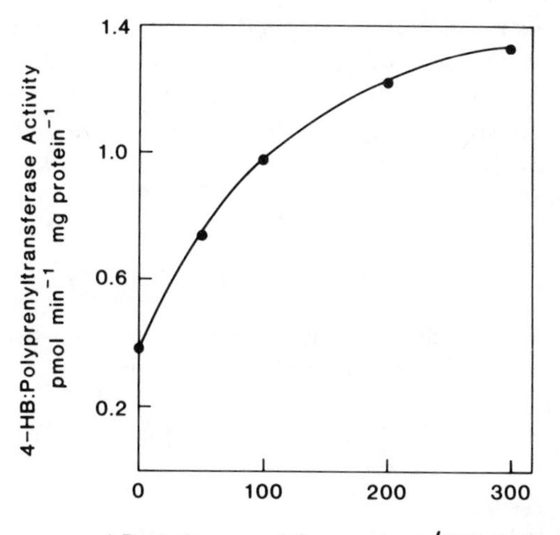

FIG. 1. Effect of increasing concentration of postmicrosomal supernatant on 4-hydroxybenzoate polyprenyltransferase activity from rat liver. All assay tubes contained 6 nmol of NPP and 0.55 mg of inner mitochondrial membrane fragments. The protein concentration of postmicrosomal supernatant was 37.0 mg/ml. All other conditions are as described in the assay procedure.

thiol groups are essential for catalytic activity.[13] Among other benzoic acid derivatives, 4-amino- and 4-chlorobenzoic acids were strong inhibitors. 4-Aminobenzoic acid inhibited by virtue of its ability to compete with 4-hydroxybenzoate as a substrate for prenylation by the enzyme.[13] Structural requirements for the aromatic substrate include a substituent in the 4 position which is an electron donor coupled with a strong electron attracting group in the one position, e.g., a carboxyl group generates sufficient electron density in the 3 position to enable a nucleophilic attack on the oxygen-carbon bond of polyprenyl pyrophosphate.

Addition of postmicrosomal supernatant of rat liver to whole mitochondria as well as to isolated mitochondrial membrane fragments from the same tissue stimulated the synthesis of polyprenyl 4-hydroxybenzoate by 4- to 5-fold.[1,14] The degree of stimulation of 4-hydroxybenzoate polyprenyltransferase activity increased with the amount of postmicrosomal supernatant added and then reached a plateau (Fig. 1). The postmitochondrial supernatant also contains a protein which facilitates the transport of solanesol pyrophosphate into the mitochondria. It is not yet determined whether the same protein is involved in stimulation of enzyme activity and in the transport of solanesyl pyrophosphate and stimulation of 4-hydroxybenzoate polyprenyltransferase activity.[14] *Micrococcus luteus* also contains a similar stimulatory factor.[14]

Enzyme Stability. Isolated mitochondria could be stored frozen for 2–3 weeks prior to preparation of membrane fragment without any loss in activity. With aged mitochondria, the results were very consistent. Once inner mitochondrial fragments are prepared, the enzyme loses its activity after two cycles of freezing and thawing. If stored in several aliquots, the enzyme is stable for 2 weeks.

Acknowledgment

We wish to acknowledge support by the National Science Foundation, PCM-820-4817, which aided in the conduct of some of the work reported here.

[13] S. S. Alam, A. M. D. Nambudiri, and H. Rudney, *Arch. Biochem. Biophys.* **171,** 183 (1975).

[14] A. Gupta, B. C. Paton, S. Ranganathan, and H. Rudney, *Biochem. Biophys. Res. Commun.* **119,** 1109 (1984).

[38] Dimethylallylpyrophosphate: L-Tryptophan Dimethylallyltransferase

By HANS C. RILLING

There are two different types of prenyltransferase reactions. The 1′-4 reaction, which is the basic polymerization reaction of polyisoprenoid biosynthesis, has been extensively studied and is the topic of several chapters in this volume.[1-3] The other prenyl transfer reaction entails the alkylation of a nonprenyl moiety by an allylic pyrophosphate. This is a 1′-aryl condensation and has been reviewed recently.[4] Few of these enzymes have been examined in anything but crude extracts. However, the enzyme that catalyzes prenylation of aspulvinone by dimethylallyl pyrophosphate to yield a family of aspulvinones has been extensively purified by Takahashi *et al.* from *Aspergillis terreus*,[5,6] and dimethylallyl pyrophosphate : L-tryptophan dimethylallyltransferase has been purified to homogeneity. This enzyme catalyzes the first reaction in a biosynthetic pathway of the ergot alkyloids in *Claviceps* sp. An extensive purification of this protein was reported by Lee, Floss, and Heinstein[7] and more recently a homogeneous preparation of this enzyme was obtained by a different and more convenient procedure.[8]

Growth of Claviceps. Claviceps sp. strain SD58 was provided by Dr. P. Heinstein (Department of Medicinal Chemistry, Purdue Univ.). The fungus is grown on a semisynthetic media consisting of 25 g sucrose, 25 g mannitol, 0.1 g KH_2PO_4, 0.3 g $MgSO_4 \cdot 2H_2O$, 10 mg $FeSO_4 \cdot 7H_2O$, 4 mg $ZnSO_4 \cdot 7H_2O$, 3 g yeast extract (Difco), and 5.4 g succinic acid in 1 liter distilled water. The pH is adjusted to 5.4 with NH_4OH. Culture conditions are important. Low form mold culture flasks or Roux bottles are used, and the depth of the media is 1–2 cm. The flasks are innoculated with *Claviceps* which were grown in shake culture. However, stationary culture conditions are used for growing *Claviceps* for the purpose of enzyme

[1] C. M. Allen, this volume [31].
[2] G. Barnard, this volume [18].
[3] H. C. Rilling, this volume [14].
[4] C. D. Poulder and H. C. Rilling, *in* "Biosynthesis of Isoprenoid Compounds" (J. W. Porter and S. L. Spurgeon, eds.), Vol. 1, p. 162. Wiley, New York, 1981.
[5] I. Takahashi, N. Ojima, K. Ogura, and S. Seto, *Biochemistry* **17,** 2696 (1978).
[6] I. Takahashi, N. Ojima, K. Ogura, and S. Seto, this volume [36].
[7] S. L. Lee, H. G. Floss, and P. Heinstein, *Arch. Biochem. Biophys.* **177,** 84 (1976).
[8] W. A. Cress, L. T. Chayet, and H. C. Rilling, *J. Biol. Chem.* **256,** 10917 (1981).

ISBN 0-12-182010-6

isolation. The cultures at 25° are fully grown at 1 week but are not harvested until 2 weeks. At that time, the medium is wine red. After harvesting the mycelium is washed with water, lyophilized, and stored at −20°.

Enzyme Assays. Two assays are used. The first is a modification of that of Lee *et al.*[7] The final composition of the assay mixture was 1 m*M* dimethylallyl pyrophosphate (0.5 mCi/mmol), 1 m*M* tryptophan, 20 m*M* $CaCl_2$, 10 m*M* Tris–HCl, pH 7.5, and 20 m*M* 2-mercaptoethanol in a total volume of 0.1 ml. The preparation of [^{3}H]allylic pyrophosphates is presented elsewhere in this volume.[9] Incubations were for 30 min at 30°. After incubation, the reaction is stopped by dilution with distilled water to 3 ml, and the incubation mixture is transferred to a 2-ml column of Bio-Rad exchange resin AG 50W-X4, hydrogen form (200 to 400 mesh). After the column is washed three times with 10 ml of distilled water, the product is eluted into a glass scintillation vial with 5 ml of $NH_4OH : H_2O$: methanol (2 : 3 : 5). The solvent was removed under a stream of nitrogen in a heated water bath, and 1 ml of water and 10 ml of Triton X-100 scintillation mixture [Triton X-100, (500 ml), toluene (1 liter), Omnifluor (8.5 g)] were added to each vial. Radioactivity was determined by liquid scintillation spectrometry.

The second, a continuous spectrophotometric procedure, has been devised by Woodside and Poulter.[10] These assays are conducted with substrate concentrations at least equal to twice K_m for each substrate. The UV spectra are obtained on a Gilford Model 2600 UV spectrophotometer. A standard assay is according to the following procedure: the assay sample was prepared by mixing L-tryptophan, dimethylallyl pyrophosphate, and buffer together at 0°. Enzyme is then added, and the mixture is incubated at 30° for 10 min. Ca^{2+} is then added and the mixture is transferred to the spectrophotometer. The reaction is followed as a function time at 298 nm. The assay mixture in 0.5 ml total volume contains 0.4 m*M* L-tryptophan, 0.5 m*M* dimethylallyl pyrophosphate, 20 m*M* in Ca^{2+}, 10 m*M* Tris pH 7.5, and enzyme.

The extinction coefficient of the product is estimated to be 8×10^4 at 298 nm by a comparison of this assay procedure with the one utilizing radioisotopic substrates. An accurate knowledge of this extinction coefficient awaits the chemical synthesis and spectrophotometric analysis of dimethylallyl tryptophan.

Enzyme Purification. The buffer for the initial extraction of the protein is 20 m*M* Tris–HCl, 20 m*M* 2-mercaptoethanol, 20 m*M* thioglycolate, 10% (v/v) glycerol, and 20 m*M* $CaCl_2$ (pH 8). In all other purification steps

[9] V. J. Davidson, A. B. Woodside, and C. D. Poulter, this volume [15].

[10] A. B. Woodside and C. D. Poulter, unpublished.

and procedures, buffer II, which contains 10 mM Tris–HCl, 20 mM 2-mercaptoethanol, and 10% (v/v) glycerol (pH 8) is used. Protein is determined by the method of Bradford using bovine serum albumin as a standard.[11]

All purification procedures are at 4°. Lyophilized *Claviceps* SD58 mycelium is ground overnight in a ball mill. Extraction buffer is added to the powdered mycelium in the ball mill (20 ml/g of lyophilized mycelium), and the mill is run for an additional 20 min. The suspension thus obtained is blended for 1 to 2 min in a Waring blender and then centrifuged for 30 min at 16,300 *g*. The supernatant fluid is filtered through glass wool and ammonium sulfate is added to the supernatant to 30% saturation while the pH is maintained at 8.0 by the addition of dilute NH_4OH. After centrifugation, the supernatant fluid is brought to 45% of saturation of ammonium sulfate. The pellet obtained by centrifugation is dispersed in extraction buffer without $CaCl_2$ and dialyzed overnight against buffer II. The dialyzate is cleared by centrifugation at 85,500 *g* for 30 min and then mixed with *n*-butyl-Sepharose which was prepared by reacting CNBr-activated Sepharose with *n*-butylamine. One milliliter of *n*-butyl-Sepharose is used for each g of lyophilized *Claviceps*, and the mixture is stirred for 15 min. After the *n*-butyl-Sepharose is washed on a Büchner funnel with a total volume of 500 ml of buffer II containing 100 mM KCl, the slurry of Sepharose is poured into a column (2.5 × 18 cm), and the column eluted with a linear gradient of 0 to 500 mM KCl in buffer II (600 ml total volume). Ten-milliliter fractions are collected. Enzymatically active fractions are pooled and the protein is precipitated with ammonium sulfate at 50% saturation. The protein is collected by centrifugation at 27,000 *g* for 20 min and then dissolved in buffer II and dialyzed overnight. The dialyzed protein is applied to a 1.5 × 10 cm column of DE-52 which is washed with buffer II until the absorbance (280 nm) of the column effluent falls to less than 0.1. The column is eluted with a 600-ml linear gradient of 0 to 300 mM KCl in buffer II. Ten-milliliter fractions are collected and active fractions pooled as before, and enzyme is precipitated with ammonium sulfate at 50% saturation. The precipitate is collected by centrifugation and dissolved in buffer II containing 0.1 M KCl. The protein is then chromatographed on Sephacryl S-200 superfine (1.5 × 87 cm column) with buffer II containing 0.1 M KCl. The enzymatically active fractions are analyzed by electrophoresis on polyacrylamide gels and the homogeneous fractions pooled.

If the procedure yields at least 20 mg of protein at this point, the protein in the pooled fractions can be crystallized. To accomplish this,

[11] M. M. Bradford, *Anal. Biochem.* **72,** 248 (1976).

TABLE I
PURIFICATION OF DIMETHYLALLYL TRYPTOPHAN SYNTHETASE[a]

Fraction	Units (nmol/min)	Protein (mg)	Specific activity (nmol/min/mg protein)	Yield (%)	Purification (fold)
Crude supernatant	6881	2710	2.5	100	
30–45% ammonium sulfate	4101	620	6.6	60	2.6
n-Butyl-Sepharose	3202	75	42.4	47	16.7
DE-52	1704	18	95	25	37.4
Sephacryl S-200	1960	5.6	350	28	138

[a] Reproduced from the *Journal of Biological Chemistry*.[8]

protein is precipitated with ammonium sulfate at 50% saturation. The pellet thus obtained is triturated with 0.5 ml of 40% saturated ammonium sulfate and allowed to stand 5 min in an ice bath. The suspension is centrifuged and the precipitate is extracted with a graduated series of 38, 36, 34, 32, 30, 28, and 26% of cold saturation ammonium sulfate solutions. The supernatant from each extraction is warmed to room temperature and left overnight. Small (approximately 10 μm in length) needlelike crystals are observed in the 32 and 30% saturation fractions. The best crystals are obtained in the 32% saturation fractions. Crystals can also be obtained from the protein after the DE-52 step, however, they are not homogeneous as judged by polyacrylamide gel electrophoresis. The purification procedure is summarized in Table I.

Properties of the Enzyme. The enzyme does not require a metal ion for activity as do the 1′-4 (polymerizing) prenyltransferases. Thus, it is possible that the protein might contain a tightly bound metal ion. To test this possibility, the enzyme was treated with a variety of chelating agents (Table II). All tested inhibited the enzyme in the presence of excess Ca^{2+}, and many were more effective after a 12-hr preincubation with the protein. A number of divalent cations were added back to DMAT synthetase that had been inactivated with 1.0 m*M* EDTA. These ions included Zn^{2+}, Ca^{2+}, Mg^{2+}, Mn^{2+}, and Cu^{2+}. None restored activity to a significant extent. Thus, it is apparent that the enzyme contains a divalent cation which is most likely involved in catalysis. However, the identity of the cation is not certain. Curiously enough, when calcium is omitted from the incubation mixture, EDTA does not inhibit the reaction.

Kinetics. A marked deviation from linearity found for this enzyme in Lineweaver–Burk plots of the kinetic data, which revealed a difference between DMAT synthetase and the other prenyltransferases in that this

TABLE II
EFFECTS OF CHELATORS ON DIMETHYLALLYL TRYPTOPHAN SYNTHETASE ACTIVITY IN THE PRESENCE OF 20 mM Ca^{2+}

	Inhibition (%)					
	30-min incubation			12-hr preincubation + 30-min incubation		
Chelator	0.1 mM	1 mM	10 mM	0.1 mM	1 mM	10 mM
8-Mercaptoquinoline	31	100		93	100	
8-Hydroxyquinoline		29		36	50	
2,2′-Bipyridine	8	17		60	96	
Nitrilotriacetic acid		24	100		15	100
Ethylenediamine-tetraacetic acid		100	100		100	100
Dithiothreitol		25	4		25	96
Acetylacetone		0	0		0	90
1,10-Phenanthroline		73	100		95	100

[a] Reproduced from the *Journal of Biological Chemistry*.[8]

enzyme displays cooperative behavior. Since this enzyme catalyzes the first committed step in the synthesis of the ergot alkaloids, it is not surprising to find that it is regulated. Hill plots show that the cooperativity is mixed, indicating that as the concentration of the two substrates is varied negative, positive, and absence of cooperativity are observed. Minimum Hill coefficients of 0.37 and 0.75 for DMAPP and tryptophan, respectively, are observed, demonstrating negative cooperativity, while positive cooperativity is indicated by maximum Hill coefficients of 1.5 for dimethylallyl pyrophosphate and 1.8 for tryptophan. The mixed cooperativity found with this enzyme is in marked contrast to the other prenyltransferases which display normal Michaelian kinetics. In the presence of 20 mM Ca^{2+}, the Lineweaver–Burk plot of dimethylallyl tryptophan synthetase is linear, indicating the loss of cooperativity. This is interpreted as deregulation of the enzyme which now proceeds by Michaelian kinetics.

The Michaelis constants for the 1′-4 prenyltransferase are in the range of 0.1 μM for isopentenyl, dimethylallyl, and geranyl pyrophosphates, while the $S_{[0.5]}$ values for DMAPP and trytophan are 316 and 200 μM, respectively, for DMAT synthetase. In contrast, the deregulated enzyme in the presence of 20 mM Ca^{2+}, has Michaelis constants of 7 and 9 μM for dimethylallyl pyrophosphate and tryptophan.

Calcium was found to be a positive allosteric effector with a Hill coefficient of 2.8 and $S_{[0.5]}$ of 4.5 mM. A concentration of 20 mM Ca^{2+} completely deregulates the enzyme while 30 mM Ca^{2+} ion is inhibitory. This may be due to calcium interaction with dimethylallyl pyrophosphate to give a precipitate. Since calcium is not essential for catalytic activity, it most likely does not have a catalytic function and is probably involved only in regulation.

[39] Assay and Partial Purification of the Cytokinin Biosynthetic Enzyme Dimethylallylpyrophosphate : 5′-AMP Transferase

By N. G. HOMMES, D. E. AKIYOSHI, and R. O. MORRIS

Introduction

The route by which cytokinins are synthesized in plants has long been obscure, but in 1979 Chen and Melitz[1] showed that habituated tobacco callus tissue contained an enzyme activity which transferred the dimethylallyl group from dimethylallyl pyrophosphate to the exocyclic amino group of 5′-AMP to give the cytokinin dimethylallyladenosine 5′-phosphate (iPA-5′P).[1a] The enzyme, dimethylallyl pyrophosphate:5′-AMP dimethylallyltransferase (DMA transferase), had a strict requirement for 5′-AMP and could not utilize adenine or adenosine. Subsequently[2] adenosine was proposed as a substrate. This has not been confirmed however, and recent work[3] suggests that in plant crown gall tumors at least, an enzyme is present which is similar in its substrate requirements to that described by Chen and Melitz. It is possibly the source of the high levels of cytokinin in crown gall tissue.[4]

[1] C. M. Chen and D. K. Melitz, *FEBS Lett.* **107,** 15 (1979).

[1a] Abbreviations: dimethylallyladenosine (isopentenyladenosine), iPA; dimethylallyl pyrophosphate, DMAPP; dithiothreitol, DTT; iPa 5′-phosphate, iPA-5′-P; octadecyl silica, ODS; phenylmethylsulfonyl fluoride, PMSF; poly(vinylpyrrolidone), PVP.

[2] N. Nishinari and K. Syono, *Z. Pflanzenphysiol.* **99,** 383 (1980).

[3] R. O. Morris, D. E. Akiyoshi, E. M. S. MacDonald, J. W. Morris, D. A. Regier, and J. B. Zaerr, *in* "Plant Growth Substances" (P. F. Wareing, ed.), p. 175. Academic Press, London, 1982.

[4] D. E. Akiyoshi, R. O. Morris, R. Hinz, B. S. Mischke, T. Kosuge, D. J. Garfinkel, M. P. Gordon, and E. W. Nester, *Proc. Natl. Acad. Sci. U.S.A.* **80,** 407 (1982).

ISBN 0-12-182010-6

This article describes the techniques required for the growth of crown gall tumor tissues (a rich source of the enzyme), the preparation of dimethylallyl pyrophosphate, an assay method for DMA transferase and a protocol for isolation and partial purification of the enzyme. The assay relies upon the immunoaffinity procedures described in the accompanying article by MacDonald and Morris.[5]

Growth of Plant Tumors

The most convenient source of DMA transferase has been found to be the tobacco crown gall tumor line (15955/01) described by Gelvin *et al.*[6] The tumor was incited originally in *Nicotiana tabacum* cv. Xanthi by infection with *Agrobacterium tumefaciens,* strain 15955, and has been cloned. It grows readily upon Murashige and Skoog medium[7] in the absence of added phytohormones and is stable in culture. Under high light intensities it forms an undifferentiated mass of dark green callus.

Reagents and Supplies

Murashige and Skoog salt mixture and Phytagar (Gibco, Grand Island, NY)

Plastic petri dishes (100 × 25 mm, Lab-Tek, Naperville, IL)

Procedure. The medium is composed of Murashige and Skoog salts (4.3 g/liter), sucrose (3% w/v), and phytagar (0.75% w/v) and is supplemented with thiamine (1.2 mg/liter) and *myo*-inositol (100 mg/liter). It is adjusted to pH 5.7 with sodium hydroxide, autoclaved for 20 min, poured under sterile conditions into plastic petri dishes to a depth of 1 cm and allowed to solidify in a stream of sterile air. The tumor tissues are maintained at 25° in continuous light (high intensity Growlux fluorescent lights) and are subcultured (1–2 g pieces) at intervals of approximately 3 weeks. No phytohormones need be added to the medium. Tissues should be used immediately after harvest for enzyme preparation.

Synthesis of Dimethylallylpyrophosphate

The synthesis of DMAPP was originally described by Cornforth and Popják.[8] It is reproduced here with additional detail including an alternative method for characterization of the product.

[5] E. M. S. MacDonald and R. O. Morris, this volume [40].

[6] S. B. Gelvin, M. F. Thomashow, J. C. McPherson, M. P. Gordon, and E. W. Nester, *Proc. Natl. Acad. Sci. U.S.A.* **79,** 76 (1982).

[7] T. Murashige and F. Skoog, *Physiol. Plant.* **15,** 473 (1962).

[8] R. H. Cornforth and G. Popják, this series, Vol. 15, p. 385.

Reagents Required

3,3-Dimethylallyl alcohol (3-methylbut-2-en-1-ol, prenyl alcohol, Rhodia Co., Freeport, TX)
Acetonitrile (HPLC grade): dried over calcium hydride for 48 hr, redistilled, and stored under anhydrous conditions
Trichloroacetonitrile: distilled before use and stored at −20° under anhydrous conditions
Triethylamine: dried over KOH pellets for 48 hr, distilled in a stream of nitrogen, and stored at −20°
TLC plates: polyethyleneimine (PEI)-cellulose (J. T. Baker)
TLC spray reagent: perchloric acid (60% w/v) : hydrochloric acid (1 *M*) : ammonium molybdate (4% w/v aqueous) : water = 5 : 10 : 25 : 60
0.1 *M* lithium chloride
0.1 *M* ammonium hydroxide
0.1 *M* ammonium bicarbonate, adjusted to pH 8.4
0.5 *M* ammonium bicarbonate (diluted from 1 *M*)
30 m*M* ammonium bicarbonate (diluted from 1 *M*)
DEAE-Sephadex (Sigma, A-125-25) equilibrated with 1 *M* ammonium bicarbonate, pH 8.4

Procedure. Procedures in this section must be carried out in dry vessels under strictly anhydrous conditions.

Bis(triethylammonium) phosphate. Phosphoric acid (85–88%, 10 ml) is dried under high vacuum over phosphorus pentoxide for 24 hr. Five grams of the anhydrous acid (0.05 mol) is dissolved in a mixture of dry acetonitrile (25 ml) and dry triethylamine (0.1 mol, 10 g). Bis(triethylammonium) phosphate crystallizes from solution on standing at 25°. After 24 hr the crystals are collected on a sintered glass filter, washed rapidly with 20 ml of cold dry acetonitrile, and used in the next stage of the reaction or stored over phosphorus pentoxide. The subsequent reaction is facilitated if the crystals are not filtered to dryness.

Dimethylallylpyrophosphate synthesis. A saturated solution of Bis-(triethylammonium) phosphate (7 g, wet weight) is prepared in dry acetonitrile (200 ml) and the excess crystals are allowed to settle. A mixture of trichloroacetonitrile (90 mmol, 9 ml) and 3,3-dimethylallyl alcohol (10 mmol, 1 ml) is prepared and the solution of bis(triethylammonium) phosphate is added slowly over a period of 3 hr at 25° with stirring. Upon completion of the addition, the remaining solid bis(triethylammonium) phosphate is added over the next 2 hr and the mixture is stirred for an additional 2 hr. It is then diluted with 1 liter of ether and extracted three times with 300 ml 0.1 *M* ammonium hydroxide. The pooled aqueous phases are back-extracted three times with 500 ml ether, evaporated to dryness at 45° *in vacuo* in a Büchi rotary evaporator, dissolved in 10 ml of water, lyophilized, and stored at −20°.

The progress of the reaction may be followed by TLC on PEI-cellulose plates developed in 0.5 *M* lithium chloride. Phosphate esters separate in this system on the basis of charge, triphosphates having low mobility and monophosphates having high mobility. They are rendered visible by spraying with acidic molybdate.[9] The plate is sprayed, dried with warm air, and exposed to ultraviolet light until uniformly light blue. Any background color is eliminated by exposure to ammonia vapor. Organic phosphates appear as blue spots on a yellow background. The reaction mixture contains at least 5 phosphate-containing compounds with R_fs of 0.8, 0.72, 0.55, 0.15, and 0.05. DMAPP has a retention time of 0.15.

Purification on DEAE-Sephadex. The DEAE-Sephadex is equilibrated in 1 *M* ammonium bicarbonate, pH 8.4, and washed by sequential decantation with 5 vol of 1 *M*, 0.5 *M*, and 30 m*M* ammonium bicarbonate. A 1.5 × 22 cm column is poured and washed with 100 ml 30 m*M* ammonium bicarbonate. The crude mixture of phosphates is dissolved in 100 ml 30 m*M* ammonium bicarbonate and applied at a flow rate of 0.5 ml/min. The column is eluted with a 400 ml linear gradient (30 m*M* to 1 *M*) of ammonium bicarbonate at the same flow rate. Fractions are collected and assayed for DMAPP by TLC. It elutes between 100 and 160 ml. Peak fractions are pooled, lyophilized, and stored at −20°. Yield of DMAPP is approximately 10%.

Enzyme Extraction and Purification

Reagents and Buffers

Buffer A: 50 m*M* Tris–HCl, 20 m*M* KCl, 10 m*M* $MgCl_2$, 20% v/v glycerol. Adjust to pH 7.00 at 25° with sodium hydroxide
100 m*M* Dithiothreitol (DTT)
200 mg/ml phenylmethylsulfonyl fluoride (PMSF) in 95% ethanol
Poly(vinylpyrrolidone) (PVP) and Amberlite XAD-4: washed according to the procedures of Loomis *et al.*[10,11] and stored in 0.1% sodium azide

Procedure

Preparation of Crude Extract. All operations are performed at or below 4°. The extraction buffer is prepared from Buffer A and contains in

[9] R. S. Bandurski and B. Axelrod, *J. Biol. Chem.* **193,** 405 (1951).
[10] W. D. Loomis, this series, Vol. 13, p. 555.
[11] W. D. Loomis, J. D. Lile, R. D. Sandström, and A. J. Burbott, *Phytochemistry* **18,** 1049 (1979).

addition, 0.5 m*M* DTT and 25 μg/ml PMSF. A convenient scale of preparation employs 40 g 15955/01 crown gall callus. Freshly havested tissue is frozen in liquid nitrogen, ground to a fine powder in a blender or chilled mortar with liquid nitrogen, and transferred immediately to a slurry of 40 g Amberlite XAD-4 and 20 g PVP suspended in 150 ml extraction buffer. The XAD-4 and PVP should be preequilibrated in extraction buffer overnight prior to use. Nitrogen is bubbled through the mixture for 15 min. The slurry is then poured onto a layer of 20 g PVP (previously equilibrated in extraction buffer) in a sintered glass filter. The supernatant is collected by vacuum filtration.

Ammonium Sulfate Precipitation. Solid ammonium sulfate is added slowly to the crude supernatant to 25% saturation. After 30 min the solution is centrifuged at 15,000 *g*/30 min, the pellet is discarded, and the ammonium sulfate concentration of the supernatant is raised to 80% saturation. After 1 hr the precipitated protein is collected by centrifugation at 15,000 *g*/30 min, redissolved in 20 ml Buffer A, and dialyzed against three 700 ml portions of Buffer A. Denatured protein is removed by centrifugation at 15,000 *g*/15 min.

DEAE-Cellulose Chromatography. DEAE-cellulose (DE-52, Whatman Corporation) is precycled according to the manufacturer's instructions, washed with water, and suspended in 10 volumes of 1 *M* Tris–HCl, pH 7.0. The pH of the stirred suspension is adjusted with 1 *M* HCl until there is no further change (3 hr). The DE-52 is washed with 100 volumes of 50 m*M* Tris–HCl, pH 7.0 and 20 volumes of Buffer A. A column (11 × 1.5 cm) is poured, washed with 50 ml of Buffer A, loaded with the dialyzed ammonium sulfate fraction at 0.5 ml/min, and washed with 60 ml of Buffer A. The enzyme is eluted with Buffer A containing 0.3 *M* KCl. The bulk of the activity elutes within 30 ml, coincident with the peak of absorbance at 280 nm. All fractions are pooled and stored frozen at −70°. Activity is stable for ~2 months.

Enzyme Assay

Principle. A satisfactory assay can be developed by measuring the amount of iPA-5′P produced upon incubating the enzyme with dimethylallyl pyrophosphate and radiolabeled 5′-AMP. Because cytokinins are present at extremely low levels in plant tissues, it might be anticipated that the enzymes which synthesize them are present at low levels or have low turnover numbers. Indeed, the best preparations of the enzyme which have been obtained to date synthesize 10 fmol IPA-5′-P/mg protein/hr.[12]

[12] D. E. Akiyoshi and R. O. Morris, unpublished results.

Estimation of the miniscule amounts of IPA-5′-P which are produced requires both high specific activity 5′-[^{3}H]AMP (19.5 Ci/mmol) and a stringent purification procedure for iPA-5′-P. The immunoaffinity protocol described in the accompanying article[5] satisfies this last requirement. After incubation of the enzyme with DMAPP and 5′-[^{3}H]AMP, the iPA and iPA-5′-P are collected on ODS, dephosphorylated, and the total iPA is then collected by passage through coupled DEAE-cellulose and immobilized anti-iPA antibody columns. An HPLC step completes the procedure.

Reagents and Solutions

The following stock buffers are useful:

100 m*M* HEPES, pH 7.0
100 m*M* $MgCl_2$
100 m*M* KF
100 mg/ml DMAPP; stored at −70°
20 m*M* Tris–HCl, pH 7.0
20 m*M* sodium phosphate, pH 7.0
20 m*M* Tris acetate, pH 9.2
40 m*M* acetic acid, adjusted to pH 7.0 with redistilled triethylamine
Buffer B: 20 m*M* Tris acetate, pH 4.9
12 m*M* $MgCl_2$
Buffer C: 40 m*M* phosphate, pH 7.0; 0.5 *M* NaCl; 2% (v/v) dimethyl sulfoxide.
Buffer D: 40 m*M* acetic acid, adjusted to pH 3.35 with redistilled triethylamine

Procedure

Precautions. Because cytokinins have a very high affinity for untreated glass surfaces, all reactions must be carried out in silanized glassware or, preferably, polypropylene microfuge tubes. It is essential for the success of the assay to select microfuge tubes which do not strongly absorb cytokinins. We have found that Labcon tubes (VWR Scientific #20170-273 or 20170-320) are suitable. New lots of microfuge tubes may be tested by measuring their ability to bind [^{3}H]zeatin trialcohol.[5]

Assay. In a final volume of 6 ml, the assay mixture contains 30 m*M* HEPES, pH 7.0, 20 m*M* $MgCl_2$, 5 m*M* KF, 10% v/v glycerol, 0.7 m*M* DMAPP, 2 μCi 5′-[^{3}H]AMP, and the enzyme derived from 2 g of 15955/01 tobacco crown gall callus. The mixture is incubated at 25° for 1–4 hr, 100 ng iPA is added as carrier, and the mixture is run through a short ODS column.

Purification on ODS. Aliquots of 1 ml of ODS (40 μm, Sepralyte,

Analytichem Intl., Harbor City, CA) are loaded into 3 ml disposable polypropylene syringes which have been plugged with fragments of cellulose tissue (glass wool should not be used) and washed sequentially with 10 ml portions of 2-propanol, methanol, and 40 m*M* triethylammonium acetate, pH 7.0 (TEA-HoAc). The assay mixture is applied at 0.5 ml/min and the column is washed with 5 ml buffer. Excess buffer is blown from the column by brief application of nitrogen pressure and any iPA and iPA-5′-P are eluted slowly with 2 ml methanol. The eluate is dried *in vacuo* at 25°.

Dephosphorylation. The dried sample is suspended in 500 μl of Buffer B, 1 mg wheat germ acid phosphatase is added, and the mixture is incubated at 37° for 1 hr. The pH is restored to neutrality by the addition of 500 μl of 20 m*M* Tris-acetate, pH 9.2.

Immunoaffinity chromatography. DEAE-cellulose (DE-52, Whatman) is precycled according to the manufacturer's instructions, washed with water, equilibrated with 1 *M* Tris–HCl, pH 7.0, until no further pH changes occur, and washed with 20 volumes of 30 m*M* Tris–HCl, pH 7.0. Anti-iPA antibody is prepared and covalently attached to cellulose as described elsewhere in this volume.[5] Columns of DE-52 and anti-iPA antibody-cellulose (1 ml each) are poured in polypropylene syringes and are linked together with the DE-52 column above the antibody column. The neutralized, phosphatase-treated, sample is run directly onto the DE-52 column at 0.5 ml/min and washed with 10 ml of 40 m*M* TEA-HoAc, pH 7.0. The DE-52 column is disconnected, and the antibody column is washed with 10 ml portions of 20 m*M* sodium phosphate, pH 7.0, Buffer C, 40 m*M* TEA-HoAc, pH 7.0, and 5 ml of water. Excess water is removed by nitrogen pressure, the cytokinins are eluted with 2 ml of methanol, and dried *in vacuo*.

HPLC. A rapid modification of existing HPLC procedures[13] may be used to analyze the radioactive iPA present. It employs a short ODS column (Altex; 3 μm, 4.6 × 75 mm), and a eluant of Buffer D increasing in acetonitrile concentration from 10 to 12% over 5 min and from 12 to 37% over 7 min at a flow rate of 1.6 ml/min. The sample is dissolved in 100 μl of a 50% mixture of methanol and Buffer D, applied to the column, and the radioactivity of the iPA fraction is estimated by liquid scintillation counting.

Further Purification and Properties

The properties of the crown gall enzyme are similar to those of the enzyme isolated from habituated tobacco tissue by Chen and Melitz.[1] The

[13] E. M. S. MacDonald, D. E. Akiyoshi, and R. O. Morris, *J. Chromatogr.* **214,** 101 (1981).

crown gall enzyme utilizes DMAPP but not isopentenyl pyrophosphate as side chain donor.[12] It utilizes 5'-AMP but not adenine or adenosine as side chain acceptor.[12] Chen and Melitz[1] were able to purify the DMA transferase of habituated tobacco callus to homogeneity. In contrast, the crown gall tumor DMA transferase appears to be somewhat more unstable and has so far only been purified to the DEAE-cellulose stage. It is conceivable that the two enzymes are, in fact, different. If, as seems likely, the DMA transferase in crown gall tumors is coded for by the *tmr* locus of T-DNA, differences might be expected.

Acknowledgments

We wish to thank Mrs. E. M. S. MacDonald for valuable technical assistance. This research was supported by Grants PCM 81-04514 and PCM 83-03371 from the National Science Foundation and by the Science and Education Administration of the U.S. Department of Agriculture under Grant 83-CRCR-1-1249 from the Competitive Research Grants Office.

[40] Isolation of Cytokinins by Immunoaffinity Chromatography and Analysis by High-Performance Liquid Chromatography Radioimmunoassay

By E. M. S. MacDonald and R. O. Morris

Cytokinins are 6-(3-methylbut-2-enylamino)purine derivatives, the main representatives of which are isopentenyladenine (iP),[1] zeatin (Z), dihydrozeatin (diHZ), and their respective ribosides (iPA, ZR, and dHZR). They occur naturally as highly active plant cell division regulators, usually at levels below 30 ng/g. Consequently, their analysis in plant extracts, which normally contain a plethora of small organic molecules, presents formidable difficulties.

Early isolation procedures[1a] called for extraction of large quantities of plant material with 1-butanol or ethyl acetate, chromatography of the extract, followed by bioassay of the column eluate. The procedures were

[1] Abbreviations: Octadecyl silica, ODS; gas–liquid chromatography–mass spectrometry, GLC-MS; zeatin, Z; bovine serum albumin, BSA; zeatin riboside, ZR; isopentenyladenine, iP; isopentenyladenosine, iPA; dihydrozeatin, dHZ; dihydrozeatin riboside, dHZR; phosphate-buffered saline, PBS.

[1a] T. Murashige and F. Skoog, *Physiol. Plant.* **15,** 473 (1962).

ISBN 0-12-182010-6

extremely labor-intensive, time-consuming (average bioassay time, 40 days), imprecise, and relatively insensitive (limit about 10 ng). The introduction of mass spectrometric methods[2] in which the cytokinins were converted to volatile derivatives and subjected to GLC-MS, improved the precision but not the sensitivity.

More recent procedures have utilized RIA, either of crude extracts[3] or of extracts after fractionation by HPLC.[4] The technique can detect cytokinins at low pg levels, making it possible to analyze mg samples of plant tissues. Problems remain in that the HPLC-RIA procedure is still labor-intensive, while the application of RIA directly to unfractionated extracts of certain plants has, in our hands, given spuriously high values.

This article describes single-pass immunoaffinity chromatographic purification of cytokinins from plant extracts prior to HPLC. Antibodies directed against cytokinins are immobilized on cellulose and used to recover the cytokinins from crude plant extracts. The technique is extremely rapid, permits recoveries of 80%, may be applied to extremely small tissue samples, and achieves purification to near homogeneity in a single step.

Isolation Protocol

The technique has been described briefly by Morris *et al.*[5] and Akiyoshi *et al.*[6] After homogenization of the sample in methanol, removal of the debris by centrifugation, and reduction of the methanol concentration by dilution with an excess of buffer, cytokinin 5′-phosphates are converted into free nucleosides by treatment with acid phosphatase. The sample is then passed through a small DEAE-cellulose column in order to remove pigments and anionic materials. The eluate, which contains the cytokinins, is run directly onto an immunoaffinity column containing anti-cytokinin antibodies covalently linked to cellulose. Immunoaffinity columns may be constructed from monoclonal antibodies against cytokinins if they are available and have the appropriate specificities and affinities.[7] If not, polyclonal antibodies raised in rabbits may be used,

[2] R. E. Summons, C. C. Duke, J. V. Eichholzer, B. Entsch, D. S. Letham, J. K. MacLeod, and C. W. Parker, *Biomed. Mass Spectrom.* **6,** 407 (1979).

[3] E. W. Weiler, *Planta* **149,** 155 (1980).

[4] E. M. S. MacDonald, D. E. Akiyoshi, and R. O. Morris, *J. Chromatogr.* **214,** 101 (1981).

[5] R. O. Morris, D. E. Akiyoshi, E. M. S. MacDonald, J. W. Morris, D. A. Regier, and J. B. Zaerr, *in* "Plant Growth Substances" (P. F. Wareing, ed.), p. 175. Academic Press, London, 1982.

[6] D. E. Akiyoshi, R. O. Morris, R. Hinz, B. S. Mischke, T. Kosuge, D. J. Garfinkel, M. P. Gordon, and E. W. Nester, *Proc. Natl. Acad. Sci. U.S.A.* **80,** 407 (1982).

[7] E. J. Trione and R. O. Morris, unpublished results.

provided they are subjected first to conventional purification or to affinity purification on cytokinin–polylysine–agarose.

After extensive washing of the immunoaffinity column, the cytokinins are eluted with methanol and fractionated by HPLC on ODS. Individual fractions from HPLC may be analyzed by RIA if desired, but, in general, the samples from most plant extracts are sufficiently pure to allow quantitation by integration of the UV-absorbance trace.

Procedure

Preparation of Immunoaffinity Columns

This section describes preparation of cytokinin–BSA conjugates for immunization, the immunization procedure, preparation, and use of cytokinin–polylysine–agarose columns for antibody purification, and finally, attachment of the purified antibody to cellulose.

Reagents and Solutions

- Bovine serum albumin, Fraction V
- Freund's complete and incomplete adjuvants
- Poly(L-lysine) agarose [poly(L-lysine) MW range, 4–15,000, Sigma, P 6893]
- TEA buffer: 40 m*M* triethylammonium acetate, pH 3.4 (40 m*M* acetic acid adjusted to pH 3.4 with redistilled triethylamine)
- 50 m*M* Na_2CO_3–$NaHCO_3$, pH 8.0
- 0.14 *M* NaCl
- Phosphate buffered saline (PBS): 50 m*M* sodium phosphate, pH 7.0; 0.14 *M* NaCl; 1 m*M* EDTA)
- PBS–methanol: 20% v/v methanol
- PBS-NaCl: 1 *M* NaCl
- PBS–azide: 0.1% sodium azide
- Glycine–ether–HCl: 0.1 *M* glycine–HCl, pH 2.2, saturated with ether
- 1 *M* Na_2CO_3
- 1 *M* Tris–HCl, pH 8.0
- Microcrystalline cellulose, TLC grade
- Tetrahydrofuran
- 5 *M* NaOH
- 0.1 *M* $NaHCO_3$, pH 8.5
- 1 *M* ethanolamine hydrochloride, pH 8.5
- DEAE-cellulose (DE-52, Whatman)
- 2 *M* ammonium acetate, pH 6.5

40 mM ammonium acetate, pH 6.5

Octadecyl silica (Sep-paks, Waters Assoc. or Sepralyte, Analytichem Intl.)

Preparation of Cytokinin–Protein Conjugates for Immunization

Cytokinin ribosides (iPA, ZR, or dHZR) are converted into the corresponding dialdehydes[8] which then form Schiff's bases with the ε-amino groups of lysine residues of BSA. The complexes are stabilized by reduction with sodium borohydride. Three separate conjugates are required: for ZR, iPA, and dHZR. Before use, the purity of the cytokinins should be checked by HPLC. Some commercial samples of ZR contain as much as 10% of the cis isomer.

The cytokinin riboside (0.13 mmol) is dissolved in 10 ml 50% aqueous methanol, 0.9 mmol sodium periodate is added, and the mixture is incubated at 4° in the dark for 1 hr. Bovine serum albumin (2 μmol, 120 mg) is added, the pH is adjusted to 9.3 with 0.3 M K_2CO_3, and the incubation is continued for 1 hr. An excess of sodium borohydride (3.6 mmol, 135 mg) is added and allowed to react overnight.

The pH is adjusted to 5.5 with 1 M formic acid, readjusted after 5 min to pH 8.5 with 1 M K_2CO_3, the conjugate is dialyzed exhaustively against PBS, and concentrated to about 8 mg/ml by ultrafiltration (Millex CX-10 ultrafilter, Millipore Corp., Bedford, MA). Difference spectra show a typical cytokinin absorbance peak at 268 nm equivalent to binding 15–17 mol of cytokinin/mol BSA. Conjugates should be stored frozen at −20°.

Immunization Schedule. Each conjugate is dissolved in 0.14 M NaCl at 4 mg/ml, mixed with an equal volume of Freund's complete adjuvant, and used to raise antiserum in New Zealand white rabbits. The first injection of 1 mg conjugate is distributed intramuscularly and subcutaneously. Two subsequent weekly injections of 1 mg are given in Freund's incomplete adjuvant, followed by a rest period of 5 weeks and by a final injection of 0.5 mg. Animals are bled at the end of the tenth week, the blood is allowed to clot for 1 hr at 25°, overnight at 4°, centrifuged 12,000 g/10 min, and the serum is stored at −20°. Crude serum can be used directly for RIA if the titer is sufficiently high, but must be further purified for use in immunoaffinity chromatography. Titer may be determined by serially diluting the antiserum in PBS and measuring the binding of [^{3}H]cytokinin trialcohol using the RIA procedure described below.

Preparation of Cytokinin–Polylysine–Agarose Columns

Prior to beginning the preparation, a 0.5 ml ODS column is pretreated sequentially with 10 ml portions of 2-propanol, methanol, TEA buffer,

[8] B. F. Erlanger and S. M. Beiser, *Proc. Natl. Acad. Sci. U.S.A.* **52,** 68 (1964).

and 20 ml water. Cytokinin riboside (6 μmol) is dissolved in 200 μl 50% aqueous methanol and 12 μmol solid $NaIO_4$ is added. The mixture is shaken at 0° in the dark for 30 min. It is diluted with 2 ml water, and run through the pretreated ODS column. The column is washed with water (5 ml), purged briefly with nitrogen, and the cytokinin dialdehyde is eluted with methanol (3 ml). The eluate is lyophilized, dissolved immediately in 200 μl of water, and added to 20 ml poly(L-lysine) agarose slurried in a minimum volume of water. The pH is adjusted to 9.3 by the addition of 1 *M* Na_2CO_3 and the slurry is tumbled slowly in the dark at 4° for 1 hr. Solid $NaBH_4$ (10 mg) is added and the mixture is tumbled for a further 16 hr. The suspension is then adjusted to pH 5.5 with 1 *M* formic acid, readjusted after 1 hr to pH 8.5 with 1 *M* NH_4OH, and washed with 50 ml portions of 50 m*M* Na_2CO_3–$NaHCO_3$, pH 8.0, PBS, PBS–NaCl, PBS–methanol, PBS, glycine–ether–HCl, and PBS.

The washes with PBS-methanol and glycine-ether-HCl are most critical since they remove adsorbed cytokinin trialcohol. Store the product in PBS–azide at 4°.

Affinity Purification of Cytokinin-Specific Antibodies

The cytokinin–polylysine–agarose (5 ml) is packed into a 10 ml polypropylene syringe jacketed at 37° and prewashed with 5 volumes of PBS, PBS–NaCl, glycine–ether–HCl, and PBS. In order to determine the column capacity, several 5-ml aliquots of serum whose titer is known may be applied at 1 ml/min, the effluents are collected, and their titers are determined.

For normal serum purification, an amount of antiserum equivalent to 50% of the column capacity is applied at 1 ml/min, washed with PBS–NaCl until the A_{280} reading approaches that of PBS–NaCl, and eluted at 0.3 ml/min with glycine–ether–HCl. Fractions (1 ml) are collected directly into 350 μl ice cold 1 *M* Tris–HCl pH 8.0. Each fraction is mixed rapidly with the Tris–HCl as it is eluted. An aliquot may be removed from each fraction at this point to determine [^{3}H]cytokinin trialcohol binding. Active antibody elutes in the first 30 ml, but the bulk elutes with the majority of the protein within the first 7 ml.

The active fractions are pooled, the ether is removed rapidly *in vacuo* at 25° (using a rotary evaporator), and concentrated by use of a Millex CX-10 submersible filter. It is essential that, during the elution, the ether is removed from the eluted fractions as rapidly as possible and that the pH is raised to between 7.0 and 8.0 to minimize denaturation. Some denatured protein is inevitably produced by this method and should be removed by centrifugation at 15,000 *g*/15 min. After a final dialysis against 5 changes of PBS, the purified immunoglobulin may be coupled at once to cellulose or it may be stored at −70° in the presence of an equal volume of glycerol. The cytokinin–polylysine–agarose column, after washing with 10–20 volumes of PBS, may be reused indefinitely.

Antibodies surviving this treatment appear to be relatively resistant to denaturation in that affinity columns prepared from them may be used many times. The columns can withstand repeated exposure to methanol at 37° and still retain substantial activity.

Binding Cytokinin Antibody to Cellulose

The procedure of Cuatrecasas[9] is modified for use with cellulose. Cyanogen bromide is used; consequently all operations should be carried out in an efficient fume hood. Microcrystalline cellulose (TLC grade, 20 g) is washed extensively with 0.1 *M* NaCl and distilled water. The cellulose is then suspended in 200 ml water in a beaker equipped with a magnetic stirrer, pH meter electrode, and thermometer, and cooled in ice. Cyanogen bromide (10 g) is dissolved in 5 ml tetrahydrofuran, added to the cellulose suspension, and the pH is raised to and maintained at 11.0 ± 0.2 by continuous addition of 5 *M* NaOH. The temperature should be held below 30° by the addition of ice if necessary. After 10 min, the consumption of alkali decreases (about 200 ml will have been consumed) whereupon the activated cellulose is filtered rapidly on a Büchner funnel and washed with 1 liter ice water. The moist powder is added immediately, at 4°, to a solution of 20 mg of the affinity-purified antibody in a minimum volume of 0.1 *M* sodium bicarbonate, pH 8.5 at 4°. After 24 hr, any remaining imidocarbonate linkages are deactivated by a 4 hr treatment at 25° with 100 ml 1 *M* ethanolamine hydrochloride, pH 8.5. The antibody-cellulose is washed with PBS and stored in PBS–glycerol (1 : 2) at −20°. Cytokinin capacity is approximately 100 ng/ml.

Note: The filtrate from CNBr activation contains large amounts of cyanide which may be oxidized to cyanate by treatment with a sodium hydroxide–sodium hypochlorite solution before disposal.

Immunoaffinity Purification of Cytokinins

Reagents and Solutions

2 *M* ammonium acetate, pH 6.5
40 m*M* ammonium acetate, pH 6.5
200 μg/ml diethyldithiocarbamate
40 m*M* ammonium acetate pH 6.5 containing 0.5 *M* NaCl and 2% dimethyl sulfoxide

[9] P. Cuatrecasas, *J. Biol. Chem.* **245**, 3059 (1970).

Analytical Precautions

Cytokinins have a high affinity for untreated glass surfaces and for many plastics, and are most readily handled in silanized glass or polypropylene ware. Glassware may be silanized by sequential treatments with 5% (v/v) dimethyldichlorosilane in hexane, hexane, methanol, and water. It is then over-dried. In analyzing subnanogram levels of cytokinins it is necessary to exclude rigorously from the laboratory any synthetic cytokinins at more than microgram levels in order to prevent accidental contamination. All glassware should be treated with strong oxidizing agents before being silanized; polypropylene syringes and tubes should be discarded after use.

Procedure

DEAE-Cellulose. (DE-52, Whatman) is cycled according to the manufacturer's directions, stirred for 3 hr with 2 *M* ammonium acetate, pH 6.5 (adjusting the pH periodically if necessary), and finally washed with 40 m*M* ammonium acetate pH 6.5. It may be stored in this buffer at 4° with the addition of 0.1% sodium azide.

Column Preparation. DE-52 and antibody-cellulose are packed separately into polypropylene syringes, the outlets of which have been plugged with fragments of cellulose tissue. Glass wool should not be used. The antibody-cellulose column should be jacketed at 37° and the two columns should be connected in series with the DE-52 outlet leading directly to the top of the immunoaffinity column. Prior to use both columns are equilibrated with 40 m*M* ammonium acetate, pH 6.4.

Preparation of Plant Extracts. Fresh plant tissue (1 g) is placed in a silanized Corex centrifuge tube and homogenized (Polytron, speed setting 10, 1 min) in 2 ml of methanol containing 400 mg of diethyldithiocarbamate. High specific activity [^{3}H]cytokinin may be added at this point for later determination of recovery. The homogenate is purged briefly with nitrogen, incubated at 4° in the dark for 30 min, centrifuged at 20,000 *g*/5 min, and the supernatant is diluted into 9 volumes of 40 m*M* ammonium acetate, pH 6.5. Wheat germ acid phosphatase (1 mg) is then added to convert the cytokinin 5′-phosphates to free cytokinins. The mixture is incubated at 25°/30 min in the dark, and is run onto the DE-52 column at a flow rate of 2–3 ml/min, and then directly through the immunoaffinity column. The DE-52 column is washed with 3 vol of ammonium acetate 40 m*M*, pH 6.5 and then detached from the immunoaffinity column, which is then washed with a further 10 ml of ammonium acetate followed by 3 ml of 40 m*M* ammonium acetate pH 6.5 containing 0.5 *M* NaCl and 2% dimethyl sulfoxide. The immunoaffinity column is finally washed with

5 ml of water, purged briefly with nitrogen, and eluted with 3–6 ml methanol at 0.5 ml/min. The methanolic extract is lyophilized and subjected to HPLC and RIA as described below.

High-Performance Liquid Chromatography

Reagents, Columns, and Equipment

TEA Buffer: 40 m*M* acetic acid, adjusted to pH 3.4 with distilled triethylamine

Organic phase: equal volumes of methanol and acetonitrile (HPLC grade)

Column: ODS (Altex, 5 μm, 250 $\times$ 4.6 mm)

HPLC UV Detector: a detector should be chosen having a flow cell volume of less than 20 μl and adequate sensitivity and baseline stability. Insensitivity to refractive index changes is also required

Analytical Precautions

Optimum resolution of cytokinins on ODS requires that the pH of the aqueous buffer should lie between 3.3 and 3.5. Complete removal of spurious organic contaminants from the aqueous buffer is also essential when analyses of cytokinins are contemplated in the ng range. This may be achieved by passage of the buffer through a 1 $\times$ 5 cm column of 40 μm ODS, followed by filtration through a prewashed 0.2 μm filter.

Some memory effects are always seen on ODS columns when analyses are performed at the nanogram level. Microgram amounts of cytokinin standards should never be applied to the columns. An ideal standard contains either *cis*-zeatin, *cis*-zeatin riboside, or dihydro-iPA which are not normally present at significant levels in plants. Samples are collected for RIA assay in polypropylene microfuge tubes. Not all polypropylene tubes are suitable. Some strongly absorb cytokinins and cannot be used. We have found that Labcon tubes (VWR Scientific #20170-273 or 20170-320) are suitable. New lots of microfuge tubes should be checked by measuring their ability to bind [^{3}H]zeatin trialcohol.

Sample Application and Elution

The ODS column is cycled five times (using a 10 min program) between 40 m*M* TEA buffer containing 10% organic phase and 100% organic phase in order to reduce the UV background signal. After equilibration in 40 m*M* TEA–10% organic phase for 30 min, the sample is injected in a maximum volume of 50 μl methanol. Using a flow rate of 1 ml/min, a linear gradient is initiated after 1 min to 25% organic phase over 1 hr and is then increased to 50% over the next 15 min, and finally, to 100% over 2 min. The column may be recycled back to 10% organic phase, and after 30 min equilibration, can accept another sample.

Fraction Collection and Evaporation

Fractions (0.5 ml) are collected in polypropylene microfuge tubes and evaporated to dryness *in vacuo* using a Savant centrifugal lyophilizer. Lyophilization should be at the lowest temperature possible, since removal of the solvent at elevated temperatures or at ambient pressure gives rise to acidic conditions which cause complete destruction of the cytokinins present.

Radioimmunoassay

Reagents and Solutions

Phosphate-buffered saline (PBS): prepared as described above
PBS–gelatin: 1 mg/ml gelatin in PBS (dissolved at 100°/10 min)
PBS–ovalbumin: 10 mg/ml ovalbumin in PBS (centrifuged to remove particulate material)
90% saturated ammonium sulfate
50% saturated ammonium sulfate

Preparation of [^{3}H]Cytokinin Trialcohols

The method of Weiler[3] is followed with some modification. *trans*-Zeatin riboside should be purified by preparative HPLC on ODS in order to remove traces of the cis-isomer. The cytokinin (1.5 μmol) in 100 μl 50% aqueous methanol is oxidized to the corresponding dialdehyde with 3 μmol sodium periodate for 45 min in the dark at 4° and is separated immediately from excess periodate by adsorption onto a 0.5 ml ODS column preconditioned with methanol and TEA buffer as described above. After washing the column with 5 ml of water and purging it with nitrogen, the dialdehyde is eluted with 3 ml methanol. The eluate is lyophilized, dissolved in 20 μl methanol, and added to 100 μl 1 *M* sodium bicarbonate, pH 9.3. Tritiated sodium borohydride (50 mCi, 8.9 Ci/mmol) is dissolved in 25 μl 0.5 *M* NaOH and added to the dialdehyde. After 30 min, excess borohydride is decomposed by the addition of glacial acetic acid to pH 5.0 (in an adequate fume hood) and the [^{3}H]cytokinin trialcohol is isolated on a 0.5 ml ODS column, which is washed extensively with

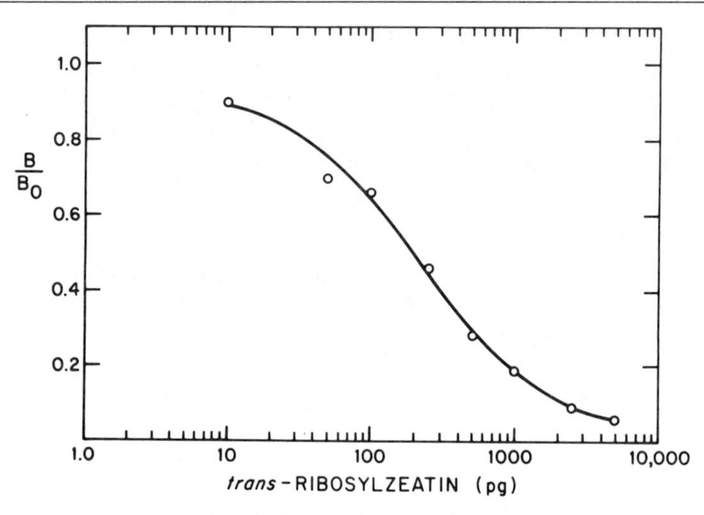

FIG. 1. Standard curve for *trans*-ribosylzeatin.

water to remove ^3H_2O, and finally eluted with methanol. Analytical HPLC on ODS indicates the presence of only one major radioactive component having a retention time appropriate to the cytokinin trialcohol. The specific activity is about 2 Ci/mmol. The solution is stored at −70° prior to use. It should not be concentrated.

Radioimmunoassay Procedure

Reactions are performed in 1.5 ml polypropylene microfuge tubes. Additions are made in the order given: 250 μl PBS–gelatin, 50 μl PBS–ovalbumin, 50 μl [^{3}H]cytokinin trialcohol, 50 μl standard or unknown cytokinin sample. After mixing, diluted antiserum (50 μl) is added in an amount which is sufficient to bind 50% of the [^{3}H]cytokinin trialcohol in the absence of added competitor. Following incubation at 20° for 20 min, the antibody complex is precipitated by the addition of 600 μl 90% saturated ammonium sulfate. The mixture is allowed to stand at 25° for 10 min and the precipitate is collected by centrifugation (Eppendorf centrifuge, 30 sec). The pellet is washed with 400 μl 50% saturated ammonium sulfate, suspended in 130 μl methanol, and after 30 min, 1 ml of scintillation cocktail is added. Radioactivity is determined by scintillation counting. *Note:* The presence of even 5% methanol will seriously affect the antibody–antigen interaction. Dimethyl sulfoxide can be present at 5% without causing problems.

Standard Curve Construction

Cytokinin standards should be assayed concurrently with unknowns, and the appropriate calibration curves constructed. The standards are dissolved in methanol or dimethyl sulfoxide at 1 mg/ml, and serially diluted in methanol directly into the assay tubes to give the desired amount in each tube. After removal of methanol *in vacuo,* PBS–gelatin is added and the RIA performed as described above. A typical standard curve obtained by this method is illustrated in Fig. 1. B_0 is the amount of radioactivity in the antibody precipitate in the absence of any cytokinins; other B values are obtained in the presence of the cytokinin standards at various levels. The detection limit with current antibodies and cytokinin trialcohols is approximately 5 pg.

Applications

Immunoaffinity Column Properties

Studies of the optimum pH, salt, temperature, and flow parameters of polyclonal immunoaffinity columns[10] have indicated that they are capable of operation over a fairly wide pH range of conditions. Their effectiveness at bending cytokinins is good between pH 5 and 8 and is not affected by salt concentrations up to 0.5 M. Operating temperature and flow rates are obviously interrelated. Significant bending of cytokinins does not occur below 25°, but at 37° recoveries are excellent (80%). The columns have been operated as high as 50° and still remain active. The most surprising property is their ability to withstand denaturing agents. In one set of experiments, cytokinins were repeatedly applied to and eluted (with methanol at 37°) from a single column. After an initial drop in capacity of 50%, the column retained constant capacity for the next 17 cycles.

The effectiveness of the immunoaffinity procedure in purifying cytokinins from complex extracts is illustrated by the data in Fig. 2 which compare a conventional purification of cytokinins on ODS with immunoaffinity purification. Figure 2A and B represent the HPLC UV-absorbance and RIA traces derived from the analysis of a sample of cytokinins isolated from 1 g of tobacco crown gall tissue by conventional extraction and passage through DE-52 and ODS. The sample contained approximately 400 ng of ZR but the UV-absorbing material effectively obscured any signal due to the presence of the cytokinin. Figures 2C and D show the corresponding HPLC traces from a sample processed by the immu-

[10] E. M. S. MacDonald and R. O. Morris, unpublished results.

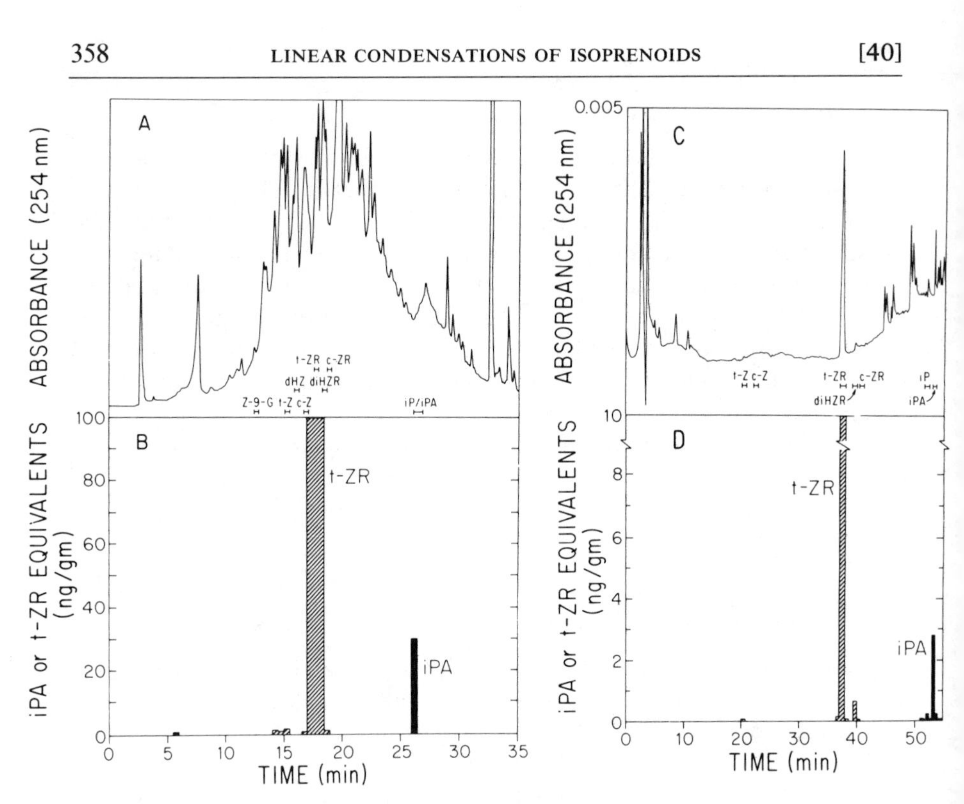

FIG. 2. Purification of cytokinins from complex extracts.

noaffinity protocol. In this case, the cytokinins constitute the predominant UV-absorbing material present in the chromatogram. The ZR peak is apparently homogeneous since estimates of its level obtained by integration of the UV-absorbance trace and from RIA agree within experimental error.

The technique shows sufficient promise that its use may well be considered for the analysis of other substances present at low levels in highly contaminated samples from which conventional purification procedures would have difficulty in removing the analyte.

Acknowledgments

This research was supported by Grants PCM 81-04514 and PCM 83-03371 from the National Science Foundation and by the Science and Education Administration of the U.S. Department of Agriculture under Grant 83-CRCR-1-1249 from the Competitive Research Grants Office.

[41] Squalene Synthetase

By WILLIAM S. AGNEW

Introduction

Squalene synthetase is the term used to designate the enzyme or enzyme complex which, in the presence of NADPH or NADH and Mg^{2+} or Mn^{2+}, condenses two molecules of all-*trans*-farnesyl pyrophosphate to form squalene.[1] Squalene synthetases are associated with the subcellular membranes in mammalian liver and in yeast. Although the enzyme and its products might initially seem to be of only moderate interest, pursuit of the reaction mechanisms and the underlying enzymology in very fundamental ways has stimulated a remarkable chapter in biochemical research. This has included the systematic resolution of all of the stereochemical transformations in the reactions between mevalonate and squalene.[2] Remarkably, there still remain major questions regarding the nature of squalene synthetase and its mechanisms of action.

Among the issues left unresolved is the observation that, although there is little doubt that squalene synthesis is mediated by two catalytic steps, apparently by two enzyme active sites, the reactions nevertheless do not occur independently.[3-9] It would appear that the product of the first reaction (and first active site) may be preferentially transferred to the second active site, without involvement of soluble carrier proteins.[10] Further, the reducing substrate of the second reaction (NADPH, NADH) markedly stimulates the turnover of the first reaction, an effect not mediated by fluctuations of the bulk pool of the water insoluble intermediate, presqualene pyrophosphate.[9,10] Further, given the requirement of soluble carrier proteins for squalene epoxidase, the enzyme may normally trans-

[1] G. Popják and W. S. Agnew, *Mol. Cell. Biochem.* **27,** 97 (1979).
[2] G. Popják, *Harvey Lect.* **65,** 127 (1971).
[3] G. Popják, DeW. S. Goodman, J. W. Cornforth, R. H. Cornforth, and R. Ryhage, *J. Biol. Chem.* **236,** 1934 (1961).
[4] H. C. Rilling, *J. Biol. Chem.* **241,** 3233 (1966).
[5] G. Popják, J. Edmond, K. Clifford, and V. Williams, *J. Biol. Chem.* **244,** 1897 (1969).
[6] W. W. Epstein and H. C. Rilling, *J. Biol. Chem.* **245,** 4597 (1970).
[7] J. Edmond, G. Popják, S.-M. Wong, and V. P. Williams, *J. Biol. Chem.* **246,** 6254 (1971).
[8] E. J. Corey and R. P. Volante, *J. Am. Chem. Soc.* **98,** 1291 (1976).
[9] F. Muscio, J. P. Carlson, L. Kuehl, and H. C. Rilling, *J. Biol. Chem.* **249,** 3746 (1974).
[10] W. S. Agnew and G. Popják, *J. Biol. Chem.* **253,** 4566 (1978).

ISBN 0-12-182010-6

fer the product squalene to a carrier protein. Thus, the usual product of the synthetase appears to be squalene complexed to a protein. These observations suggest that the intricacy of the reaction mechanisms may be matched by that of the protein itself.

This chapter briefly describes methods pertinent to the assay of the reactions in squalene synthesis. Further, the enzyme may be solubilized and depleted of membrane lipids, with loss of catalytic activities.[11] When reconstituted with the appropriate lipids, normal catalytic behavior can be restored. Methods for solubilization, delipidation, and reconstitution are given. As background to the information provided here, the reader is directed to earlier contributions in this series.[12,13]

Overall synthesis of squalene requires the head-to-head reductive condensation of two farnesyl pyrophosphates (Fig. 1). The two pyrophosphoryl moieties are eliminated as anions, and the pro-*S* hydrogen from C-1 of one of the two farnesyl groups is lost as a proton. This is replaced, with retention of absolute configuration, by the B-side (pro-*S*) hydrogen of NADPH (or NADH) as a hydride ion. The hydrogens around C-1 of the second farnesyl group are retained, but with inversion of absolute configuration.

The complexity of these reactions suggested at an early stage that there should exist one or more stable intermediates in the condensation.[3] Such an intermediate was first observed,[4] and can most readily be demonstrated, when the reducing substrate is omitted from the incubation. When farnesyl pyrophosphate is added in the absence of NADPH the 30-carbon substituted cyclopropylcarbinyl pyrophosphate, presqualene pyrophosphate accumulates.[4-7] In the presence of NADPH or NADH this compound is converted to squalene. Several lines of evidence lend support to the conclusion that presqualene pyrophosphate is not only a competent, but also an obligatory intermediate.[4-10] This would require the existence of two enzymes, or two distinct active sites on an enzyme complex.

Catalytic Characteristics of Squalene Synthetase

Most work on squalene synthetases has been carried out with the mammalian liver microsomal enzyme or yeast microsomal enzyme. The enzymes from these two sources appear to be similar, though the yeast membranes contain lower levels of membrane bound phosphatases which can attack substrate or intermediates.

[11] W. S. Agnew and G. Popják, *J. Biol. Chem.* **253,** 4574 (1978).

[12] R. H. Cornforth and G. Popják, this series, Vol. 15, p. 359.

[13] G. Popják, this series, Vol. 15, p. 393.

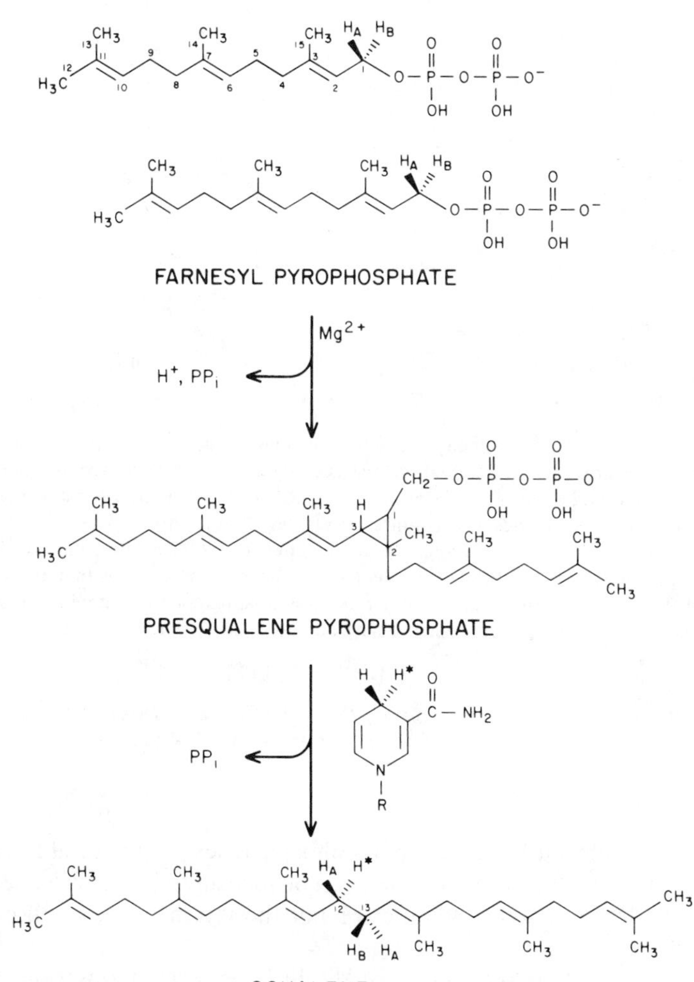

FIG. 1. Reactions catalyzed by squalene synthetase: In the presence of reducing pyridine nucleotide, preferably NADPH, squalene is formed with the indicated stereochemical rearrangements of hydrogens on the C-1 carbon of farnesyl pyrophosphate. In the absence of reducing cofactor the rate of the condensation reaction is lower, and all of the product accumulates as presqualene pyrophosphate.

The substrate requirements for overall squalene synthesis are farnesyl pyrophosphate, Mg^{2+}, and either NADPH (preferred substrate) or NADH. If the reduced nucleotide is omitted, only the condensation reaction occurs, with concomitant accumulation of presqualene pyrophosphate as well as the presqualene monophosphate and presqualene alcohol

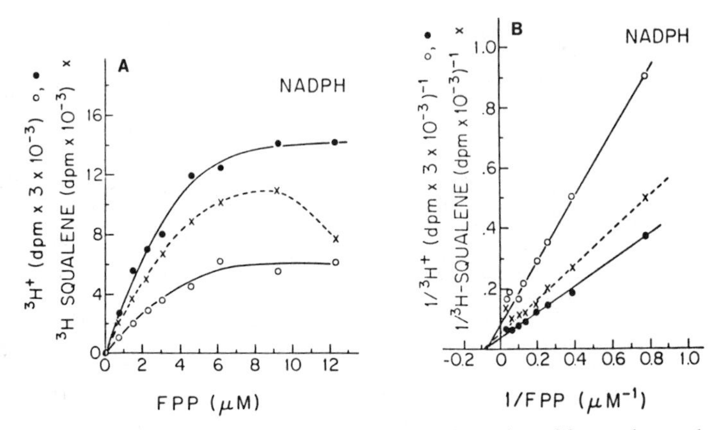

FIG. 2. (A) Substrate vs velocity curve for the condensation of farnesyl pyrophosphate, monitored as ejected $^3H^+$, and overall synthesis of squalene. Farnesyl pyrophosphate is the varied substrate and the reaction was in 1 m*M* NADPH. Note the pronounce substrate inhibition of squalene synthesis. The inhibitory effects generally are not observed at concentrations below 25 μM farnesyl pyrophosphate, and may result from detergent-like character of this compound. (B) Double reciprocal plots of data from A. Note that the effect of NADPH on proton release results from an increase in apparent V_{max} rather than to alterations in substrate affinity. Proton release may be described by reaction kinetics of the form

$$v = V_{max}[FPP]^2/[K_1[FPP] + K_2[FPP] + [FPP]^2]$$

which, after cancelling common substrate terms, looks like a simple Michaelis–Menten equation. These kinetics may be explained by a Ping-Pong–Bi-Bi reaction mechanism (10 and refs therein.)

which are produced by endogenous phosphatases. With suitably labeled substrate the condensation reaction can be measured by the release of the 1-*pro-S* hydrogen as H^+, whether or not the intermediate is allowed to accumulate.

When NADPH is present, normally little presqualene pyrophosphate accumulates, so there is nearly stoichiometric release of protons and overall synthesis of squalene. Interestingly, NADH is slightly less effective in supporting stoichiometric conversion *in vitro*. Farnesyl pyrophosphate, however, is a substrate inhibitor of the second, although not the first reaction. Thus high levels of farnesyl pyrophosphate ($\geq 25\ \mu M$) inhibit appearance of squalene, but do not slow proton release. Under these conditions a net accumulation of presqualene pyrophosphate may be detected.[9] This inhibition may be due to the similarity of farnesyl pyrophosphate to presqualene pyrophosphate, representing an authentic competitive inhibition, or may be due to the detergent properties of farnesyl pyrophosphate, with deleterious effects on the coupling of the reactions.

In addition, the rate of condensation, measured by proton release, is stimulated by addition of NADPH or NADH. The enhancement is due to an increase in the maximal enzyme velocity and does not reflect altered substrate affinity (Fig. 2). These observations seem consistent with two distinct enzyme active sites, and with the interpretation that the enzymes form an aggregate or complex in the membrane, with the water insoluble intermediate preferentially transferred from one active site to the other, with no accumulating pool of intermediate. In keeping with this it has been observed that the stimulation of proton release by NADPH is not blocked by allowing a pool of presqualene pyrophosphate to accumulate during a preincubation. Further, the rate of squalene synthesis seems more closely correlated with the rate of the condensation reaction than the amount of presqualene pyrophosphate present. This suggests that substrate "delivered" from the active site of the first reaction may be preferred over presqualene pyrophosphate accumulated in the membranes. Although these interpretations remain speculative, they are instructive in considering conditions for assay of the reactions and for solubilization and reconstitution of the enzyme. The enzyme may be solubilized with the bile salt sodium deoxycholate. The soluble form is catalytically less active and thermally labile. Activity together with all of the properties suggesting an enzyme complex can be restored by removal of detergent in the presence of appropriate membrane lipids.

Methods

Choice of Substrates for Assays

Squalene synthesis can be measured with the immediate precursor presqualene pyrophosphate as substrate, or the penultimate precursor farnesyl pyrophosphate. Presqualene pyrophosphate presents the difficulty that, as a water-insoluble compound, it must be introduced as an emulsion with detergents which in themselves may alter enzyme behavior. Further, exogenous presqualene pyrophosphate seems to be less preferred as a substrate than that generated by the condensation reaction. Farnesyl pyrophosphate provides a water-soluble substrate for both steps in squalene synthesis, although its detergent properties may make determinations of its concentration ambiguous. The substrate may be present exclusively as monomers, as micelles, or may partition between membrane and solution, depending on factors not easily analyzed. These considerations compromise detailed interpretations of kinetic mechanisms of the enzyme. Nevertheless, as a substrate farnesyl pyrophosphate gives reproducible results, if its concentration does not exceed ~25 μM.

In the formation of presqualene pyrophosphate the 1-*pro-S* hydrogen is eliminated. Thus, when farnesyl pyrophosphate labeled with tritium at carbon 1 is used, the products presqualene pyrophosphate and squalene will have specific activities 1.5 that of the substrate. One tritium is released as $^3H^+$ for every three in squalene or presqualene. Thus, this substrate allows measurement of the condensation reaction whether or not presqualene pyrophosphate accumulates or is converted further to squalene. The effects of pyridine nucleotide on the condensation reaction can be unambiguously monitored. Inhibitors may easily be tested for their site of action. Thus most questions about the enzyme may be addressed with [1-$^3H_2$]farnesyl pyrophosphate and a labeled presqualene pyrophosphates as substrates.

[1,3,5-$^3H_6$]Farnesyl pyrophosphate is commercially available. This may be used to assay presqualene pyrophosphate synthesis or squalene synthesis, corrections being made in each case for loss of isotope during condensation. It may also be used for measuring the proton released during condensation as described below. However, in this procedure failure to remove all of the alcohols produced by phosphatase action can lead to overestimation of reaction rates, an error magnified by the higher specific activity of the alcohol compared to the released proton.

The most useful substrate for the assays described here is [1$^3H_2$]farnesyl pyrophosphate. Chemical synthesis of [1-$^3H_2$]farnesyl pyrophosphate is described elsewhere in this series.[12,14] An alternative, if prenyltransferase is available, is to couple, enzymatically, commercially available [1-$^3H_2$]isopentenyl pyrophosphate to unlabeled geranyl pyrophosphate.[13] This allows quantitative synthesis of all-trans product of the same specific activity as the starting isopentenyl pyrophosphate. The product may be isolated by preparative thin layer chromatography on silica gel HF plates with isopropanol : ammonia : water (6 : 3 : 1) as solvent. After scanning to locate farnesyl pyrophosphate, using the commerically available material as a standard, the substrate may be eluted from the gel with aqueous ammonia, and concentrated by rotary evaporation and stored in aqueous ammonia, in the cold for subsequent use in assays. For the proton release assay, specific activities of at least 100 Ci/mol are desired.

Presqualene pyrophosphate may be prepared as a substrate for squalene synthetase from farnesyl pyrophosphate labeled either with ^3H or ^{14}C, and isolated as follows. The incubation medium contains phosphate buffer, pH 7.4, 10 m*M*, $MgCl_2$, 5 m*M*, labeled farnesyl pyrophosphate 0.2–0.4 m*M*, mixed with 1% (v/v) 1-butanol, equilibrated under

[14] V. J. Davisson, A. B. Woodside, and C. D. Poulter, this volume [15].

nitrogen at 30°. To this is added a yeast membrane fraction prepared as described below, 2.5 mg protein/ml final volume. After incubation for 1 hr and the reaction is stopped by addition of 0.25 volumes 1-butanol. The mixture is heated to 60° and then cooled in an ice bath. To this suspension is added 20 g of $(NH_4)_2SO_4$. Following vigorous mixing and the phases are separated by centrifugation. Three or four additional extractions with butanol are carried out until the majority of the radioactivity is removed. The butanol extract is then concentrated by rotary evaporation. Butanol is removed as an azeotrope with aqueous ammonia by constant addition of aqueous ammonia to a volume which is about half of the residual butanol. After the butanol has been removed, methanol containing 10 m*M* ammonia is added. Several cycles of distillation are used to remove the aqueous phase. This is repeated until a thick, turbid material is produced (1–3 ml). The product is applied to a column 40 × 2 cm of DEAE cellulose (Bio-Rad Cellex D) equilibrated with methanol containing 80 m*M* ammonium formate. The column is eluted with a linear gradient of 0 to 0.3 *N* ammonia in 80 m*M* ammonium formate in methanol (800 ml). Presqualene pyrophosphate elutes as the major component between 0.1 and 0.16 *N* ammonia. At lower concentrations of ammonia there are often three small peaks of radioactivity corresponding to farnesenes and to a breakdown product of presqualene pyrophosphate. Trailing the major peak which is presqualene pyrophosphate is a smaller peak of residual farnesyl pyrophosphate. The resolution of these can be decided by thin layer chromatography as described by Popják,[13] with authentic farnesyl pyrophosphate as a standard. The product can be concentrated by rotary evaporation and stored under nitrogen in ammoniacal methanol in the cold.

Incubation Conditions

$[1\text{-}^3H_2]$Farnesyl pyrophosphate as substrate. For measuring squalene synthesis, proton release, or presqualene pyrophosphate synthesis 180 μl of a standard reaction mixture is placed in a 10 ml calibrated, stoppered glass centrifuge tube, equilibrated with nitrogen. This can be stored at 4° for several hours before use.

The standard reaction mixture consists of potassium phosphate buffer, 55 m*M*, pH 7.4, KF, 11 m*M*, NADPH, 1 m*M* (may be omitted if overall squalene synthesis is not desired), $[1\text{-}^3H_2]$farnesyl pyrophosphate (>100 Ci/mol) 27.5 μ*M* (5.5 μ*M*), $MgCl_2$ 5.5 m*M* (added last to avoid precipitation), and ~1 mg unlabeled squalene (solvent free).

For incubation, each tube is brought to 30° in a water bath and the reaction started by addition of 20 μl of enzyme solution (50–150 μg of protein). After incubation for 5–10 min the reactions are stopped by addition of 50 μl of 40% KOH in the case of the proton assay, or with 100 μl of 40% KOH if squalene formation is to be assayed. The samples are worked up as described below.

Mention should be made of several specific considerations concerning the selection of reaction conditions. Phosphate buffer and KF are used to reduce the breakdown of substrate or of the product presqualene pyrophosphate by a membrane bound, nonspecific phosphatase. The substrate concentration for 50% maximal activity ($S_{0.5}$) is 10–12 μM for farnesyl pyrophosphate. However, the reaction is very nearly maximal at ~25 μM. Addition of a higher concentration of farnesyl pyrophosphate inhibits squalene synthesis but not proton release. Thus for maximal activity of the overall reaction, and for stoichiometric appearance of squalene and proton release 25 μM is the upper concentration limit. At the levels of enzyme activity normally used (50–150 μU, where 1 μU catalyzes formation of 1 pmol of squalene per minute) the reaction will consume approximately 10% of the substrate in 8–10 min. Rates are linear with time and nearly linear with protein concentration under these conditions. NADPH (K_m ~70 mM) may be substituted for by NADH although the former is the preferred substrate ($V_{max} \cong 0.4$ of that with NADPH). NADPH produces slightly higher rates of squalene synthesis and better stoichiometry between appearance of squalene and proton release. The coenzymes stimulate proton release by increasing the apparent V_{max} of the reaction by 2- to 3-fold, leaving the $S_{0.5}$ for farnesyl pyrophosphate unchanged.

Presqualene Pyrophosphate as Substrate

The incubation should be as above except that presqualene pyrophosphate is added as an emulsion in low concentrations of detergent such as Tween-80. The amount of detergent the enzyme will tolerate depends on the preparation being assayed. Microsomal enzyme may function well with 0.25% (w/v) final detergent, while reconstituted enzyme recovered from a Sephadex G-200 column, as described below, may be inhibited at this level. In general pilot assays should be carried out to optimize the detergent carrier concentration for the sample to be tested. Because of deleterious effects of the detergent on enzyme activity and because of uncertainties in the efficiency of delivering presqualene pyrophosphate to the active site, these assays should be considered to be semiquantitative.

Work Up of Assays

Proton Release

The proton assay, first used by Rilling and co-workers[6,9] is based on the equilibration of the tritium expelled from one of the two condensing farnesyl pyrophosphates with protons and the primary hydroxyl of methanol. A fixed volume of methanol is added to an assay mixture, which is equilibrated and dehydrated and the methanol is then distilled. From the specific activity of the methanol and the volumes of solvent the number of nmol of released $^3H^+$ is calculated. The procedure is not rapid, but has the redeeming quality of being reproducible, and may be conducted with standard laboratory microdistillation equipment.

The proton release assay is as follows. As described above, the reaction is stopped by addition of 50 μl of 40% KOH. The tube is extracted three times with ~1.5 ml of light petroleum ether to remove volatile alcohols generated by phosphatases. The last few microliters of solvent are removed with a stream of nitrogen. Two milliliters of methanol is added to the sample, and the stoppered tube is heated for >15 min in a sand bath. After cooling 425 mg of anhydrous $MgSO_4$ (freshly dried in an oven at >100°) is added (170 mg/0.10 ml of aqueous sample). The samples then are stored at room temperature >6 hr or, preferably, overnight. The samples are vortexed to break up caked precipitate, centrifuged in a desk top centrifuge, and the supernatant placed, together with one or two boiling chips, in a pear-shaped flask of a standard microdistillation apparatus held in a 60° sand bath. The distillation arm is cooled with circulating tap water, and the distilled methanol collected in a chilled 10 ml calibrated centrifuge tube. Although not necessary, a light vacuum may facilitate distillation. Four such apparatuses may be set up for simultaneous use in a small sand bath. It normally takes 10–15 min to distill 0.7 ml of methanol, of which 0.5 ml is taken for determination of radioactivity by liquid scintillation spectrometry.

The radioactivity in the methanol is related to that in H_2O by the relationship

$$^3\text{H released (dpm)} = K(\text{dpm/0.5 ml methanol}) \quad (1)$$

where K is 6.25, calculated as

$$K = [V_m H_d^M + V_s H_d^w]/V_e H_d^M \quad (2)$$

In this equation V_m is volume of methanol added to the sample; V_s is volume of the incubation mixture; V_e is volume of methanol distilled and counted; H_d^M is mg atoms of dissociable hydrogen per ml of methanol (24.67 at 20°); and H_d^w is mg atoms of dissociable hydrogen per ml of water (111.0 at 20°).

In experiments in which optimal stoichiometries between proton release and squalene synthesis are to be demonstrated, care should be taken to keep farnesyl pyrophosphate concentrations below ~25 μM as mentioned above. Higher levels inhibit the second reaction, while having no

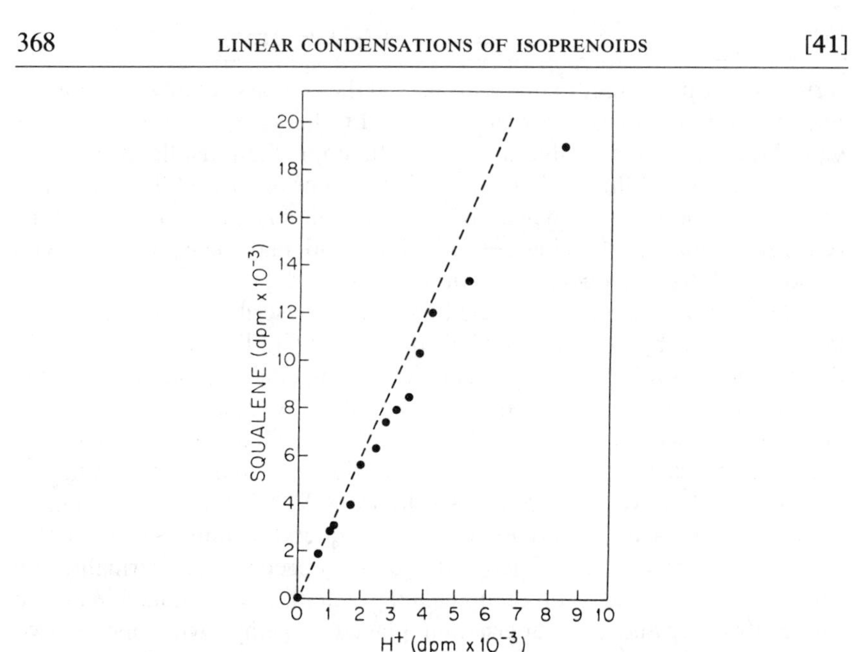

FIG. 3. Stoichiometry of squalene synthesis and proton release when measured by the procedures described. Theoretically there should be three times the level of label recovered in squalene as in H^+ if the stoichiometry between initial and overall reactions were perfect (dashed line).

effect on the first. In the presence of NADPH nearly stoichiometric conversion is observed, with 0.85–1.0 equivalents of squalene recovered for every equivalent of proton released (Fig. 3). As previously indicated, it appears that the rate of squalene synthesis is generally a function of the rate of proton release, rather than the amount of presqualene pyrophosphate which has accumulated.[10]

Squalene Assays

As described above, assays to be analyzed for squalene synthesis are quenched by addition of 0.10 ml of 40% KOH. To these tubes 0.10 ml of 95% ethanol is added. Each tube is blanketed with nitrogen and saponified for 30 min at 60–70° in a sand bath. The unsaponified lipids are extracted three times from the cooled samples with 1.5 ml of light petroleum ether. The pooled petroleum ether fractions are chromatographed on a column of activated alumina (1 g) in a glass column stoppered with glass wool. The hydrocarbons are eluted with an additional 15 ml of petroleum ether, collected in glass scintillation vials. The solvent is removed by evaporation with a stream of nitrogen. If the vial is suspended in a warm water bath (~45°) the time for removing solvent is halved.

Special Considerations

It is important that the reactions not be halted by boiling or by addition of ethanol. These procedures result in variable and low recoveries. Duplicate assays stopped by boiling can vary by more than 2-fold. Apparently these procedures denature the membranes in such a way as to trap the product, squalene. Stopping the reaction by adding base may tend to dissolve rather than precipitate the proteins, and the mild saponification further breaks up the membraneous material.

While failure to saponify results in low recoveries of product, extreme heat during saponification (>70°) should be avoided since it appears to produce farnesene hydrocarbons which are readily eluted from the silica gel columns. Even in the absence of enzyme, or with heat denatured enzyme, too vigorous a saponification results in high backgrounds and contributes to erratic determinations.

Yeast Squalene Synthetase

Preparation of the Microsomal Enzyme

Yeast cells (*Saccharomyces cerevisiae*) are a convenient source of enzyme. One pound blocks may be obtained from a local baker, and kept stored in a refrigerator for 1–2 weeks. Usually 100 g is suspended, with stirring, in a final volume of 250 ml of 0.10 m*M* Tris–HCl buffer at 4°. The cell suspension may be broken in a needle-valve type homogenizer such as a Manton-Gaulin homogenizer, run at 7000 psi through about 80 compression cycles. The homogenate is then cooled for 20 min with stirring in an ice-bath and rehomogenized for an additional 60 compression cycles. The homogenate is cooled and centrifuged at 20,000 *g* for 30 min at 4° in the GSA rotor of a Sorval preparative ultracentrifuge. The supernatant is decanted through six layers of cheesecloth. This supernatant (s_{20}), usually between 190 and 240 ml, is centrifuged at 100,000 *g* for 1 hr. The supernatant is discarded and the pellets resuspended in 100 ml of 1 m*M* EDTA, 10 m*M* $MgCl_2$, 0.10 *M* Tris–HCl buffer, pH 7.6 and centrifuged again at 100,000 *g* for 1 hr. The supernatant is again discarded and the pellet resuspended in 200 ml of 0.1 *M* Tris–HCl buffer at pH 7.6. The suspension can be divided into aliquots and stored at −20°. Normally a sample of the membranes is thawed, dialyzed against 25 m*M* Tris–HCl buffer, 1 m*M*

EDTA at 4°, overnight. Such a suspension contains 3–5 mg protein/ml and ~1 mU squalene synthetase/mg protein.

Solubilization

Solubilization of squalene synthetase can be achieved with the bile salt, sodium deoxycholate. In the soluble state both activities are associated with particles with sedimentation coefficient 3.3 S, a Stokes radius of 40 Å, and computed molecular weight of 54,500.[11] When solubilized both catalytic steps are inhibited, although activity is restored partially by 10-fold dilution of protein into the assay medium, or by removing detergent with cholestyramine resin (Dowex 1X-2-C1). When solubilized the enzyme is extremely sensitive to thermal inactivation, losing 50% of its activity (assayed after removing detergent) in 5 min at 25°.

To solubilize the yeast microsomal enzyme, a suspension is made of 2–5 mg of protein/ml at ice bath temperature. To this is added a stock solution (10%, w/v) of sodium deoxycholate to a final concentration of 0.25%. (Solubilization is maximal between 0.2 and 0.3% deoxycholate under these conditions.) After 10 min with occasional mixing the suspension is centrifuged at 4° for 30 min at 100,000 *g* to remove debris. The supernatant contains about 30% of the initial protein and nearly all of the original enzyme activity. To assay the enzyme an aliquot of the supernatant is first treated with ~10 mg/ml cholestyramine resin to remove detergent. The dry resin is sprinkled into the solution, mixed occasionally for 15 min, and then removed by settling or low-speed centrifugation. The supernatant is kept on ice for an additional 30 min before assay. It should be noted that the enzyme so treated has reprecipitated together with the solubilized membrane phospholipids.

Lipid Depletion

The solubilized protein(s) of squalene synthetase may be separated from much or all of the membrane phospholipids by gel filtration in the presence of detergent. The lipid-depleted preparation can then be used in recombination experiments to study the activation of the two catalytic steps by the lipids. Activation does not simply result from removal of the inhibitory detergent, because without lipid supplementation, removal of deoxycholate fails to activate the enzyme.

For delipidation, chromatography over Sephadex G-200 may be used. A column, 1 × 60 cm, is packed and equilibrated at 4° with 20 m*M* Tris–HCl, pH 7.6 (measured at 4°), and 0.2% (w/v) deoxycholate. Higher ionic strength will result in gelation of the detergent. The column is normally run with a hydrostatic head of 20 cm and a flow rate of 5–6 ml/hr. Fractions of 0.8–1.2 ml are collected. The column should be calibrated with standard proteins in the absence of detergent, because while the filtration properties of the gel are not altered by the detergent, several commonly used molecular weight standards bind to deoxycholate micelles. The solubilized squalene synthetase (both reactions) elutes as a particle with 40 Å Stokes radius.

A sample of soluble squalene synthetase (1 ml, 2–5 mg protein) is delivered to the column which is then run overnight at 4°. During the elution, the enzyme is largely resolved from mixed detergent–lipid micelles. In initial testing of the effectiveness of the procedure, it is useful to assay alternating fractions with and without lipid supplementation. This may be done by adding an aliquot containing 200 μg of pure phosphatidylcholine dispersed in 0.2% deoxycholate to odd or even numbered tubes in the fraction collector. After collecting the fractions, 10 mg of cholestyramine resin is added to each tube, followed by occasional mixing for 15 min. The resin is removed by settling or low-speed centrifugation, and the fractions assayed as described above. It is recommended that the fractions not be assayed less than 45 min after adding the cholestyramine. The results from such an experiment are somewhat variable in that enzyme activity remaining in the peak from the column may be very low (less than 5% of the lipid-supplemented level) to 30–40% of the augmented level in cases of poor separation. In the latter instances, fractions toward the trailing end of the peak generally have higher activity, reflecting incomplete separation from the membrane lipids.

Reconstitution

When solubilized in deoxycholate, the enzyme is thermally unstable, catalytically inactive, and is dispersed as one or more protein species of molecular weight 54,500. If separated from membrane lipids, removing of detergent no longer results in recovery of catalytic activity. However, if supplementary lipids are added before removing detergent (specifically phosphatidylcholine or phosphatidylethanolamine), both enzymatic activities can be recovered and reasonable stoichiometries between proton release and overall squalene synthetase may be demonstrated. The ability of the pyridine nucleotides to stimulate proton release is also recovered.[11] In addition the preparation regains thermal stability and the enzymatic activity and at this point is sedimented by centrifugation at 100,000 *g* for 1 hr. Thus, procedures for recombining the soluble, lipid-depleted protein with membrane lipid may be useful for investigation of mechanisms of lipid activation, for further examination of the question of the separability of the two putative enzymes comprising squalene synthetase, examina-

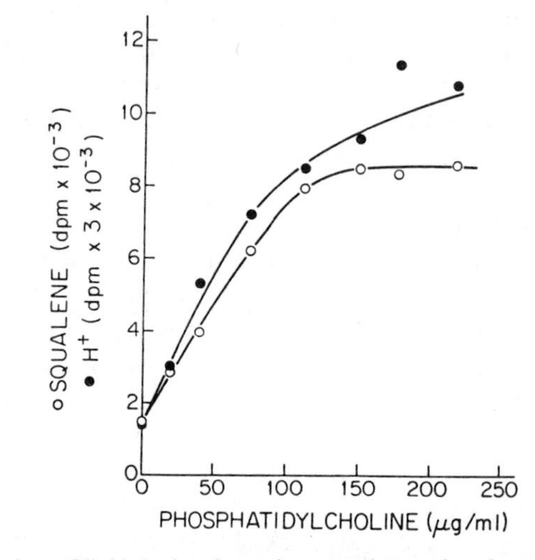

FIG. 4. Activation of lipid-depleted squalene synthetase by phosphatidylcholine. Both proton release and overall squalene synthesis are activated in parallel by addition of the lipid before removal of detergent with cholestyramine resin. Stoichiometry of conversion is nearly perfect, and the reconstituted preparation exhibits the NADPH and NADH stimulation of H^+ release found in the original membrane bound enzyme (not shown).

tion of the coupling of the two stages of the reaction and pyridine nucleotide stimulation.

In general the method for reconstitution of the depleted enzyme with one or another pure phospholipids is the same as used for the column fractions. Pooled, lipid depleted material removed from the Sephadex G-200 column fractions is stored at 0°. If the above procedures are used this material will have 300–600 μg protein/ml. To 0.30-ml aliquots of this material may be added 50 μl of lipid suspended in deoxycholate. After 20–30 min for equilibration 5 mg of cholestyramine resin is added, and the samples mixed occasionally for 45 min. The resin may be removed by settling or low-speed centrifugation and the supernatant assayed by standard procedures. Because some purification has occurred during fractionation on the Sephadex column, 10–20 μg of protein per assay is sufficient. It should be emphasized that the reconstitution procedure described here is rapid and convenient, but should not preclude the use of the slower approach of equilibrium dialysis which may provide a more controlled reformation of lipid–protein aggregates in the form of small liposomes.

Lipids which have been shown to activate the lipid-depleted material include phosphatidylcholine and phosphatidyl ethanolamine. Lipids which are negatively charged at neutral pH, phosphatidylserine and phosphatidylinositol have not been found to activate, and may inhibit the residual activity. It appears that permitting at least 45 min to pass between removing detergent and assaying the enzyme may result in improved stoichiometries of proton release and squalene synthesis, although this has not been systematically analyzed. The ability of the reducing cofactor to enhance the condensation reaction is also restored on reconstitution. This suggests that, if there are two enzyme proteins of ~55,000 Da separately in solution, they may reassociate during or after reconstitution.

Acknowledgments

The author acknowledges being supported by NIH-NINCDS Grant NS17928 and a grant from the National Multiple Sclerosis Society during the writing of this article.

[42] Formation of 12-*cis*-Dehydrosqualene Catalyzed by Squalene Synthetase

By TOKUZO NISHINO and HIROHIKO KATSUKI

Upon incubation of a microsomal fraction from *Saccharomyces cerevisiae* with [^{14}C]farnesyl pyrophosphate in the presence of Mn^{2+}, radioactivity is incorporated into 12-*cis*-dehydrosqualene.[1,2] Dehydrosqualene formation is observed also with a microsomal fraction of *Rhodotorula glutinis*[3] or mammalian microsomes.[2] When the reaction is performed in the presence of NADPH, squalene is formed instead of dehydrosqualene. With a squalene synthetase preparation which is solubilized and partially purified from *S. cerevisiae* using a Sephacryl S-200 column by a modification[2] of the method of Agnew and Popják,[4] the involvement of squalene synthetase in the dehydrosqualene formation was established. Dehydrosqualene is presumed to be formed through elimination of a proton from Cα of an allylic cation, an intermediate in squalene formation in the presence of NADPH.

[1] T. Nishino, H. Takatsuji, S. Hata, and H. Katsuki, *Biochem. Biophys. Res. Commun.* **85,** 867 (1978).
[2] H. Takatsuji, T. Nishino, K. Izui, and H. Katsuki, *J. Biochem.* (*Tokyo*) **91,** 911 (1982).
[3] T. Nishino, N. Suzuki, H. Takatsuji, and H. Katsuki, *Mem. Fac. Sci., Kyoto Univ.* **36,** 67 (1981).
[4] W. S. Agnew and G. Popják, *J. Biol. Chem.* **253,** 4578 (1978).

ISBN 0-12-182010-6

Assay Methods

Principle. When crude microsomal preparations including squalene synthetase from which cofactors such as NADPH or NADP have been removed are incubated with farnesyl pyrophosphate and Mn^{2+} in the absence of NADPH, dehydrosqualene is formed instead of squalene. The formation of dehydrosqualene can be easily detected by reversed phase TLC in which dehydrosqualene is separated from squalene. Presqualene pyrophosphate which is formed under the above conditions[5] is not extracted in the petroleum ether used for extraction of dehydrosqualene.

Reagents. [^{14}C]Farnesyl pyrophosphate can be prepared by incubation of [1-^{14}C]isopentenyl pyrophosphate (specific radioactivity, 56 Ci/mol, Amersham) with geranyl pyrophosphate and the 105,000 *g* supernatant of cell homogenates of *R. glutinis*[2] or *S. cerevisiae*[6] after preincubation with iodoacetamide for inactivation of isopentenyl pyrophosphate isomerase.

- Tris–HCl buffer, 0.1 *M*, pH 7.4
- $MgCl_2$, 0.1 *M*
- $MnCl_2$, 0.1 *M*
- [1-^{14}C]Farnesyl pyrophosphate (sp. act., more than 50 Ci/mol), 0.1 m*M*
- KF, 1 *M*
- KOH, 6 *M*
- Petroleum ether

Procedure. In a 15-ml screw-capped test tube are placed 0.1 ml of Tris buffer, 50 μl of $MgCl_2$, 20 μl of $MnCl_2$, 10 μl of KF, 10 μl of [^{14}C]farnesyl pyrophosphate, the microsomal preparation, and water (final volume 1.0 ml). When the crude microsomal preparation is to be used instead of the partially purified one, it should be passed through a Sephadex G-50 column to remove cofactors such as NADPH, NADP and metal ions. Reaction is for 120 min at 30°. The reaction is stopped by the addition of 0.5 ml of KOH, and the mixture is heated for 60 min at 80°. Nonpolar lipids including dehydrosqualene are extracted with petroleum ether.

Product analysis. The lipids are subjected to TLC analysis on kieselguhr plate (Merck) impregnated with liquid paraffin using 95% acetone saturated with liquid paraffin as solvent (R_f of dehydrosqualene, 0.39; R_f of squalene, 0.25). In radio-gas chromatography using a column of 2% Dexisil 300 GC or 1.5% SE-30, dehydrosqualene has a retention time relative to squalene of 1.40 or 1.23, respectively.

[5] H. C. Rilling, *J. Biol. Chem.* **241,** 3233 (1966).

[6] T. Nishino, N. Suzuki, and H. Katsuki, *J. Biochem.* (*Tokyo*) **92,** 1731 (1982).

Properties

Presqualene pyrophosphate (also isopentenyl pyrophosphate when the 15,000 *g* supernatant is used) can be used as a substrate instead of farnesyl pyrophosphate. For the dehydrosqualene formation, Mn^{2+} is more effective than Mg^{2+}. This is presumed to be due to the action of Mn^{2+} which makes the cleavage of the C–O bond of presqualene pyrophosphate easier.[6] The ratio of squalene to dehydrosqualene formed under optimal conditions, in the presence or absence of NADPH, is about 200 : 1. Presqualene alcohol and several species of C_{30}-alcohols[2] with unidentified structures, which are presumed to be derived from the cation intermediates postulated by Rilling, Poulter, and their collaborators,[7,8] are formed besides dehydrosqualene.

[7] H. C. Rilling, C. D. Poulter, W. W. Epstein, and B. Larsen, *J. Am. Chem. Soc.* **93,** 1783 (1971).
[8] C. D. Poulter, O. J. Muscio, and R. J. Goodfellow, *Biochemistry* **13,** 1530 (1974).

[43] Squalene Epoxidase from Rat Liver Microsomes

By TERUO ONO and YOH IMAI

$$\text{Squalene} + O_2 + \text{NADPH} + H^+ \longrightarrow \text{2,3-Oxidosqualene} + \text{NADP}^+ + H_2O$$

The mixed-function oxidases of liver microsomes which catalyze the hydroxylation of a wide variety of lipophilic substrates have been shown to consist of a NADPH–cytochrome *P*-450 reductase, cytochrome *P*-450 isozymes, and phosphatidylcholine.[1] The squalene epoxidase [EC 1.14.99.7, squalene monooxygenase (2,3-epoxidizing)] system is comprised of the terminal oxidase which is distinct from hemoproteins such as cytochrome *P*-450 isozymes,[2] and a flavoprotein identical with NADPH–cytochrome P-450 reductase.[3]

NADPH–Cytochrome *c* (*P*-450) Reductase [EC 1.6.2.4]

Assay Method

Assay of NADPH–cytochrome c reductase activity is performed with an Hitachi 124 spectrophotometer. For determining activity the enzyme

[1] M. J. Coon, this series, Vol. 52, p. 200.
[2] T. Ono, K. Nakazono, and H. Kosaka, *Biochim. Biophys. Acta* **709,** 84 (1982).
[3] T. Ono, S. Ozasa, F. Hasegawa, and Y. Imai, *Biochim. Biophys. Acta* **486,** 401 (1977).

ISBN 0-12-182010-6

(2–10 μg of protein) is added to 0.1 *M* potassium phosphate buffer, pH 7.4, which contains 0.02 μmol of cytochrome *c* (Boehringer, horse heart) and 0.1 μmol of NADPH (Sigma). The final volume is 1.0 ml. NADPH is added to initiate the reaction. One unit of the activity represents 1 μmol of cytochrome *c* reduced per min.

Preparation of NADPH–Cytochrome c (P-450) Reductase

The NADPH–cytochrome *c* (*P*-450) reductase fraction eluted in the first DEAE-cellulose chromatography is further purified on a 2′,5′-ADP Sepharose column according to Yasukochi and Masters.[4] The specific activity of the purified reductase is 45–55 μmol cytochrome *c* reduced per min per mg of protein.

Squalene Epoxidase

Assay Method

Principle. The activity of squalene epoxidase can be determined in the reconstituted system of NADPH–cytochrome *P*-450 reductase and squalene epoxidase by measuring the formation of radioactive 2,3-oxidosqualene from [^{14}C]squalene.[5]

Reagents

[^{14}C]Squalene (1000–2000 cpm/nmol) in benzene, 1 m*M*
FAD, 1 m*M*
NADPH, 1 m*M*
Tween-80 in acetone, 0.25%
Tris–HCl buffer, 0.5 *M*, pH 7.4
Potassium hydroxide in methanol, 10%

Procedure. The substrate, either [11-^{14}C]squalene purchased from CEA, France or [^{14}C]squalene prepared enzymatically from [2-^{14}C]mevalonolactone is purified by alumina column chromatography as described by Popják.[6] The specific activity of the labeled squalene can be adjusted by mixing with unlabeled squalene which was purified through the thiourea clathrate.[6] The assay incubation is carried out at 37° in a Pyrex culture tube (12 ml). Ten microliters of [^{14}C]squalene and 20 μl of Tween-80 are mixed, and the solvents are evaporated under nitrogen. Along with substrate and detergent, the reaction mixtures contain in a volume of 0.5

[4] Y. Yasukochi and B. S. S. Masters, *J. Biol. Chem.* **251,** 5337 (1976).
[5] S. Yamamoto and K. Bloch, *J. Biol. Chem.* **245,** 1670 (1970).
[6] G. Popják, this series, Vol. 15, p. 442.

ml, 20 mM Tris–HCl buffer, pH 7.4, 0.01 mM FAD, 0.2 unit NADPH–cytochrome *c* (*P*-450) reductase, squalene epoxidase (1–100 μg of protein), Triton X-100 (0.05–0.3%), and 1 mM NADPH. NADPH is added to initiate the reaction. Mixtures are incubated at 37° for 30 min. The reaction is stopped by the addition of 0.5 ml of 10% KOH-methanol and the nonsaponifiable lipids are extracted twice with 3 ml of petroleum ether. The extract is evaporated under nitrogen. The residue is dissolved in a small amount of diethyl ether and applied to a silica gel thin layer plate (Kiesel gel 60 F_{254}, Merck, 2 × 10 cm). The plate is developed with ethylacetate/benzene (0.5 : 99.5, v/v).[5] R_f values are 0.91 for squalene, 0.54 for 2,3-oxidosqualene, 0.21 for lanosterol, and 0.13 for cholesterol.[5,7] Plates are divided into three bands, squalene, epoxide, and sterol regions, and each band is scraped into a scintillation vial. Radioactivity is determined by liquid scintillation spectrometry. Controls for nonenzymatic squalene epoxidation are performed by omitting addition of the enzyme.

Purification Procedure

Wistar strain rats, weighing 200–300 g, are fed Oriental chow from Oriental Co. and water *ad libitum* without any pretreatment with chemical inducers of cytochrome *P*-450 isozymes. Rats are fasted overnight prior to sacrifice. Rat livers are perfused with 0.1 M Tris–HCl buffer, pH 7.4, containing 1 mM EDTA. After mincing, the tissue is homogenized in a Potter-Elvehjem homogenizer with a loose fitting Teflon pestle in the presence of 2 volumes of buffer. The homogenate is centrifuged at 10,000 *g* for 10 min and the supernatant is recentrifuged at 78,750 *g* for 90 min in a Beckman Model 12-65B centrifuge (rotor 35). The sedimented material is dispersed in a buffer volume equal to that of 10,000 *g* supernatant and further centrifuged at 78,750 *g* for 90 min. The sediment is taken up in a volume of buffer equivalent to one-third the original liver weight and homogenized. The microsomal suspension has a protein concentration of 30–40 mg/ml. Alternatively active microsomes can also be prepared with 0.25 M sucrose or 1.15% KCl as suspending media.

All buffers contained 0.5% Triton X-100, 1 mM EDTA, and 1 mM dithiothreitol throughout the purification procedures unless otherwise stated.

Step 1. Solubilization of Microsomes.[8] Squalene epoxidase is a membrane-bound enzyme. For solubilization, 100 ml of a microsomal suspension (~3–4 g of protein) in 0.1 M Tris–HCl buffer, pH 7.4, containing 1

[7] J. A. Nelson, S. R. Steckbeck, and T. A. Spencer, *J. Biol. Chem.* **256,** 1067 (1981).
[8] T. Ono and K. Bloch, *J. Biol. Chem.* **250,** 1571 (1975).

m*M* EDTA is mixed with 25 ml of 10% Triton X-100 (final concentration of detergent, 2%). The mixture is gently stirred for 20 min at 4°, diluted with 375 ml of 1 m*M* EDTA and 1 m*M* dithiothreitol to lower the concentration of detergent to 0.5%, and centrifuged at 78,750 *g* for 90 min.

Step 2. DEAE-Cellulose Chromatography.[8] The 78,750 *g* supernatant (~460 ml) resulting from the above solubilization procedure is applied to a DEAE-cellulose column (4.6 × 21 cm) previously equilibrated with 20 m*M* Tris–HCl, pH 7.4. After washing the column with 1.5 bed volumes of equilibration buffer, protein is eluted with a linear gradient consisting of 1200 ml of 20 m*M* Tris–HCl pH 7.4 and 1200 ml of the same buffer plus 0.5 *M* KCl and 20-ml fractions are collected. Squalene epoxidase activity is recovered at about 0.1 *M* KCl and NADPH–cytochrome *c* (*P*-450) reductase is recovered at about 0.25 *M* KCl. Fractions positive for squalene epoxidase and NADPH–cytochrome *c* reductase are combined and dialyzed overnight against 6 liters of 40 m*M* potassium phosphate, pH 7.4 (buffer A) and 20 m*M* Tris–HCl, pH 7.4 (buffer B), respectively. NADPH–cytochrome *c* (*P*-450) reductase fraction is further purified according to Yasukochi and Masters[4] as mentioned above.

Step 3. Alumina Cγ gel Fractionation. The DEAE-cellulose fraction (~300–400 ml), which is contaminated with NADH-ferricyanide reductase and cytochrome b_5 is treated with alumina Cγ gel (aged, Sigma) by a batch method. One gram by wet weight of alumina Cγ gel is suspended in 40 ml of the enzyme solution. The mixture is gently stirred for 60 min and then centrifuged at 10,000 *g* for 20 min. The precipitates are washed twice with 150 ml of 70 m*M* buffer A by centrifugation. Finally the sedimented material is suspended in 150 ml of 200 m*M* buffer A and extracted for 60 min. Unextracted material is removed by centrifugation for 20 min at 10,000 *g* and 4°. The supernatant thus obtained is dialyzed overnight against 4 liters of 40 m*M* buffer A.

Step 4. Hydroxylapatite Chromatography. The dialyzed alumina Cγ gel fraction is applied to a hydroxylapatite column (1.8 × 8.7 cm) previously equilibrated with 50 m*M* buffer A and then eluted with a linear gradient formed from 150 ml each of 50 m*M* buffer A and of 300 m*M* buffer A. Squalene epoxidase is eluted at phosphate buffer concentration between 100 to 150 m*M*. Fractions positive for squalene epoxidase fraction are pooled and dialyzed against 4 liters of 20 m*M* buffer A, pH 7.2.

Step 5. CM-Sephadex C-50 Chromatography. The dialyzed hydroxylapatite fraction is applied to a CM-Sephadex C-50 column (bed volume, 25 ml) and then eluted with a linear gradient consisting of 30 m*M* buffer A, pH 7.2 (100 ml) and 300 m*M* buffer A, pH 7.2 (100 ml). The fractions positive for squalene epoxidase activity are pooled and dialyzed against 50 m*M* buffer B, pH 7.4.

PURIFICATION OF SQUALENE EPOXIDASE FROM RAT LIVER MICROSOMES

Steps	Protein (mg)	Squalene epoxidase: Total activity (nmol/30 min)	Specific activity (nmol/30 min/mg)
DEAE-cellulose	810	1053	1.3
Alumina Cγ gel	172	826	4.8
Hydroxylapatite	52	484	9.3
CM-Sephadex C-50	14	287	20.5
Blue Sepharose 4B	1.24	107	86.2
Chromatofocusing	0.24	45	185.8

Step 6. Cibacron Blue Sepharose 4B Chromatography.[9] The CM-Sephadex C-50 fraction is applied to a Cibacron Blue Sepharose 4B column (bed volume, 10 ml) previously equilibrated with 50 m*M* buffer B. After extensive washing of the column with 0.2 *M* KCl–50 m*M* buffer B, squalene epoxidase is eluted with 1.0 *M* KCl–50 m*M* buffer B. The Blue Sepharose 4B fraction is dialyzed against 50 m*M* buffer B and stored at −70° until use.

Step 7. Chromatofocusing.[2,10] About 12 ml (~1.5 mg of protein) of the Blue Sepharose fraction are applied to a column of chromatofocusing gel PBE 94 (Pharmacia) (1.2 × 8 cm), previously equilibrated with 25 m*M* ethanolamine acetate buffer (pH 9.4), containing 0.5% Triton X-100 and 0.1 m*M* dithiothreitol. The pH gradient elution is achieved by Polybuffer 96 (Pharmacia), diluted 10-fold with distilled water and titrated with 1.0 *M* acetic acid to pH 6.0, containing 0.5% Triton X-100 and 0.1 m*M* dithiothreitol. Four-milliliter fractions are collected. The fractions positive for squalene epoxidase activity are pooled as the final enzyme preparation and stored at −70°. The results of a typical experiment on the isolation of squalene 2,3-monooxygenase from rat liver microsomes are shown in the table.

Purity. Based on polyacrylamide gel electrophoresis in Triton X-100 and in SDS, the squalene epoxidase is nearly homogeneous (90–95% pure) after purification through step 7. The final yield is about 4.2%, with specific activity of 186 nmol/mg of protein/30 min for 2,3-oxidosqualene under the standard assay conditions. The minimal molecular weight is about

[9] T. Ono, K. Takahashi, S. Odani, H. Konno, and Y. Imai, *Biochem. Biophys. Res. Commun.* **96,** 522 (1980).

[10] T. Ono, S. Odani, and K. Nakazono, *in* "Oxygenase and Oxygen Metabolism" (M. Nozaki, S. Yamomoto, Y. Ishimura, M. J. Coon, L. Ernster, and R. W. Estabrook, eds.), p. 637. Academic Press, New York, 1982.

51,000. Chromatofocused enzyme solutions (2–4 μM) showed no distinct absorption spectrum in the visible regions. CO-reduced minus oxidized difference spectrometry also failed to detect the presence of a cytochrome *P*-450 isozymes.

Properties.[2] The enzyme activity was linearly proportional to a protein concentration of between 2 to 50 μg/0.5 ml in the presence of 0.2 unit of NADPH–cytochrome *P*-450 reductase. FAD is required for enzyme activity.[11] A double reciprocal plot of velocity of reaction against substrate concentration and FAD indicated apparent K_m values for squalene and FAD of 13 and 5 μM, respectively. The specific activity was estimated to be approximately 10 nmol 2,3-oxidosqualene formed per nmol enzyme per 30 min, indicating a turnover number of 0.33.

The enzyme is stable for several weeks at −70° in 50 m*M* Tris–HCl buffer, pH 7.4, containing 0.5% Triton X-100. After freezing and thawing once, squalene epoxidase activity is decreased by 80% when compared with control.

SKF 525 A at concentration up to 0.1 m*M* and metyrapone at 50 μM did not inhibit product formation. Similarly no inhibition of squalene epoxidation is noted when substrates for cytochrome *P*-450, such as benzphetamine at 0.2 m*M* and aniline at 0.1 m*M*, are included in the reaction medium. Agents that modify the steady-state concentration of oxygen reduction products, such as sodium azide (0.5 m*M*), mannitol (20 m*M*), catalase (1000 IU), or superoxide dismutase (10 IU), are without effect on the squalene epoxidase activity.[2]

[11] T. Ono, T. Okayasu, K. Kameda, and Y. Imai, *in* "Biochemical Aspects of Nutrition" (K. Yagi, ed.), p. 115. Jpn. Sci. Soc. Press, Tokyo, 1979.

Section III

Cyclization Reactions

[44] Monoterpene and Sesquiterpene Cyclases

By RODNEY CROTEAU and DAVID E. CANE

The cyclic monoterpenes (C_{10}) and sesquiterpenes (C_{15}) constitute two large groups of natural products synthesized and accumulated in distinct glandular structures by some 50 families of higher plants or excreted by more than 60 fungal species, particularly those belonging to the *Fungi imperfecti* or *Basidiomycetes,* as well as by a limited number of *Streptomyces*. Unlike higher plants, which produce both monoterpenes and sesquiterpenes, microorganisms appear to be largely restricted to the synthesis of sesquiterpenes, in spite of a few scattered reports of the occurrence of acyclic monoterpenes in fungi. Various species of *Laurencia* (red algae) and marine sponges are also prolific producers of a wide variety of both halogenated and nonhalogenated sesquiterpenes. Phytoalexins, which include a rich spectrum of cyclic and acyclic sesquiterpenes, represent a special class of inducible plant metabolites which are absent in healthy plant tissue but which are secreted in response to external challenge, particularly tissue damage or exposure to fungi. In all these organisms, the characteristic monoterpenes and sesquiterpenes are derived from the cyclization of one of two acyclic precursors, geranyl pyrophosphate (monoterpenes) or farnesyl pyrophosphate (sesquiterpenes)[1] (Fig. 1). The initially formed cyclization products, consisting of one or more rings, are generated as olefins or simple alcohol or ether derivatives. Subsequent transformations, frequently oxidative and including, in some cases, extensive carbon–carbon bond cleavage, ultimately give rise to the vast number of mono- and sesquiterpenoid metabolites. The crucial cyclization reactions by which the parent monoterpene and sesquiterpene carbon skeletons are generated are catalyzed by enzymes collectively known as cyclases. Multiple cyclases, each producing a different skeletal arrangement from the same acyclic precursor, often occur in higher plants while single cyclases which synthesize a variety of skeletal types are also known.[2–4] Individual cyclases, each generating a simple derivative or positional isomer of the same skeletal type, have been described, as have

[1] For recent comprehensive reviews of monoterpene and sesquiterpene biosynthesis, respectively, see (a) R. Croteau, *in* "Biosynthesis of Isoprenoid Compounds" (J. W. Porter and S. L. Spurgeon, eds.), p. 225. Wiley, New York, 1981 and (b) D. E. Cane, *in* "Biosynthesis of Isoprenoid Compounds" (J. W. Porter and S. L. Spurgeon, eds.), p. 283. Wiley, New York, 1981.

[2] R. Croteau and F. Karp, *Arch. Biochem. Biophys.* **179,** 257 (1977).

[3] R. Croteau and F. Karp, *Arch. Biochem. Biophys.* **198,** 512 (1979).

[4] H. Gambliel and R. Croteau, *J. Biol. Chem.* **259,** 740 (1984).

ISBN 0-12-182010-6

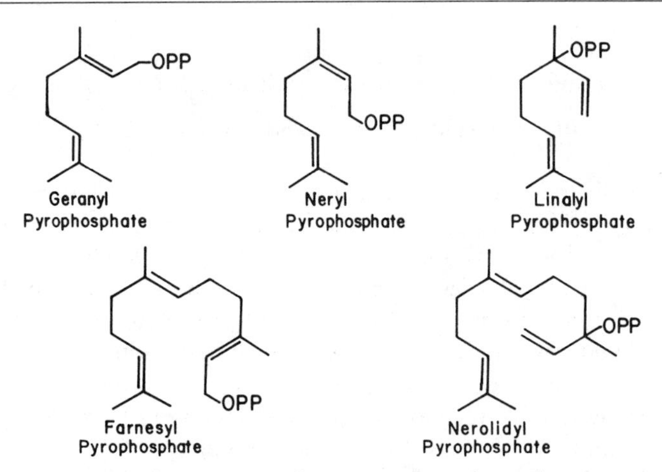

FIG. 1. Acyclic precursors of monoterpenes and sesquiterpenes.

distinct cyclases catalyzing the synthesis of enantiomeric products.[5] Most sesquiterpenoid-synthesizing microorganisms produce a rather limited range of sesquiterpene skeletal types and, by implication, contain a relatively small contingent of cyclases. The fungi as a class, however, synthesize sesquiterpenes of over 45 discrete structural types.[6] The number of monoterpene and sesquiterpene cyclases in nature is presently uncertain, yet studies on the known examples from a limited number of organisms[1] suggest that the total is at least 200 to 300. The obvious diversity of this enzyme class precludes even a brief overview of the entire field. Nonetheless, it is possible to draw some broad generalizations about the nature of these enzymes which will precede a description of the most relevant isolation, purification, and assay methods. Since cyclases from *Pinus* and *Citrus* are described by Cori (this volume [45]), we have restricted our coverage to the cyclases of herbaceous plant species and microorganisms with which we have had direct experience.

General Considerations

Cyclases, with apparently rare exception,[7] are operationally soluble enzymes possessing molecular weights in the 50,000–100,000 range.

[5] R. Croteau, *in* "Isopentenoids in Plants" (W. D. Nes, G. Fuller, and L. S. Tsai, eds.), p. 31. Dekker, New York, 1984.

[6] For an extensive review of monoterpenoid and sesquiterpenoid fungal metabolites see W. B. Turner, "Fungal Metabolites," pp. 218–235; 368–370. Academic Press, New York, 1971; W. B. Turner and D. C. Aldridge, "Fungal Metabolites II," pp. 225–272, 519–526. Academic Press, New York, 1983.

[7] C. Bernard-Dagan, C. Pauly, A. Marpeau, M. Gleizes, J. Carde, and P. Baradat, *Physiol. Veg.* **20,** 775 (1982).

Based on relatively few examples from higher plant sources,[4,8] they appear to be rather hydrophobic and to possess relatively low p*I* values. The only cofactor required is a divalent metal ion, Mg^{2+} or Mn^{2+} generally being preferred. In spite of earlier speculations, there is no evidence which has withstood experimental scrutiny for the involvement of redox coenzymes.[9] Most monoterpene cyclases can utilize geranyl pyrophosphate, neryl pyrophosphate, and linalyl pyrophosphate (Fig. 1) as acyclic precursors, without detectible interconversion among these substrates or preliminary conversion to other free intermediates. Assessment of the relative efficiency of cyclization of these three allylic pyrophosphates is often hindered by the presence of competing phosphohydrolases, and earlier conclusions based on data in which this complication was ignored may be erroneous. In preparations of (+)-bornyl pyrophosphate cyclase from which phosphatases were removed, geranyl pyrophosphate was shown to be a more efficient substrate than neryl pyrophosphate, based on comparison of the V/K_m values for each substrate.[3] In pinene cyclase preparations, where corrections were made for substrate loss due to competing phosphatases, kinetic constants for the three acyclic precursors were roughly equivalent.[4] Michaelis constants for all three substrates with all known cyclases are in the 1–20 μM range.

Geranyl pyrophosphate, the first C_{10} compound to arise in the classical isoprenoid pathway, is assumed to be the universal precursor of monoterpenes. Steric constraints imposed by the trans-2,3-double bond, however, prevent direct cyclization of this substrate and necessitate preliminary conversion to a bound intermediate competent to cyclize. Since geranyl pyrophosphate is efficiently cyclized without formation of free intermediates, it is clear that monoterpene cyclases are capable of catalyzing both the required isomerization and cyclization reactions, the overall process being essentially irreversible in all cases.[9] The mechanism of cyclization of geranyl pyrophosphate is considered to involve initial ionization of the pyrophosphate moiety, in which the divalent cation is presumed to assist, followed by isomerization to a linalyl intermediate (i.e., linalyl pyrophosphate, the corresponding ion pair, or other bound equivalent), with rotation about the C-2–C-3 single bound and subsequent cyclization of the cisoid conformer.[10,11] The proposed mechanism, which is completely con-

[8] A. J. Poulose and R. Croteau, *Arch. Biochem. Biophys.* **191,** 400 (1978).

[9] Considerable evidence has ruled out earlier redox schemes for the isomerization of geranyl pyrophosphate to neryl pyrophosphate via the corresponding aldehydes [see R. Croteau and M. Felton, *Arch. Biochem. Biophys.* **207,** 460 (1981)], and of *trans,trans*-farnesyl pyrophosphate to the cis,trans-isomer by similar means (see refs. 10 and 26 as well as P. Anastasis, I. Freer, C. Gilmore, H. Mackie, K. Overton, and S. Swanson, *J. Chem. Soc., Chem. Commun.*, p. 268 (1982).

[10] D. E. Cane, *Tetrahedron* **36,** 1109 (1980).

[11] D. E. Cane, R. Iyengar, and M.-S. Shiao, *J. Am. Chem. Soc.* **103,** 914 (1981).

sistent with the results of numerous model studies of terpenoid cyclizations,[12] readily accounts for the direct cyclization of both linalyl pyrophosphate and neryl pyrophosphate for which there is no steric obstruction. It is not yet clear, however, whether these latter acyclic precursors are in fact true intermediates of the cyclization process or simply efficient substrate analogs.

A similar mechanistic scheme can be applied to the cyclization of *trans,trans*-farnesyl pyrophosphate to any of the approximately 200 known sesquiterpene carbon skeletons, with the additional variable that intramolecular electrophilic attack of the initially generated allylic cation may take place on either the central or distal double bond of the farnesyl chain. Formation of six-membered rings is believed to occur by a mechanism completely analogous to that mediated by the monoterpene cyclases, with prior isomerization of the *trans*-allylic pyrophosphate most likely involving the corresponding tertiary allylic isomer, nerolidyl pyrophosphate.[10,11] An isomerase which catalyzes the conversion of *trans, trans*-farnesyl pyrophosphate to nerolidyl pyrophosphate has in fact been demonstrated in extracts of the fungus *Gibberella fujikuroi* and the mechanism of the reaction has been studied in detail.[11] Formation of 10- or 11-membered rings by initial electrophilic addition to the distal double bond of farnesyl pyrophosphate may or may not involve prior isomerization of the trans-2,3-bond, depending on the geometry of the eventually formed products. For example, cyclization of *trans,trans*-farnesyl pyrophosphate to the macrocyclic sesquiterpene, all-*trans*-humulene, by sage leaf extracts, apparently does not require initial double bond isomerization,[13] whereas strong, albeit indirect, evidence has been advanced for the intermediacy of nerolidyl pyrophosphate in the formation of longifolene and sativene via an initially generated *cis,trans*-germacradienyl cation.[14] On the other hand, the role of nerolidyl pyrophosphate as an intermediate in sesquiterpene cyclizations has yet to be explicitly demonstrated; neither has the possible involvement of the *cis,trans*-farnesyl pyrophosphate in cyclization processes been rigorously defined experimentally.

The monoterpenoid and sesquiterpenoid cyclases are unusual in that a small number of acyclic substrates, geranyl and farnesyl pyrophosphate, and presumably the corresponding tertiary allylic pyrophosphates, linalyl and nerolidyl pyrophosphate, serve as universal precursors of some 200–

[12] S. Gotfredsen, J. P. Obrecht, and D. Arigoni, *Chimia* **31**(2), 62 (1977); S. Winstein, G. Valkanas, and C. F. Wilcox, *J. Am. Chem. Soc.* **94,** 2286 (1972); K. Stephan, *J. Prakt. Chem.* **58,** 109 (1898); C. D. Poulter and C.-H. R. King, *J. Am. Chem. Soc.* **104,** 1420, 1422 (1982).

[13] R. Croteau and A. Gundy, *Biochem. Biophys.* **233,** 838 (1984).

[14] D. Arigoni, *Pure Appl. Chem.* **41,** 219 (1975).

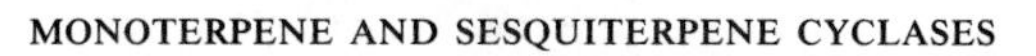

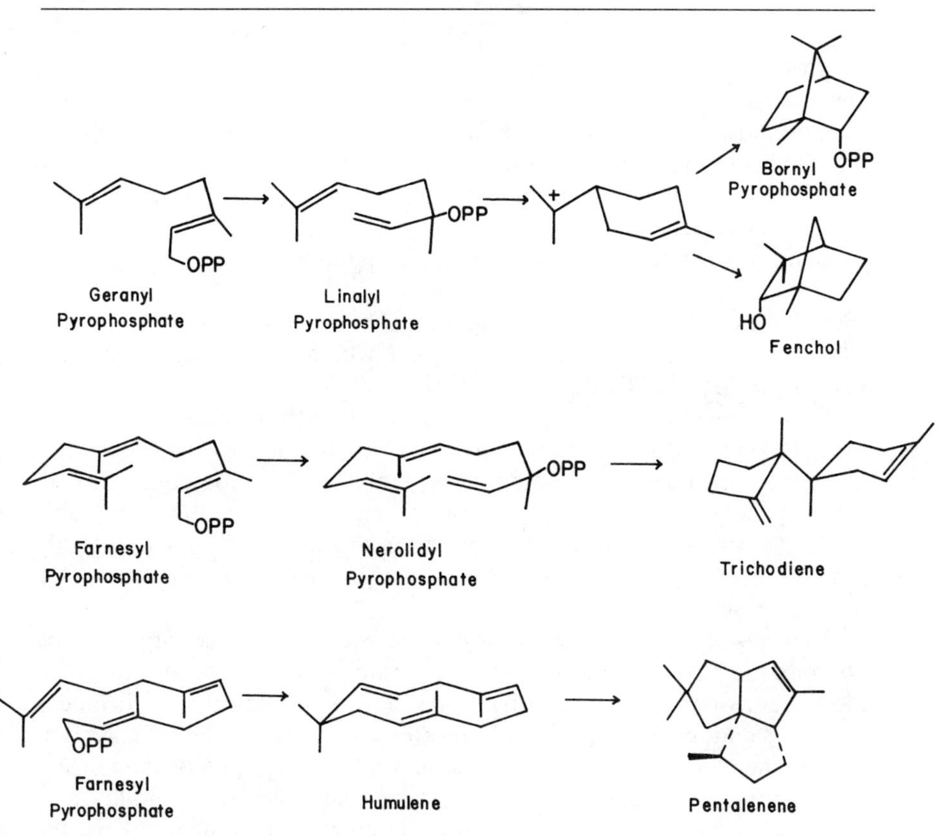

FIG. 2. Presumed conformations of acyclic precursors leading to cyclized terpenoids.

300 cyclized products under the catalysis of a nearly equal number of distinct cyclases. Not only the substrates, but the cyclization mechanisms as well, are believed to be general. In considering the role of the enzyme in these transformations, it is essential to evaluate the relative importance of rate enhancement and of product specificity. It is generally agreed that the enzyme catalyzes the initial ionization of the pyrophosphate moiety to generate an allylic cation, in a manner completely analogous to the action of prenyltransferase. The subsequent course of the cyclization reaction critically depends upon the precise folding of the acyclic geranyl or farnesyl substrate. Stereoelectronic considerations require that the reacting double bonds of the precursor be aligned in roughly parallel planes so as to allow effective overlap of the appropriate $2p$ orbitals (Fig. 2). Taking into account the further conformational constraints imposed by the presence of two or three π-systems within a chain of only 8 or 12 carbon

atoms, respectively, it becomes apparent that a small number of substrate conformations can account for the majority of known monoterpene and sesquiterpene skeletons. It is not known, however, in what way the enzyme enforces the required conformation on the reacting substrate, or what factors determine formation of distinct products arising from ostensibly identical substrate conformations. Moreover, any model of substrate-catalyst interactions must take into account the substantial changes in charge distribution, hybridization, configuration, and bonding which characterize the cyclization process and its concomitant Wagner–Meerwein rearrangements, hydride migrations, and deprotonation-reprotonation events. It is not as yet understood whether the enzyme actually lowers the ΔH of activation for any of these fundamental bond transformations, or simply forces the selection of a single reaction channel by precise control of substrate conformation and the positioning of counterions. It is also relevant to note that terpenoid cyclases, in common with other cyclization and linear condensation enzymes, have the remarkable ability to survive the transient generation at their active sites of highly reactive electrophilic species involving charge separations of up to 5–10 Å in some cases.

In terms of isolation, purification, and assay, the cyclases appear to be no more (or less) intractable than other proteins; yet, the tissues in which they occur often contain relatively high levels of interfering substances (e.g., oils and resins) and enzymes (proteases, phenol oxidases, and competing phosphohydrolases and cyclases) which can greatly complicate these operations. Such problems must be dealt with by the addition of complexing agents, adsorbents, and inhibitors and, ultimately, by the purification of the cyclase of interest. Far greater limitations in working with these enzymes are imposed by the facts that the cyclases, like most enzymes involved in the biosynthesis of natural products, do not occur in very high intracellular concentration and the reactions that they catalyze are rather slow.[15] Highly sensitive radiochemical assays of substrate consumption or product formation are thus required, as is the need to utilize as the enzyme source the organism or tissue at the developmental stage which affords the greatest enrichment of the relevant cyclase.

Finally, it should be pointed out that the parent cyclic products are often actively metabolized *in vivo,* undergoing extensive oxidative or reductive secondary transformations to afford the final product mixtures commonly associated with the organism. The initial cyclic product may therefore be only a minor *accumulated* product which bears limited re-

[15] D. E. Cane, *in* "Enzyme Chemistry, Impact and Applications" (C. J. Suckling, ed.), pp. 196–231. Chapman & Hall, London, 1984.

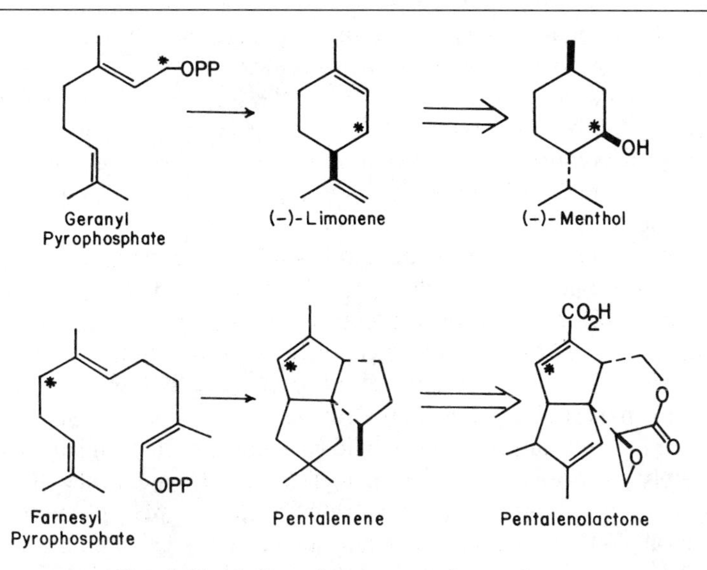

FIG. 3. Metabolism of cyclase reaction products.

semblance to its ultimate derivative(s). Thus, for example, (−)-limonene and pentalenene are the major cyclization products of cell-free preparations from *Mentha piperita*[16] and *Streptomyces* UC5319,[17] respectively, yet both compounds occur in but trace amounts *in vivo,* being largely converted to (−)-menthol and pentalenolactone (Fig. 3). Preliminary knowledge of biosynthetic pathways (and acquisition of the relevant chemical standards) is therefore essential to a successful search for cyclase activities. The lack of convenient sources of many sesquiterpene hydrocarbons can impose an important logistical burden on the study of the relevant cyclases. In fact, in some cases it has been necessary to prepare the cyclization products by total synthesis in order to assure adequate supplies of cyclic sesquiterpenes for enzyme isolation and purification studies.

Assay Methods

General Principles. Cyclic products derived from ^3H- or ^{14}C-labeled acyclic precursors[18] in the presence of cyclase and divalent cation are

[16] R. Kjonaas and R. Croteau, *Arch. Biochem. Biophys.* **220,** 79 (1983).

[17] D. E. Cane and A. M. Tillman, *J. Am. Chem. Soc.* **105,** 122 (1983).

[18] Procedures for the preparation of a number of labeled substrates useful for the study of monoterpene and sesquiterpene cyclizations have been described: [1-$^3H_2$]geranyl pyrophosphate,[18a] [8,9-^{14}C]geranyl pyrophosphate,[18b] [1-^{18}O]geranyl pyrophosphate,[18b] [1-α-

extracted and measured by radioassay after suitable chromatographic purification. Gas chromatographic measurement of product formation has been employed in a few instances,[19] but the technique is not sufficiently sensitive for most applications.[20] Since the mixture and types of cyclases present in a given extract are unique to each higher plant or microbial source, the details of incubation conditions and chromatographic separations must be tailored to each specific application. A general approach, applicable to most monoterpene and sesquiterpene cyclase preparations in which multiple products are formed, is provided here, as are some special precautions relevant to the assay of this class of enzyme.

Special Precautions. Since many crude preparations of mono- and sesquiterpene cyclases contain multiple cyclase activities as well as competing enzymes such as phosphohydrolases[21] and oxidases,[22] it is essential to determine the nature of all cyclic products formed and to define as completely as possible the alternate fates of the substrate. To obtain meaningful comparative data (substrate specificity, pH optimum, etc.) it is necessary, at minimum, to estimate substrate losses by determining the level of recoverable substrate after the assay. Various inhibitors of competing activities (as described below) often have been employed in assaying crude extracts. The lack of influence of such substances on cyclase activity cannot be assumed, however, and must be determined empirically.

The allylic pyrophosphate substrates utilized in cyclase assays[18] are also prone to acid and metal ion-catalyzed solvolysis[23] and the appropri-

---

^{32}P]geranyl pyrophosphate,[18b] [1-β-^{32}P]geranyl pyrophosphate,[18b] [1-$^{3}H_2$]farnesyl pyrophosphate,[11] [12,13-^{14}C]farnesyl pyrophosphate,[11] [12,13-^{14}C]nerolidyl pyrophosphate,[11] *E*-[1-^{3}H]nerolidyl pyrophosphate,[11] [1-^{18}O]farnesyl pyrophosphate,[11] [8-$^{3}H_2$]farnesyl pyrophosphate,[17] and [9-$^{3}H_2$]farnesyl pyrophosphate.[18c]

[18a] R. Croteau and F. Karp, *Arch. Biochem. Biophys.* **176,** 734 (1976).

[18b] D. E. Cane, A. Saito, R. Croteau, J. Shaskus, and M. Felton, *J. Am. Chem. Soc.* **104,** 5831 (1982).

[18c] D. E. Cane, C. Abel, and A. M. Tillman, *Bioorg. Chem.* **12,** 312 (1984).

[19] R. Croteau, A. J. Burbott, and W. D. Loomis, *Biochem. Biophys. Res. Commun.* **50,** 1006 (1973).

[20] Activities are usually measured in nmol/hr-mg protein in crude extracts, and may reach μmol/hr-mg protein levels in partially purified preparations.

[21] R. Croteau and F. Karp, *Arch. Biochem. Biophys.* **198,** 523 (1979).

[22] D. V. Banthorpe, G. A. Bucknall, J. A. Gutowski, and M. G. Rowan, *Phytochemistry* **16,** 355 (1977).

[23] D. S. Goodman and G. Popjak, *J. Lipid Res.* **1,** 286 (1960); W. Rittersdorf and F. Cramer, *Tetrahedron* **24,** 43 (1068); B. K. Tidd, *J. Chem. Soc. B* p. 1168 (1971); C. George-Nascimento, R. Pont-Lezica, and O. Cori, *Biochem. Biophys. Res. Commun.* **45,** 119 (1971); D. M. Brems and H. C. Rilling, *J. Am. Chem. Soc.* **99,** 8351 (1977); M. V. Vial, C. Rojas, G. Portilla, L. Chayet, L. M. Perez, O. Cori, and C. A. Bunton, *Tetrahedron* **37,** 2351 (1981).

ate control experiments should be included in each set of assays.[24] Such substrates, especially at high specific activity, also undergo degradation on storage by additional mechanisms that are only poorly understood.[25]

In as much as cyclic terpenoids are generated in nmole amounts in most assays, and they are generally volatile and often labile (e.g., to oxidative decomposition or acid-catalyzed isomerization reactions), unlabeled carrier compounds should be added as soon as possible to minimize degradation of the labeled product and losses on handling.

Incubation Procedures. Cyclases are generally assayed in buffers of moderate ionic strength (10 to 50 mM) in the pH range of 6 to 7. Sodium or potassium phosphate buffers, or combinations of sodium phosphate and Good's buffers such as Mes or Tricine, have frequently been employed,[2–4,8,17,26] with acetate and Tris-based buffers finding occasional use.[11,27] Polyols (5 to 20% sucrose, sorbitol, or glycerol) and thiol reagents such as dithioerythritol (0.5 to 1 mM) or mercaptoethanol (5 to 10 mM) markedly improve enzyme stability, and these additions to the buffer, as well as the appropriate divalent cation, are now routine.[28] The reaction is

[24] To obtain an accurate assessment of metal ion-dependent solvolysis, the divalent cation should be added only after boiling the enzyme control. The pH of the solution also should be rechecked after boiling.

[25] 1-^{3}H-Labeled substrates (at the 300 Ci/mol level) prepared by the Cramer and Bohm pyrophosphorylation procedure [*Angew. Chem.* **71,** 775 (1959)], and after treatment with inorganic pyrophosphatase, conversion to the ammonium salt, and purification by ion-exchange chromatography on *O*-diethylaminoethyl cellulose (see ref. 18a), can be frozen in 10 mM $(NH_4)_2CO_3$ containing 0.001% BHT and stored under paraffin oil in plastic vials at $-20°$ with less than 10% decomposition in a year. The purity of the substrate should be verified periodically by combination of adsorption thin-layer chromatography on buffered silica gel H or partition thin-layer chromatography on cellulose MN 300, in both cases developed with chloroform : methanol : 0.1 M $(NH_4)_2HPO_4$, pH 6.8 [for further details and R_f values see S. S. Sofer and H. C. Rilling, *J. Lipid Res.* **10,** 183 (1969) and ref. 25a], as well as by analysis of the products obtained on enzymatic and acid hydrolysis (see ref. 23 and footnote 34). Simply counting an aliquot of the hexane extract used to remove the paraffin oil overlay from stored samples will often reveal the extent of decomposition. Lack of radioactivity in such extracts, however, does not indicate that the substrate is intact, since we have on occasion also documented very polar substances and ^3H_2O (from [1-^{3}H]geranyl pyrophosphate) as decomposition products.

[25a] R. Croteau and R. C. Ronald, *in* "Chromatography: Fundamentals and Applications of Chromatographic Electrophoretic Methods. Part B: Applications" (E. Heftmann, ed.), p. 147. Am. Elsevier, New York, 1983.

[26] D. E. Cane, S. Swanson, and P. N. Murthy, *J. Am. Chem. Soc.* **103,** 2136 (1981).

[27] R. Croteau, M. Felton, and R. C. Ronald, *Arch. Biochem. Biophys.* **200,** 534 (1980).

[28] Optimum levels must be determined empirically. In the absence of polyol and thiol reagent cyclase activity in crude preparations may not survive more than 24 hr at 0–4°. Most cyclases are also extremely sensitive to thiol-directed reagents such as *p*-hydroxymercuribenzoate (see, for example, refs. 2–4,8 and 27). With the cyclases examined thus far, K_m for Mg^{2+} is in the 1 to 5 mM range; for Mn^{2+}, in the 0.1 to 0.5 mM range.

initiated by the addition of substrate (generally at the 10 to 50 μM level) and incubation carried out from one to several hours at 27–30°. Since many terpenoid cyclases are thermally labile, or may be subject to competing degradation processes, especially in crude extracts, it is important that the length of the assay period be small compared to the enzyme half-life. Teflon-lined screw-capped vials of 7 ml capacity are ideal for these incubations, and 1 ml is a convenient assay volume for μg to mg levels of protein. Phosphatase inhibitors, such as the sodium or ammonium salts of fluoride, molybdate, tungstate, and vanadate(V) (0.5 to 5.0 mM) are sometimes added to minimize substrate losses during the assay of crude preparations.[21,29] In our hands, 1 mM sodium molybdate has proven to be most effective for this purpose, providing maximum inhibition of phosphohydrolases with minimum adverse influence on cyclase activity. Phosphate-based buffers (5–20 mM) also provide a moderate degree of inhibition of phosphatases,[21] with little effect on cyclases. In some cases it may be possible to exploit the apparent differences in K_m for cyclase and contaminating phosphatase activities by judicious choice of substrate concentrations.

If adequate precautions have been taken during the extraction and preparation of the enzyme (see below), phenolases and other oxidases cause little apparent interference in the cyclase assay. The addition of inhibitors such as diethyldithiocarbamate and mercaptobenzthiazole[30] to the assay media or the maintenance of anaerobic conditions have provided only marginal improvement of cyclase activity in our experience.

Isolation of Products. If a complex mixture of cyclic products is anticipated in the assay, it is generally advisable to carry out a group separation of hydrocarbons from oxygenated terpenes as the initial step of the product analysis. This is readily accomplished by selective adsorbtion of oxygenated products on silicic acid. The ice-cooled reaction mixture is initially extracted with an equal amount of pentane, and after vigorous shaking (followed by centrifugation, if necessary, to facilitate separation of phases) the pentane extract is eluted through a 0.5 × 3 cm column of silicic acid.[31] The extraction is repeated with another portion of pentane, and the column washed with this extract and with a further 0.5 ml of pentane. The combined pentane eluate contains the terpenoid hydrocarbons, which do not bind to silicic acid under these conditions. The origi-

[29] D. V. Banthorpe, A. R. Chaudhry, and S. Doonan, *Z. Pflanzenphysiol.* **76,** 143 (1975).

[30] W. D. Loomis, this series, Vol. 31, p. 528.

[31] Columns can be conveniently prepared in glass wool-plugged Pasteur pipets. Both Mallinckrodt SilicAR CC-7 and Baker silica gel (40–140 mesh) have been used. For some olefin cyclases[4,16] the assay can be run with a pentane overlay to improve the recovery of volatile products; however, not all cyclases tolerate this procedure.

nal reaction mixture is then reextracted with an equal portion of diethyl ether, and the ether extract passed through the same silicic acid column to elute the oxygenated products. Complete elution of the oxygenated products is ensured by washing the column with an additional 1.5 ml of ether. This procedure allows essentially quantitative separation of even relatively nonpolar oxygen-containing products such as 1,8-cineole from monoterpene dienes and trienes (limonene, myrcene, etc.) and allows complete recovery of compounds at least as polar as geraniol, liberated from geranyl pyrophosphate by endogenous phosphohydrolases.[32]

To assay the formation of cyclic phosphorylated products and to assess the residual level of phosphorylated substrate in the assay,[33] any traces of ether are removed from the extracted reaction mixture under a stream of nitrogen followed by the addition of 3 units of wheat germ acid phosphatase and 2 units of potato apyrase in 1.0 ml of 0.1 *M* sodium acetate, pH 5 (or 3 units of calf intestine alkaline phosphatase and 2 units of apyrase in 1.0 ml of 0.1 *M* Tris–Cl, pH 8.5). The vial is resealed and incubated for 2 hr at 30°.[34] The chilled reaction mixture is then extracted twice with 1.5-ml portions of ether to recover the liberated alcohols, and

[32] Even relatively polar compounds such as geranic acid and diepoxygeraniol, formed by the action of undefined dehydrogenases and oxidases, have been quantitatively recovered by this method.

[33] Since pyrophosphate hydrolases may be present in plant and fungal extracts, it may be necessary to determine how much substrate has been converted to the catalytically inactive monophosphate. For the purpose of determining the distribution of phosphorylated products, the aqueous phase remaining after the extraction of ether-soluble products is adjusted to a pH of 9 with NH_4OH and lyophilized. The labeled residue is dissolved in the minimum quantity of water and the separation of monophosphate and pyrophosphate esters achieved by either adsorption, partition or ion exchange chromatography (see note 25). For the rapid estimation of monophosphate, extraction of the aqueous phase with water-saturated *n*-butanol suffices, but the method must be calibrated for the efficiency of extraction of the product of interest (see ref. 21).

[34] Under the conditions outlined, fresh preparations of wheat germ acid phosphatase (Sigma Type 1) or calf intestine alkaline phosphatase (Sigma Type 1S) alone are capable of hydrolyzing ~100 nmol of geranyl, farnesyl, and bornyl pyrophosphate to the corresponding alcohols. However, preparations with the same acid or alkaline phosphatase activity (as measured by hydrolysis of *p*-nitrophenyl phosphate) have been shown to vary severalfold in ability to hydrolyze terpenyl pyrophosphates, suggesting that the ability to hydrolyze terpenyl pyrophosphates to the corresponding alcohols is dependent, at least in part, on the presence of a pyrophosphate hydrolase in these commercial preparations. Potato apyrase (Sigma Grade 1) is thus routinely added to effect the terpenyl pyrophosphate to monophosphate conversion and ensure complete hydrolysis [see G. Del Campo, J. Puente, M. A. Valenzuela, A. Traverso-Cori, and O. Cori, *Biochem. J.* **167,** 525 (1977) for a description of this enzyme]. In our hands, acid phosphatase has been found to be more efficient than alkaline phosphatase for the complete hydrolysis of cyclic terpenyl pyrophosphate esters to the corresponding alcohols. In any event, the procedure should be calibrated with authentic standard under the conditions of the cyclase assay.

the combined extract dried by passage through a 0.5 × 3 cm column of anhydrous Na_2SO_4. In preparation for subsequent radiochromatographic analysis, carrier standards are added at this point to the various extracts containing hydrocarbons, oxygen-containing products, and alcohols liberated by phosphatase treatment,[35] and the samples, after radioassay of an aliquot, are concentrated to a convenient volume under a stream of N_2 at 0°.

If only one or a few cyclic products is anticipated in an assay, as is generally the case for microbial cyclase systems,[11,26] or if preliminary work has revealed it to be unnecessary to examine the complete spectrum of products, it is often sufficient to carry out only the initial partitioning step (if only hydrocarbons are required) or to substitute a simple ether extraction for the above sequence. In the latter instance, the separation of hydrocarbons from oxygenated products must be carried out in a subsequent chromatographic step. In assaying for synthesis of saturated products (1,8-cineole, borneol, patchouliol, etc.), we have found it convenient to treat the extracts with OsO_4 to convert double bonds of olefinic products to the corresponding diols, thereby simplifying subsequent chromatographic analysis.[36] Further modifications can be made to tailor the extraction procedure to the recovery of a specific product, the primary consideration being the elimination of interference from other labeled products. For example, when assaying for bornyl pyrophosphate synthetase,[3] the aqueous phase remaining after extraction of ether-soluble products is treated with 1.0 ml of 0.1 *M* sodium acetate, pH 5, and heated on a steam bath for several minutes to solvolyze residual substrate and drive off the volatile products, and to hydrolyze simultaneously bornyl pyrophosphate to bornyl phosphate. Removal of coagulated protein and subsequent incubation with acid phosphatase releases essentially only borneol which can be radioassayed after ether extraction.

Radiochromatographic Separation of Products. A number of chromatographic procedures can be employed for the separation of individual components in mixtures of terpene hydrocarbons, oxygenated terpenoids, and phosphatase-derived products; thin-layer chromatography and gas–liquid chromatography are particularly well-suited for most routine

[35] As much carrier is added as is compatible with the subsequent separation and radioassay (e.g., 5 to 10 mg of each component for thin-layer chromatography; 3 to 5 mg of each for gas–liquid chromatography).

[36] After the addition of the relevant standard, the ether extracts are treated with a few drops of pyridine containing an excess of OsO_4. After mixing and incubation for ~30 min, saturated aqueous Na_2SO_3 is added to decompose the osmate esters followed by a volume of pentane equivalent to that of the ether, vigorous mixing and separation of phases. Monoterpene triols and more polar products remain in the aqueous phase, and only diols (derived from monoenes) are carried through to the subsequent step.

assays. Thin-layer chromatography on silica gel G (developed with hexane containing up to 5% ether) resolves several monoterpene and sesquiterpene olefins. Argentation chromatography on 12% $AgNO_3$-silica gel G (developed with hexane : benzene : ether, 50/50/1, v/v/v) is far superior for separating complex olefin mixtures.[37] Thin-layer chromatography on silica gel G using hexane containing ethyl acetate (5 to 30%, depending on the polarity of the products) as the developing solvent is suitable for the separation of many oxygen-containing terpenoids.[38] At concentrations of ethyl acetate exceeding 5%, essentially all terpene hydrocarbons migrate as a single band. Up to six samples can be analyzed on a single 20 × 20 cm plate, and the products readily located by spraying with a 0.2% ethanolic solution of 2,7-dichlorofluorescein and visualization under UV light. Products of interest may be radioassayed by scraping the appropriate section of gel directly into a scintillation vial containing fluor, or the powder transferred to a glass wool-plugged Pasteur pipet and the product eluted with dry ether or other solvent for subsequent radioassay or further analysis. Thin-layer chromatography radioscanners (e.g., of the Berthold-type) are unsuitable for the assay of most monoterpene and sesquiterpene cyclases since the evaporative losses of products during the scanning manipulations are variable and unacceptably high.

Assays should be calibrated with labeled standard or, if a standard is not available, by running the initially purified sample back through the entire procedure to define the limits of maximum loss. Alternatively, the losses in handling can be determined gas chromatographically, although this is less convenient. The recovery of monoterpenes is in the range of 70 to 95% depending on the nature of the product and the type of assay.[2,18a,37] Without added carrier the bulk of the monoterpene olefins generated in an assay may be lost during isolation and separation. The recovery of sesquiterpenes is correspondingly higher (e.g., [^{3}H]humulene diluted with 5 mg of carrier was recovered in ca. 90% yield after concentration of an extract and thin-layer chromatography[13]). Thin-layer chromatography of radiochemical levels of product leads not only to volatility losses but can result in contamination of other chromatographically well-defined labeled products which have been diluted with labeled standard. It may thus be advisable in some instances to add carrier even for labeled side products which may not be of significance in a given assay.

A particularly powerful technique for the separation and analysis of complex mixtures of isoprenoids is radio-gas–liquid chromatography, whereby individual components separated on the chromatographic column are determined directly in a heated flow-through proportional

[37] H. Gambliel and R. Croteau, *J. Biol. Chem.* **257,** 2335 (1982).

[38] J. Battaile, R. L. Dunning, and W. D. Loomis, *Biochim. Biophys. Acta* **51,** 538 (1961).

counter, or are similarly monitored in an ambient flow-through counter after conversion to $^{14}CO_2$ and/or $^{3}H_2$ in a combustion-reduction train.[39] Retention factors on methyl silicone oil, Apiezon, and Carbowax 20M have been compiled for many of the common monoterpenes and sesquiterpenes.[25a,40,41] Using 12 ft × 1/8 in. stainless-steel columns containing 15% liquid phase on 80/100 mesh Chromosorb W-HP and temperature programming, even crude product mixtures containing both hydrocarbons and oxygenated products can be separated and analyzed at a sensitivity of a few hundred dpm per component of injected sample. Two-channel recording with integration of both chromatograph and counter output readily allows assignment of radioactivity to the appropriate component eluted, and permits simple correction for handling losses based on single or multiple internal standards. [^{3}H]Toluene and [^{14}C]toluene are utilized for external calibration of the counting system. Figure 4 illustrates a radio-gas–liquid chromatogram of the monoterpene olefin fraction isolated from a cell-free extract of *Salvia officinalis* which had been incubated with [1-^{3}H]geranyl pyrophosphate.

For the separation of phosphorylated products, thin-layer chromatography is often employed, either in the adsorption (silica gel H) or partition (cellulose) mode as previously noted.[25] Ion exchange chromatography finds occasional use, particularly on the semipreparative scale. Thus bornyl pyrophosphate was first isolated from large scale enzyme incubations by column chromatography on *O*-diethylaminoethyl cellulose using a 10 to 50 m*M* $(NH_4)_2CO_3$ gradient (the pyrophosphate ester elutes at ~33 m*M*).[42]

[39] The only commercially available instrument at this writing is the Packard 894 Gas Proportional Counter. For maximum radiosensitivity and resolution several modifications to the instrument are necessary. The separate oxidation and reduction furnaces and associated plumbing are removed and placed end-to-end as close to the exit port of the gas chromatograph as possible. A Gow-Mac 550P Thermal Conductivity Gas Chromatograph is ideal for this purpose in that no effluent splitter is required (as with a flame ionization detector) and the entire heated gas stream exits directly from the detector at the rear of the instrument, thus eliminating variations in split-ratio with chromatographic conditions and the need for auxiliary transfer heaters. The two quartz furnace tubes must be replaced with a single longer tube (7 mm o.d. × 30 cm) containing cupric oxide in the front third, a quartz glass plug, and methanol-washed steel wool (extra fine) in the rear two-thirds. The plumbing from the rear septum (1/16-in. stainless steel tubing) to the proportional counter is kept to the minimum length, and, when counting only $^{3}H_2$, a CO_2 adsorbent trap (6 mm i.d. × 15 cm) is placed in line to sequester CO_2 and minimize quenching. With such a system, as little as 200 dpm ^{3}H exiting over a 10-sec interval can be detected.

[40] W. Jennings and T. Shibamoto, "Qualitative Analysis of Flavor and Fragrance Volatiles by Glass Capillary Gas Chromatography." Academic Press, New York, 1980.

[41] A. A. Swiger and R. M. Silverstein, "Monoterpenes." Aldrich Chemical Co., Milwaukee, Wisconsin, 1981.

[42] R. Croteau and F. Karp, *Arch. Biochem. Biophys.* **184,** 77 (1977).

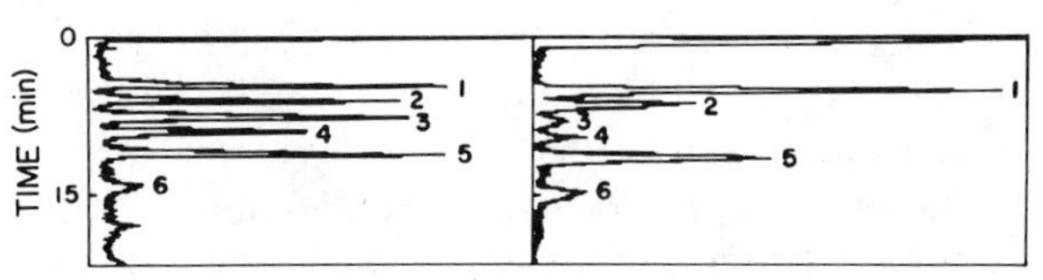

FIG. 4. Reconstructed radio-gas chromatogram of the monoterpene olefin fraction obtained by incubating a partially purified enzyme preparation from *Salvia officinalis* leaves with [1-$^3H_2$]geranyl pyrophosphate in the presence of $MgCl_2$. The tracing on the left is the thermal conductivity detector response obtained from coinjected standards of α-pinene (1), camphene (2), β-pinene (3), myrcene (4), limonene (5), and terpinolene (6). The tracing on the right is the response of the radioactivity monitor attached to the gas chromatograph (~5000 dpm injected). The chromatographic column was 12 ft × 1/8 in. o.d. stainless steel containing 15% Carbowax 20M on 80/100 mesh Chromosorb W-HP and programmed from 90 to 140° at 5°/min at a He flow rate of 55 ml/min. Outputs of both thermal conductivity and radioactivity detectors were plotted and integrated with a Sica 7000a chromatogram processor.

Product Identification and Determination of Labeling Pattern. Since chromatographic coincidence is insufficient to establish the identity or radiochemical homogeneity of a labeled product, additional criteria must be applied for confirmation of structure, at least initially. Seldom is enough biosynthetic material available to permit spectroscopic analysis, and thus cocrystallization to constant specific radioactivity of the suspected product (or a suitable derivative) with an authentic standard is most often employed for this purpose. Olefins are generally converted to crystalline halides or nitrosohalides, or oxidized to alcohols or carbonyl compounds and recrystallized as the respective esters, urethanes, or conventional carbonyl derivatives. Classical methods for the preparation of numerous derivatives of terpenoid olefins, alcohols, aldehydes, and ketones have been described.[43]

Many plant species are known to produce enantiomeric monoterpenes, and it therefore may be necessary to determine the stereochemical composition of the cyclic products generated in assays of cell-free extracts. Sage, for example, produces (+)- and (−)-limonene, (+)- and (−)-α-pinene, and (+)- and (−)-camphene, as well as (+)- and (−)-bornyl pyrophosphate, each enantiomer being synthesized by separate cyclases.[4,5] Conversion of enantiomers to diastereomeric derivatives, followed by fractional crystallization or chromatographic separation, has been employed to resolve α- and β-pinene,[37] camphor,[42] and fenchone.[44]

[43] F. S. Sterrett, *in* "The Essential Oils" (E. Guenther, ed.), Vol. 2 (reprinted), p. 769. Robert E. Krieger, Huntington, New York, 1975.

[44] R. Croteau, M. Felton, and R. C. Ronald, *Arch. Biochem. Biophys.* **200,** 524 (1980).

Direct separation of enantiomers by gas–liquid chromatography on chiral liquid phases (the technique is feasible only on capillary columns[45]) offers a possible alternative, assuming that sufficient biosynthetic product is available for mass detection or that newly developed capillary radio-gas chromatographic methods will be used.[46] Radiochemical cocrystallization offers an unequivocal, but tedious approach to the resolution of labeled enantiomers when optically pure carriers are available.[4]

Finally, it is necessary to establish the labeling patterns of products derived from specifically labeled precursors in order to confirm basic cyclization and rearrangement schemes. Failure to follow such fundamental precautions has too often led to claims which have later been withdrawn in the light of more careful work. A detailed description of the procedures which have been used for the degradation of terpenoid metabolites is well beyond the scope of this chapter. However, routes for determining the location of label in camphor, fenchone, α-terpinene, γ-terpinene, limonene, cineole, and α- and β-pinene from 1-^{3}H-labeled acyclic precursors have been described (see 1a and 5 for lead refs.). Similarly, routes for the degradation of a number of sesquiterpenes derived from specifically labeled [^{3}H]- and [^{14}C]farnesyl pyrophosphate and other precursors have been described.[1b,11,17,26]

Extraction of Cyclases

Sources of Cyclases. Any extensive study of higher plant monoterpene and sesquiterpene cyclases requires a regular, relatively large, and consistent supply of tissue. Most such studies have been carried out with common herbs (sage, thyme, fennel, etc.) of which seed is commercially available, or with locally obtained wild species (e.g., tansy) where seed collection is feasible. To ensure reproducibility, the plants must be propagated under controlled conditions, and where possible, long warm day (30°, 16 hr), cool night (15°) growth conditions should be employed with high light intensity (15,000 lux) in order to maximize monoterpene and sesquiterpene production and enhance endogenous cyclase levels.[47] Studies of the synthesis of monoterpenes as a function of leaf development have revealed that cyclization processes (though not necessarily secondary transformations) essentially cease at full leaf expansion because of the diminution of cyclase levels.[48] Resins, phenolics, and other interfering

[45] W. A. Konig, W. Francke, and I. Benecke, *J. Chromatogr.* **239,** 227 (1982).

[46] D. Gross, H. Gutekunst, A. Blaser, and H. Hambock, *J. Chromatogr.* **198,** 389 (1980).

[47] See, for example, A. J. Burbott, and W. D. Loomis, *Plant Physiol.* **42,** 20 (1967) for a discussion of the influence of environment on monoterpene production.

[48] R. Croteau, M. Felton, F. Karp, and R. Kjonaas, *Plant Physiol.* **67,** 820 (1981); R. Croteau, *ibid.* **59,** 519 (1977); see also ref. 18a.

substances also accumulate during leaf development and thus further complicate the isolation of cyclases from mature tissue. Young expanding leaves are the most suitable source for the isolation of monoterpene cyclases,[49] and probably sesquiterpene cyclases as well, although this question has been less thoroughly explored.

Generally speaking, microbial cultures present fewer problems of reproducibility in growth and metabolic activity, provided careful control is exercised over growth conditions, including nutrients, aeration, size, and age of inocula, etc. On the other hand, many strains of fungi and *Streptomyces* can show enormous variability in the levels of secondary metabolic pathways and it is therefore crucial to utilize one of the standard microbiological techniques of strain maintenance such as lyophilization of spore suspensions or storage of vegetative inocula, mycelial plugs, or spore suspensions mixed with glycerol at liquid nitrogen temperature. As a general rule most terpenoid pathways, in common with the majority of secondary metabolic pathways in microorganisms, are not expressed until either late in the replicative or tropophase or early in the stationary or idiophase. The timing of cell harvesting is therefore likely to be critical to efficient recovery of cyclase activity. Preliminary determination of metabolite levels as a function of time is essential, with optimum harvesting time likely to coincide with the early phase of the production curve. Since in most cases one will actually be measuring the excretion of more oxidized metabolites derived from the cyclization product of interest, it is important to bear in mind that induction of the relevant sesquiterpene cyclase may occur significantly earlier than the first appearance of more oxidized metabolic products. On the other hand, since the parent sesquiterpenes themselves rarely accumulate in easily measured concentrations, it is usually more convenient to assay their more highly oxidized metabolic derivatives when working with the intact organism. Nonetheless, many of the production media which have been optimized to give the best yields of final metabolite frequently contain insoluble components ($CaCO_3$, soy grits, corn gluten meal) which interfere with effective cell harvesting and enzyme isolation. It will therefore often be necessary to utilize soluble culture media, keeping in mind that the apparent decreases in fermentation yields may not reflect a decrease in actual titers of the cyclase of interest. The use of early idiophase cultures may also avoid

[49] Most of our experimental plants are raised from seed in a greenhouse equipped with supplemental lighting and temperature control. Possible seasonal and development variation are minimized by using only expanding leaves from relatively young plants (less than 3 weeks postgermination). As experimental plants need not be maintained for long periods, commercial peat pellets with a single application of slow-release fertilizer provide a simple and convenient growth media.

interference by late-stage redox enzymes or problems associated with autolysis of senescent cells.

Special Precautions. A major obstacle in the isolation of cyclases from higher plants is the presence of resins and phenolic substances (along with phenol oxidases) which inactivate cyclases and other plant enzymes through both covalent and noncovalent interactions. The problem is particularly severe in essential oil bearing plants, and has been discussed in detail by Loomis[30,50,51] and other investigators.[52,53] Numerous techniques have been devised to prevent, or at least to minimize, such destructive interactions through the use of reducing agents, chelators, and various adsorbent substances.[30,50–53] In our hands, the use of high ionic strength acidic buffers, multiple reducing agents (metabisulfite and ascorbate), thiol protecting reagents, and both polyvinylpolypyrrolidone and polystyrene as adsorbents has been found to be particularly effective in maximizing extractable cyclase activity. Thiol reagents, polyols, and proline are useful in stabilizing activity during extraction and purification, especially during the early stages of manipulation when the enzyme must be handled in dilute form and where reduced, hydrogen-bonded phenolics may still be present.

In microbial extracts, an additional problem frequently arises from contamination with endogeneous proteases which are often produced in substantial amounts by these organisms. Besides the use of standard protease inhibitors, it has also been found that successive washing of freshly harvested cells with cold 1 *M* KCl and 0.8 *M* NaCl solutions is effective in removing extracellular proteases prior to cell rupture.[54] Finally, in preparative scale incubations, when yields of product are the most important factor, the degradation of sensitive cyclases may sometimes be retarded by carrying out incubations at somewhat lower temperatures (24°) thereby offsetting the slightly slower rate of substrate turnover by a prolongation of catalytic activity.

Based on our experience with a number of systems, we provide below some general protocols for the extraction of monoterpene and sesquiterpene cyclases from higher plants and microbial sources.

Protocol for Cyclase Extraction from Whole Leaves. Immature leaves, after washing with distilled water, are homogenized in an ice-filled

[50] W. D. Loomis and J. Battaile, *Phytochemistry* **5,** 423 (1966).

[51] W. D. Loomis, J. D. Lile, R. P. Sandstorm, and A. J. Burbott, *Phytochemistry* **18,** 1049 (1979).

[52] J. W. Anderson, *Phytochemistry* **7,** 1973 (1968).

[53] M. J. C. Rhodes, *in* "Regulation of Enzyme Synthesis and Activity in Higher Plants" (H. Smith, ed.), p. 245. Academic Press, London, 1977.

[54] H. Grisebach and B. Kniep, *Eur. J. Biochem.* **105,** 139 (1980).

glass homogenizer (Ten-Broeck) with a slurry (4 mg/g tissue) consisting of an equal weight of insoluble polyvinylpolypyrrolidone[55] in cold 0.1 *M* sodium phosphate buffer, pH 6.5, containing 0.25 *M* sucrose, 50 m*M* $Na_2S_2O_5$, 10 m*M* sodium ascorbate, and 1.0 m*M* dithioerythritol. The homogenate is then slurried at 0–4° with an equal tissue weight of hydrated XAD-4 polystyrene resin[56] for 5 to 10 min and filtered through four layers of cheesecloth. Sufficient extraction buffer to make up the original volume is used to rinse the residual adsorbents and cell debris, and the combined filtrate centrifuged at 27,000 *g* for 20 min (pellet discarded), then at 105,000 *g* for 1 hr to provide the soluble supernatant, containing 2–5 mg protein/ml, depending on the tissue. After concentration and adjustment to assay conditions, this S_{105} fraction is used as the enzyme source.

Protocol for Cyclase Extraction from Leaf Epidermis. Since the epidermal oil glands constitute the major site of monoterpene biosynthesis in herbaceous plants,[48,57] a procedure for the selective removal of these glands, or of the epidermis itself, can provide extracts enriched in cyclase(s) and free of bulk proteins and other substances derived from the mesophyll. For exploratory studies, the leaf epidermis can be manually removed from the mesophyll by gentle brushing with a soft bristle toothbrush leaves submerged in extracting buffer.[58] Homogenization of the transparent epidermal fragments, followed by centrifugation, affords soluble preparations containing the bulk of the cyclase activity of the leaf, yet relatively low levels of competing phosphatases, phenolases, phenolics, and other vacuolar constituents. The technique is applicable to many leaf types, but is of limited preparative value. The following is an automated procedure for the preparation of epidermis extracts from sage leaves which can be readily scaled in either direction and modified for other tissue types.

From 5 to 35 g of leaves are placed in a Virtis[59] No. 5130 finned homogenizing flask and the following additions made for each gram of tissue:

[55] Insoluble polyvinylpolypyrrolidone (Polyclar AT) is available from the GAF Corp., New York. For the preparation of this material for use as an adsorbent see ref. 30.

[56] Amberlite XAD-4 polystyrene beads are available from Rohm and Haas, Philadelphia, PA. For the preparation of this material for use as an adsorbent see ref. 51. The powdered resin can be added directly during the homogenization step, but there appears to be little advantage to this procedure.

[57] R. Croteau and M. A. Johnson, *in* "Biology and Chemistry of Plant Trichomes" (P. Healey, E. Rodriguez, and I. Meta, eds.), p. 133. Plenum, New York, 1984.

[58] R. Croteau and J. N. Winters, *Plant Physiol.* **69,** 975 (1982).

[59] The Virtis Co., Gardiner, New York 12525. Stainless steel beads are available from Small Parts Inc., Miami, Florida 33138, and Pyrex beads from Corning Glass Works, Corning, New York 14830.

40 3/16″ stainless steel beads[59]
2 g ground glass (sieved through 1.5 mm mesh)
1 g insoluble polyvinylpolypyrrolidone (Polyclar AT)
10 ml 10 mM Mes–5 mM sodium phosphate buffer, pH 6.5, containing 10 mM $Na_2S_2O_5$
10 mM sodium ascorbate, 0.5 mM dithioerythritol, 100 mM L-proline, and 10% (v/v) glycerol

The flask is sealed with a rubber stopper and rotated in a horizontal position at ~100 rpm[60] for 20 min at 0–4°. One gram of XAD-4 polystyrene resin per g of tissue is then added, and the flask resealed and rotated an additional 10 min. The contents are vacuum filtered through eight layers of cheesecloth (with 90–95% recovery of extraction buffer), and the filtrate centrifuged directly at 105,000 g for 75 min to afford the soluble protein fraction. Homogenization prior to centrifugation does not increase levels of soluble cyclase activity, since the tumbling procedure alone is sufficient both to shear off the glands and to rupture the secretory cells. The yield of cyclase activity from these epidermis preparations is routinely 30 to 40% of a whole leaf homogenate; yet these extracts are essentially devoid of mesophyll-derived material. (The leaves remaining after epidermis extraction are intact and somewhat brighter green in color due to the removal of the nonphotosynthetic epidermis.) Preliminary studies with several other species (mint, thyme, tansy) indicate that the procedure can be applied to more robust tissue by increasing the proportion of steel beads, and to more fragile tissue by decreasing the proportion of steel beads or by substituting 3 or 5-mm Pyrex glass beads.[59]

Assay of cyclase activity in the crude extracts is generally unreliable due to the great dilution of enzyme and/or high phosphate concentration and presence of other inhibitory substances. Preliminary protein concentration and adjustment to more suitable assay conditions are required. For large-scale preparations, ultrafiltration (Amicon PM 30) and desalting with buffer change on a Sephadex G-25 columns are preferred for speed. For small-scale preparations, protein concentration with dry Sephadex G-25-300 is very effective.[4] Osmotic concentration and $(NH_4)_2SO_4$ precipitation with subsequent dialysis have been employed with varying degrees of success.[2,3,8,27,42,61]

Protocol for Cyclase Extraction from Fungal Cultures. Trichodiene Synthetase.[26,62] The mycelium from 2.4 liters of a 4-day-old suspension

[60] A rock polisher or horizontal rotary evaporator is used for this purpose. In the latter case, a coupling shaft is affixed to the stopper.

[61] R. Croteau and F. Karp, *Biochem. Biophys. Res. Commun.* **72,** 440 (1976).

[62] R. Evans and J. R. Hanson, *J. Chem. Soc., Perkin Trans. 1* p. 326 (1975).

culture of *Trichothecium roseum* ATCC 8685, grown as previously described,[63] is collected by filtration and washed thoroughly with glass-distilled water. (As a rule, cultures which have been growing between 84 and 108 hr give the best yields of cyclase.) The mycelium is suspended in 30 ml of cold, 0.1 *M* potassium phosphate buffer, pH 7.0, and partially homogenized in a blender before being disrupted by passage through a prechilled French press at 15,000–18,000 psi and 4°. The extract is separated from broken cell debris by centrifugation at 4°, first at 10,000 *g* for 20 min and then at 17,000 *g* for 90 min. The resultant supernatant (200 ml, ~10 mg protein/ml[64]) is mixed with an additional 10 ml of 0.1 *M* phosphate buffer, pH 7.0 (total phosphate buffer, 4.0 mmol) along with 2.0 mmol of $MgCl_2$, 0.8 mmol of $MnCl_2$, 0.1 mmol of dithiothreitol, and 10 μmol of [1-^{3}H, 12,13-^{14}C]farnesyl pyrophosphate (1.0×10^6 dpm ^{14}C) and the mixture incubated for 2 hr at 30°. The enzyme reaction is quenched by addition of 12 *N* NaOH (25 ml) and 95% ethanol (25 ml), allowed to stand for 30 min, and then extracted with 120 ml of *n*-pentane containing 5 mg of carrier trichodiene. After purification by preparative thin-layer chromatography (silica gel, hexane), the recovered trichodiene ^{14}C activity (1.8×10^4 dpm ^{14}C) corresponds to the formation of 0.05 nmol of trichodiene/mg protein/hr.

Protocol for Cyclase Extraction from Streptomyces Cultures. Pentalenene Synthetase. A vegetative inoculum of *Streptomyces* UC5319 is grown as previously described[65] for 72 hr and then transferred in 1-ml aliquots to 24 500-ml Delong flasks each containing 100 ml of culture medium consisting of 2.0 g Black Strap molasses, 20 g dextrin, 2.0 g NaCl, 1.25 g $CaCO_3$, 4.17 g casein hydrolyzate, and 3.33 g of Bacto dextrose per liter distilled water, pH adjusted to pH 7.2. After inoculation the flasks are shaken (300 rpm, 28°) for 60 hr,[66] following which the culture broth is combined and the cells collected by centrifugation (8 min, 6000 *g*, 0°). The supernatant is decanted and the cells washed successively with cold glass-distilled water, 1 *M* potassium chloride, and 0.8 *M* sodium chloride solutions. After each washing the supernatant is discarded. The washed cells are combined and made into a slurry with ~25 ml of 50 m*M* potassium phosphate buffer, pH 7.2, containing 5 m*M* dithioerythritol, 1.0 m*M* EDTA, and 10% (v/v) glycerol.

The cells in buffer are ruptured by rapid stirring with 0.1–0.15-mm

[63] G. G. Freeman, *J. Gen. Microbiol.* **12,** 213 (1955).

[64] O. A. Bessey, O. H. Lowry, and R. H. Love, *J. Biol. Chem.* **180,** 755 (1949).

[65] D. E. Cane, T. Rossi, A. M. Tillman, and J. P. Pachlatko, *J. Am. Chem. Soc.* **103,** 1838 (1981).

[66] Cells harvested after 84 hr showed little or no cyclase activity.

glass beads in a 50-ml jacketed Bead-Beater cell.[67] A 15 sec on, 15 sec off, cycle and the ice/alcohol cooling jacket reduces the risk of a deleterious temperature increase. The total cycle time is 6 min.

The homogeneous slurry from the beater is decanted into small centrifuge tubes and centrifuged at 14,500 *g* to deposit cell-wall material and the glass beads. Polyvinylpolypyrrolidone (PVPP)(3 mg/ml) is then added to the supernatant and the mixture centrifuged (15 min, 14,500 *g*). The supernatant is decanted from the PVPP and centrifuged again (60 min, 47,800 *g*) to give 28 ml of enzyme solution (1.4 mg protein/ml[68]). A portion (21 ml) of the clear supernatant obtained from the final centrifugation is treated with finely ground ammonium sulfate (344 mg/ml). The ammonium sulfate is added slowly over 40 min with stirring at 0°. After the addition, the solution is stirred for an additional 60 min and then centrifuged (60 min, 47,800 *g*) to precipitate the protein pellet.[69] The pellet is resuspended in 5.0 ml of phosphate buffer containing dithioerythritol, EDTA, and glycerol, as above (2.8 mg protein/ml); $MgCl_2$ (75 μl of a 1 *M* solution) is added, and the solution is purged with nitrogen. The enzyme reaction is initiated by addition of labeled farnesyl pyrophosphate (final concentration, 20–100 μM) and the incubation is carried out at 24° for 2 hr.[70] The incubation is terminated by the addition of acetone (2 ml) and the organic products are extracted (3×) into pentane. Carrier (±)-pentalenene (3 mg) is added at this stage. The pentane extract is purified by flash chromatography (silica 230–400 mesh, column 1 × 15 cm, pentane, fraction size 4 ml). Radioactive pentalenene, free from farnesol, is isolated in the third or fourth fraction. Based on the activity of the recovered pentalenene, the rate of pentalenene formation typically corresponds to 1 nmol/mg protein/hr.

[67] Biospec Products, Bartlesville, Oklahoma 74003. This is a benchtop version of more sophisticated cell-disruption units such as the Dyno-Mill, Impandex, Inc., Maywood, New Jersey 07607.

[68] Bio-Rad Protein Assay, Bio-Rad Laboratories, Richmond, California 94804. Cf. M. Bradford, *Anal. Biochem.* **72,** 248 (1976) and T. Spector, *ibid.* **83,** 773 (1977).

[69] The pellet typically contains 50% of the total protein. Greater than 90% of the contaminating phosphatase activity remains in the ammonium sulfate supernatant.

[70] Incubation of the enzyme with 3 μM [1-$^3H_2$]farnesyl pyrophosphate for differing times at 30° showed that 55% of the pentalenene was formed in the first 5 min and that there was no significant conversion after 30 min. Depletion of substrate was not the only cause of this termination of turnover, as adding extra substrate after 15 min did not increase the yield. When the incubation was repeated at 24°, a 25% lower turnover was observed (3 μM substrate, 30 min incubation), but addition of extra substrate after 15 min gave an overall better conversion. These experiments suggest that even though the enzyme may be less active at 24° it retains activity longer and for a given amount of protein converts more substrate to product.

For smaller scale enzyme assays, incubations can be carried out on 1-ml volumes of enzyme solution which are quenched with acetone (0.5 ml) after 30 min. Synthetic pentalenene (0.2 mg) in pentane (1 ml) is added and the two layers mixed thoroughly. A portion of the pentane extract is withdrawn and passed directly down a 2 cm silica column in a pipet (thin-layer chromatography grade silica, pretreated with 1 ml of pentane). The eluate is collected and an additional 1 ml of pentane passed down the column and combined with the initial fraction. This combined volume (1.5 ml) contains all the pentalenene and no farnesol.

Cyclase Purification. As mentioned earlier, a major impediment to a serious program aimed at cyclase purification is the extremely small quantities of these enzymes which are available even from large scale preparations. Nonetheless, several monoterpene cyclases, including those for the synthesis of 1,8-cineole,[2] (+)-bornyl pyrophosphate,[3] γ-terpinene,[8] (−)-*endo*-fenchol,[27] and (+)-α-pinene,[4] have been partially purified (10- to 20-fold) by combination of gel filtration and chromatography on *O*-diethylaminoethyl cellulose and hydroxylapatite. The major emphasis of these earlier attempts at purification was the removal of competing phosphatases and an assessment of the number and types of cyclases present. An additional chromatofocusing step allowed an olefin cyclase from *S. officinalis,* capable of synthesizing (−)-α-pinene, (−)-β-pinene, (−)-limonene, and (−)-camphene, to be obtained in electrophoretically homogeneous form, but in amounts too small to be of practical significance.[4]

Hydrophobic chromatography on geranic acid-substituted agarose was employed in the purification of γ-terpinene cyclase,[8] but this and similar hydrophobic matrices (octyl and phenyl Sepharose) were of little value in the purification of other monoterpene cyclases, which bind tenaciously and can only be eluted with considerable loss of activity. Likewise, cation exchange techniques have been of little value for the purification of monoterpene cyclases in that these enzymes are unstable at the rather low pH needed to effect binding. Thus far only a few sesquiterpene cyclases have been purified to the extent of a gel filtration and hydroxylapatite or ion exchange chromatography step.[11,13]

Storage stability varies enormously with the particular cyclase and degree of purification. Freezing in pH 6.5 buffer containing 25% glycerol, 5 m*M* sodium phosphate, and 1 m*M* dithioerythritol appears to be superior to lyophilization for long-term storage of the few monoterpene cyclases examined. Sesquiterpene cyclase preparations from *Gibberella fujikuroi,* lyophilized in Tris buffer containing glycerol and dithioerythritol, can be stored at 4° for a month without loss of activity.[11] The presence of polyols and thiol reagents also markedly improve cyclase stability during purification.

[45] Carbocyclases from *Citrus limonum*

By Osvaldo Cori and María Cecilia Rojas

Carbocyclases[1] form cyclic monoterpenes from allylic pyrophosphates (neryl, geranyl, or linalyl pyrophosphate).[4]

$$\underset{\text{allylic pyrophosphate}}{C_{10}H_{17}\text{-OPOPO}} \xrightarrow{M^{2+}} \underset{\text{cyclic monoterpenes}}{C_{10}H_{16}} + POP^- + H^+$$

This reaction was first demonstrated *in vitro* with crude extracts from *Pinus radiata* seedlings.[5] Enzyme preparations, crude or purified have been obtained from *Salvia, Mentha, Foenilicum,*[6-8] and various species of citrus.[9,10] Mono and bicyclic monoterpenes have been found as products. More than one substrate is utilized or several products are formed and evidence points to the existence of several enzymes.

The most plausible mechanism of the reaction catalyzed by carbocyclases is the intramolecular addition of an olefinic double bond to a positive charge generated by the leaving of a pyrophosphate group, followed by regiospecific proton elimination. It is the intramolecular counterpart of the reaction catalyzed by prenyltransferase.[11-13]

[1] We will use the term "carbocyclase" rather than "cyclase" to designate these enzymes. The name, "Carboligase, C-C, intramolecular, pyrophosphate eliminating" is still too cumbersome, and also inadequate if product and substrate specificities are not clarified. The term "carbocyclase" also distinguishes these enzymes from more specific cyclases (epoxysqualene cyclase) or from those that form cyclic ethers like cineol or tertiary cyclic alcohols like α-terpineol.[2,3]

[2] R. Croteau and F. Karp, *Arch. Biochem. Biophys.* **179,** 257 (1977).

[3] R. Croteau and F. Karp, *Arch. Biochem. Biophys.* **176,** 734 (1976).

[4] Abbreviations: GPP, geranylpyrophosphate (C_{10}, $\Delta^{2,6}$; 2*E*); NPP, neryl pyrophosphate (C_{10}, $\Delta^{2,6}$; 2*Z*); LPP, linalyl pyrophosphate (C_{10}, $\Delta^{1,6}$; tertiary); POP, inorganic pyrophosphate ($H_3P_2O_7^-$); TLC, thin-layer chromatography; GLC, gas chromatography; M, bivalent cation (Mn^{2+}, Mg^{2+}, Co^{2+}).

[5] O. Cori, *Arch. Biochem. Biophys.* **135,** 415 (1969).

[6] H. Gambliel and R. Croteau, *J. Biol. Chem.* **257,** 2335 (1982).

[7] R. Kjonaas and R. Croteau, *Arch. Biochem. Biophys.* **220,** 79 (1983).

[8] R. Croteau, M. Felton, and R. C. Ronald, *Arch. Biochem. Biophys.* **200,** 534 (1980).

[9] C. George-Nascimento and O. Cori, *Phytochemistry* **10,** 1803 (1971).

[10] M. C. Rojas, L. Chayet, G. Portilla, and O. Cori, *Arch. Biochem. Biophys.* **222,** 389 (1983).

[11] O. Cori, *Phytochemistry* **22,** 331 (1983).

[12] O. Cori, G. Portilla, and L. Chayet, *Arch. Biol. Med. Exp.* **15,** 357 (1982).

[13] R. Croteau, *in* "Biosynthesis of Isoprenoid Compounds" (J. W. Porter and S. L. Spurgeon, eds.), p. 227. Wiley, New York, 1981.

ISBN 0-12-182010-6

Although formation of labeled cyclic monoterpenes from any precursor implies the participation of a cyclase, direct evidence of the conversion of phosphorylated substrates into cyclic products is scarce. The activity of extracts ranges in the pmol/min/mg, and purifications achieved are not spectacular. Table I summarizes relevant data of direct evidence with cell-free extracts or purified enzyme preparations from different plant species. Whenever tested, NPP, GPP, or LPP have been substrates, although enzymes from different sources differentiate between precursors.[3,5]

The present chapter describes the partial purification and properties of carbocyclase from the flavedo of *Citrus limonum*.[14] This is a convenient and abundant source and the enzyme obtained is reasonably stable.

Assay

The assay is based on the conversion of radioactive substrates (1-^{3}H or 2-^{14}C) into hexane-soluble products and the separation of hydrocarbons from alcohols by adsorption on silicic acid. The hydrocarbons may be identified by TLC or radio-GLC.[15]

The following standard mixture is incubated in conical glass stoppered tubes for 30 min at 30°:

Total volume, 0.5 to 2 ml
TES-NH_3 buffer pH 7.0, 100 m*M*
2-Mercaptoethanol, 20 m*M*
$MnSO_4$, 3 m*M*
Substrates ([1-^{3}H]GPP, [1-^{3}H]NPP, or [1-^{3}H]LPP), 0.004–0.15 m*M*
Protein concentration, 5–600 μg/ml

Substrates and protein concentration must be adjusted, according to the level of purification.[10,16,17] When LPP is used as substrate, incubation time must be only 10 min. Initial rates are obtained under these conditions.

The reaction is quenched by cooling to 0° and vigorous shaking with 2 ml of hexane, which extracts prenols and hydrocarbons. Allylic alcohols are the hydrolysis products of phosphatase activity.[18] Total radioactivity

[14] O. Cori, L. Chayet, L. M. Pérez, M. C. Rojas, G. Portilla, L. Holuigue, and L. A. Fernández, *Arch. Med. Biol. Exp.* **14,** 129 (1981).

[15] A. Karmen, this series. Vol. 14, p. 465.

[16] L. Chayet, M. C. Rojas, E. Cardemil, A. M. Jabalquinto, J. R. Vicuña, and O. Cori, *Arch. Biochem. Biophys.* **180,** 318 (1977).

[17] G. Portilla, M. C. Rojas, L. Chayet, and O. Cori, *Arch. Biochem. Biophys.* **218,** 614 (1982).

[18] L. M. Pérez, G. Taucher, and O. Cori, *Phytochemistry* **19,** 183 (1980).

TABLE I

Comparison of Cyclases from Different Plant Tissues

Source	Effective substrates tested	Major products	Highest specific activity (nmol/min/mg protein)	Purification over extract	V_{max}/K_m[m]
Pinus radiata D Don[a]	NPP only	α-Pinene, β-pinene	0.0003	*n*	
Citrus sinensis[b]	NPP, GPP	α-Pinene, limonene	0.21	*n*	
Citrus limonum[c]	NPP, GPP, LPP	α- and β-Pinene, limonene	7.0	50	0.25 (2.5)
Mentha piperita[d]	NPP, GPP	Limonene	1.5	*n*	
Foenilicum vulgare[e]	GPP, NPP	1-endo fenchol	0.01	10	
Salvia officinalis[f]	NPP, GPP, LPP	α- and β-Pinene, limonene, camphene	0.016	*n*	
Salvia officinalis[g]	NPP, GPP	Bornyl-PP	0.16	>3.5	0.14 (20)
Salvia officinalis[h]	NPP, LPP, GPP	1,8-Cineole	0.03	17	
Salvia officinalis[i]	NPP only	α-Terpineol, hydrocarbons	0.02	11	
Salvia officinalis[j]	NPP	Borneol, camphor	0.003	—	
Thymus vulgaris[k]	NPP, GPP, terpinyl-PP	γ-Terpinene	0.04	10	0.065 (18)
Thymus vulgaris[l]	γ-Terpinene, *p*-cymene	*p*-Cymene, thymol	—	*n*	

[a] O. Cori, *Arch. Biochem. Biophys.* **135,** 416 (1969).

[b] C. George-Nascimento and O. Cori, *Phytochemistry* **10,** 1803 (1971).

[c] M. C. Rojas, L. Chayet, G. Portilla, and O. Cori, *Arch. Biochem. Biophys.* **222,** 389 (1983).

[d] R. Kjonaas and R. Croteau, *Arch. Biochem. Biophys.* **220,** 79 (1983).

[e] R. Croteau, M. Felton, and R. C. Ronald, *Arch. Biochem. Biophys.* **200,** 534 (1980).

[f] H. Gambliel and R. Croteau, *J. Biol. Chem.* **257,** 2335 (1982).

[g] R. Croteau and F. Karp, *Arch. Biochem. Biophys.* **198,** 512 (1979).

[h] R. Croteau and F. Karp, *Arch. Biochem. Biophys.* **179,** 257 (1977).

[i] R. Croteau and F. Karp, *Arch. Biochem. Biophys.* **176,** 734 (1976).

[j] R. Croteau and F. Karp, *Biochem. Biophys. Res. Commun.* **72,** 440 (1976).

[k] A. J. Poulose and R. Croteau, *Arch. Biochem. Biophys.* **191,** 400 (1978).

[l] A. J. Poulose and R. Croteau, *Arch. Biochem. Biophys.* **187,** 307 (1978).

[m] Numbers in parentheses are the $(V_{max}/K_m)_{GPP}/(V_{max}K_m)_{NPP}$ values.

[n] Only activity of extract reported.

in this fraction ("Free lipids")[19] is counted in an aliquot by scintillation spectrometry as described below. Hydrocarbons are separated from prenols by the following procedure in another aliquot: 350–400 mg portions of silicic acid (100 mesh) are weighed into centrifuge tubes, activated for at least 1 hr at 110°, and allowed to cool in a desiccator prior to use. An aliquot (1.5 ml) of the hexane extract is added and vigorously shaked on a Vortex stirrer. After centrifuging for 1–2 min at 1000 *g* two-thirds of the supernatant which contains only hydrocarbons ("Hydrocarbons fraction") is pipetted into a scintillation vial and counted by conventional methods.[15]

Controls performed with radioactive prenols show that they are retained to the extent of 99% on silicic acid. They may be recovered by extraction with 2 ml of ether. If the hydrocarbon fraction contains less than 2% of the total radioactivity of the "Free lipids" fraction, the results are considered negative. Controls with boiled enzyme must always be performed, since Mn^{2+} catalyzes the nonenzymatic formation of limonene, myrcene, or ocymenes from allylic pyrophosphates.[17,20] This becomes relevant at low enzyme activities.

Identification of Products

Reaction products may be identified in the "Hydrocarbon fraction" by GLC or TLC.

Gas Chromatography. Suitable column packing for stainless steel or copper columns (300 × 0.64 cm) is either 0.83% ethylene glycol adipate coated on Chromosorb G-AW 60/70 mesh or Chromosorb W 60/80 mesh coated with 20% Apiezon. The operating temperature for hydrocarbon identification is 40° for polyethylene glycol and 120° for Apiezon; injector and thermal conductivity detector temperature should not exceed 170 and 210°, respectively. The effluent of the gas chromatograph is introduced by means of a heated connecting tube (150°) into a heated Geiger counter for gas phase (Biospan 4998, Nuclear Chicago, Des Plaines, IL). This instrument, which does not destroy the sample, is more suitable for the volatile products of carbocyclases than those based on combustion of the sample.[15] In the latter instruments, CuO is readily consumed by the combustion of the solvent which cannot be eliminated without loss of the volatile radioactive products.[21]

[19] E. Beytía, P. Valenzuela, and O. Cori, *Arch. Biochem. Biophys.* **129,** 346 (1969).

[20] M. V. Vial, M. C. Rojas, G. Portilla, L. Chayet, L. M. Pérez, O. Cori, and C. A. Bunton, *Tetrahedron* **37,** 2351 (1981).

[21] H. A. Massaldi, and C. J. King, *J. Food Sci.* **39,** 438 (1974).

The hexane extract containing the terpene hydrocarbons is added 0.10 to 0.3 mg of authentic carriers and injected into the GLC instrument. Mass and radioactivity peaks are simultaneously recorded with a two channel instrument. Radioactivity is estimated by measuring peak area.

Thin-Layer Chromatography. Aliquots of the hexane extract containing the radioactive products plus added carriers are applied on argentated Silica gel G plates (12.5% $AgNO_3$) and developed in benzene. The hydrocarbons are localized by means of iodine vapor. The plate is scraped in 0.5 cm portions into scintillation vials and counted. GLC is the preferable procedure to identify terpene hydrocarbons, since their high volatility makes TLC of qualitative value only.[16]

Chemical Synthesis of Radioactive Substrates

Carbocyclases utilize GPP, NPP, and LPP.[10,17] They may be synthesized as ^{14}C- or ^{3}H-labeled molecules by well established procedures. However many details are crucial for satisfactory yield and purity of these precursors, and they are detailed in this section.

Radioactive neryl, geranyl, and linalyl pyrophosphates are prepared by phosphorylation of the corresponding alcohols [1-^{3}H]nerol, [1-^{3}H]geraniol, and [1-^{3}H]linalool.

Oxidation of Geraniol and Nerol with MnO_2. Nerol and geraniol are obtained commercially (Aldrich) and further purified by distillation under reduced pressure in a spinning band column.

MnO_2 is prepared from $KMnO_4$ and $MnSO_4$,[22] and thoroughly washed with hot water. The solid obtained after centrifugation is activated by heating at 110° for 42 hr and then partially deactivated for 30 hr at room temperature. MnO_2 thus prepared oxidizes nerol or geraniol at a rate that allows recovery of neral or geraniol without cross contamination. Fully active MnO_2 forms two nonidentified oxidation products; if oxidation is too slow cis–trans isomerization of the aldehydes occurs.

The oxidation reaction is carried out at room temperature by stirring 2.3 g of activated MnO_2, with 0.92 mmol of geraniol or nerol dissolved in 10 ml of *n*-hexane. The reaction should be carried out in darkness to avoid *E-Z* isomerization of the newly formed aldehydes. Samples of the reaction mixture are taken at different time intervals and centrifuged to analyze the supernatant by GLC. The reaction is carried out to only 80% of completion (30 to 60 min). This prevents the formation of side products.

Reduction of the Aldehydes with $NaB^{3}H_4$. The reaction is performed at room temperature in isopropanol : hexane = 2 : 1 (v/v). Eight milliliters

[22] E. F. Pratt and J. F. Van de Castle, *J. Org. Chem.* **26,** 2973 (1961).

of hexane solution containing 0.612 mmol of geranial or neral is mixed with 13 mg of NaB^3H_4 (228 mCi/mmol) and 9 ml of isopropanol. At different time intervals, an aliquot of the reaction mixture is added to 1 ml of 0.05% NH_4OH, extracted with hexane, and analyzed for radioactive alcohols by GLC, as described. The reaction is completed in approximately 4 hr and stopped by adding 0.05% NH_4OH.

When higher concentrations of NaB^3H_4 and larger proportions of isopropanol are used the reduction product is a mixture of geraniol, nerol, and citronellol. Some batches of NaB^3H_4 contain an impurity which causes extensive *E-Z* isomerization of the allylic alcohols, under conditions where unlabeled borohydride yields a single product.[23]

After washing the hexane phase with 0.05% NH_4OH, unlabeled alcohol is added to obtain a specific radioactivity of 4–10 × 10^7 dpm/μmol.

The solvent is evaporated by distillation. Water in the remaining alcohol is eliminated by adding benzene and blowing out the solvents with a stream of nitrogen. The operation is repeated 2–3 times.

Rearrangement of [1-^{3}H]Geraniol. Tritiated linalool is prepared by rearrangement of [1-^{3}H]geraniol in 1 : 1 acetone/HCl 0.1 *M* at 37°. Of [1-^{3}H]linalool 60% is obtained after 62 hr of incubation and separated from the other rearrangement products (hydrocarbons, nerol, α-terpineol) and unreacted geraniol by chromatography on an Adsorbosil column.[17]

The column (1.5 × 11 cm) is packed with Adsorbosil-5 (Applied Science Laboratories) and equilibrated with 1% ethyl acetate in hexane. The rearranged alcohols are applied to the column in a final volume of 2 ml, and washed with 40 ml of the equilibration solvent. [1-^{3}H]Linalool is eluted with a gradient of ethyl acetate from 1 to 5% (v/v) in hexane. Fractions of 2 ml are collected and performance of the column is monitored by GLC analysis. Uncontaminated linalool emerges between 2 and 3.5% ethyl acetate.

Phosphorylation of the Alcohols. Primary alcohols are phosphorylated with bis(triethylammonium)phosphate.[24]

Phosphorylation of [1-^{3}H]Linalool. The tertiary isomer linalool is phosphorylated less efficiently than nerol or geraniol. Addition of new portions of the phosphorylating agent improves the yield.[25]

[1-^{3}H]Linalool (0.2 mmol) is stirred with 0.18 ml of trichloroacetonitrile for 40 min, and 5 ml of a solution of 0.23 *M* bis(triethylammonium) phosphate in acetonitrile[24] is added dropwise. After 24 hr of reaction, only 10% of the alcohol has been phosphorylated.

[23] L. Chayet, L. Holuigue, and O. Cori, unpublished observations (1981).
[24] G. Popják, this series, Vol. 15, p. 386.
[25] G. Portilla and O. Cori, unpublished results (1982).

Solid bis(triethylammonium) phosphate (150 mg) is added to the reaction mixture and stirred at room temperature for additional 24 hr. This operation increases the phosphorylation yield to 26% of the initial linalool. The reaction is stopped, and the aqueous phase is first extracted with hexane and then with ether as described.[24] Addition of KOH to obtain a pH of 10 avoids decomposition of linalyl pyrophosphate during the separation procedure.

Tritiated linalool that has not been phosphorylated can be recovered from the ether phase and may be recycled, but trichloroacetamide must be eliminated from the ether phase. The solvent is evaporated and linalool is extracted with hexane from the trichloroacetamide crystals. In each phosphorylation cycle additional 10% of the radioactive alcohol is transformed into water soluble compounds.

Elimination of Inorganic Phosphates. The aqueous phase obtained after phosphorylation of nerol, geraniol, or linalool contains orthophosphate inorganic pyrophosphate, and triphosphate in addition to the phosphate and pyrophosphate esters of these prenols. The aqueous phase is lyophilized and the phosphate and pyrophosphate esters are extracted with *n*-propanol : NH_4OH = 2 : 1 v/v. Inorganic phosphates remain in the insoluble residue. The residue is separated by centrifugation and the solvent is evaporated from the supernatant.[20]

Resolution of the Phosphate Esters. Primary allylic phosphomonoesteres and pyrophosphate esters are separated by chromatography on DEAE-Sephadex[26] with a linear gradient (50–650 m*M*) of triethylammonium bicarbonate pH 7.4 as eluting solvent.[27] Since linalyl derivatives are very unstable at this pH, their separation is achieved under different conditions on a silica gel 60 column[28] eluted with a linear gradient of propanol : NH_3.

Purification of the Isolated Allylic Pyrophosphates. The tritiated pyrophosphate substrates purified by the methods described above still contain impurities that are visible as an oily residue after freeze drying. These compounds are eliminated by precipitation of the phosphates with LiCl.

The fractions containing the radioactive substrates are concentrated to 200 m*M* and LiCl is added to a final concentration of 3.3 *M*. The white precipitate which appears after 24 hr at 4° is washed with cold acetone and ether. In the purification of LPP, KOH must be added before concentration.

The solubility of the lithium salt of NPP, GPP, and LPP is low, and Li^+ may remain associated with the substrate under enzymatic assay condi-

[26] M. Oster and C. A. West, *Arch. Biochem. Biophys.* **127,** 112 (1968).
[27] M. Smith and H. G. Khorana, this series, Vol. 6, p. 651.
[28] Y. Gafni and I. Schechter, *Anal. Biochem.* **92,** 248 (1959).

tions thus affecting the results obtained. For this reason Li^+ is exchanged with K^+ by chromatography on a Chelex column before use.

Chelex Na^+ (100 mesh, 0.7 meq/ml) is washed with 1 *M* HCl, 1 *M* KOH, 0.5 *M* EDTA pH 7.0, and water to obtain a conductivity of 10^{-6} mho. The Li^+ salt of the pyrophosphates is applied to the column and eluted with distilled water.[29]

Identity and purity of the synthetic substrates is established by TLC, by the ratio of alcohol to phosphorus[30] and by GLC identification of the alcohol obtained by enzymatic hydrolysis.[31]

The final cross contamination of primary pyrophosphates resulting from isomerization during the preparation is less than 5%. No radioactive contaminants were detected in LPP (limit 1%).

When stored for 50 days in aqueous solution pH 10 at 4°, 7% of the LPP is rearranged. No rearrangement of NPP or GPP was observed, but unidentified decomposition products appear when freeze dried and stored for extended periods.

Enzyme Purification

A partially purified preparation of carbocyclase with a specific activity of 4–7 nmol/min/mg is obtained from the flavedo of *Citrus limonum* by the following procedure (Table II).

Preparation of the Extract. Ripe lemons are collected in winter and early spring and stored at 4°. Carbocyclase specific activity of the extracts obtained during this period is about 40 times higher than that obtained from fruits collected in summer or fall.[16]

Lemons are washed with distilled water and 200 g of flavedo are grated over 225 ml of 0.1 *M* Tris–HCl pH 7.8, 0.02 *M* 2-mercaptoethanol buffer kept at 0°.

The grated flavedo suspension is gently stirred with a glass rod for 20 min and then filtered through Nylon cloth. This procedure is sufficient to extract carbocyclase from the flavedo and more vigorous grinding does not improve the yield. The filtrate is centrifuged for 30 min at 30,000 *g* at 0° and filtered through glass wool to separate the floating lipid layer from the supernatant. This extract contains 2–3 mg of protein per ml, determined by turbidimetry[32] and its final pH is about 6.0. For recovery reasons, we recommend to continue the purification procedure to completion of the following step.

[29] G. Portilla, V. Calvo, and O. Cori, unpublished results (1983).

[30] L. Ernster, R. Zettenström, and O. Lindberg, *Acta Chem. Scand.* **6,** 804 (1952).

[31] E. Cardemil and O. Cori, *J. Label. Compd.* **9,** 15 (1974).

[32] E. Stadtman, G. D. Novelli, and F. Lipmann, *J. Biol. Chem.* **191,** 365 (1951).

TABLE II
PARTIAL PURIFICATION OF CARBOCYCLASE FROM *Citrus limonum*

Fraction	Total protein (mg)	Substrate	Specific activity[a] (U/mg)	Total units	Half-life[b] (days)	Stability[c] at $-18°$, freeze dried	Ratio of cyclase/phosphatase activity
Extract	104	NPP	0.16	16.6	2.2	*d*	0.17
		GPP	0.22	22.9			0.63
PEG 10–20%	17.8	NPP	0.59	10.5	0.9	4 years	0.56
		GPP	1.03	18.3			1.7
DEAE-Sephadex	2.3	NPP	6.78	15.6	2.2	6 months	1.7
		GPP	10.3	23.6			5.9

[a] One unit = 1 nmol/min at 30°.
[b] Stored at 4°.
[c] No decrease in specific activity observed.
[d] The instability of the extract to freeze drying and storage did not allow any accurate estimate.

Polyethylene Glycol (PEG) Fractionation. The procedure is carried out at 0°. The pH of the extract is adjusted to 8.0 with 10 *M* NaOH before precipitation.

PEG 4000 (Fluka A. G., Buchs, Switzerland)[33] is finely ground in a mortar and added with continuous stirring to the extract until 10% w/v concentration is reached. The solution is equilibrated by stirring for 20 min, centrifuged 20 min at 30,000 *g* and 0°, and the protein precipitate is discarded. A final 20% w/v concentration is obtained by adding solid PEG to the supernatant. The precipitate is separated by centrifugation for 20 min at 30,000 *g* and 0°.

The protein pellet containing 60–80% of the carbocyclase activity of the extract is dissolved in 0.1 *M* Tris–HCl pH 7.8, containing 0.02 *M* 2-mercaptoethanol and freeze-dried. Since the enzyme may thus be stored for over 4 years at $-18°$, this is a convenient stage to interrupt the process and accumulate enzyme for the next step. PEG fractionation eliminates 70% of the phosphatase activity present in the extract and a considerable amount of the pigments that absorb at 320 nm (tannins, flavonoids, and others). PEG carried over does not interfere with the assay, provided its concentration in the incubation mixture is below 10% (v/w).

DEAE Sephadex A-25 Chromatography. DEAE-Sephadex A-25-OH^- is treated with 0.01 *M* EDTA and then equilibrated with 0.05 *M* Tris–HCl pH 7.0, 0.05 *M* KCl, 2 m*M* EDTA, and 1 m*M* dithiothreitol buffer.[10]

[33] PEG from other sources causes partial loss of activity or changes in the precipitation pattern. Crystallization of PEG from acetone does not improve this problem.

The freeze-dried PEG pellet is dissolved in cold water to a final concentration of 15–30 mg of protein per ml. After centrifugation at 30,000 *g* for 20 min, to eliminate insoluble material the solution is dialyzed with stirring for 2 hr at 4° against four changes of 100 volumes of a solution of 0.05 *M* Tris–HCl pH 7.0, 0.05 *M* KCl, 2 m*M* EDTA, and 0.1 m*M* dithiothreitol and then centrifuged at 30,000 *g* and 0° for 20 min.

The protein treated as described above (366 mg) is applied to a column (2.1 × 37 cm = 122 ml bed volume) and washed with four column volumes of the equilibration buffer.

Carbocyclase is eluted as a broad peak with a linear gradient of 0.05 to 0.25 *M* KCl in the same buffer. Fractions are assayed individually and those containing carbocyclase (usually between 0.1 and 0.2 *M* KCl) are concentrated by ultrafiltration in a PM-30 Amicon membrane to obtain a protein concentration of 0.8 mg/ml. The concentrated enzyme solution is freeze-dried and stored at −18°. There is no loss of activity for at least 6 months.

The flavedo extract may be prepared with a buffer which contains in addition to Tris and mercaptoethanol, 6 m*M* $MnSO_4$ plus 0.03 *N,N'*-diethyl dithiocarbamate. Although a brown precipitate forms, the yield of total units in the extract is increased by a factor of two to five. PEG fractionation is not affected by these two components, but the elution pattern in ion exchange chromatography on DEAE-Sephadex is altered. About 30% of cyclase activity is not retained in the column. It behaves normally if rechromatographed on a second DEAE-Sephadex-Cl column. The Mn^{2+} ions added at the extract stage are eliminated in steps 2 and 3. This procedure, although more laborious, allows a higher total yield of carbocyclase units, since the initial extracts exhibit a higher activity.

Properties of Carbocyclase

Effect of pH. Maximum activities are obtained at pH values between 6.0 and 7.0 with inflections at pH 5.6 and 7.5. A 3-fold increase of activity with GPP as substrate is observed at pH 8.0 when Tris buffer is substituted by Tricine.[16] This effect is not observed with NPP as substrate. At pH 5.0 and 30°, the half life of carbocyclase is 19 min.

Substrate Specificity and Kinetic Parameters. Most carbocylases utilize both NPP and GPP. Enzymes from *Pinus radiata*,[5] *Salvia*,[6] and peppermint[7] have been reported to utilize exclusively or preferentially NPP. The tertiary isomer LPP is utilized by *Citrus*,[17] *Salvia*,[6] and *Mentha*[5] enzymes to form cyclic monoterpenes. Table III shows the substrate specificity of the enzyme from *C. limonum* for three allylic pyrophosphates.[10,17] Allylic phosphomonoesters and citronellyl pyrophosphate are not substrates.[10]

TABLE III
KINETIC PARAMETERS AND CROSS INHIBITORS OF CARBOCYCLASE FROM *Citrus limonum*[a]

Substrate	K_m (μM)	V_{max}/K_M (min^{-1} mg^{-1})	$I_{0.5}/K_m$	
			NPP	GPP
NPP	1.0	0.1	—	18.2
GPP	2.9	0.25	2.3	—
LPP	0.4	2.75	6.0	8.5

[a] Adapted from M. C. Rojas, L. Chayet, G. Portilla, and O. Cori, *Arch. Biochem. Biophys.* **222,** 389 (1983), and G. Portilla, M. C. Rojas, L. Chayet, and O. Cori, *Arch. Biochem. Biophys.* **218,** 614 (1982), by permission of Academic Press.

The ability of the enzyme to utilize one or another substrate does not vary with purification, but varies with aging or heating.

Product Specificity. Carbocyclases from *C. limonum* produce α-pinene, β-pinene, and limonene independently of the substrate used. Some preparations form in addition γ-terpinene or sabinene, but these components are not always present.[10,16] The distribution pattern of the three main hydrocarbons changes somewhat during the last purification step but the formation of α-pinene is greatly diminished by aging at $-18°$.[10,34] The distribution pattern is slightly different with LPP as substrate.[17] All these data support the existence of several enzymes.

Inhibitors. Carbocyclase substrates show cross inhibition (Table III).

Citronellyl pyrophosphate, the 2,3-saturated analog of GPP and NPP has the same I_{50}/K_m for both substrates. Allylic phosphomonoesters, IPP, inorganic pyrophosphate, and triphosphate are ineffective inhibitors; inhibition by DMAPP is more effective with GPP than with NPP as substrate.[10,12,17]

Requirement of Metal Ions. Carbocyclases from *C. limonum* have an absolute Mn^{2+} requirement. Mg^{2+} and Co^{2+} may substitute it with about 2% efficiency. There are slight changes in product distribution patterns with different ions. Reaction rate correlates with the calculated concentration of the R-PP$(M^{2+})_2$ complex.[10,20,35]

Inactivators. Carbocyclase require reduced sulfhydryl groups for catalytic activity. It is reversibly inactivated by *p*-chloromercuribenzoate or

[34] O. Cori, L. Chayet, L. M. Pérez, M. de la Fuente, M. C. Rojas, G. Portilla, and M. V. Vial, *Mol. Biol. Biochem. Biophys.* **32,** 97 (1980).

[35] M. C. Rojas, L. Chayet, and O. Cori, unpublished observations (1983).

dithiobis(nitrobenzoic acid). The enzyme is also inactivated by benzyl bromide and protected from this inactivation by the Mn^{2+} complexes of GPP or NPP, but not by free metal or uncomplexed substrate.[14]

Acknowledgments

The authors are very indebted to Dr. Liliana Chayet and Gloria Portilla for allowing us to use their unpublished results and for their interest in this publication. The research supporting this communication was financed by Dep. de Desarrollo de la Investigación, U. de Chile. Mr. Gonzalo Pérez generously supplied citrus fruits from selected trees of "Huertos de Betania," Mallarauco, Chile.

[46] Isolation and Identification of Diterpenes from Termite Soldiers

By GLENN D. PRESTWICH

Distribution and Importance of Diterpenes

Chemical defense has evolved numerous times in the termite families Rhinotermitidae and Termitidae (Isoptera).[1-3] Two defense strategies are employed by specialized soldiers chemically armed with terpenoid secretions: (1) biting, with the addition of an oily, toxic or irritating secretion from a frontal gland reservoir, or (2) ejection of an irritating glue-like substance from an elongated frontal tube (the "nasus"). Whereas mono- and sesquiterpenoid defense secretions are widely distributed throughout termite subfamilies, diterpenes are restricted to two groups of higher termite genera.[4] Higher terpenoids are rare among arthropods in general,[5] which cannot synthesize steroids *de novo* since they lack the ability to couple two farnesyl pyrophosphate units.[6] Only among the Homoptera

[1] G. D. Prestwich, *J. Chem. Ecol.* **5,** 459 (1979).

[2] G. D. Prestwich, *Annu. Rev. Entomol.* **29,** 201 (1984).

[3] G. D. Prestwich, *Annu. Rev. Ecol. Syst.* **14,** 287 (1983).

[4] A single exception is *Reticulitermes lucifugus,* in which geranyllinalool has been identified [R. Baker, A. H. Parton, and P. E. Howse, *Experientia* **38,** 297 (1981)].

[5] M. S. Blum, "Chemical Defenses of Arthropods." Academic Press, New York, 1981.

[6] R. D. Goodfellow, H. E. Radtke, Y. S. Huang, and G. C. K. Liu, *Insect Biochem.* **3,** 61 (1973).

ISBN 0-12-182010-6

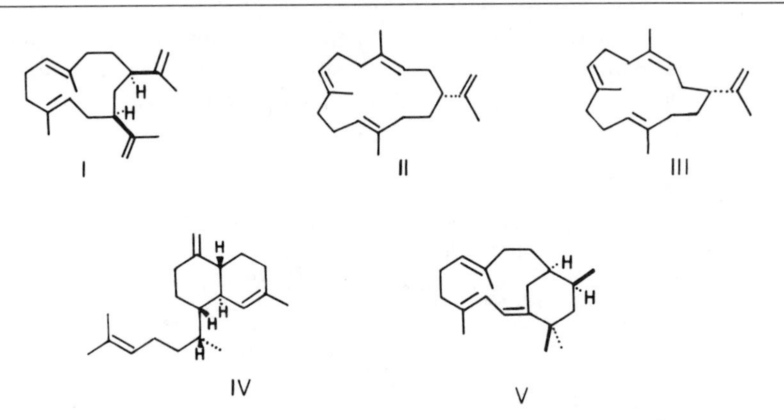

FIG. 1. Diterpenes (**I–V**) of *Cubitermes* and *Crenetermes* species.

(aphids, wax scale, and lac insects) are sesterterpenes[7,8] found. *De novo* biosynthesis of diterpenes has been demonstrated only in the glue-squirting nasute termite soldiers.[9] Further characterization of the conversion of geranylgeranyl pyrophosphate to cembrene-A and then to trivervitanes is now in progress.

Two closely related genera of humivorous African termites, *Cubitermes* and *Crenetermes* (Termitidae, Termitinae), have biting soldiers with diterpene hydrocarbons in cephalic reservoirs.[10,11] Five out of the 17 structures are now known (Fig. 1, **I–V**), and they exemplify novel skeletons, often irregular isoprenoids, with unusual biosynthetic pathways.

A second group of related genera are the pantropical nasute termites (Termitidae, Nasutitermitinae), which are the most advanced, abundant, and diverse subfamily of termites.[1,2] The soldiers eject viscous, lipophilic solutions of oxygenated, dome-shaped diterpenes dissolved in monoterpenes, and there is considerable inter- and intraspecific variation in the compositions of these sticky secretions.[3,12] The diterpenes appear to be derived biogenetically from cembrene-A (**II**), a known termite defense secretion (see above) and trail pheromone[13,14] for several nasute species. Some 50 derivatives of 5 carbon skeletons are currently known for these

[7] K. S. Brown, *Chem. Soc. Rev.* **4,** 263 (1974).

[8] F. Miyamoto, H. Naoki, T. Takemoto, and Y. Naya, *Tetrahedron* **35,** 1913 (1979).

[9] G. D. Prestwich, R. W. Jones, and M. S. Collins, *Insect. Biochem.* **11,** 331 (1981).

[10] G. D. Prestwich, *J. Chem. Ecol.* **10,** 1219 (1984).

[11] D. F. Wiemer, J. Meinwald, G. D. Prestwich, B. A. Solheim, and J. Clardy, *J. Org. Chem.* **45,** 191 (1980).

[12] G. D. Prestwich, *Biochem. Syst. Ecol.* **7,** 211 (1979).

[13] A. J. Birch, W. V. Brown, J. E. T. Corrie, and B. P. Moore, *J. Chem. Soc., Perkin Trans. 1* p. 2653 (1972).

[14] P. G. McDowell and G. Oloo, *J. Chem. Ecol.* **10,** 835 (1984).

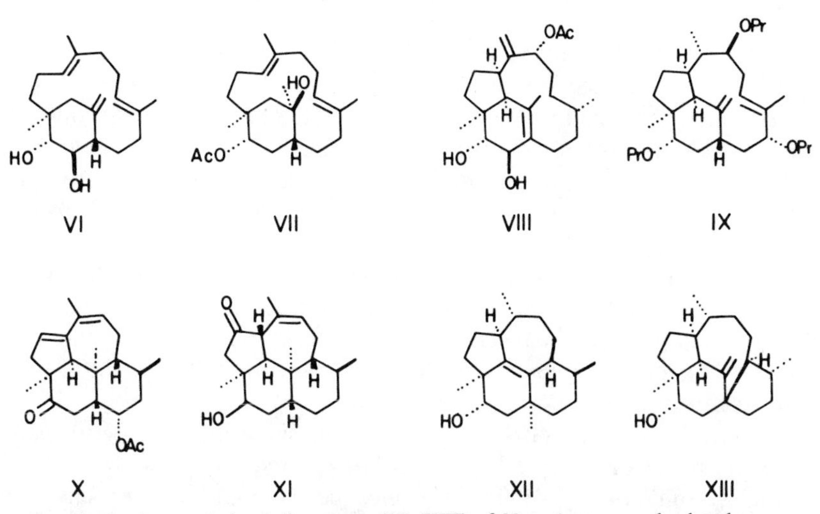

FIG. 2. Cembrene-derived diterpenes (**VI–XIII**) of *Nasutitermes* and related genera.

diterpenes (Fig. 2). The bicyclic secotrinervitanes (e.g., **VI, VII**)[15,16] and the tricyclic trinervitanes (e.g., **VIII, IX**),[1,17,18] exhibit features of the absolute stereochemistry and olefin geometry found in cembrene-A (**II**). The kempanes[1,19,20] (e.g., **X, XI**), the rippertane[21] (**XII**), with a 1,2-shifted angular methyl group, and the spirocyclic longipane[22] (**XIII**) each possess tetracyclic arrays in which the former cembrene-A derivation is somewhat obscured by the extensive transannular cyclizations which occur during biogenesis.

Isolation of Diterpenes

Nasute termite soldiers are individually removed from the colonies, cooled to 0°, and decapitated. The heads are crushed under distilled hex-

[15] J. C. Braekman, D. Daloze, A. Dupont, J. Pasteels, B. Tursch, J. P. Declerq, G. Germain, and M. van Meerssche, *Tetrahedron Lett.* **21,** 2761 (1980).

[16] J. C. Braekman, D. Daloze, A. Dupont, J. M. Pasteels, P. Lefeuve, C. Bordereau, J. P. Declerq, and M. van Meerssche, *Tetrahedron* **39,** 4237 (1983).

[17] G. D. Prestwich, S. P. Tanis, F. G. Pilkiewicz, I. Miura, and K. Nakanishi, *J. Am. Chem. Soc.* **98,** 6062 (1976).

[18] G. D. Prestwich, S. G. Spanton, S. H. Goh, and Y. P. Tho, *Tetrahedron Lett.* **22,** 1563 (1981).

[19] G. D. Prestwich, B. A. Solheim, J. Clardy, F. G. Pilkiewicz, I. Miura, S. P. Tanis, and K. Nakanishi, *J. Am. Chem. Soc.* **99,** 8082 (1977).

[20] G. D. Prestwich, J. W. Lauher, and M. S. Collins, *Tetrahedron Lett.* p. 3827 (1979).

[21] G. D. Prestwich, S. G. Spanton, J. W. Lauher, and J. Vrkoč, *J. Am. Chem. Soc.* **102,** 6825 (1980).

[22] S. H. Goh, C. H. Chuah, Y. P. Tho, and G. D. Prestwich, *J. Chem. Ecol.* **10,** 929 (1984).

ane and the crude extract is dried and freed from polar impurities by passage through a Pasteur pipet containing 1–2 cm of Florisil [elution with 1 : 1 (v/v) diethyl ether (or ethyl acetate)-hexane]. Solvents are removed *in vacuo* (30°/20 mm) using a Buchi Rotovapor-R. The resulting viscous oils can then be evaporatively distilled (75°/0.1 mm) using a Buchi kugelrohrofen to separate the less volatile diterpenoid compounds (residue) from the volatile monoterpenes and sesquiterpenes (distillate, trapped in dry-ice cooled receiving flask). To avoid isolation of extractable artifacts, the "milked" secretion is also expressed directly from the nasus onto a glass surface or is collected from the fontanelle with a microcapillary. Spectral and chromatographic analysis of these materials must match those of extracts. Nasute termite soldiers have 10–25% of the dry weight of their heads as defense secretion.

The residue is then chromatographed on silica gel or Florisil using increasing percentages of ethyl acetate in hexane. We now prefer flash chromatography on 230–400 mesh silica gel G under N_2 pressure[23] to effect the initial separations of diterpenes with different degrees of hydroxylation and/or esterification. Fractions are monitored by TLC on silica gel plates as described below. Unresolved compounds of like polarity can be separated further by semipreparative HPLC. Thus, the trinervitane diols (e.g., **VIII**) are separated on μPorasil with 10% ethyl acetate–hexane (1.0 ml/min)[24] or on Ultrasphere silica with 2–10% ethyl acetate in hexane.[25] Rippertenol (**XII**) has been separated from trinervitadienols on 10 μm Lichrosorb SI-60 with 10% ethyl acetate in hexane.[21] Many of the hydroxylated (but not esterified) diterpenes can be crystallized from methanol containing a few percent water.

Chromatographic separations are conveniently monitored by TLC using Machery-Nagel polygram Sil G/UV254 4 × 8-cm sheets and eluting with 25% (v/v) ethyl acetate in hexane. Compounds were visualized by first observing with UV (254 nm) and then by spraying with a vanillin reagent followed by heating with a hot air blower. The vanillin reagent is prepared by dissolving 9.0 g vanillin in 300 ml absolute ethanol and adding 1.5 ml concentrated sulfuric acid. With this system, most nasute diterpenes were easily resolved, and each exhibited a characteristic color when heated in the presence of the vanillin reagent.[25]

Gas chromatographic analysis can be performed on a variety of packed columns, particularly the nonpolar and medium-polarity silicones such as OV-1, OV-101, and OV-17 at 200–250°. Some compounds, such as the tripropionate **IX** require higher temperatures (up to 300°) to elute, while others, such as the β,γ unsaturated ketone **XI,** decompose on the

[23] W. C. Still, M. Kahn, and A. Mitra, *J. Org. Chem.* **43,** 2923 (1978).

[24] G. D. Prestwich and M. S. Collins, *Biochem. Syst. Ecol.* **9,** 83 (1981).

[25] C. H. Chuah, S. H. Goh, G. D. Prestwich, and Y. P. Tho, *J. Chem. Ecol.* **9,** 347 (1983).

column. Most recently, we have succeeded in using capillary gas chromatography to separate the nasute diterpenes. Metal support coated open tubular (SCOT) columns, glass wall coated open tubular (WCOT) columns and fused silica columns can cause extensive decomposition of hydroxylated diterpenes. However, we have found the durably bonded phases with high temperature stability (>300°) such as DB-5 (30 m × 0.25 mm, 0.25 μm film) and DB-1701 (15 m × 0.25 mm, 0.15 μm film) (J & W Scientific Co.) will provide extremely high resolution separations of the trimethylsilyl (TMS) derivatives. A normal WCOT 50-m fused silica OV-101 column (250–270°) can also be used if the Durabond columns are unavailable. Thus, an aliquot of the hexane extract (1–10 head equivalents is warmed at 80° for 1 hr with 100 μl of TriSil (Pierce Chemical Co.). Shorter times give incomplete silylation of the neopentyl-type C-3 hydroxyl groups. The solvent is evaporated with N_2 and 50 μl of hexane is added and the supernatant analyzed at 180–220° on OV-101 or DB-5 fused silica capillaries (Fig. 3). Again, some problems occur with heavy or labile structures like (**IX**) and (**XI**).

The diterpene hydrocarbons of *Cubitermes* are obtained analogously. However, all 17 structures coelute on silica gel and must be separated by sequential argentation chromatography and preparative GLC. Thus, cubitene (**I**) was easily purified by prep GLC on 15% FFAP or 6% Carbowax columns.[26] Medium pressure LC on 5% $AgNO_3$ on 30 μm silica gel G gave partial separation of (**II**), (**III**), (**IV**), and (**V**), but preparative GC on 15% FFAP or 20% QF-1 was needed for final purification.[10,11,27] No packed GLC column could resolve the 17 diterpenes. However, a 25-m SCOT column of FFAP gave respectable resolution of ten different isomers in *C. ugandensis*. The best resolution is achieved with a bonded phase, Durawax DX-4 (30 m × 0.25 mm, 0.25 μm film) at 170° (Fig. 4).[10]

Identification of Diterpenes

Four out of five nasute diterpene carbon skeletons[15,17–21] and two out of the five *Cubitermes* diterpene hydrocarbon structures[11,26] were determined by X-ray crystallography. In many cases, derivatization was required to obtain suitable crystals. The rippertenol skeleton (**XII**) was determined on the epoxy acetate derivative,[21] the tripropionate structure (**IX**) was obtained on the triol monohydrate,[18] the kempene structure (**XI**) was determined as the bromobenzoate.[20] The bifloratriene (**IV**) was determined on the bromobenzoate of one of the diastereomeric primary alcohols obtained by hydroboration of the *exo*-methylene group.[11] With heavy

[26] G. D. Prestwich, D. F. Wiemer, J. Meinwald, and J. Clardy, *J. Am. Chem. Soc.* **100,** 2560 (1978).

[27] D. F. Wiemer, J. Meinwald, G. D. Prestwich, and I. Miura, *J. Org. Chem.* **44,** 3950 (1979).

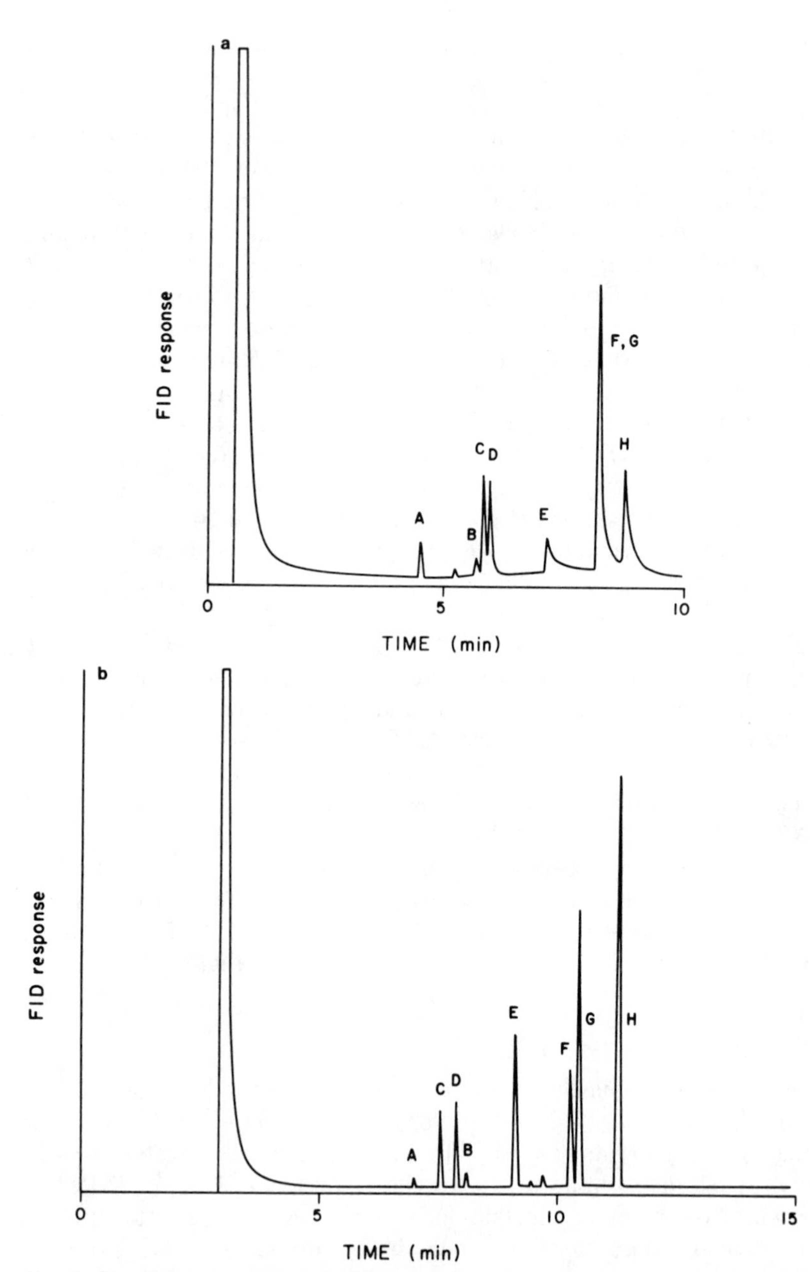

FIG. 3. Gas chromatograms of diterpenes of *Nasutitermes ephratae:* (a) 18-m OV-101 WCOT glass capillary analysis of native diterpenes (200–220° at 4°/min) and (b) 50-m OV-101 fused silica analysis of diterpene TMS ethers (250–270° at 2°/min). Letters indicate structures as follows: A, unidentified; B, unidentified; C, trinervita-1(15),8(19)-dien-3α-ol; D, trinervita-1(15),8(19)-dien-2β-ol; E, 2-oxotrinervita-1(15),8(19)-diene-3α-ol; F, trinervita-1(15),8(19)-diene-2α,3α-diol; G, trinervita-1(15),8(19)-diene-2α,3β-diol; and H, trinervita-1(15),8(19)-diene-2β,3a-diol.

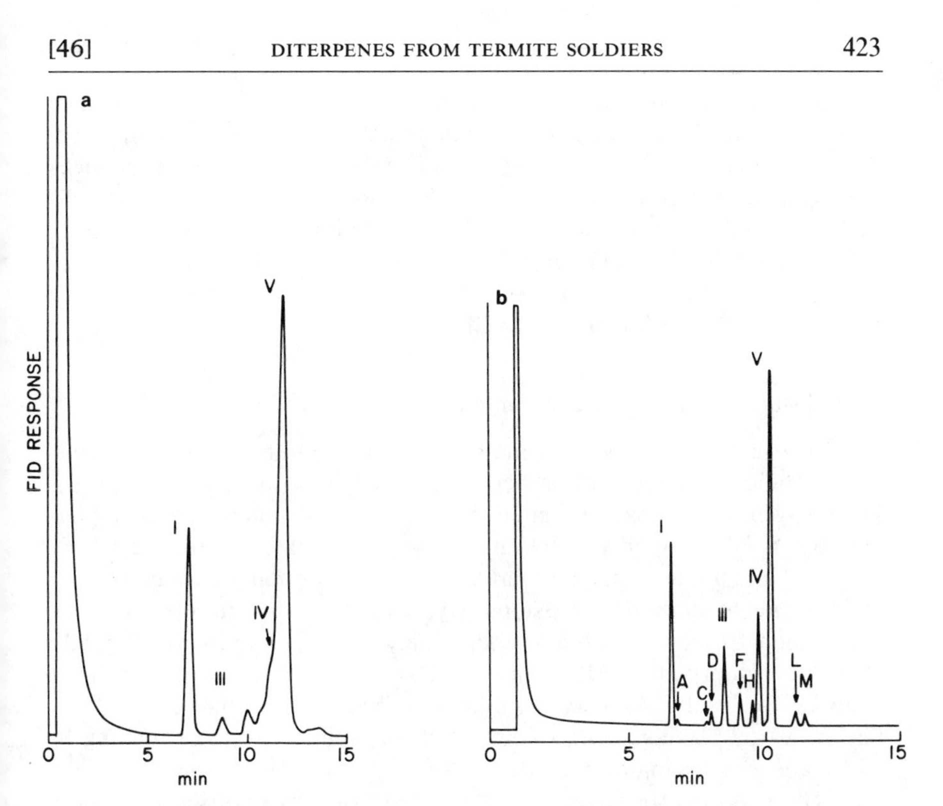

FIG. 4. Gas chromatograms of *Cubitermes ugandensis* diterpenes on (a) 2 m × 2 mm i.d. 6% Carbowax 20M packed column, and (b) 30 m × 0.25 mm i.d. Durawax DX-4 fused silica capillary column.

atom-containing derivatives, absolute configurations were obtained by anomalous dispersion studies. For nonheavy atom structures, absolute configurations were obtained by chiroptical methods, primarily circular dichroism. Diene helicity was used to established the kempene (**X**) stereochemistry, while the twist of the β,γ-unsaturated ketone (**XI**) revealed the absolute stereochemistry of this chromophore. The exiton chirality method, using a $Pr(dpm)_3$ complex of the $2\beta,3\alpha$-diol of (**VIII**), gave the absolute configuration of the first trinervitene studied.[17] All termite compounds examined to date derive from the *R*-(−)-cembrene-A structure, as expected for a terrestrial organism.[28]

The majority of the 60 or more diterpenes now characterized from termites have been determined by spectral methods, of which the most informative are ^1H and ^{13}C NMR spectroscopy and mass spectrometry.

[28] A. J. Weinheimer, C. W. J. Chiang, and J. A. Matson, *Fortschr. Chem. Org. Naturst.* **36,** 286 (1978).

Prominent early contributions to the nasute diterpenes includes the use of partially relaxed ^{13}C FTNMR (Fourier transform NMR) and both selective and off-resonance proton decoupling techniques to assign resonances and determine stereochemistry.[15–22,24,29–31] More recently, the powerful techniques of two-dimensional-FTNMR, including 2D-J resolved, 2D ^{1}H correlation (COSY), 2D ^{1}H-^{13}C correlation, and 2D-NOE pulse sequences, have been employed in *de novo* structure determination of the cubugene (**V**)[32] and longipenol (**XIII**)[33] structures.

Quantification of Diterpene Variation

The use of chemical systematics to study within colony variation, intraspecific population differences and interspecific diversity requires a large data base. We have initiated the collection of such data and emphasize the need for high resolution GLC analysis of individual termites, colonies, and populations to obtain data suitable for computerized numerical taxonomic methods of cluster analysis and discriminant function analysis.[3] The GLC data are obtained separately for monoterpenes (DB-5, 80–100°) and diterpene TMS ethers (DB-5, 180–220°). Individual head analyses (at least 3 colonies/population and at least 5 heads/colony) are performed by placing single soldier heads in capillary tubes containing 10 μl hexane and sealing them in the field for later analysis at the laboratory site. The head is crushed with a wire, 5 μl of extract is used for 2 or 3 injections for monoterpene analysis, while the remaining 5 μl is freed of solvent treated with TriSil, and analyzed. An internal standard such as octadecanol is valuable for calculation of relative retention times and standardization of chromatographic of runs from week to week.

Twenty populations of *Reticulitermes lucifugus* were examined for variation in geranyllinalool content. Eight polychemic types were identified and correlated with enzyme polymorphisms in the EST3 and ACPH2 loci.[34] In *Cubitermes,* diterpene fingerprints have been noted for seven East African species.[10] Each of two African *Trinervitermes* species had

[29] J. Vrkoč, M. Budešinsky, and P. Sedmera, *Collect. Czech. Chem. Commun.* **43,** 2478 (1978).

[30] R. Baker and S. Walmsley, *Tetrahedron* **38,** 1899 (1982).

[31] A. Dupont, J. C. Braekman, C. Daloze, J. M. Pasteels, and B. Tursch, *Bull. Soc. Chim. Belg.* **90,** 485 (1981).

[32] M. Tempesta, J. Pawlak, T. Iwashita, K. Nakanishi, Y. Naya, and G. D. Prestwich, *J. Org. Chem.* **49,** 2077 (1984).

[33] G. D. Prestwich, M. Tempesta, and C. Turner, *Tetrahedron Lett.* **25,** 1531 (1984).

[34] A. H. Parton, P. E. Howse, J.-L. Clement, and R. Baker, *in* "Biosystematics of Social Insects" (P. E. Howse and J.-L. Clement, eds.), p. 193. Academic Press, New York, 1981.

three allopatric populations with distinct secretion compositions (monoterpenes, diterpenes).[3] Four *Nasutitermes* species in Central American rainforests show remarkable examples of convergence to a common secretion type,[3] although rigorous cluster analysis allows segregation of species and geographic populations.[35] In contrast, mound-building nasutes of Venezuelan savannahs show extensive divergence from the common type to form many idiosyncratic chemical subpopulations.[3] The systematic analysis of diterpenes from termites promises exciting insights into chemistry, biochemistry, and evolutionary relationships in these primitive social insects.

Acknowledgments

I thank the NSF (DEB-7823257, CHE-7925081, and CHE-8304012) for grants in support of our research in this field. I am also grateful to the Alfred P. Sloan Foundation and the Camille and Henry Dreyfus Foundation for research awards. Most of all, I thank my many co-workers, whose names appear in the references and who helped to secure both termites and structures.

[35] T. Gush, M.Sc. Thesis, State University of New York at Stony Brook (1983).

Author Index

Numbers in parentheses are footnote reference numbers and indicate that an author's work is referred to although the name is not cited in the text.

C

D

E

F

G

H

I

J

K

L

M

N

O

P

Q

R

S

T

U

V

W

Y

Subject Index

A

B

D

G

H

I

M

R

S